RADIOACTIVE RELEASES IN THE ENVIRONMENT: IMPACT AND ASSESSMENT

RADIOACTIVE RELEASES IN THE ENVIRONMENT: IMPACT AND ASSESSMENT

John R. Cooper
National Radiological Protection Board, Chilton, UK

Keith Randle
University of Birmingham, UK

Ranjeet S. Sokhi
University of Hertfordshire, UK

JOHN WILEY & SONS, LTD

Telephone (+44) 1243 779777

Email (for orders and customer service enquiries): cs-books@wiley.co.uk
Visit our Home Page on www.wileyeurope.com or www.wiley.com

Other Wiley Editorial Offices

John Wiley & Sons, Inc., 111 River Street, Hoboken, NJ 07030, USA

Jossey-Bass, 989 Market Street, San Francisco, CA 94103-1741, USA

Wiley-VCH Verlag GmbH, Boschstr. 12, D-69469 Weinheim, Germany

John Wiley & Sons Australia Ltd, 33 Park Road, Milton, Queensland 4064, Australia

John Wiley & Sons (Asia) Pte Ltd, 2 Clementi Loop #02-01, Jin Xing Distripark, Singapore 129809

John Wiley & Sons Canada Ltd, 22 Worcester Road, Etobicoke, Ontario, Canada M9W 1L1

Library of Congress Cataloging-in-Publication Data

Cooper, John R.
Radioactive releases in the environment : impact and assessment / John R. Cooper,
Keith Randle, Ranjeet S. Sokhi.
p. cm.
Includes bibliographical references and index.
ISBN 0-471-89923-2 (cloth : alk. paper) – ISBN 0-471-89924-0 (pbk. : alk. paper)
1. Radioactive pollution. 2. Radioactivity – Measurement. I. Randle, Keith. II. Sokhi, Ranjeet S.
III. Title.

TD196.R3 .C663 2002
363.17′99 – dc21 2002033145

British Library Cataloguing in Publication Data

A catalogue record for this book is available from the British Library

ISBN 0 471 89923 2 (Cloth)
ISBN 0 471 89924 0 (Paper)

Typeset in 10/12pt Times by Kolam Information Services Pvt. Ltd, Pondicherry, India.

This book is dedicated to our families for their support,
patience and inspiration

Contents

Preface

Radiation, like taxes, is always with us. It is frequently not realized that the very fabric of our planet is radioactive and that the earth has been continuously bombarded with particles from extraterrestrial sources which induce radioactive nuclides in the surrounding atmosphere. Thus, all life from its very beginnings has been bathed in a pervasive stream of ionizing radiation emanating from the very environment in which it exists and, indeed, internally from the materials from which it is constructed. In addition, the last century has seen an additional anthropogenic contribution to this background radioactivity from medical diagnostic and clinical uses of radiation, the operation of nuclear power plants, and the fall-out from the testing of nuclear weapons. These latter sources make only a minor contribution to the overall background radioactivity but tend to figure disproportionately in the public imagination. Nuclear accidents, such as Three Mile Island and especially, Chernobyl, of course tend to reinforce this assumption, even though they contributed only a relatively small fraction to the overall dose received by total populations remote from the accident spots.

It is the purpose of this present book to explore both this natural radioactivity and also that arising from the various anthropogenic sources. Our overall aim has been to try and draw together into one text the somewhat disparate strands of knowledge that are necessarily required in a field encompassing many different scientific disciplines as does environmental radioactivity. It is hoped that each chapter can stand more or less on its own and thus be understood without excessive reference to earlier chapters. Thus, for those readers with the necessary background knowledge and possible expertise in one or more areas of environmental radioactivity and radiation counting but who wish to broaden the scope of their knowledge this text should provide them with a useful starting point. At the same time, the earlier chapters provide an introduction to the basics of radioactive theory, the interactions of radiation with matter, the biological effects of radiation and approaches to radiation protection. Other chapters also cover the basics of radiation counting and nuclear instrumentation, together with the most important methods of measuring environmental radiation, and we have also included chapters on obtaining samples and necessary preparation procedures prior to counting. The final chapters provide an introduction to modelling the dispersion of radionuclides in the environment and on methods for conducting the assessment of radiation doses. In this way, we also hope that this book can be used as a suitable text in advanced undergraduate or graduate courses in environmental science which involve aspects of radioactivity. To this end, a bibliography at the end of the book directs the interested reader to texts that deal with the various subject areas in much greater depth and detail. Finally, we believe that this text will also be a guide to those scientists who have to provide radiological data or information for legislative and related purposes and to those involved in the more general area of environmental analysis and who wish to monitor the local or large-scale environment.

The authors would like to acknowledge the help and support of Lakhumal Luhana, Hongjun Mao, Julie Kedward, Gina Cooper and Ginny Randle in the preparation of the manuscript. In addition, we are indebted to Bruce Gilliver for his critical reading of Chapter 11.

John R. Cooper, Keith Randle and Ranjeet S. Sokhi

1

Introduction to Environmental Radioactivity

Several modern industries, particularly in the area of power production and medicine, involve radiation and may discharge radioactive materials to the environment. This present book is about the effects of these discharges and the basis for their control. This chapter sets the scene by briefly describing the history of the discovery of radiation, its uses and the radiation exposures that we may all experience. It also provides a 'route map' for the rest of the book.

1.1 THE DISCOVERY AND EARLY HISTORY OF RADIOACTIVITY

In 1895, a German physicist, Wilhelm Röntgen, made an epoch-making discovery. While studying the cathode rays formed by electrical discharges through tubes containing a rarefied gas, he noticed that across the room a screen coated with barium platinocyanide gently glowed. The rays responsible for this phenomenon were quickly shown to pass through materials opaque to ordinary light. Importantly, the rays, now called X-rays, would expose a photographic plate, enabling, for example, photographs of bone fractures to be made.

This discovery provoked a wave of scientific and public interest. One scientist who was excited by the discovery was Antoine Henri Becquerel, a French physicist. He had been investigating fluorescence and phosphorescence and thought that fluorescing substances might produce X-rays. He used potassium uranyl sulfate in his experiments because these crystals were unaffected by exposure to air. Photographic plates were wrapped in black paper with the crystals being placed on top. The entire device was then exposed to bright sunlight for several hours before the photographic plate was developed. The photographic plate showed a faint outline of the crystals, which Becquerel ascribed to rays that were emitted during fluorescence of the crystals. Becquerel attempted to repeat these experiments but the weather was overcast and he placed the photographic plates together with the crystals away in a drawer after only a brief exposure to sunlight. The weather remained overcast during the following days and eventually Becquerel developed the plates expecting to see only a very faint image. He was surprised to see that the image was dark and well defined – rays were being emitted from the crystals without any prior exposure to sunlight! Becquerel quickly showed that the phenomenon was related to the uranium content of the crystals. The uranium emitted penetrating radiation continuously and without any form of initiation – the phenomenon of *radioactivity* had been discovered but it prompted little immediate interest (further details of Becquerel's work are given in Allisy, 1996).

One person who was interested was Marie Curie; she discovered that another heavy element, thorium, also emitted radiation. Early on, together with her husband Pierre, Marie Curie discovered that pitchblende and the mineral chalcocite emitted more radiation than could be accounted for by their uranium content. Further investigations on pitchblende by the Curies led to the identification of two other radioactive elements, i.e. polonium and radium. The significance of these discoveries was quickly recognized and in 1903 the Curies shared the Nobel Prize for physics with Becquerel for their work on radioactivity. Marie Curie was later awarded the Nobel Prize for chemistry for her work on radium and polonium – she is one of the few people to have been awarded two Nobel Prizes. Becquerel gave his name to the unit of radioactivity – the *becquerel*, symbol *Bq* (one becquerel is one nuclear disintegration per second).

Not only had the phenomenon of radioactivity been discovered but it could be said that environmental radioactivity had *also* been discovered. Uranium, thorium, polonium and radium are all naturally occurring elements. Work on radioactivity developed and expanded rapidly following the early discoveries of Becquerel and the Curies. In 1899, Rutherford showed that one type of radiation comprised positively charged particles and these were later identified as helium nuclei (alpha particles). In 1900, the French scientist Villard characterized another type of radiation – the uncharged, penetrating gamma rays. The laws of radioactive decay were established by 1903 by Rutherford and Soddy, and by 1912 the concept of isotopes had been developed to explain why radioelements exist with more than one atomic weight. The first artificial radioactive isotopes were produced in 1934.

Radioactivity decay and the decay laws are explained in detail in Chapter 2. For the moment, it is important to realize that atoms of the same element may have different atomic weights and some types of atoms are unstable. An unstable atom can transform spontaneously into an atom of another element and, in so doing, emit radiation. As this process involves the transformation of atoms, the number of the original atoms must decrease with time. The time taken for the number to fall by half is called the *half-life*. It is a unique and unalterable property of each radionuclide. Half-lives can range from fractions of a second to millions of years.

The three main types of ionizing radiation were quickly identified as alpha particles (two protons plus two neutrons), beta particles (a negatively charged electron), and gamma rays (uncharged, energetic photons). The characteristics of ionizing radiation are described in Chapter 2. It should be emphasized that this book is about ionizing radiation, namely radiation of sufficient energy to displace electrons from atoms or molecules and so produce ions (see Chapter 3). Ionizing radiation should be distinguished from non-ionizing radiation, e.g. UV light and radio waves, neither of which have enough energy to cause ionization. In this book, the word 'radiation' always refers to ionizing radiation.

The possible beneficial effects of radiation were quickly exploited. Radioactivity was soon recognized as a more concentrated form of energy than had been hitherto known. Radium was used in the treatment of solid cancers where the radiation often destroyed the tumour tissue. It became to be viewed as an almost miraculous substance and at one time it was more valuable than gold.

Fascination with radium and radioactivity lead to a plethora of 'quack uses', including belts, pads, face creams, etc., all of which were claimed to have therapeutic properties. One of the most significant legitimate uses was in the production of luminous dials – a use that continued until fairly recently. In the early days, the dials were often painted by young women who licked the brushes to obtain a fine point; consequently they took some of the radioactive material into their bodies!

The fascination with radioactivity was tempered when it slowly became clear that it could have fatal effects. Over a number of years some people who had worked with radioactive materials experienced ill health and died. Sufferers included some of the radium dial painters. In July 1934, at the age of 66, Madame Curie died of a blood disorder that was almost certainly brought on by her work with radioactivity. Recognition of possible harmful effects led, in 1928, to the setting up of the International X-ray and Radium Protection Committee that was to advise on standards of protection (see Chapter 4 for a discussion of the biological effects of ionizing radiation and of radiological protection).

Nevertheless, we live in an environment pervaded by radiation of natural origin. Artificial uses of radioactivity can bring about many benefits, ranging from nuclear medicine to the production of electricity by nuclear power. It is the purpose of this book to present the reader with a balanced view of the sources of radiation exposure, the behaviour of radionuclides in the environment, the effects on man and the basic principles of control. The remainder of this chapter gives an overview of the sources of radiation exposure; many of the topics covered are amplified in later chapters.

1.2 SOURCES OF RADIATION

We cannot sense radiation but it is a continual feature of human existence. Natural radiation sources are the most significant contributors to exposure for the vast majority of individuals.

1.2.1 Natural Sources of Radiation

Studies since the early 1900s have characterized the natural radiation environment. Historical reviews are contained in Kathren (1984) and Glasstone (1979). On Earth, we are exposed to ionizing radiation of both terrestrial and extra-terrestrial origins. Radiation from outside the Earth's atmosphere is called cosmic radiation and is described first, followed by other natural sources.

Cosmic radiation

During the first decade of the twentieth century, experiments suggested a hitherto unknown source of radiation that was thought to be entirely gamma radiation from terrestrial sources. An Austrian physicist, V F Hess, was investigating the attenuation of terrestrial gamma radiation by the atmosphere and conducted experiments with radiation detectors in balloons; he found, to his surprise, that his instruments registered increasing radiation levels with increasing altitude. He concluded, in 1912, that penetrating radiation was coming from extraterrestrial sources: cosmic radiation had been discovered.

Cosmic radiation has two main sources, i.e. galactic and solar. Before they interact with the Earth's atmosphere, both types of radiation consist mainly of protons together with some helium and other heavier ions. Galactic cosmic rays are produced outside of the solar system possibly because of stellar flares, supernova explosions, etc. However, there is no generally accepted theory as to their generation. The energy of galactic cosmic rays averages about 10^4 MeV, with maximum energies of up to around 10^{14} MeV.

The other origin of cosmic radiation is the Sun. Particles of low energy are emitted continuously by the Sun but more energetic ones are emitted during solar events such as magnetic disturbances or flares. Thus, emission of solar cosmic radiation tends to follow the 11 year solar cycle with a maximum during increased solar activity and a minimum in the 'quiet' period.

Solar cosmic rays are usually of considerably lower energy than those of galactic origin and energies are typically between 1 and 100 MeV. They are of less significance at the Earth's surface because most have insufficient energy to penetrate the Earth's magnetic field.

On entering the atmosphere, energetic particles, primarily protons, may interact with nuclei of atmospheric gases (nitrogen, oxygen and argon) in a variety of nuclear reactions. The products include neutrons, protons, muons, pions and kaons, together with radioisotopes such as ^3H (tritium) and ^7Be (see *cosmogenic radionuclides*, p. 11 below). The secondary particles may be energetic enough to initiate further reactions leading to a cascade of events. Cosmic radiation at the Earth's surface comprises directly ionizing secondary particles, together with a small neutron component. Exposure to cosmic rays increases with altitude and hence air travel increases exposure to cosmic radiation. There is also a relatively smaller variation with latitude – the Earth's magnetic field causes a greater flux of low-energy protons to reach the top of the atmosphere at the poles than at equatorial regions.

Terrestrial sources of radiation

The work of the Curies showed that radioactive materials existed naturally in the environment. Natural radionuclides can be divided into two groups depending on their origin. The first and the most important group is the *primordial radionuclides*. These have been present since the origin of the Earth. The second group is the *cosmogenic radionuclides*. They are continuously produced in the upper atmosphere by the action of cosmic rays. Concentrations of primordial and cosmogenic radionuclides in environmental materials and foodstuffs are given in Tables 1.1 to 1.5. The information is taken from two published reviews by Bradley (1993) and The United Nations Scientific Committee on the Effects of Atomic Radiation (UNSCEAR, 2000a).

Table 1.1 Concentrations of naturally occurring radionuclides in air (Bq m^{-3}).

Nuclide	Indoor	Outdoor	Reference[a]
^{238}U		1.2×10^{-8}–1.8×10^{-5}	1
	1×10^{-6}	1×10^{-6}	2
^{235}U	5×10^{-8}	5×10^{-8}	2
^{232}Th		1.4×10^{-8}–8.7×10^{-7}	1
	5×10^{-7}	5×10^{-7}	2
^{230}Th		1.5×10^{-8}–1.4×10^{-6}	1
	5×10^{-7}	5×10^{-7}	2
^{226}Ra		8×10^{-7}–3.2×10^{-5}	1
	1×10^{-6}	1×10^{-6}	2
^{222}Rn	3–100 000	0.3–48	1
^{220}Rn	0.6–88		1
^{214}Pb	1.1–98		1
^{212}Pb	5.9×10^{-2}–6.6		1
^{210}Pb		4.4×10^{-6}–3.3×10^{-3}	1
	5×10^{-4}	5×10^{-4}	2
^{214}Bi	0.5–92		1
^{212}Bi	0.11–1.5		
^{210}Bi		5.6×10^{-5}–4.2×10^{-4}	1
^{218}Po	0.93–113	1.9–10	1
^{210}Po		2.2×10^{-6}–2.6×10^{-4}	1
	5×10^{-5}	5×10^{-5}	2

[a]References: 1, Bradley (1993), where the values given are for over land; 2, UNSCEAR (2000a).

Primordial radionuclides This group of radionuclides is responsible for most of the radiation exposure of the vast majority of individuals. The principal radionuclides are ^{40}K (half-life, 1.28×10^9 years), ^{232}Th (half-life, 1.41×10^{10} years) and ^{238}U (half-life, 4.47×10^9 years). There are two other primordial radionuclides of lesser importance for human exposure, i.e. ^{87}Rb (half-life, 4.7×10^{10} years) and ^{235}U (half-life, 7.04×10^8 years). The uranium and thorium radionuclides are the starting points for decay chains of several radionuclides. Members of these decay chains may be important in their own right, either as contributors to human exposure or for other radioactive properties. The decay chains of ^{238}U, ^{232}Th and ^{235}U are shown in Appendix 3. The general properties of decay chains are discussed in Chapter 2, but for the moment it is sufficient to understand that a radioactive isotope decays to another and that one to another, etc., until a stable isotope is reached. The two decay series headed by ^{238}U and ^{232}Th are the most important, while the series headed by ^{235}U is of lesser importance. The primordial radionuclides are listed in Table 1.6.

Table 1.2 Concentrations of naturally occurring radionuclides in various waters (Bq l^{-1}).[a]

Nuclide	Rainwater	Groundwater	Surface water	Drinking water	Spa/mineral water[b]	Seawater	Water supplies[c]
^{238}U	—	2.0×10^{-7}–185	$< 1.0 \times 10^{-3}$–5.6	9.4×10^{-5}–7.7×10^{-2}	$< 1 \times 10^{-3}$–0.93	4.9×10^{-4}–4.4×10^{-2}	1×10^{-3}
^{234}U	—	—	4.5×10^{-3}–2.7	5.6×10^{-4}–0.12	$< 1 \times 10^{-3}$–0.2	4.6×10^{-2}–4.8×10^{-2}	1×10^{-3}
^{232}Th	—	—	0–5×10^{-2}	—	—	4.4×10^{-8}–2.9×10^{-5}	5×10^{-5}
^{230}Th	—	—	2.6×10^{-3}–5.0	—	—	2×10^{-6}–5.2×10^{-5}	1×10^{-4}
^{228}Th	—	—	3.5×10^{-3}–1.9×10^{-2}	—	—	7.4×10^{-6}–1.1×10^{-4}	5×10^{-5}
^{228}Ra	—	1.5×10^{-2}–0.21	5×10^{-4}	—	—	2×10^{-4}	5×10^{-4}
^{226}Ra	—	7.4×10^{-5}–56	$< 1.0 \times 10^{-3}$–3.0	0–4.0	0–2.2	1×10^{-3}–1.6×10^{-3}	5×10^{-4}
^{224}Ra	—	0–7.4×10^{-2}	4×10^{-3}–5×10^{-2}	0–6.3×10^{-2}	—	1.5×10^{-4}	—
^{222}Rn	400	0–1.1×10^{4}	0–9.3×10^{2}	0–2.5×10^{4}	0–1.8×10^{3}	7×10^{-4}–1×10^{-3}	—
^{220}Rn	—	—	—	—	—	1.5×10^{-4}	—
^{210}Pb	3.7×10^{-3}–0.86	$< 4.9 \times 10^{-4}$–51	1.7×10^{-3}–0.25	1.1×10^{-4}–0.26	0–8.5	3.7×10^{-4}–3×10^{-3}	1×10^{-2}
^{210}Po	3.7×10^{-3}–0.15	0–4.8×10^{-2}	5.6×10^{-4}–2.6×10^{-3}	2.2×10^{-4}–0.56	2.0×10^{-4}–0.10	1.9×10^{-4}–8.1×10^{-3}	5×10^{-3}
^{40}K	1.6×10^{-2}	1.3×10^{-2}–0.16	0–0.85	$< 3.0 \times 10^{-2}$–3.6	3.1×10^{-2}–3.7	13	—

[a] From Bradley (1993).
[b] Including bottled water.
[c] Reference values for water supplies taken from UNSCEAR (2000a).

Table 1.3 Concentrations of naturally occurring radionuclides in soils, rocks and sediments (Bq kg^{-1}, dry weight).[a]

Nuclide	Soils	Rocks	Sediments
^{238}U	2^{b}–9.8×10^3 [c]	0.5–4.2×10^6 [c, d]	2–1×10^4
^{234}U	55–9.7×10^3 [c]	29–3.5×10^3 [c]	18
^{232}Th	1–180	2–280	1–130
^{230}Th	17–160[c]	< 11–26[c]	35
^{228}Th	3.7–7.4	—	—
^{228}Ra	5–185	—	—
^{226}Ra	2.6–200	30–1.7×10^3	3–1.3×10^4 [e]
^{210}Pb	8.5–230	—	20–700
^{210}Po	10–51	—	—
^{40}K	0–3.2×10^3	4–40	19–1.6×10^3

[a] From Bradley (1993).
[b] As collected.
[c] From an area of known enhanced natural radioactivity.
[d] Bulk uraniferous vein material.
[e] May include mine waste.

A primordial radionuclide that causes widespread and inescapable exposure but not necessarily the highest is the beta/gamma emitter, ^{40}K. This is present in natural potassium in an effectively fixed proportion of 0.012%. Potassium is a biologically essential element and its body concentration is under homeostatic control; thus, exposure to ^{40}K is unavoidable but does vary with age and sex in line with corresponding changes in the concentration of potassium. The presence of ^{40}K in rocks and soils is also a source of external exposure. Concentrations of ^{40}K in soils range from zero for some soils in the Scottish islands to over 3200 Bq kg^{-1} dry weight, again in the Scottish islands. UNSCEAR estimates the worldwide average concentration of ^{40}K in soil at 420 Bq kg^{-1} (UNSCEAR, 2000a). Concentrations in rocks are less well studied but appear to be lower: concentrations of between 4 and 40 Bq kg^{-1} have been reported for phosphate-bearing rocks.

The decay chain headed by ^{238}U is important. ^{238}U is a ubiquitous radionuclide. It is present in most environmental materials and it has been widely studied in rocks and soils in connection with uranium prospecting. Levels of ^{238}U are generally in the range of 2 to 300 Bq kg^{-1} dry soil. The world-wide average reported by UNSCEAR is 33 Bq kg^{-1} dry weight. In areas where uraniferous rocks such as granite are found, soil levels range up to 400 Bq kg^{-1}. Some silts and sandstones have concentrations up to 30 000 Bq kg^{-1} and uraniferous rocks have levels around 4000 Bq kg^{-1} with up to 2×10^6 Bq kg^{-1} in bulk vein material. Levels in seawater are typically about 0.04 Bq l^{-1}.

Some members of the ^{238}U chain can be taken up by foodstuffs and in some situations can contribute significantly to human exposure (see Table 1.4). With the exception of ^{40}K, the highest concentration of any one radionuclide in foodstuff is ^{210}Po in fish and shellfish. Levels of around 300 Bq kg^{-1} dry weight have been measured in mussel flesh from Cumbria, UK and France. Even higher levels of 100 to 7000 Bq kg^{-1} dry weight have been measured in whole shrimps from the North

Table 1.4 Concentrations of naturally occurring radionuclides in foodstuffs (Bq kg^{-1}, fresh weight).[a]

Nuclide	Meat and meat products	Fish and shellfish	Vegetables	Fruit	Cereals, grains and cereal products	Dairy produce	Fats and oils	Sugars	Beverages	Others
^{238}U	7.8×10^{-4}–4.9×10^{-3} [b]	2.5×10^{-3}–1.9	9.0×10^{-4}–10[c]	3.7×10^{-4}–3.7[c]	3×10^{-3}–1.5[c]	1.2×10^{-4}–1.7×10^{-2}	2.5×10^{-2}	2.5×10^{-3}	6.2×10^{-2}–7.4×10^{-2}[d]	2.5×10^{-3}–1.1
^{234}U	7.7×10^{-4}–1.8×10^{-3}	1.8×10^{-2}–2.2	1.1×10^{-3}–3.1×10^{-2}	5.5×10^{-4}–2.7×10^{-3}	2.5×10^{-3}–3.0×10^{-2}	9.7×10^{-4}–1.0×10^{-3}	—	—	—	—
^{232}Th	2.7×10^{-4}–2.0×10^{-3}	1.2×10^{-3}–3.0×10^{-2}	8.0×10^{-5}–0.38 [d]	1.2×10^{-4}–5.6×10^{-4}	1.0×10^{-4}–1.2×10^{-2}[d]	2.7×10^{-4}–3.0×10^{-3}	—	—	—	—
^{230}Th	4.7×10^{-4}–3.0×10^{-3}	1.2×10^{-3}–2.9×10^{-2}	2.0×10^{-4}–3.2×10^{-2}	1.2×10^{-4}–6.9×10^{-4}	9.3×10^{-4}–1.0×10^{-2}	3.7×10^{-4}–9.8×10^{-4}	—	—	—	—
^{228}Th	2.2×10^{-2}–9.3×10^{-2} [c]	5.6×10^{-2}–0.7[c]	—	2.2×10^{-2} [c]	0.18–2.3[c]	5.6×10^{-2} [c]	—	7.0×10^{-2}[c]	< 0.18–6.8	3.0×10^{-2}–44
^{226}Ra	0–0.12	8.5×10^{-3}[b] –2.1c	2.2×10^{-3}[c]–0.84[e]	6.3×10^{-3}–10	7.4×10^{-4}[c] – 7.8[f]	< 3.7×10^{-4}–0.26[c]	3.7×10^{-3}–0.18	2.4×10^{-2}–8.9×10^{-2}	5.6×10^{-3}[g] – 15[c]	1.4×10^{-2}–130
^{210}Pb	1.7×10^{-2}–24	7.4×10^{-3}–790	7.8×10^{-3}–3.3	8.5×10^{-3}–7.4×10^{-2}	3.3×10^{-2}–1.1	3.5×10^{-2}–8.8×10^{-2}[b]	0.11[b]	4.1×10^{-2}–7.4×10^{-2} [b]	1.5×10^{-5}–2.0×10^{-2} [g]	0.11[b]
^{210}Po	3.7×10^{-2}–120	1.5×10^{-2}–6.7×10^{3} [d]	3.7×10^{-2}–7.4	3.4×10^{-2}–0.14 [c]	1.5×10^{-2}–0.37 [c]	3.3×10^{-3}–37	3.0×10^{-2}–0.22	1.8×10^{-2}–0.15 [b]	1.5×10^{-5}–3.7×10^{-2} [g]	—
^{40}K	60–120[b]	34–170	30–240	23–140	26–120	7.0–56	0–18	0.15–120	14–1000[g]	39–240

[a]From Bradley (1993).
[b]Prepared for consumption.
[c]As purchased.
[d]Dry weight.
[e]Form an area of known enhanced natural activity.
[f]Including decay products.
[g]Units of Bq l^{-1}.

Table 1.5 Reference concentrations of natural radionuclides in food given by UNSCEAR (2000a) (mBq kg^{-1}).[a]

Radionuclide	Meat products	Fish products	Leafy vegetables	Roots and fruits	Grain products	Milk products
^{238}U	2	30	20	3	20	1
^{234}U	2	30	20	3	20	1
^{232}Th	1	—	15	0.5	3	0.3
^{230}Th	2	—	20	0.5	10	0.5
^{228}Th	1	—	15	0.5	3	0.3
^{228}Ra	10	—	40	20	60	5
^{226}Ra	15	100	50	30	80	5
^{210}Pb	80	200	30	25	100	40
^{210}Po	60	2000	30	30	60	60

[a]1 mBq = 0.001 Bq.

Table 1.6 Primordial radionuclides.

Nuclide	Half-life (years)	Mode of decay
^{40}K	71.3×10^{9}	Beta
^{50}V	6×10^{14}	Beta
^{87}Rb	4.7×10^{10}	Beta
^{113}Cd	9×10^{15}	Beta
^{115}In	5×10^{14}	Beta
^{123}Te	1.2×10^{13}	EC[a]
^{138}La	1.1×10^{11}	Beta
^{142}Ce	$> 5 \times 10^{16}$	Alpha
^{144}Nd	2.1×10^{15}	Alpha
^{147}Sm	1.1×10^{11}	Alpha
^{148}Sm	8×10^{15}	Alpha
^{152}Gd	1.1×10^{14}	Alpha
^{156}Dy	2×10^{14}	Alpha
^{176}Lu	2.7×10^{10}	Beta
^{174}Hf	2×10^{15}	Alpha
^{180}Ta	$> 1.6 \times 10^{13}$	Beta
^{187}Re	5×10^{10}	Beta
^{190}Pt	7×10^{11}	Alpha
^{204}Pb	1.4×10^{17}	Alpha
^{238}U[b]	—	—
^{235}U[c]	—	—
^{232}Th[d]	—	—

[a]EC, electron capture.
[b]Plus decay chain (see Appendix 3, Table 2 and Figure 1).
[c]Plus decay chain (see Appendix 3, Table 2 and Figure 3).
[d]Plus decay chain (see Appendix 3, Table 2 and Figure 2).

Atlantic. Particular radionuclides appear to concentrate in certain foodstuffs; for example, ^{226}Ra is found in significant levels in dry tea leaves and Brazil nuts with concentrations of 20 Bq kg^{-1} and 100 Bq kg^{-1}, respectively.

The nuclide ^{226}Ra is a member of the ^{238}U decay chain. All of the ^{226}Ra now present has ultimately been produced from decay of ^{238}U via the decay of a number of other radionuclides. ^{226}Ra decays to a radioactive gas, ^{222}Rn, which is a significant contributor to human exposure. ^{222}Rn is commonly referred to as radon. The radioactive gas seeps from the ground into the atmosphere. Out of doors, it merely disperses and decays, but indoors it builds up. The problem is sometimes exacerbated by the fact that houses are often under slight negative pressure and so the radon can be sucked out of the ground. It is not the radon gas itself that irradiates people but some of the short-lived daughter products (see Appendix 3, Figure 2 and Table 2). Exposure to these may cause lung cancer. Radon levels in houses depend upon the underlying geology and in particular on the prevailing concentration of ^{238}U or, more strictly, that of the immediate precursor, ^{226}Ra. There have been extensive programmes in many countries to measure indoor levels of radon and to identify geographical areas of concern since the recognition of radon as a public health issue. In the UK, the average indoor concentration of radon is 20 Bq m^{-3}, but the range is very wide – from 5 to 5000 Bq m^{-3}. Outdoor levels are about 4 Bq m^{-3}. UNSCEAR estimates the world-wide population weighted average radon level at about 40 Bq m^{-3}, with an outdoor concentration of about 10 Bq m^{-3} in continental areas – lower concentrations are observed in coastal areas. However, significantly higher concentrations are seen in some parts of the world; for example, concentrations up to 100 000 Bq m^{-3} have been measured indoors in some parts of Germany and France. Steps have been taken in many countries to reduce radon levels in the most affected areas; these are discussed in Chapter 4. Enhanced levels of radon may be found in caves and mines.

^{232}Th is the parent of the other major decay chain (see Appendix 3, Figure 1 and Table 2). This radionuclide generally occurs in soils at similar levels to ^{238}U and UNSCEAR estimates a world-wide soil average of 45 Bq kg^{-1}. There is a wide range in observed levels. Levels up to 180 Bq kg^{-1} have been reported for Cornish soils. Some of the highest levels occur in heavy mineral sands, in particular, those bearing the mineral monazite. Such sands occur in various parts of the world, most notably at Kerala on the Arabian Sea coast of India and Esprito Santo on the coast of Brazil. At Kerala, some of the more radioactive areas of sand have ^{232}Th concentrations of up to 7000 Bq kg^{-1}. Levels of ^{40}K and ^{226}Ra are also elevated, with concentrations of 100 Bq kg^{-1} and 1000 Bq kg^{-1}, respectively.

One member of the ^{232}Th decay chain is the radioactive gas, ^{220}Rn. However, unlike its counterpart in the ^{238}U chain, ^{222}Rn, it is not a significant source of human exposure as its half-life, 56 s, is too short to allow it to build up significantly in air spaces. There is not such an extensive database of measurements as for ^{222}Rn, but indoor levels in the UK appear to be in the range of 0.6 to 2 Bq m^{-3}. Levels up to 40 Bq m^{-3} have been reported in Australia, while levels in some parts of Austria are twice as high. It is thought that because of the short half-life, the building materials are more important as a source of ^{220}Rn than are the underlying strata.

Cosmogenic radionuclides Cosmic rays are a form of ionizing radiation that originate outside of the Earth's atmosphere (see above). They produce a range of radionuclides in the atmosphere, biosphere and lithosphere by various nuclear reactions; these radionuclides are called the cosmogenic radionuclides.

The four most important cosmogenic radionuclides for human exposure are tritium, ^{7}Be, ^{14}C and ^{22}Na, although a much wider range is produced (see Table 1.7). By far the most significant of these radionuclides is ^{14}C. This radionuclide is produced in the upper atmosphere by interaction between the neutrons generated by cosmic rays and natural, non-radioactive, ^{14}N. This general process is termed *activation*. The annual production rate of natural ^{14}C is around 1200 to 1400 TBq (10^{12} Bq) a year. The specific activity of ^{14}C is 230 Bq kg^{-1} of carbon. Most of the other cosmogenic radionuclides are produced when the nuclei of atoms in the atmosphere are split by collision with the high-energy particles. The process is generally known as *spallation*. The global inventory of important cosmogenic radionuclides is given in Table 1.8.

The constant generation of ^{14}C by cosmic rays is used to advantage in the technique of *radiocarbon dating*. This technique was developed at Chicago University by W F Libby and colleagues in 1947. It can be used to date materials containing organic carbon and has application in fields such as archaeology, climatology and oceanography. The principle is that through metabolic activity the specific activity of the ^{14}C nuclide in a living organism will equal that of the atmosphere. However, when the organism dies the ^{14}C will decay but will not be replaced by ^{14}C from the atmosphere. Thus, from a knowledge of the half-life of ^{14}C (5730 ± 40 years) and from measurements of the specific activity of the ^{14}C in the material in question, an estimate can be made of the time since the organism died.

The method is suited to a variety of organic materials, such as wood, but its accuracy will be affected by any variation in the atmospheric levels of ^{14}C, as well as by contamination. The half-life of ^{14}C also limits the dating period to around 50 000 years.

Table 1.7 Cosmogenic radionuclides.

Radionuclide	Half-life (years)[a]	Mode of decay[b,c]
^{3}H	12.26	Beta
^{7}Be	0.15	EC
^{10}Be	1.6×10^{6}	Beta
^{14}C	5.73×10^{3}	Beta
^{22}Na	2.6	EC
^{26}Al	7.4×10^{5}	EC
^{32}Si	280	Beta
^{32}P	0.04	Beta
^{33}P	0.07	Beta
^{35}S	0.24	Beta
^{36}Cl	3.01×10^{5}	Beta
^{39}Ar	269	Beta
^{81}Kr	2.29×10^{5}	EC

[a]To two decimal places.
[b]Primary decay mode.
[c]EC, electron capture.

Table 1.8 Global inventory of important cosmogenic radionuclides (from UNSCEAR, 2000a).

Radionuclide	Inventory (PBq)[a]
^{3}H	1.3×10^{3}
^{7}Be	413
^{10}Be	230
^{14}C	1.275×10^{4}
^{22}Na	0.4

[a] 1 PBq = 10^{15} Bq.

1.2.2 Man–Made Sources of Radionuclides

A number of human activities contribute to the levels of radionuclides in the environment. Some activities result in the release of naturally occurring radionuclides to the accessible environment, e.g. producing phosphoric acid from phosphate rocks. Other activities produce novel radionuclides, with these include generating electricity by nuclear power and nuclear weapons testing. The latter activities are called sources of anthropogenic radionuclides.

Anthropogenic radionuclides

Advances in the understanding of the structure of the atom in the first decades of the 20th century led to the realization that energy could be released by splitting or fissioning uranium nuclei. Fission is induced by the absorption of a neutron. Energy can also be produced by combining two small nuclei – a process known as *fusion*. Energy is produced as a result of the binding together of the protons and neutrons in the nucleus in both processes. The binding energy may be different for different elements. In general, the most stable isotope of each element has a mass that is less than the summed masses of the protons and neutrons it contains. The difference is greatest for the medium-sized nuclei, i.e. from about strontium (mass number, 90) to cerium (mass number, 140). It is smallest for nuclei with mass numbers less than 20 and for nuclei with mass numbers greater than 210. Thus, in the fission of uranium into two smaller nuclei, the summed masses of the products (including the neutrons) will be less than that of the uranium nucleus; the difference will be manifest as energy. In fusion, the nucleus produced has a smaller mass than the two nuclei that have been combined – with a resultant release of energy.

Fission was the first nuclear process to be developed and is the basis for all current nuclear power reactors. The general principle can be illustrated as follows:

$$^{235}\text{U} + \text{neutron} \longrightarrow \text{fission products} + 2\text{–}3 \text{ neutrons} + \text{energy}$$

The fission products are radioactive isotopes. There are two key features of nuclear fission. First, very large amounts of energy can be produced. The fission of one kg of ^{235}U produces 18.7 million kilowatt hours of energy. Secondly, the process itself releases two to three neutrons per fission of ^{235}U. This leads to the possibility of a self-sustaining series of nuclear fissions – a chain reaction.

Only some isotopes of heavy elements are easily fissioned – the most important being ^{233}U, ^{235}U and ^{239}Pu. Naturally occurring uranium contains 0.71% of ^{235}U, with the remainder being almost entirely the non-fissile ^{238}U. A very small number of fissions occurs naturally in uranium but because of the low proportion of ^{235}U, a chain reaction is not initiated. However, if neutrons are slowed down by, say, elastic collisions with light nuclei such as hydrogen or carbon, the probability of inducing fission can be increased by orders of magnitude. This formed the basis for the first nuclear 'pile' and is the basis for nearly all power reactors. The slowing down of the neutrons is called *moderation* and it is achieved by a *moderator*.

The first artificial nuclear chain reaction occurred in 1942 in a graphite-moderated natural uranium reactor at the University of Chicago under the direction of the Italian physicist Enrico Fermi. The first large-scale nuclear reactors were built at Hanford in Washington State, USA. They were designed and operated for the production of plutonium for nuclear weapons. The reactors were again fuelled with natural uranium with a graphite moderator to slow down the neutrons.

Self-sustaining fission chain reactions may not be entirely man-made phenomena on Earth. In the 1970s, it was noticed that uranium from the Oklo mine in Eastern Gabon was depleted in ^{235}U. Some samples contained 0.44% of this nuclide – around 60% of the normal level (0.71%). Trace levels of fission products were also found. These observations have led to the conclusion that a natural self-sustained chain reaction had occurred at the site around two thousand million years ago which had used up some of the ^{235}U. At that time the proportion of ^{235}U is estimated from radioactive decay rates to have been around 3% (^{235}U has a half-life of 7.04×10^8 years, while ^{238}U has a half-life of 4.5×10^9 years). Thus, the likelihood of a chain reaction is increased, but a moderator to slow down the reactors would have been required; this could have been water, either as water of crystallization or from groundwater seeping into the one body.

The natural Oklo reactor is thought to have been functional on and off for about 150 000 years; during this time it generated about the same amount of energy as a modern reactor would generate in four years of operation.

The two principal uses of nuclear energy are nuclear weapons and generation of electricity. Both of these applications have contributed to environmental levels of radiation. The uses of nuclear power are discussed in detail in Chapter 6, while a brief summary is given below.

Nuclear weapons tests Nuclear power reactors rely on the controlled, relatively slow, release of nuclear energy. A nuclear bomb tries to achieve the opposite, i.e. a very rapid release of energy. A fission bomb depends upon a supercritical mass of fissile material being brought together almost instantaneously so that a rapid chain reaction occurs, so releasing enormous quantities of energy. The atomic bombs used on Japan were fission bombs: a ^{235}U device was exploded over Hiroshima, followed by a ^{239}Pu device exploded over Nagasaki. Each weapon was equivalent to about 20 thousand tonnes of (2,4,6-trinitrotoluene (TNT)). The fusion process is also used in nuclear weapons. In thermonuclear devices (hydrogen bombs), the high temperatures generated by a fission explosion are used to cause fusion of various hydrogen isotopes with a release of vast amounts of energy. As yet, it has not proved possible to control the fusion reaction to the extent that the energy can be harnessed for electricity generation.

The desire to develop ever more powerful nuclear weapons led to the testing of such weapons. Tests in the 1940s and 1950s were usually conducted in the atmosphere, with consequent dispersion of radioactive material. Over 2400 nuclear detonations have been documented, of which about 540 were atmospheric. The total yield of the atmospheric tests was 440 megatonnes (Mt) (UNSCEAR, 2000a). Most of these tests were conducted between 1954–1958 and 1961–1962. Testing in the atmosphere largely ceased after the signing of the Limited Nuclear Testing Ban Treaty in 1963. However, some tests were carried out afterwards by non-signatories to the Treaty, in particular France and China. China conducted the last atmospheric test in 1980.

The largest number of atmosphere tests, 219, was carried out by the former USSR. The United States conducted 197, and the UK carried out 21 tests. France conducted four tests before the 1963 limited test ban. Between 1963 and 1980, France and China have together conducted 64 tests.

The atmospheric tests injected substantial amounts of radionuclides into the atmosphere. Typically, a fireball rises rapidly perhaps at a speed of some hundreds of miles an hour but slowing as it cools. When it reaches the top of the troposphere at around 50000 ft, the fireball will spread out, so leading to the characteristic mushroom shape. If the explosion was sufficiently close to the ground, then particles of soil and other debris will be entrained in the plume. Soon after detonation, as the fireball begins to cool, particles will form containing materials with relatively high melting points, e.g. oxides of iron and aluminium. These particles have diameters in the range 0.4 – 4 μm and will contain radionuclides with the higher melting points. The more volatile radionuclides coalesce later into particles of size less than 0.4 μm. The larger particles undergo gravitational settling, returning to earth as local fall-out within a hundred miles or so. The smaller particles, containing most of the volatile radionuclides such as radiocaesium, radiostrontium and radioiodine, remain in the atmosphere for much longer and may return to Earth thousands of miles from the point of detonation. Particles injected into the stratosphere by the explosion produce widespread fall-out but this is confined largely to the hemisphere in which the explosion occurred: mixing of air masses between the hemispheres occurs on timescales longer than the time taken for the particles to settle. Stratospheric airflows in each hemisphere are predominantly from the equator to the pole. Air rises into the stratosphere at the equator and returns in the temperate and polar zones. This leads to the characteristic increase in global fall-out at higher latitudes.

The amounts of radionuclides injected into the atmosphere by the weapons tests are huge. Many of the radionuclides are short-lived and decay away quickly, but some are long-lived. The quantities of important long-lived radionuclides injected into the atmosphere during weapons testing are given in Table 1.9. Comparisons help to provide a perspective on the numbers and the following ones are of interest. First, during the early 1960s, when weapons testing was at its highest, the activity of ^{14}C per unit mass of carbon in the atmosphere was increased to about twice the cosmogenic level by ^{14}C produced by weapons testing. Secondly, the amount of ^{137}Cs released into the atmosphere from weapons testing was about twenty times that released from the Chernobyl reactor accident.

Table 1.9 Radionuclides produced and globally dispersed by nuclear weapons testing in the atmosphere (from UNSCEAR, 2000a).[a]

Radionuclide	Half-life (years)[b]	Total (EBq)[c]
^{3}H	12.32	186
^{14}C	5730	0.21
^{54}Mn	0.86	4.0
^{55}Fe	2.73	1.5
^{89}Sr	0.14	117
^{90}Sr	28.78	0.62
^{91}Y	0.16	120
^{95}Zr	0.18	148
^{103}Ru	0.11	247
^{106}Ru	1.02	12.2
^{125}Sb	2.76	0.741
^{131}I	0.02	675
^{137}Cs	30.07	0.948
^{140}Ba	0.03	759
^{141}Ce	0.09	263
^{144}Ce	0.78	30.7
^{239}Pu	24110	0.006 52
^{240}Pu	6563	0.004 35
^{241}Pu	14.35	0.142

[a]Excludes local fall-out.
[b]To two decimal places.
[c]1 EBq = 10^{18} Bq.

Nuclear power The first electricity generating nuclear power reactors started operation in the mid 1950s and a general expansion in the use of nuclear power followed: the number of power reactors increased from 77 to 438 over the period 1970 to 2001 and this was accompanied by an increase in the energy generated from 9 to 351 GW(e) y.[1] As of 1999, nuclear power is used in 32 countries and generates around 17% of the world's total electricity. In some countries, the percentage is much higher; for example, in France the nuclear power contribution was around 75%, while in Belgium it was approximately 56%. France is the country with the largest contribution from nuclear power. However, the United States has the largest nuclear power programme with 104 operating plants providing about 20% of its electricity requirements in 1999 (Nuclear Engineering International, 2001).

The use of nuclear electricity for generating power is expanding in some parts of the world. In 1995, four new nuclear power stations were connected to the electricity distribution grid: one each in the UK, India, the Republic of Korea and Ukraine. There are a further 39 nuclear power plants under construction world-wide and when these are completed the number of countries operating nuclear power stations will increase by three, i.e. Cuba, the Islamic Republic of Iran and Romania.

Two nuclear power plants were shut down in 1995, one in Canada and one in Germany. Some countries currently operating nuclear power stations, for

[1] 'GW(e) y' is a unit of energy used in the electrical power industry. One gigawatt-year of electricity (GW(e) y) is the amount of energy produced by a generator giving 10^9 watts of electricity for one year.

example, the UK, have no plans at present to replace existing plants when they close down.

A controlled nuclear fission chain reaction takes place inside a nuclear reactor. The heat produced is usually used to generate steam to produce electricity. Worldwide, several different types of reactor are in use. The main differences are in the moderator and the coolant. The most common type of nuclear reactor is the pressurized-water reactor (PWR) in which ordinary water (termed 'light water'), under pressure to prevent boiling, takes heat away from the reactor core to heat exchangers. A secondary circuit of water is heated in the heat exchangers to provide steam which drives electricity-generating turbines. The steam is then condensed and returned to the heat exchangers. The water coolant also acts as the moderator. Over half of the reactor units in operation in 1996 were PWRs.

The second major reactor type is the boiling-water reactor (BWR). These also use light water as coolant and moderator. The water coolant is, however, allowed to boil and generate steam directly to power electricity-generating turbines. The steam is cooled and the water is returned to the reactor unit as coolant.

There are other, more minor, types of power reactor. Some use carbon dioxide as coolant and graphite as the moderator, while others use 'heavy water' as coolant and moderator. The various types of nuclear reactor are briefly described in Chapter 6.

During normal operation, nuclear power reactors discharge very low levels of radioactive waste to the environment as liquid and gaseous effluents. These wastes arise because during the operation of a reactor, radioactive fission products are generated in the fuel and activation products are formed in structural and cladding materials. Coolant becomes contaminated as fission products diffuse into it from fuel that has defective cladding and corrosion particles also become activated as they pass through the core. These processes may require the coolant to be replaced or treated to remove contaminants; thus, radioactive wastes arise. Discharges may also occur from fuel storage ponds (see below). All reactors have effluent treatment systems to keep discharges to a minimum. The discharges of radionuclides from nuclear power reactors are summarized in Table 1.10. These data are taken from the 1993 UNSCEAR and 2000 UNSCEAR reports (UNSCEAR, 1993, 2000a).

Table 1.10 Estimated total releases[a] of radionuclides from nuclear power reactors up to 1997 (inclusive).

Radionuclide	Release (TBq)[b,c]
Noble gases	922 400
^{3}H (atmospheric)	117 622
^{14}C	1984.8
^{131}I	45.7
Particulates	122.0
^{3}H (liquid)	150 888
Other (liquid)	837.9

[a]Releases during normal operations (thus excluding accidents).
[b]$1\ TBq = 10^{12}\ Bq$.
[c]Discharges to 1989 taken from UNSCEAR (1993) and discharges from 1990 to 1997 taken from UNSCEAR (2000a).

Nuclear fuel in a reactor does not last indefinitely; ^{235}U is used up and there is a buildup of fission products that absorb neutrons and inhibit the chain reaction. Thus, fuel requires replacement and the spent fuel has to be dealt with adequately. In a typical 1000 MW PWR with 200 fuel elements, around one third of the fuel elements are replaced each year. The spent fuel is highly radioactive and is still generating heat. For cooling at the power station it is placed in a purpose-built facility, which is normally a water pool. Following a period of cooling, one of two things will ultimately happen to the spent fuel: chemical dissolution and treatment to recover unused ^{235}U and potentially useful ^{239}Pu (reprocessing) or disposal following appropriate conditioning. Reprocessing also produces solid waste in the form of vitrified fission products. Currently, both vitrified fission products and spent nuclear fuel are being stored pending construction of appropriate disposal facilities.

Commercial reprocessing is at present only undertaken at two sites, i.e. Sellafield in the UK and Cap de la Hague in France. Japan also intends to build a commercial reprocessing plant, but plans to operate plants in the USA and in Germany were abandoned. The economics of reprocessing are complex and are linked to the price of uranium – if uranium is cheap, the incentive to reprocess is diminished. Furthermore, concerns have been voiced over the possible illegal use of recovered ^{239}Pu in weapons manufacture. The discharges to the environment from reprocessing are summarized in Table 1.11. The alternative, direct disposal of spent fuel, is being pursued in some countries including the USA, Sweden, Canada and Finland. The discharges from reprocessing can appear high when compared with those from power generation (see Table 1.10), particularly if account is taken of the fact that only around 5 to 10% of spent nuclear fuel is reprocessed. The summed figures in Tables 1.10 and 1.11, however, hide some trends with time. In particular, significant changes have occurred in the discharges of some radionuclides from reprocessing plants. Discharges of ^{137}Cs, ^{90}Sr and ^{106}Ru have decreased significantly between the 1970s and the 1990s. Discharges of many other radionuclides have also decreased over this time period, particularly when normalized to the power that had been produced from the fuel (see UNSCEAR, 2000a).

Table 1.11 Estimated total releases from fuel reprocessing plants up to 1997 (inclusive) (from UNSCEAR, 2000a).

Radionuclide	Release (TBq)[a]	
	Atmospheric	Liquid
^{3}H	10 852	126 012
^{14}C	589	242
^{85}Kr	3 246 000	—
^{90}Sr	—	6620
^{106}Ru	—	14 800
^{129}I	0.92	14.9
^{131}I	4.9	—
^{137}Cs	4.4	41 903

[a] 1 TBq = 10^{12} Bq.

Fuel is fabricated from uranium. PWRs and BWRs use uranium oxide enriched in ^{235}U. Some other types of reactors can use unenriched uranium metal. Uranium is found in several parts of the world, including Africa, Canada, the USA, Australia and parts of Eastern Europe. Typically, uranium-bearing ore contains around 0.2% U_3O_8. After the ore is mined it is transported to nearby mills where the U_3O_8 is separated from the residue by using either acid or alkali leaching. The waste products from this process contain almost all of the radioactive decay products of the original uranium (see Appendix 3, Figure 2 and Table 2). Of particular significance is ^{226}Ra, which acts as a long-term source of the radioactive gas, ^{222}Rn. This is an example of a human activity that causes natural radionuclides to be released into man's environment (see above).

Enrichment in ^{235}U, if required, is undertaken by centrifugation or, in the past, gaseous diffusion. It is at this stage that the uranium recovered by reprocessing would be used.

Accidents at nuclear installations have also contributed to environmental levels of radionuclides in some areas of the world in particular. The most significant accident was the Chernobyl reactor accident of 1986 (see Section 6.4.3). The Chernobyl reactors were of a type largely peculiar to the USSR. They were graphite moderated and cooled by water contained in pressure tubes surrounding each fuel element. Water boils in the upper part of the tube to produce steam that is fed directly to electricity-generating turbines. The accident was caused when, for complex reasons, the operators reduced the flow of water through the core, thus leading to steam voids which in turn made the reactor produce more power. The reactor eventually went 'superprompt critical' and achieved 100 times normal power levels within 4 s. Fuel was ruptured and a steam explosion occurred which resulted in the 1000 tonne reactor cover being displaced and all cooling channels being cut off. After 2 or 3 s, a second explosion occurred and the graphite moderator caught fire. A release of radionuclides occurred over 10 days.

Man-made sources of natural radionuclides

A number of human activities release natural radionuclides to the environment as wastes or by-products. One such activity, that has been mentioned above, is the mining and milling of uranium ore. Other such activities include combustion of fossil fuels, production of phosphate-containing materials from phosphate rock, and mining and milling of mineral sands. These activities are described in Chapter 7.

All fossil fuels contain natural radionuclides to a greater or lesser extent. Typical concentrations of natural radionuclides in coal are: ^{40}K, 50 Bq kg^{-1}; ^{238}U, 20 Bq kg^{-1}; ^{232}Th, 20 Bq kg^{-1}. Some coals have higher concentrations, e.g. coal mined in China. These natural radionuclides may be released to the environment during combustion. The extent of release will depend upon factors such as the combustion temperature and the efficacy of off-gas treatment systems. Other fossil fuels also contain natural radionuclides. Radon is present in natural gas, although decay during storage and distribution results in a reduction in the radon concentration when the gas is eventually burned.

Phosphate rock is a starting point for a number of important materials, including some fertilizers and detergents. The main producers of phosphate rock are China, Morocco, the former Soviet Union and the USA. Phosphate rock contains natural radionuclides. Levels of ^{40}K and ^{232}Th are generally similar to those in soils. However, concentrations of ^{238}U and decay products tend to be elevated. Processing the rock releases significant amounts of natural radionuclides. Important processing plants operate in Europe. Waste products containing natural radionuclides are often discharged in liquid effluents and significant local elevations in concentrations of ^{210}Po in some aquatic foodstuffs have been observed. The phosphate fertilizer product also contains some natural radionuclides; concentrations of ^{238}U and decay products may be up to 50 times higher than normal soil levels.

Mineral sands originate from erosion of inland rocks. These represent an important source of titanium and zirconium. The sands may contain elevated concentrations of ^{232}Th and ^{238}U, together with their respective decay products. Releases of these radionuclides may occur during processing of the sands but enrichment can also occur in some of the products. Mineral sands are mined in Australia, Bangladesh, Indonesia, Malaysia, Thailand and Vietnam.

Miscellaneous sources

There are a number of other minor sources of radionuclides to the environment. These include consumer products containing radionuclides and the use of radionuclides in industry, research and medicine.

Some consumer products deliberately contain radionuclides. These include luminous dials and smoke detectors.

The use of ^{226}Ra to luminize the dials of clocks, watches, etc. has been mentioned earlier. From the 1960s onwards, this use of ^{226}Ra has been progressively discontinued to be replaced by the weak beta emitters, tritium and ^{247}Pm. Light sources containing gaseous tritium have also been used to provide low-level illumination in consumer products such as fishing floats and telephone dials.

The artificial alpha-emitting radionuclide ^{241}Am is used in some smoke detectors. Levels are typically about 40 kBq per detector. These have become increasingly popular in some countries.

Small amounts of radionuclides will be released to the environment following disposal of consumer products containing radionuclides. This is, however, generally considered to be a minor source of radiation exposure. Furthermore, the deliberate addition of radionuclides to consumer products should only be undertaken within radiological protection criteria (see Chapter 4) and many countries have legislation to this end.

Radionuclides are used for various purposes in industry, medicine and research. Isotopes such as ^{14}C, ^{99}Tc and ^{131}I find uses as tracers in research and medical diagnosis. Large ^{60}Co and ^{137}Cs sources are used in radiotherapy. Industrial uses include thickness gauging, weld testing and leak testing, with ^{137}Cs and ^{85}Kr being among the radionuclides used. Sources containing these radionuclides are produced in plants in several western countries. Estimates of annual production rates and of release rates to the environment are given in Table 1.12.

Table 1.12 Production estimates for radioisotopes used in medical educational and industrial applications (from UNSCEAR, 1993).

Radionuclide	Estimated annual global production and release (PBq)[a]
^{3}H	0.13
^{14}C	0.05
^{85}Kr	0.02
^{123}I	0.7
^{125}I	0.06
^{131}I	0.3
^{133}Xe	2.6

[a]1 PBq = 10^{15} Bq.

1.3 SUMMARY AND OVERVIEW OF BOOK

Radioactivity is a property of atoms. Radioactive atoms are called radionuclides. Man's environment is pervaded by ionizing radiation of predominantly natural origin but man's activities can increase radiation levels either by acting on natural sources or by producing artificial radionuclides. Some naturally occurring radionuclides have been present since the origin of the Earth and these are responsible for most of the radiation exposure of the vast majority of individuals; other naturally occurring radionuclides are produced by the action of cosmic rays on atoms in the Earth's upper atmosphere.

Three important types of ionizing radiation are emitted by radionuclides, namely α-radiation, β-radiation and γ-radiation. Identical radionuclides will emit radiation of the same types and energies, and at a characteristic rate. The process whereby radionuclides emit ionizing radiation is called radioactive decay. The physical properties of ionizing radiation and the principles governing radioactive decay are described in Chapter 2. Ionizing radiation can penetrate into matter, thus causing damage by interacting with the atoms and molecules of the medium. This forms the topic of Chapter 3. If the medium is living tissue, damage to cells can occur. Very large doses of radiation will cause wholesale cell death, resulting in gross tissue damage that could lead to death of the organism. Lower doses may also be harmful – these do not cause the immediate damage of high doses but instead act to increase the likelihood of contracting cancer. Thus, exposure to ionizing radiation can have health implications, which is why we are concerned about it and, to a large extent, is why this book has been written. For protection purposes, at low doses, the damage radiation does to human tissue is assumed to be proportional both to the energy deposited in the tissue and to how that energy is deposited. The unit of radiation dose, the sievert (named after a Swedish scientist; symbol, Sv) takes both of these factors into account, as well as the radiation sensitivity of the particular tissue being irradiated. When account is also taken of the differing radiosensitivities of different tissues, a quantity called effective dose can be estimated. This quantity is also measured in sieverts and provides a common basis for comparing exposures to different radionuclides, each of which may irradiate different tissues with different types of radiation (see Chapter 4 for a more complete description of effective dose). Sieverts are relatively large units and sub-multiples such as the millisievert (one thousandth of a sievert) and the microsievert (one millionth of a sievert) are more commonly used. The effective dose that an average member of the world's population would receive in a year is given in Table 1.13, together with an indication of the possible ranges. This table is a baseline comparator for many of the radiation doses discussed later in this book. There is another unit, the gray (named after a British scientist; symbol, Gy) that refers only to the energy deposited in the tissue. The biological effects of ionizing radiation and the bases for controlling radiation doses to members of the public are described in Chapter 4.

The fact that ionizing radiation interacts with matter provides a means for its detection; Chapters 8 to 12 deal with the techniques and procedures involved in measuring radionuclide and radiation levels in the environment. The various types of detectors for ionizing radiation, together with their advantages and disadvantages, are described in Chapter 8. Included in this chapter is a description of the associated detector and counting electronics. Measurement techniques employing these detectors, including energy and efficiency calibrations, are the subject of Chapter 9. Broadly speaking, these techniques can be broken down into those that require chemical separations in order to isolate radioisotopes of a particular element and those that do not. Specific methods for determining the environmentally important gas radon and for analysing gamma-ray spectra are also included here. Quantification of activity levels in samples of environmental materials is but one part of the general problem of evaluating levels of radionuclides in the wider environment. In Chapter 10, the importance of sampling in order to make an overall estimate of the level of radioactivity or radiation in samples too large to measure directly is stressed. Various sampling strategies and techniques are described, along with the importance of documentation and storage of samples. Statistical treatment of results from the analyses, including confidence levels and limits of detection, is the subject of Chapter 11. Finally, in chapter 12, approaches to radioactive surveying and remote sensing are described. Although most of the techniques are based on measurements of gamma rays, a brief discussion of surveys based on alpha- and beta-emitting radionuclides is included. The major emphasis is on aerial monitoring, particularly in remote areas and examples of remote monitoring and aerial surveys are included.

For people to receive a radiation dose they have to be exposed to radiation. Some types of radiation, α- and β-radiation, are so short-ranged that in order to cause a radiation dose, radionuclides emitting them usually have to be taken into the body by being inhaled, ingested or, possibly, through wounds. Beta-radiation may have a sufficiently long range that it can also cause a radiation dose, largely to the skin, when produced outside the body. In contrast, γ-radiation has a much longer range and, consequently, radionuclides emitting this type of radiation can cause a radiation dose when they are outside the body, for example, in nearby ground or building materials. Thus, if a particular type of radionuclide is released to the environment from, say, a nuclear power station, it will give rise to radiation doses by characteristic environmental pathways that are determined both by its chemical properties and by the physical properties of the emitted radiation. These

Table 1.13 Average annual effective dose to a member of the world's population (taken largely from UNSCEAR, 2000a).

Source	Annual effective dose (mSv)	Comments
Natural background	2.4	About half of this is due to radon exposure. This can be lower in areas where little radon is produced or where it readily disperses, such as tropical islands. In areas where radon accumulates in buildings, doses can be typically 10 mSv or more. In some areas, annual doses can be several tens of mSv from radon or from external exposure to radionuclides in monazite sands
Diagnostic medicine	0.4	Many individuals will receive no dose, while others undergoing complex examinations will receive more than this figure
Atmospheric weapons testing	0.005	Higher in the northern hemisphere than in the southern hemisphere. The highest annual dose recorded from this source was about 0.15 mSv in 1963
Chernobyl accident	0.002	Dropped from an average value in the northern hemisphere of 0.04 mSv in 1986
Nuclear power production	0.0002	Individuals living near nuclear installations may receive higher doses. The most exposed individuals may receive up to 1 mSv from older Western European plants, with much lower doses from modern facilities

characteristic environmental pathways are called *exposure pathways*. An example should make this clear. ^{131}I is a fairly short-lived β emitter ($t_{1/2}$, 8 days) which can be released from reactors during a severe accident. This radionuclide is taken up by grass. If cows eat the grass, the biological properties of iodine are such that the ^{131}I will concentrate in milk, which, if drunk, will result in an intake of the radioiodine by humans. When isotopes of iodine are taken into the body, they concentrate in the thyroid gland, so giving rise to a dose to that tissue. The 'grass–cow–cow's milk' route is an exposure pathway for radioiodine. Radionuclides of other elements may have different exposure pathways. Isotopes of plutonium are not readily taken up into crops and also do not readily cross the intestinal wall; the main exposure pathway for such isotopes is often inhalation of plutonium-containing material in the air. Other examples of exposure pathways are the consumption of fish containing radionuclides from the discharges to the aquatic environment and external irradiation from γ-emitting radionuclides in soils or sediments. A schematic illustration of possible exposure pathways is shown in Figure 1.1. In many cases, the transfer of radionuclides through the environment has to be modelled, for example, because there are insufficient measurement data. Some general modelling concepts are discussed in Chapter 13. Modelling radionuclide behaviour in the atmosphere, and in the terrestrial and aquatic environments, is described in Chapters 14 and 15, respectively. The integration of modelling techniques and measurement data into a radiological impact assessment is the subject of Chapter 16; an outline scheme for assessing radiation doses to people is shown in Figure 1.2.

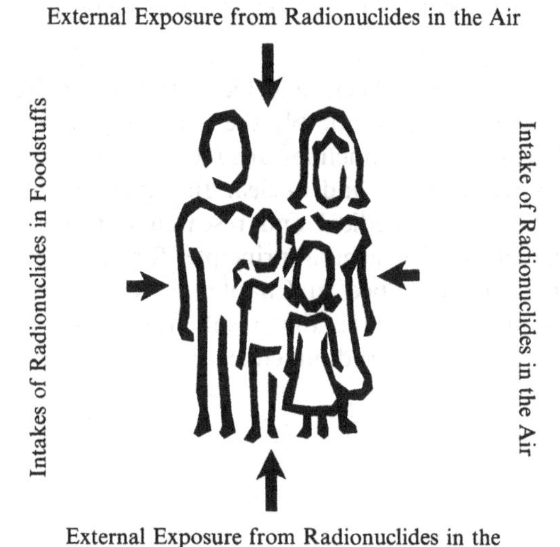

Figure 1.1 Potential radionuclide exposure pathways for humans

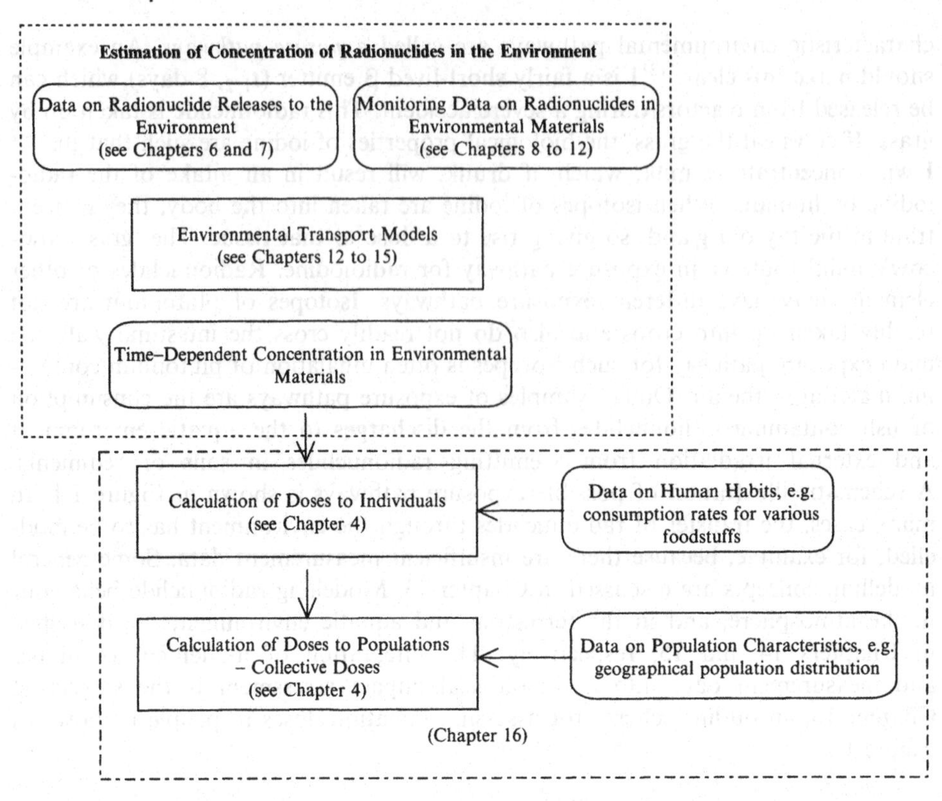

Figure 1.2 Outline scheme of a typical environmental dose assessment process

Artificial radionuclides have been injected into the environment by nuclear weapons testing and by accidents, notably the Chernobyl reactor accident in 1986. Very low levels of artificial radionuclides are released to the environment in effluents from nuclear power stations and nuclear fuel cycle facilities. The production and use of radionuclides for medical and research purposes also leads to some releases to the environment. The general principles for managing radioactive wastes from these industries are described in Chapter 5, while Chapter 6 discusses the environmental levels of radionuclides that result from man's use of nuclear power in its various forms. Releases of radionuclides from other industries are described in Chapter 7.

2

Nature of Radioactivity

The previous chapter provided you with an introductory outline of radioactivity and highlighted its importance in the context of our environment. This chapter considers the various aspects of the nature of radioactivity in more detail. In order to appreciate the concepts regarding radioactivity it is important to have a basic understanding of the structure of matter and the nomenclature adopted in this field. Essential terms, units and definitions are explained before introducing various aspects, such as stability of nuclei, radioactive decay, properties of the main types of ionizing radiation, activity of the daughter products and types of equilibria in radioactive decay.

2.1 STRUCTURE OF MATTER AND NOMENCLATURE

2.1.1 Structure of Matter

All matter that we experience in our everyday lives can be considered to be composed of three main types of particles, as follows:

- Protons
- Neutrons
- Electrons

These particles are the fundamental constituents of *atoms* which form the basic building blocks of all matter. An atom can be envisaged simplistically as a dense, positively charged nucleus surrounded by negatively charged electrons arranged in shells representing definite energy levels. The nucleus is made up of protons, which carry a positive electric charge and neutrons which carry no charge. (The nucleus of a hydrogen atom is an exception and contains one proton only and no neutrons.) Protons and neutrons, which are collectively known as *nucleons*, are held together by strong, short-ranged, attractive nuclear forces. Nuclei have dimensions of the order of 10^{-15} m, whereas atomic diameters (that is, the outer dimensions of the electron cloud) lie in the range $1\text{–}3 \times 10^{-10}$ m, about 10^5 times larger than the nuclear diameter.

Protons and neutrons have approximately the same mass and are about 1840 times heavier than the electron. The charge on the proton is equal in magnitude to that on the electron. In general, an atom will contain the same number of protons and electrons and thus, is electrically neutral. The properties of these sub-atomic particles are summarized in Table 2.1. The electrons in the outermost shells, known as *valance* electrons, take part in chemical combinations of atoms leading to elements and compounds, which form all matter.

2.1.2 Definitions and Nomenclature

Atomic numbers and neutron numbers

The number of protons in a nucleus of an atom is called the *atomic number* (Z) and characterizes that atom. That is, it determines the chemical properties of the atom and hence its position in the Periodic Table. The number of neutrons in a nucleus of an atom is called the *neutron number* (N). For example, carbon has six protons ($Z = 6$) and six neutrons ($Z = 6$) in its nucleus.

Mass numbers and atomic mass

The sum of protons and neutrons, that is, the total number of nucleons, in a given nucleus is known as its *mass number* (A). Thus, we can write:

$$A = Z + N \tag{2.1}$$

The *atomic mass* (or relative atomic mass) is the mass of an atom relative to one twelfth of the mass of an atom of carbon-12. Atomic mass is measured in terms of the *unified atomic mass unit* (u), which is defined such that the mass of carbon-12 is exactly 12 u. In terms of the conventional SI mass units:

$$1\,\mathrm{u} = 1.660\,43 \times 10^{-27}\,\mathrm{kg}$$

Nuclide

The term *nuclide* is used to specify an atom in terms of its atomic number (Z) and mass number (A). Symbolically, a nuclide is represented as ${}^{A}_{Z}\mathrm{X}$ where X is the chemical symbol of the atom. As an example, cobalt has an atomic mass of 59 and an atomic number of 27 and thus can be represented as ${}^{59}_{27}\mathrm{Co}$ ($N = 59 - 27 = 32$). As the atomic number is characteristic of the element, as is its symbol, it is common to omit the value for Z and simply represent the nuclide by its symbol and its mass number. Thus, cobalt can be represented simply as ${}^{59}\mathrm{Co}$, or sometimes written as Co-59.

Table 2.1 Major properties of the three sub-atomic particles.

Particle	Symbol	Charge (C)	Mass (kg)
Proton	p	1.602×10^{-19}	1.673×10^{-27}
Neutron	n	0	1.675×10^{-27}
Electron	e^-	-1.602×10^{-19}	9.110×10^{-31}

Isotopes and natural abundances

An element may have a number of different nuclides which are called *isotopes*. These are nuclides which have the *same* number of protons (Z) but different number of neutrons (N). As the isotopes of a given element have the same Z, they retain the chemical name of that element. For example, $^{31}_{15}P$ and $^{32}_{15}P$ are isotopes of phosphorus. Sometimes, the atomic number is omitted for convenience as it remains unchanged. In the case of phosphorus, the isotopes are indicated as ^{31}P and ^{32}P and are referred to as phosphorus-31 and phosphorus-32, respectively. Hydrogen is a special case where its three isotopes are given different names, i.e. hydrogen (^{1}H), deuterium (^{2}D) and tritium (^{3}T).

The isotopes of a given element, of course, are not all equally common. For example, 98.9% of naturally occurring carbon is the isotope ^{12}C. These percentages are known as *natural abundances*. In the case of uranium, the isotopes ^{238}U, ^{235}U and ^{234}U have abundances of 99.27, 0.72 and 0.006%, respectively.

Electronvolt

The *electron volt* (eV) is a convenient unit to express atomic and nuclear energies and is defined as the energy gained by an electron when passing through an electrical potential of one volt. Energies of atomic radiation, such as X-rays, are usually expressed in *kilo electron volts* (keV) where:

$$1\ \mathrm{keV} = 10^3\ \mathrm{eV}$$

while energies of nuclear radiation, such as gamma rays, are usually quoted in millions of electron volts (MeV). Thus:

$$1\ \mathrm{MeV} = 10^3\ \mathrm{keV} = 10^6\ \mathrm{eV}$$

The relationship between the SI units of energy, joules (J), and the electron volt is given by the following:

$$1\ \mathrm{eV} = 1.602 \times 10^{-19}\ \mathrm{J}$$

The unified atomic mass unit (u) can also be expressed in this energy unit where:

$$1\ \mathrm{u} = 931.5\ \mathrm{MeV}$$

This arises by the virtue that, in physics, there is a definite relationship between energy and mass, which allows one to be expressed in terms of the other.

2.2 STABILITY OF ATOMIC NUCLEI

The mass of a given nucleus is always less than the sum of the masses of its constituent *nucleons* (protons and neutrons). One may then ask how is this possible? Where has this mass gone? This mass difference, known as *mass excess*, arises because energy (equivalent to the mass difference) is released when nucleons are combined to form a nucleus. Conversely, it is the minimum energy that must be supplied into a nucleus in order to break it apart into its constituents. If, for example, the mass of a helium nucleus ($^4_2$He) was exactly equal to the mass of the two neutrons plus two protons, the nucleus would break apart without any input of energy. This energy (or mass) difference is referred to as the *binding energy* of the nucleus. For example, the binding energy of ^{4}He is 28.3 MeV.

It is, however, more useful to define the *average binding energy per nucleon* (binding energy of a nucleus divided by the number of its nucleons) when considering the stability of the nucleus. Figure 2.1 illustrates a plot of the average binding energy per nucleon as a function of the mass number (A) for stable nuclei. It can be seen that the nuclides of intermediate mass numbers have the largest value of binding energy per nucleon. The average binding energy per nucleon of ^{56}Fe, for example, has a value of 8.8 MeV and is one of the most stable nuclides. At higher mass number values, the curve falls slowly, indicating that larger nuclei are held together less tightly than those in the intermediate mass range.

As shown in Figure 2.1, stable nuclei tend to have equal number of protons and neutrons ($N = Z$) up to about $A = 20$–30. Beyond this, stable nuclei contain more neutrons than protons, with the neutron-to-proton ratio reaching about 1.6. This imbalance in the number of protons and neutrons has the consequence of the nuclear forces being weaker. Nuclides which are too unbalanced in this regard, and thus fall outside the stable region, are termed *unstable*. These nuclides have to undergo changes, resulting in the emission of radiation, in order to achieve a relatively more stable state. In general, nuclides with atomic number greater than 82 are unstable.

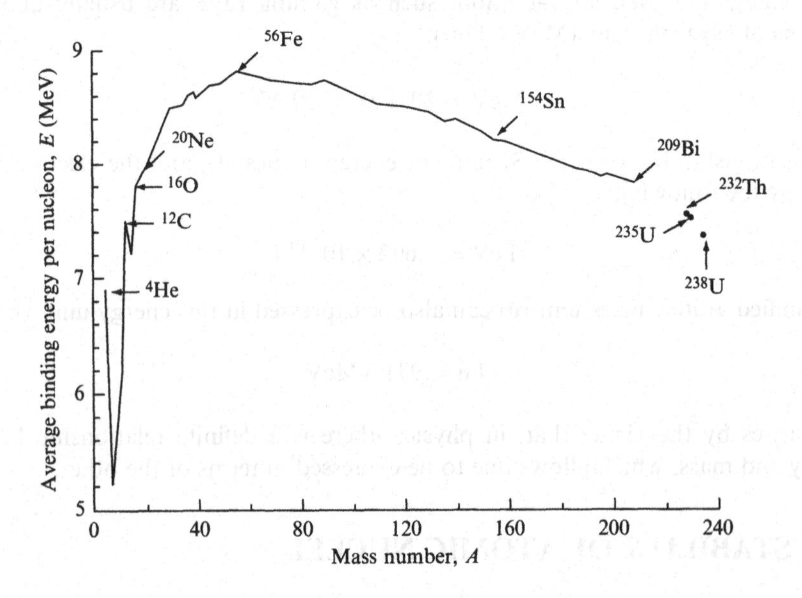

Figure 2.1 Average binding energy per nucleon as a function of mass number for stable nuclei

2.3 RADIOACTIVE DECAY

Certain nuclides, termed *radionuclides* (or radioisotopes), disintegrate by emitting radiation in order to acquire a more stable state. This phenomenon is called *radioactivity* or *radioactive decay*. The disintegration is *spontaneous* and commonly involves the emission of particles called *alpha* (α)- *and beta* (β)-*particles*. The emitting nucleus, called the *parent nucleus*, undergoes a change of atomic number and, therefore, becomes the nucleus of a different element – a process known as *transmutation*. This new nucleus is called the *daughter nucleus* or the *decay product*.

In general, the radioactive decay process can be expressed in the following form:

$$\mathrm{A} \longrightarrow \mathrm{B} + \mathrm{b} \tag{2.2}$$

where a parent radioactive nuclide, A, disintegrates to form the nuclide, B (daughter nucleus), with release of some form of radiation, b. For spontaneous radioactive decay to take place, the mass of the nucleus of the parent radionuclide must be *greater* than the combined masses of the products (daughter nucleus and the emitted particle). This mass difference is converted to kinetic energy which is shared by the products B and b. It should be noted that even if this mass difference exists, it is not a certainty that the radionuclide *will* decay or when it will decay, only that it *can* and *may* decay spontaneously.

If the daughter nucleus is in an *excited* (i.e. unstable) state when it is formed, it reaches its *ground* (or stable) state by emitting a third type of radiation called *gamma* (γ)-*rays*. This process is termed *de-excitation*.

2.4 PROPERTIES OF NUCLEAR RADIATIONS

When a radionuclide decays, the process may, therefore, result in the emission of one or more of the following types of nuclear radiations:

- Alpha (α)-radiation
- Beta (β)-radiation
- Gamma (γ)-radiation

2.4.1 Alpha Particle Radiation

This type of radiation consists of alpha (α)-particles which are identical to the nuclei of helium atoms ($Z = 2$; $N = 2$; $A = 4$). In terms of the nuclide notation, an α-particle is represented as ${}^{4}_{2}\mathrm{He}$. Although *alpha decay* is known to occur in a few of the medium-weight nuclides, such as samarium-147 (${}^{147}\mathrm{Sm}$), it is more common in the heavy nuclides with $Z \geq 83$. During this process, the mass number (A) of the radionuclide decreases by 4 and its atomic number (Z) decreases by 2. The alpha decay process, therefore, can be represented as follows:

$${}^{A}_{Z}\mathrm{X} \longrightarrow {}^{A-4}_{Z-2}\mathrm{Y} + {}^{4}_{2}\mathrm{He} + Q \tag{2.3}$$

where Q, the energy equivalent to the excess mass (see Section 2.2), is shared as kinetic energy between the decay product Y and the alpha-particle. The process thus results in a new nuclide, which may itself be radioactive. Energies of the emitted a-particles are discrete and range from about 3 to several MeV. Examples of some α-emitters are listed in Table 2.2, with additional data given later in Appendix 3.

Alpha-particles have very small ranges in matter and are stopped by a piece of paper and generally do not penetrate the skin. A 3 MeV α-particle has a range of ~ 16 mm in air and is stopped by ~ 15 μm Al foil. Interaction of radiation and matter is considered in detail in Chapter 3.

2.4.2 Beta-Particle Radiation

Beta (β)-decay is the spontaneous transformation of a neutron within the nucleus into a proton. When the number of neutrons in the nucleus is greater than that of a stable nucleus of the same mass number, a neutron is converted into a proton with the emission of a negatively charged energetic electron, called a beta (β)-particle, and another particle known as a *antineutrino*. Such radionuclides are termed *neutron-rich* and the β-emission process results in the N/Z ratio of the radionuclide changing towards a more stable configuration of the same mass number. Since a neutron changes into a proton, the mass number of the radionuclide does not change but the atomic number increases by one, hence resulting in a new nuclide. As in the case of α-radiation, this daughter nucleus may still be radioactive, that is, it may be in an excited state. This process can be represented as follows:

$$^{A}_{Z}\mathrm{X} \longrightarrow {}^{A}_{Z+1}\mathrm{Y} + \beta + \bar{\nu} \qquad (2.4)$$

where $\bar{\nu}$ represents the antineutrino. Unlike α-particles, the emitted β-particles have a continuous energy spectrum, from 0 to a maximum value, with an average energy (E') of about one-third of the maximum β-decay energy, as illustrated in Figure 2.2. Some common β-emitters are listed in Table 2.3.

Table 2.2 Some common α-emitters.

Nuclide notation	Radionuclide name	Energy of major emission (MeV)
^{241}Am[a]	Americium	5.48
^{212}Bi	Bismuth	6.10
^{239}Pu[b]	Plutonium	5.16
^{210}Po	Polonium	5.30
^{226}Ra	Radium	4.78
^{238}U	Uranium	4.20
^{232}Th	Thorium	4.01

[a]Artificially produced emitter.
[b]Pu is produced artificially but very small amounts have been found in uranium ore resulting from interactions with neutrons produced by naturally occurring fission (Eisenbud, 1987).

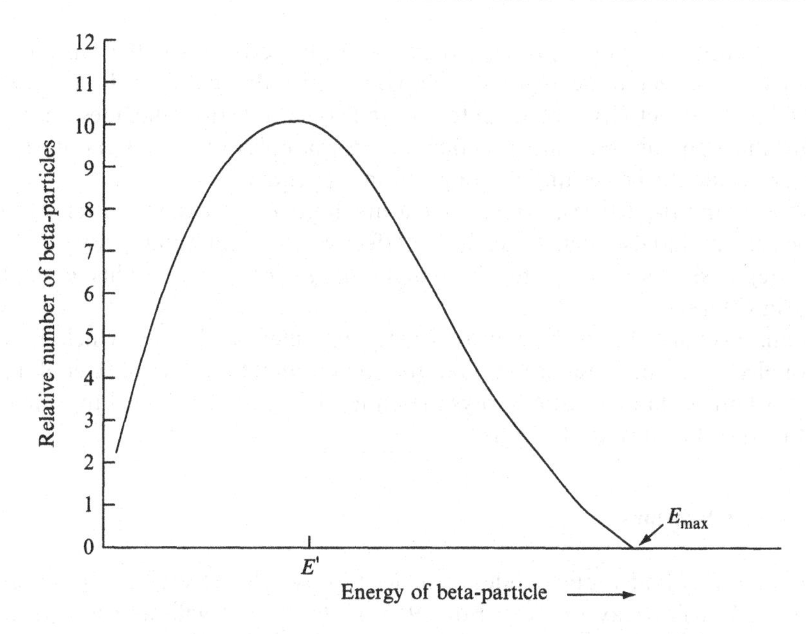

Figure 2.2 Illustration of an typical beta-particle spectrum, showing both the average energy, E', and the end-point energy, E_{max}; $E' \sim 0.33E_{max}$

Table 2.3 Some common β-emitters.

Nuclide notation	Radionuclide name	Energy (MeV)	
		Maximum	Average
^{3}H	Tritium	0.0186	0.006
^{14}C	Carbon	0.156	0.049
^{45}Ca	Calcium	0.252	0.077
^{32}P	Phosphorus	1.71	0.70
^{90}Sr	Strontium	2.27[a]	1.13[b]

[a]From ^{90}Y which is the decay product of ^{90}Sr; 0.55 MeV (maximum) β-particles are emitted by ^{90}Sr.
[b]From ^{90}Sr (0.196 MeV) plus ^{90}Y (0.93 MeV).

β-particles leave the nucleus at velocities close to that of light and penetrate further in matter than α-particles. The amount of material necessary to stop these particles depends on their maximum energy . For example, β-particles with a maximum energy of 1.71 MeV (^{32}P) will be stopped by ~ 3.5 *mm* of glass.

2.4.3 Gamma-Radiation

Gamma (γ)-radiation results from the de-excitation of the decay product nucleius, which is in an excited state, to its ground state. This type of radiation, therefore, is usually accompanied by α- and β-radiation. γ-rays are *high-energy electromagnetic radiation* and have no charge or mass. The energy of these γ-rays are characteristic of the radionuclides which produce them. This property forms the basis of γ-ray spectrometry, a technique regularly employed for identifying radionuclides and quantifying their activity (see Chapters 8 and 9). Many radionuclides emit γ-rays of more than one discrete energy value. Some examples of γ-emitters are listed in Table 2.4, while further examples are given in Appendix 3.

Unlike α- and β-particles, γ-rays have a much greater range in matter. However, high-density materials, such as lead, are effective in attenuating γ-rays and hence lead is regularly used as a shielding material against γ-rays. This is explained further in Chapter 3.

Gamma-rays should not be confused with characteristic X-rays which are also a form of electromagnetic radiation. Gamma-rays originate in the nucleus of a radionuclide, whereas characteristic X-rays are emitted as a result of rearrangements (i.e. transitions) of the atomic electrons.

2.4.4 Decay Schemes

It is common to find radionuclides that decay through more than one mode. For example, ^{40}K will decay to ^{40}Ca. For 89% of the time it will emit a β-particle of energy 1.314 MeV (maximum) and for the rest of the time it decays by emitting a positron (β^+) resulting in the emission of a photon of energy 1.460 MeV. Such a photon is characteristic of ^{40}K and is used to identify this radionuclide by gamma-spectrometry (see Chapter 9). These decay modes can be represented in the form of schematic diagrams called *decay schemes*, as illustrated in Figure 2.3 for ^{40}K.

Table 2.4 Some common γ-emitters.

Nuclide notation	Radionuclide name	Energy (MeV)
^{60}Co	Cobalt	1.17, 1.33
^{131}I	Iodine	0.364, 0.638
^{137}Cs	Caesium	0.662, 0.032
^{40}K	Potassium	1.460

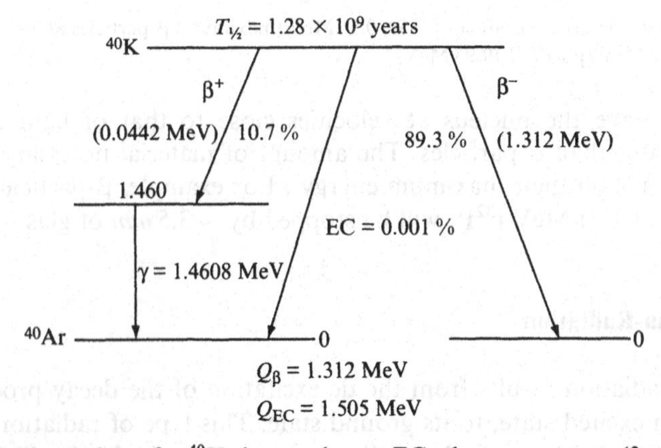

Figure 2.3 An illustration of a ^{40}K decay scheme; EC, electron capture (from OECD/NEA Data Bank, 1993)

2.5 RATE OF RADIOACTIVE DECAY AND HALF-LIFE

A sample of any radionuclide consists of a vast number of radioactive nuclei which decay in a random nature; we cannot predict exactly when a given nucleus will decay. The process, therefore, has to be treated in a statistical manner. We, thus, say that there is a *probability* that a given radioactive nucleus will decay. Over a given period of time we can predict approximately how many radioactive nuclei will disintegrate. It should be emphasized that the probability that any nucleus will decay at a particular time is *independent* of the fate of neighbouring nuclei, the chemical state of the atom and the physical conditions, such as pressure and temperature. The decay process, therefore, cannot be speeded up or slowed down.

For a large number of radioactive nuclei, the *rate of decay*, that is, the number of *disintegrations per unit time*, is *proportional to the number of parent nuclei present*. Therefore, if there are N parent nuclei present at any given time t, then the *rate of decay*, also called the *activity* of the radionuclide source, is given by the following equation:

$$\frac{dN}{dt} = -\lambda N \tag{2.5}$$

where dN/dt is the rate of decay (activity), and λ is a constant of proportionality, called the *decay constant* for a particular radionuclide.

The negative sign in Equation (2.5) is included because the number of parent nuclei present (N) are *decreasing* as the source decays with time. The activity of the radioactive source is expressed in the unit of *becquerel (Bq)*, which is equal to an activity of *one disintegration per second*. The unit of λ is s^{-1}.

The radioactive decay law can be derived from Equation (2.5), and can be expressed mathematically by the following equation:

$$N = N_0 e^{-\lambda t} \tag{2.6}$$

where N_0 is the initial number of radioactive nuclide at time $(t) = 0$.

Equation (2.6) expresses the *exponential* nature of radioactive decay, that is, the number of nuclei remaining after some time t, and therefore the activity, decreases exponentially with time. This is illustrated in Figure 2.4(a). By taking *natural* logarithms, Equation 2.6 can be transformed to a linear form, as follows:

$$\ln(N) = \ln(N_0) - \lambda t \tag{2.7}$$

This equation is illustrated in Figure 2.4(b) which shows a linear plot of ln (N) versus time (t). The gradient of this plot gives the value of the decay constant (λ) directly.

It is clear from Equation (2.6) that the *lifetime* of a radioactive source, that is, the time it would take for the radionuclides in the source to disintegrate completely ($N = 0$), is infinity. It is thus impractical to use this as a measure of how quickly a source will decay. In order to overcome this difficulty, it is usual for the rate of decay of a radioactive source to be specified in terms of its *half-life* ($T_{1/2}$). *The half-life of a radionuclide source is defined as the time taken for half of the nuclei present to decay*, that is, the time taken for N to equal $N_0/2$.

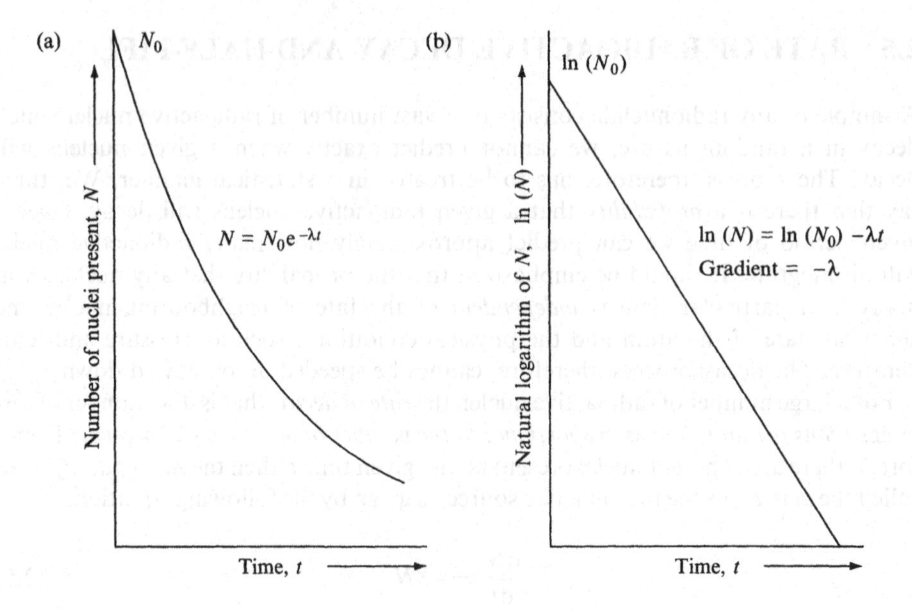

Figure 2.4 (a) A schematic representation of the exponential nature of radioactive decay. (b) The linear relationship between ln (N) and time (t), as defined by Equation (2.7). The gradient of the straight line gives the decay constant, λ

An expression for the half-life of a source in term of λ can be derived from Equation (2.6) by substituting $N_0/2$ for N, and then taking natural logarithms of both sides to give the following:

$$T_{1/2} = \frac{\ln(2)}{\lambda} = \frac{0.6931}{\lambda} \tag{2.8}$$

For example, the $T_{1/2}$ of ^{137}Cs is 30.17 years, and if at some time a source contains 1×10^{22} nuclei, then after 30.17 years, the source will contain 0.5×10^{22} radioactive nuclei. After another 30.17 years, there will be only 0.25×10^{22} nuclei, and so on. After ten half-lives, the activity drops to about a thousandth of its original value. Half-lives of radionuclide sources vary from as short as 10^{-22} s to 10^{28} s (about 10^{21} years). Examples of some common radionuclides with their respective half-lives are given in Table 2.5.

The half-life of a radioactive material has obvious implications when considering environmental hazards posed by radiation. If the half-life is short, the radionuclide may not pose a significant hazard over an extended period as its activity will decay rapidly. Over shorter periods of time, however, it may pose a considerable hazard. Long-lived radionuclides may also pose a significant hazard if taken into the body. In general, materials that are most radioactive are those with relatively short half-lives, while those with long lives tend to emit very little radiation, with the latter being of low penetrating power.

In addition to the half-life, the time-dependence of the exponential decay of a radioactive source can be characterized in terms of the *mean life*, (τ). For a radioactive source containing initially N_0 radioactive nuclei, this parameter, is the time

Table 2.5 Half-lives of some common radionuclide sources.

Nuclide notation	Radionuclide name	Half-life (years)
^{3}H	Tritium	12.26
^{14}C	Carbon	5730
^{40}K	Potassium	1.3×10^{9}
^{60}Co	Cobalt	5.27
^{90}Sr	Strontium	28.8
^{131}I	Iodine	8.04[a]
^{137}Cs	Caesium	30.17
^{210}Po	Polonium	138.38[a]
^{226}Ra	Radium	1.6×10^{3}
^{234}Th	Thorium	24.1[a]
^{238}U	Uranium	4.51×10^{9}
^{239}Pu	Plutonium	2.41×10^{4}

[a]Measured in units of days.

taken for these nuclei to decay to 1/e of their original number, that is, the time taken for N to equal N_0/e. Thus, we can write the following:

$$\frac{N}{N_0} = \frac{1}{e} = e^{-\lambda\tau} \tag{2.9}$$

Taking natural logarithms on both sides then gives:

$$\ln\left(\frac{1}{e}\right) = \ln\left(e^{-\lambda\tau}\right)$$

$$\therefore -1 = -\lambda\tau \tag{2.10}$$

$$\therefore \tau = \frac{1}{\lambda}$$

The mean life represents the average time from an arbitrary initial value to a time later when the nucleus disintegrates. Note that the difference in the magnitude of the mean life and the half-life is a factor of ln (2). In addition, whereas the half-life refers to a radioactive source containing many nuclei, the mean life refers to the decay of an individual nucleus.

2.6 ACTIVITY OF DAUGHTER PRODUCTS AND PARTIAL DECAY CONSTANTS

In order to quantify the radioactivity of a source we have to consider not only the radiation emitted by the parent nuclei but also that emitted by its *daughter products*. It is common that a nucleus will decay through more than one mode of disintegration, that is, it will produce more that one daughter product. Each mode of disintegration will have an associated *partial decay constant* (λ_i, where

$i = $ a, b, c...). The *total decay constant* (λ) can now be expressed as the sum of the partial decay constants, as follows:

$$\lambda = \lambda_a + \lambda_b + \lambda_c + \ldots = \sum_{i=a}^{n} \lambda_i \tag{2.11}$$

where n is the number of decay modes. The total activity, λN, is now given by a similar expression, namely:

$$\lambda N = \lambda_a N + \lambda_b N + \lambda_c N + \ldots = \sum_{i=a}^{n} \lambda_i N \tag{2.12}$$

The *partial activity*, $\lambda_i N$, associated with the ith mode of disintegration, can be written as follows:

$$\lambda_i N = \lambda_i N_0 \mathrm{e}^{-\lambda t} \tag{2.13}$$

where N has been substituted from Equation (2.6). In relation to the above equation, we should note that each partial activity, $\lambda_i N$, will decay at a rate that is determined by the total decay constant, λ, and not by λ_i. This is because the number of available nuclei, N, at the time, t, for each mode of disintegration is the same for all modes and the reduction of N results from the combined action of the different decay modes. Furthermore, as each λ_i is a constant, the individual partial activities are proportional to the total activity. Thus, the ratios of $\lambda_i N/\lambda N$ are constants and their sum for all i modes will equal unity, that is:

$$\sum_{i=a}^{n} \frac{\lambda_i N}{\lambda N} = \sum_{i=a}^{n} \frac{\lambda_i}{\lambda} = 1 \tag{2.14}$$

We can consider a radioactive source containing initially N_1 nuclei after time t, decaying with a decay constant of λ_1 resulting from a given type of disintegration and that the decay of the N_2 daughter product that results is characterized by the decay constant λ_2. Then the net rate of the accumulation of the daughter nuclei is given by the rate of production of these daughter nuclei minus the rate of their removal, as follows:

$$\begin{aligned} \frac{\mathrm{d}N_2}{\mathrm{d}t} &= \lambda_1 N_1 - \lambda_2 N_2 \text{ and since } N_1 = N_{10}\mathrm{e}^{-\lambda_1 t} \\ \frac{\mathrm{d}N_2}{\mathrm{d}t} &= \lambda_1 N_{10}\mathrm{e}^{-\lambda_1 t} - \lambda_2 N_2 \end{aligned} \tag{2.15}$$

where N_{10} represents the initial number of parent nuclei. The activity of the daughter product can be derived by solving the above differential equation. The derivation of this equation is beyond the scope of this present book and the interested reader is referred to Attix (1986) for further details. The final equation for the daughter product activity is given by the following:

$$\lambda_2 N_2 = \lambda_1 N_{10} \frac{\lambda_2}{\lambda_2 - \lambda_1} (e^{-\lambda_1 t} - e^{-\lambda_2 t}) \tag{2.16}$$

From this equation, one can deduce that the activity of the daughter product will be zero when $t = 0$ and ∞. As the activity of the parent nuclei is given by the following:

$$\lambda_1 N_1 = \lambda_1 N_{10} e^{-\lambda_1 t} \tag{2.17}$$

then the ratio of the daughter to parent activity is given by dividing Equation (2.16) by Equation (2.17), as follows:

$$\frac{\lambda_2 N_2}{\lambda_1 N_1} = \frac{\lambda_2}{\lambda_2 - \lambda_1} (1 - e^{-(\lambda_2 - \lambda_1)t}) \tag{2.18}$$

This equation is valid if we assume that the parent decays only to one daughter; however, in most cases other daughter products are produced and in such situations the ratio of the daughter to parent activity is given by multiplying Equation (2.18) by the fractional decay constant λ_{1a}/λ_1, where λ_{1a} is the decay constant for type-a disintegration. Thus, the ratio of the activity of daughter a to that of the parent activity is given by the following expression:

$$\frac{\lambda_2 N_2}{\lambda_1 N_1} = \frac{\lambda_{1a}}{\lambda_1} \frac{\lambda_2}{\lambda_2 - \lambda_1} (1 - e^{(\lambda_2 - \lambda_1)t}) \tag{2.19}$$

2.7 EQUILIBRIA IN RADIOACTIVE DECAY

As we have already seen, the activity of a daughter product resulting from a pure population of parent nuclei is zero when $t = 0$ and ∞, which implies that there will be a maximum activity at an intermediate time t_m. The value for t_m can be calculated from the following:

$$t_m = \frac{\ln (\lambda_2/\lambda_1)}{(\lambda_2 - \lambda_1)} \tag{2.20}$$

At t_m, the activity of the daughter product is at a maximum when starting from an initially pure population of parent nuclei. If the decay constants for the daughter and the parent are equal ($\lambda_{1a} = \lambda_1$), that is, the parent only has one daughter, then it can be shown that the activities of the parent and the daughter will also be equal at time $t = t_m$. Figure 2.5 shows schematically the relative activities of a parent and its daughter nuclei. At a time $t = 0$, the parent nuclei are pure and the activity of the daughter is zero ($N_2 = 0$), while at a time $t = t_m$ the activity of the daughter is at a maximum and occurs at the same time at which the activity curves of the daughter and parent cross.

As we have already seen, the relationship between the activities of the parent and one daughter depends on their decay constants. If we only consider a parent and one daughter, then there are three possibilities of equilibria that can exist, as follows:

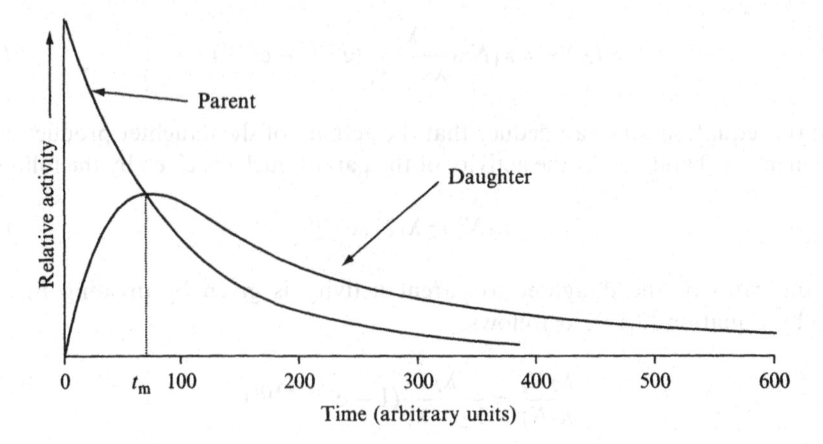

Figure 2.5 An illustration of the relative activities of a parent and daughter radioactive nuclei

(i) *Secular equilibrium*, when the half-life of the parent is much greater than that of the daughter ($\lambda_1 << \lambda_2$).

(ii) *Transient equilibrium*, when the half-life of the parent is equal to or nearly equal to (just greater than) that of the daughter ($\lambda_1 \sim \lambda_2$).

(iii) *Non-equilibrium*, when the half-life of the parent is less than that of the daughter ($\lambda_1 > \lambda_2$).

2.7.1 Secular Equilibrium

For large values of time (t), we can show, from Equation (2.18), that:

$$\frac{\lambda_2 N_2}{\lambda_1 N_1} = \frac{\lambda_2}{\lambda_2 - \lambda_1} \tag{2.21}$$

In this case, as the half-life of the parent is much greater than that of the daughter, that is, the daughter decays much quicker than does the parent and, therefore, $\lambda_1 << \lambda_2$, the equation simplifies to the following:

$$\frac{\lambda_2 N_2}{\lambda_1 N_1} = \frac{\lambda_2}{\lambda_2 - \lambda_1} = \frac{\lambda_2}{\lambda_2} = 1$$

and so:

$$\lambda_2 N_2 = \lambda_1 N_1 \tag{2.22}$$

Thus, given an initially pure parent, the activity of the daughter will increase with time until it equals the activity of the parent, at which stage daughter and the parent are at *secular equilibrium*. This stage is reached after about seven half-lives of the daughter, again assuming that at $t = 0$, $N_2 = 0$. At equilibrium, every time that a parent nucleus decays a daughter nucleus decays, that is, they decay at the same rate and thus the rate of creation of the daughter nuclei is balanced by their rate of decay. As a result, the total activity is now approximately twice the original activity and the decay now follows the decay rate of the parent (see Figure 2.6). As the half-life of the parent is very large, its decay can be neglected. You should note that the ratio of the parent to daughter nuclei is equal to the ratio of their half-lives.

2.7.2 Transient Equilibrium

For large values of time ($t >> t_m$), the ratio of the activities of the daughter and the parent becomes constant, as shown earlier in Equation (2.21), assuming that at $t = 0$, $N_2 = 0$. Thus:

$$\frac{\lambda_2 N_2}{\lambda_1 N_1} = \frac{\lambda_2}{\lambda_2 - \lambda_1} \tag{2.21}$$

In this case, where the half-life of the parent is greater, but not much greater, than the half-life of the daughter, that is, $t_{1/2(1)} > t_{1/2(2)}$, or $\lambda_1 < \lambda_2$, the activity of the daughter decreases at the same rate as that of the parent (shown in Figure 2.7). At

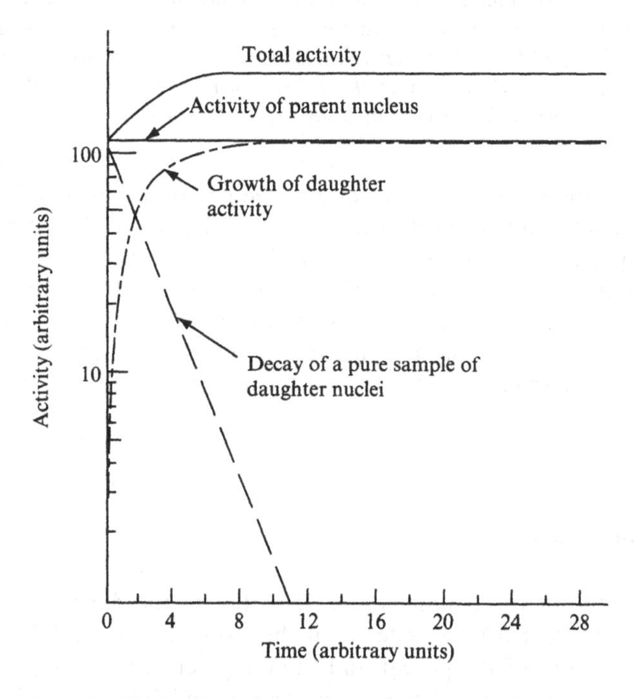

Figure 2.6 A schematic illustration of a secular equilibrium between a parent and the resulting daughter nuclei. In this case, the half-life of the parent is much greater than that of the daughter. Reproduced from Kathren, R. L., *Radioactivity in the Environment*, Figure 3.3, p. 57, Harwood Academic, 1991, with permission from Taylor & Francis

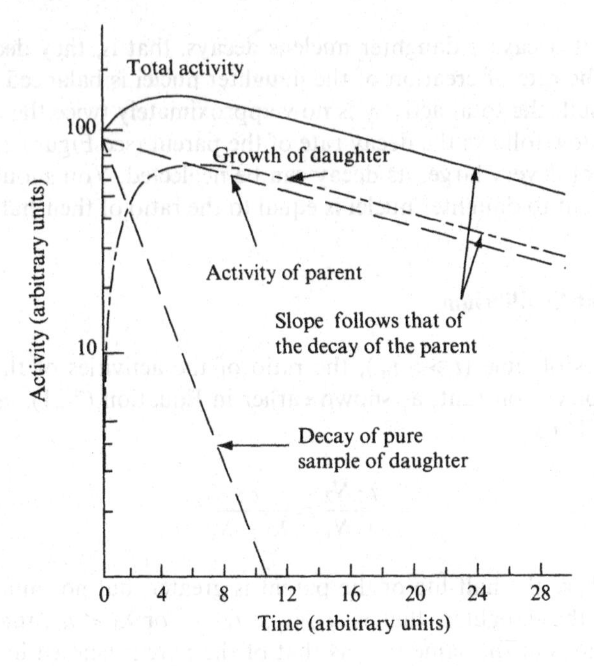

Figure 2.7 A schematic illustration of a transient equilibrium between a parent and the resulting daughter nuclei. In this case, the half-life of the parent is equal or nearly equal to (just greater than) that of the daughter. Reproduced from Kathren, R. L., *Radioactivity in the Environment*, Figure 3.4, p. 59, Harwood Academic, 1991, with permission from Taylor & Francis

this stage, the activities of the daughter and the parent are in *transient equilibrium*. Before this equilibrium is reached, the activity of the parent decreases as the activity of the daughter increases. The total activity, however, never reaches twice of the original value as in the case of secular equilibrium. As the parent decays more slowly than the daughter when equilibrium is reached, the activity of the daughter remains greater than that of the parent. At the time when the activity of the daughter, $\lambda_2 N_2$, is a maximum, i.e. $t = t_m$, the activities of the parent and daughter are equal.

2.7.3 No Equilibrium

When the half-life of the parent is less than that of the daughter ($t_{1/2(1)} < t_{1/2(2)}$, or $\lambda_1 > \lambda_2$) , that is, it decays faster than the daughter, no equilibrium is reached between the activities of the parent and the daughter. This is illustrated in Figure 2.8, which shows the activity of the parent decreasing while the number of daughter nuclei increase exponentially with time (assuming that initially the parent nuclei are pure, and $N_2 = 0$). If the parent decays to a stable daughter, then the number of daughter nuclei will exactly equal the number of parent nuclei that have decayed.

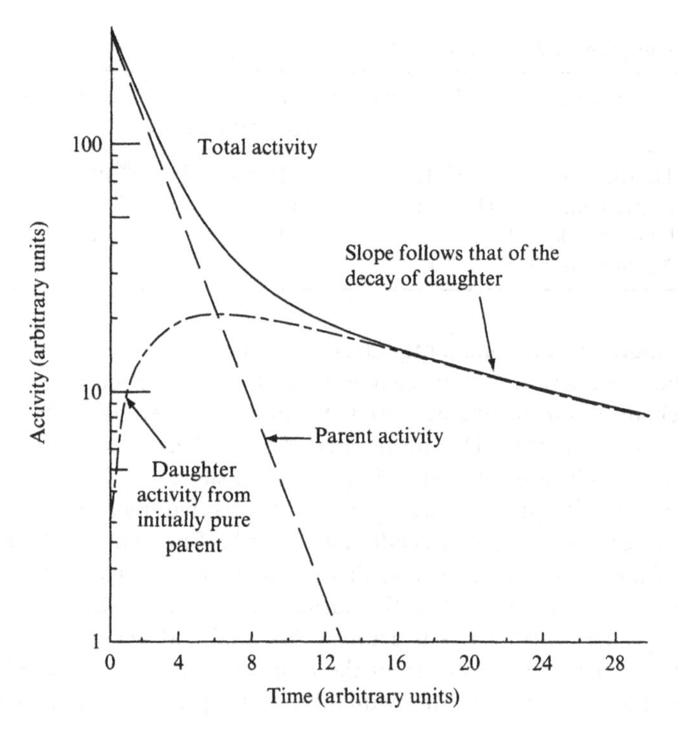

Figure 2.8 A schematic illustration of a 'no-equilibrium' situation between a parent and the resulting daughter nuclei. In this case, the half-life of the parent is less than that of the daughter. Reproduced from Kathren, R. L., *Radioactivity in the Environment*, Figure 3.5, p. 60, Harwood Academic, 1991, with permission from Taylor & Francis

2.8 RADIONUCLIDE DECAY SERIES

In nature, there are three radioactive series or chains of radionuclides. In addition to these three, there is a fourth, known as the Neptunium Series, which is artificial and not found in nature. These four series can be characterized in terms of the mass numbers of their constituents, as follows:

$$A = 4n + m \tag{2.23}$$

where A is the mass number of a member of the series, n is the largest integer divisible into A and m is the remainder. Thus, the four series are characterized by the following expressions:

$$A = 4n \tag{2.24}$$

$$A = 4n + 1 \tag{2.25}$$

$$A = 4n + 2 \tag{2.26}$$

$$A = 4n + 3 \tag{2.27}$$

Table 2.6 Basic characteristics of the four radioactive series.

Name of series	First member	Half-life of first member (years)	Last member
Thorium ($4n$)	^{232}Th	1.41×10^{10}	^{208}Pb
Neptunium ($4n + 1$)	^{241}Pu	13.2	^{209}Bi
Uranium ($4n + 2$)	^{238}U	4.51×10^{9}	^{206}Pb
Actinium ($4n + 3$)	^{235}U	7.1×10^{8}	^{207}Pb

Heavy radionuclides predominantly decay by alpha-emission, which means that there will be a reduction in the mass number by 4 (see Section 2.4) or by beta-decay, which results in an increase of the atomic number by 1, but with also an affect on the mass number. This means that the members of a series will only have mass numbers which satisfy the above four relationships.

The three series found in nature all end with a stable nuclide, i.e. an isotope of lead, and are headed by a radionuclide that has a half-life which is longer than that of the age of the earth, hence making them of primordial origin. The radionuclides in these series account for most of the natural radioactivity whose origin is terrestrial. Although the fourth series is artificial, its final end nuclide (^{209}Bi) is, in fact, found in nature. The characteristics of the four series are summarized in Table 2.6, while more detailed data on the individual series are presented in Appendix 3.

3

Interaction of Radiation with Matter

Radiation can penetrate into matter causing damage by interacting with the atoms and molecules of the medium. In order to understand the processes, we have to consider the interactions between the incident radiation and the receiving material at atomic collisional scales. This chapter considers the main processes that are important when alpha- or beta-particles or gamma-rays pass through matter. The chapter begins with an introduction to the basic processes of ionization and excitation before moving on to the interaction of gamma-rays and simple charged particles with matter. A short treatment is given of the chemical and biological processes that take place when radiation passes through cells before briefly introducing the concept of *Kinetic Energy Released per unit Mass* (Kerma).

3.1 IONIZATION AND EXCITATION

3.1.1 Ionization

When a charged particle, such as an α- or a β-particle, passes close to an atom, electrostatic forces operate between it and the orbital electrons surrounding the nucleus of the atom. If it passes close enough, one of the electrons may acquire sufficient energy to escape from the atom. This process is known as *ionization* and the atom is said to have become 'ionized'. The atom is now no longer neutral and has a net positive charge because it has lost an electron. The atom, together with the escaped electron, forms an *ion pair*. During this process, the particle loses energy, and the amount of energy lost by the particle depends on the nature of the medium through which it is passing, its initial energy and the type of particle. In air, for example, 34 eV energy is required to form an ion pair. Hence, if we have an α-emitter giving 4 MeV α-particles with 1 MBq (10^6 Bq) activity, and it transfers all this energy to air through ionization, then there will be:

$$\frac{1 \times 10^6\ (\text{Bq}) \times 4 \times 10^6\ (\text{eV})}{34\ (\text{eV})} = 1.18 \times 10^{11} \text{ ion pairs per second}$$

As the charge on each electron is 1.6×10^{-19} C, the current produced by 1.18×10^{11} ion pairs will be:

$$1.6 \times 10^{-19}(\text{C}) \times 1.18 \times 10^{11}(\text{s}^{-1}) = 1.89 \times 10^{-8}\text{A}$$

Radiation detectors, such as *Geiger-Müller monitors* (see Chapter 8), use such a process to measure radioactivity. It is this process that is mainly responsible for radiation damage of biological tissues, for example, as it can result in a breakup of a molecule or cause other changes that would not normally occur (see Section 3.5).

3.1.2 Excitation

Excitation occurs instead of ionization when insufficient energy is imparted by the incident radiation to the atomic electron to allow it to escape and the electron instead acquires a higher energy within the atom. The *excited* atom may subsequently return to its *normal (ground) state* with the emission of characteristic electromagnetic radiation, such as light in the case of outer-shell electrons. This phenomenon is exploited in the detection of radiation by scintillation counters (see Section 8.1.2).

3.1.3 Directly and Indirectly Ionizing Radiation

Radiation that can produce ionization of atoms in matter is known as *ionizing radiation*. This type of radiation can be further classified into:

- directly ionizing radiation
- indirectly ionizing radiation

Directly ionizing radiation consists of charged particles such as electrons, α-particles and β-particles. *Indirectly ionizing radiation* consists of uncharged particles, such as neutrons, and photons (electromagnetic radiation), such as γ-rays and X-rays.

The basic difference between these two types of ionizing radiation lies in their modes of interaction with matter. Directly ionizing radiation, which consists of charged particles, ionizes the atoms of the medium in small intervals along the path of the particles. As the particles impart their energy to the medium, they lose kinetic energy until they are finally stopped. The more energy they have initially, then the deeper they penetrate before being stopped. In the case of indirectly ionizing radiation, the radiation penetrates through the material without any interaction with the electrons in the medium, until, by chance, they make a collision with electrons, atoms or nuclei, which results in the liberation of *energetic charged particles*, such as electrons. The charged particles liberated are themselves directly ionizing, and it is through them that ionization and damage in the medium are produced. Thus, the basic damage is caused by charged directly ionizing particles, even when the incident radiation is indirectly ionizing.

Directly ionizing radiation, such as α- and β-particles, emitted from radionuclides have a limited energy range and are stopped in a relatively short distance, usually less than a few millimetres in the body. Indirectly ionizing radiation, on the other hand, such as γ-rays, liberate directly ionizing particles deep within a medium – much deeper than alpha- or beta-particles.

3.2 INTERACTION OF GAMMA (γ)-RAYS WITH MATTER

A beam of γ-rays passing through a substance can interact with atoms of the material either through *absorption* or through *scattering*. In the absorption process, a photon is removed as a result of interacting with an atom of the material, and in the second case the photon is deflected from the original direction of the beam. Both of these processes can be described by the term *attenuation*. The interactions responsible for these processes take place between the photons and the orbital electrons of the atom.

The probability of an interaction taking place between an incoming photon and a 'target' atom is related to the *cross-section* of the interaction. The cross-section reflects the probability of an interaction taking place, and hence, greater the probability of interaction, then larger the value of the cross-section for that particular interaction. It is obvious that the chances of a projectile striking a target will depend on the area of the target and the number of projectiles. When considering atomic interactions, the cross-section is considered to be the area associated with the atoms with which the photon (projectile) interacts. Typically, atomic cross-sections are of the order of 10^{-28} m^2. Cross-sections are often quoted in units of barns (b) where:

$$1\,\text{b} = 1 \times 10^{-28}\ \text{m}^2$$

For a photon beam incident on a material of thickness dx, as illustrated in Figure 3.1, the probability of interaction is given by the following:

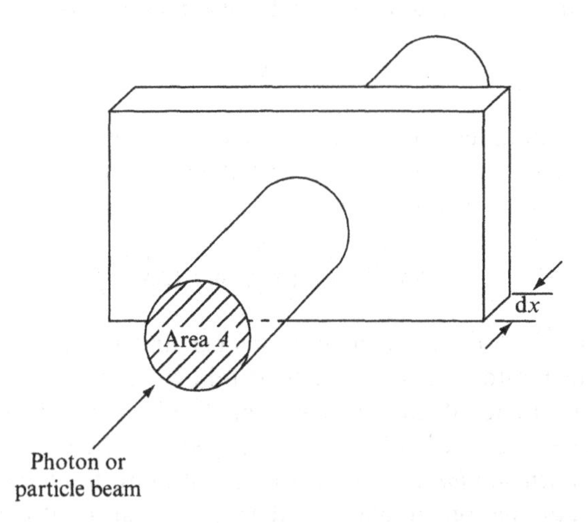

Figure 3.1 A schematic of a photon or particle beam of cross-sectional area A incident on a target of thickness dx

$$\frac{AN\sigma dx}{A} = N\sigma dx \tag{3.1}$$

where A is the area of the incident photon beam, N is the number of atoms per unit volume (m^{-3}), σ is the cross-section of the interaction (m^2) and dx is the thickness of the material. For n photons or particles, the number of interactions is given by the following:

$$nN\sigma dx \tag{3.2}$$

During these number of interactions, dn photons will be attenuated or removed from the incident beam and thus:

$$\begin{aligned} dn &= -nN\sigma dx \\ \therefore \frac{dn}{n} &= -N\sigma dx \end{aligned} \tag{3.3}$$

The n_t photons present in the transmitted beam can now be obtained by integrating Equation (3.3) to give:

$$n_t = n_o \exp(-N\sigma x) \tag{3.4}$$

In terms of the intensity of a homogeneous incoming photon beam, Equation (3.4) can be written as follows:

$$I = I_0 \exp(-\mu x) \tag{3.5}$$

where I is the intensity of the beam after passing through the material of thickness x, I_0 is the incident intensity of the incoming beam and μ is the *total linear attenuation coefficient*. The latter is a function of the number of atoms per unit volume and the cross-section, σ, of interaction and is given by the following:

$$\mu = N\sigma \tag{3.6}$$

The linear attenuation coefficient represents the fraction of X-rays or γ-rays removed from the incoming beam per unit thickness of the target material. The units of μ are m^{-1}, with this becoming clear from Equation (3.6), namely:

$$m^{-3}(N)m^2(\sigma) = m^{-1}(\mu)$$

As μ is proportional to the number of atoms per unit volume of the material with which an incident photon beam interacts, it follows that it is also proportional to the density of the material and thus will have different values for the different phases of the same material (solid, liquid or gas). As there is this dependence on density, the *mass attenuation coefficient* (μ/ρ) is a more fundamental and practical measure of the attenuation of photons than is the linear attenuation coefficient. This parameter is independent of the density of the absorber material and thus does not depend on its physical state. The units of μ/ρ are as follows:

$$\frac{\mathrm{m}^{-1}\ (\mu)}{kgm^{-3}\ (\rho)} = m^2kg^{-1}\ (\mu/\rho)$$

The total mass attenuation coefficient thus represents the fraction of X-rays or γ-rays removed from a beam of photons with a unit cross-sectional area passing through a medium of unit mass. The parameter μ/ρ is related to the cross-section of interaction through the following expression:

$$\frac{\mu}{\rho} = \frac{N\sigma}{\rho}$$

By substituting for N with the expression:

$$N = \frac{N_A\rho}{A} \tag{3.7}$$

we obtain the following:

$$\frac{\mu}{\rho} = \left(\frac{N_A\rho}{A}\right)\left(\frac{\sigma}{\rho}\right) = \frac{N_A\sigma}{A} \tag{3.8}$$

where N_A is the Avogadro constant, 6.02×10^{26} kg mol (or 6.02×10^{23} g mol) which is the number of atoms of carbon in exactly 12 kg (or 12 g) of the isotope carbon-12. The amount of substance containing this number of entities (such as atoms or molecules) is called the *mole* (mol). In the above equation, A is the atomic weight of the absorber material.

In order to maintain the dimension of the exponential term in Equation (3.5), the thickness of the absorber (x) is now replaced by the *areal density* (ρx). Thus, Equation (3.5) becomes:

$$I = I_0 \exp\left(-\frac{\mu}{\rho}\rho x\right) \tag{3.9}$$

where μ/ρ is dependent upon various interaction processes between the photon and the absorbing material.

There are five processes by which γ-rays can interact with matter. These are as follows:

- The Photoelectric Effect
- Compton (Incoherent) Scattering
- Pair Production
- Elastic (Coherent) Scattering
- Photonuclear Interaction

The extent to which γ-rays interact by each of these mechanisms is largely determined by the energy of the γ-rays and the atomic number (Z) of the material. The first three processes are the most important in the context of environmental radioactivity and are accompanied by the transfer of energy to an atomic electron which can be ejected from the atom and can then cause further ionization along its track within the absorber. An understanding of these mechanisms is important when considering the biological effects of radiation (see Chapter 4) and in radiation detection techniques such as gamma-ray spectroscopy (see Chapter 8).

3.2.1 The Photoelectric Effect

The photoelectric effect is the most important mechanism by which low-energy photons interact with matter. In this process, the γ-ray photon interacts with the atom during which the photon is completely absorbed and an inner-shell atomic electron is ejected, as shown in Figure 3.2. The vacancy created during this ionizing process is filled by an electron from higher shells, thus leading to the possible emission of X-rays. This process of X-ray production is more clearly illustrated in Figure 3.3. For example, if the K-shell electron is ejected, then the electrons in the higher shells (L and M) can fill this vacancy, so leading to the possible emission of K_α or K_β X-rays. These X-rays are characteristic of the absorber material. Thus, if a photon of energy E ejects an electron from the K shell with a binding energy E_B, then the energy of the ejected electron (E_j) will be given by the following:

$$E_j = E - E_B \tag{3.10}$$

The recoil energy of the atom is nearly zero and thus can be ignored. In competition to X-ray production, there is a finite probability that the process will result in the emission of an *Auger electron* from the outer shells, rather than X-rays from the inner shells (see, for example, Dyson, 1990 and Tertian and Claisse, 1982). This

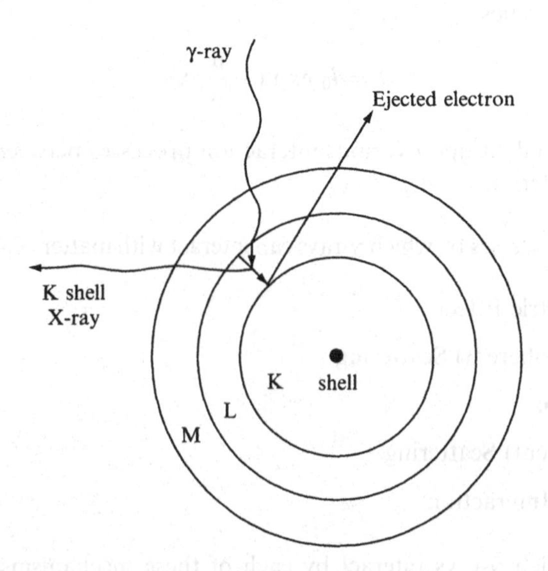

Figure 3.2 Illustration of the photoelectric absorption process where a γ-ray photon is absorbed and a characteristic X-ray is emitted

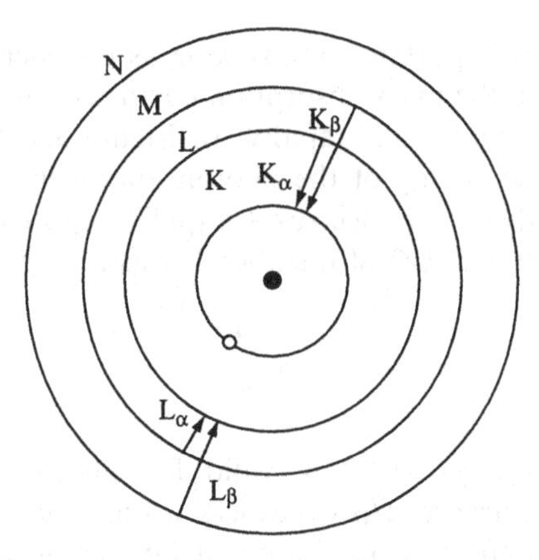

Figure 3.3 Representation of the inner electron shells of an atom and the electronic transitions that can lead to the emission of characteristic X-rays

process cannot occur unless the energy of the incident photon, E, is greater than the binding energy of the inner shell, that is:

$$E > E_B$$

The probability of the photoelectric effect taking place is greatest when:

$$E = E_B$$

and declines rapidly as the photon energy increases. As the incoming photon is removed or *absorbed* from the incident beam, the mass attenuation coefficient is referred to as the *mass absorption coefficient* (τ/ρ), where τ is the *photoelectric linear absorption coefficient*. The rapid decrease in τ/ρ with increasing E is illustrated in Figure 3.4. As the photon energy increases, the absorption due to the individual

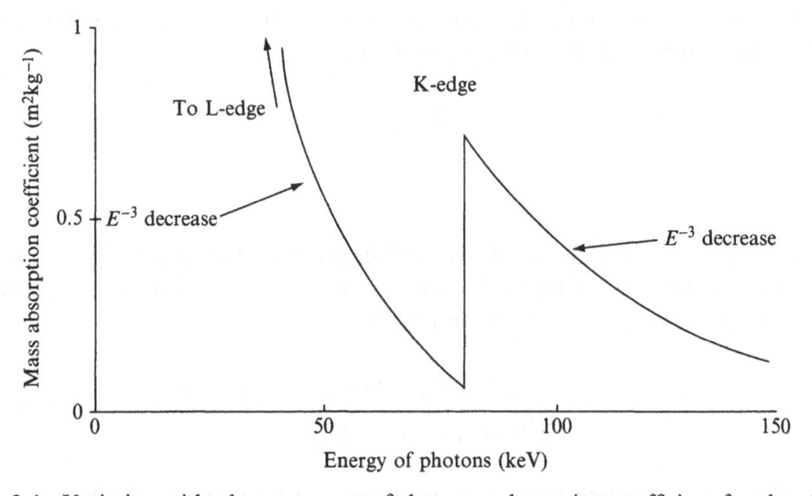

Figure 3.4 Variation with photon energy of the mass absorption coefficient for the photoelectric process

electronic shells becomes important, but as soon as the energy exceeds a particular shell binding energy then the mass absorption coefficient, which gives a measure of the probability of interaction, falls rapidly. In Figure 3.4, the 'edges' refer to the sudden rise in τ/ρ as the energy of the photon approaches the binding energy of a shell. Beyond the edges, τ/ρ decreases rapidly, approximately as $1/E^3$, and increases with atomic number (Z) of the absorber as Z^3, and thus:

$$\frac{\tau}{\rho} \propto \frac{Z^3}{E^3} \tag{3.11}$$

This proportionality is approximately valid for photon energies up to about 200 keV, but at higher energies, τ/ρ varies less strongly and decreases as $1/E^2$, and eventually as $1/E$. This effect is the dominant interaction mechanism for photons of intermediate and low energies (0.5 to 200 keV). As described by Equation (3.11), the absorption increases rapidly with the atomic number of the absorber and this is the main reason why high-Z materials, such as lead, are used for shielding against γ-rays. Such shielding materials are not only used for protection against radiation but also for the purpose of reducing background signals in techniques like γ-ray spectroscopy (see Chapters 8 and 9).

3.2.2 Compton Scattering

At higher photon energies, the significance of binding energies reduces, unlike in the case of the photoelectric effect where the absorption is greatest when $E = E_B$. Here, a photon interacts with an electron that can be considered to be 'free'. In this type of interaction, known as Compton scattering, the incident γ-ray photon is not completely absorbed but is scattered by an atomic electron in the material. The scattering process, illustrated in Figure 3.5, results in the partial absorption of the photon energy and, consequently, the photon has less energy than its initial value and the electron is ejected in the process. The energy (E_1) and momentum (p_1) of the incident photon are given by the following:

$$E_1 = h\nu_1 \tag{3.12}$$

$$p_1 = \frac{h\nu_1}{c} \tag{3.13}$$

where h is the Planck constant, ν_1 is the frequency of the incident photons, and c represents the velocity of light. The energy of the scattered photon (E_2) and the kinetic energy of the recoiled electron (E_e) are given by:

$$E_2 = h\nu_2 \tag{3.14}$$

$$E_e = h\nu_1 - h\nu_2 = h(\nu_1 - \nu_2) \tag{3.15}$$

From the law of conservation of momentum and resolving in the horizontal and vertical directions we obtain the following:

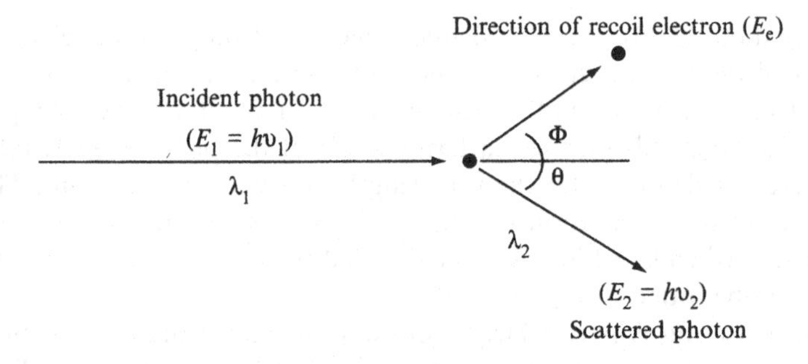

Figure 3.5 Illustration of the Compton scattering process. Here, the incoming photon interacts with an electron and is scattered at an angle θ, whereas the electron recoils forward at an angle ϕ; λ_1 and λ_2 are the wavelengths for the incident and the scattered photons, respectively

$$\frac{h\nu_1}{c} = \frac{h\nu_2}{c}\cos\theta + p_e\cos\phi \tag{3.16}$$

in the horizontal direction, and:

$$0 = \frac{h\nu_2}{c}\sin\theta - p_e\sin\phi \tag{3.17}$$

in the vertical direction: ϕ and θ represent the scattering angles for the recoil electron and the scattered photon, respectively (see Figure 3.5). The energy of the recoiled electron (E_e) is related to its momentum through the equation of special relativity, as follows:

$$p_e^2c^2 = E_e(E_e + 3m_ec^2) \tag{3.18}$$

where m_e is the mass of the electron. By eliminating ϕ and p_e from Equations (3.16) and (3.17), the equation for the energy of the scattered photon ($h\nu_2$) can be derived to give:

$$h\nu_2 = \frac{h\nu_1}{1 + \frac{h\nu_1}{m_ec^2}}(1 - \cos\theta) \tag{3.19}$$

This equation is easily transformed to give an expression for the Compton wavelength shift ($\lambda_2 - \lambda_1$) experienced by the photon as a result of this type of interaction:

$$\lambda_2 - \lambda_1 = \frac{h}{m_ec}(1 - \cos\theta) \tag{3.20}$$

The term h/m_ec ($= 2.43 \times 10^{-12}$ m) is known as the Compton wavelength. The energy of the incident photon is distributed between the recoiled electron and the scattered photon and thus depends on the scattering angle and the original energy of the photon (E_1). Compton scattering is termed an inelastic process because the incident photon loses energy in the process, although the total energy is obviously conserved.

From Equation (3.20), we can deduce that the Compton wavelength shift is independent of the wavelength of the incident photon and of the scattering material. As the Compton wavelength term ($h/m_e c$) is a constant, the Compton wavelength shift only depends on the scattering angle of the photon (θ). It follows then that for a given angle of scatter the wavelength shift will be a constant. The photon energy ($h\nu_2$), given by Equation (3.19), is a maximum when $\theta = 0$ (no scattering) and a minimum when $\theta = 180°$. A more detailed treatment of the Compton scattering process can be found in Dyson (1990).

Compton scattering is particularly important at intermediate γ-ray energies and increases linearly with the atomic number of the absorbing material. This process has particular significance in γ-ray spectroscopy where it leads to the so-called Compton edge and is a major source of background (see Chapter 8). The mass attenuation coefficient for Compton scattering (σ/ρ) (where σ is the linear attenuation coefficient and ρ is the density of the scattering medium) is proportional to the ratio of the electron density of the scattering medium and the energy of the incoming photon. The mass attenuation coefficient, therefore, decreases with increasing photon energy. The electron density can be estimated by using the following simple relationship:

Electron density = Number of atoms per unit mass × Number of electrons per atom

The first term is the Avogadro constant (N_A) divided by the mass number (A) of the medium, while the second term is simply the atomic number (Z), as the number of electrons for an atom is equal to the number of protons in the atom's nucleus. Thus, the mass attenuation coefficient for Compton scattering is given by the following:

$$\frac{\sigma}{\rho} \propto \left(\frac{N_A Z}{A}\right)\left(\frac{1}{E_1}\right) \tag{3.21}$$

For most elements, the ratio Z/A is about 0.5 and it follows that the mass attenuation is of similar magnitude for all substances. Hydrogen is a special case, where the ratio Z/A is unity as $Z = A$ (that is, the hydrogen nucleus only has one proton and no neutron) and thus σ/ρ will be twice the magnitude of the values found for most other substances.

3.2.3 Pair Production

Pair production is another major process by which photons can interact with matter. Although this process can occur in the vicinity of an electron it is much more likely to occur in the proximity of an atomic nucleus. During this process, if a high-energy γ-ray interacts with the electric field of an atomic nucleus it can create an electron and its antiparticle, the positron. As a consequence of this process the photon completely disappears through the following reaction:

$$E_\gamma = 2mc^2 + T_1 + T_2 \tag{3.22}$$

where E_γ is the gamma-ray photon energy, $T_{1,2}$ are the kinetic energies of the electron and the positron, respectively, and mc^2 is the rest-mass energy for each of the electron and positron (m is the mass of the electron or the positron and c is the speed of light). In terms of the energy units MeV, the rest mass of an electron or a positron is 0.511 MeV. A small amount of energy is imparted to the recoiling nucleus which serves to conserve the momentum of the interaction.

As each of these particles have a rest-mass equivalent of 0.511 MeV in terms of energy, Equation (3.22) can be written as follows:

$$E_\gamma \text{ (MeV)} = 1.02 + T_1 + T_2 \tag{3.23}$$

and thus at least 1.02 MeV energy is required for pair production. Any excess energy appears as kinetic energy of the electron/positron pair, as described by Equation (3.23). In practice, this process for absorbing γ-rays is only important if their energies are greater than about 1.5 MeV and becomes dominant for γ-ray energies of greater than 5 MeV.

The positron can annihilate in flight but it is more probable that as it losses its kinetic energy and slows down it will interact with an electron in the absorbing medium and produce two annihilation photons of energy 0.511 MeV (equivalent to mc^2), with each travelling in opposite directions. This process has special significance in gamma-ray spectrometry (Chapter 8). Since environmentally important radionuclides do not emit γ-rays of these energies, this process is not very significant. The mass attenuation coefficient for pair production, κ/ρ, where κ is the linear attenuation coefficient and ρ is the density of the medium, is given by:

$$\frac{\kappa}{\rho} \propto Z(E - 1.02) \tag{3.24}$$

This equation shows clearly that unlike the other attenuation processes, pair production increases with increasing photon energies. It also increases linearly with the atomic number of the attenuating substance.

3.2.4 Elastic (or Coherent) Scattering

Although this process is not significant to the subject of this present text we shall consider it briefly here for the sake of completeness. If the energy of the incoming photon is smaller than the binding energy of the electrons in the absorbing medium, then the photon can interact with one of the electrons and suffer deflection without any loss of energy. In this type of interaction, the electron does not gain significant energy to become excited or ionized but the interaction does cause the electron to oscillate at the same frequency as the incoming electromagnetic radiation (photon). The electron radiates electromagnetic radiation at the same oscillating frequency (with no overall loss of energy) and in the same phase as the incoming radiation. Thus the process is referred to as *Elastic* or *Coherent*. This process is also known as *Thomson* or *Classical scattering*. The result of this interaction is to deflect the photons in the forward direction. Thus, the incoming photon suffers a small attenuation (as a result of the slight forward deflection) although there is no absorption as no energy loss takes place.

The dependence of this process on the energy of the incoming photon (E) and the atomic number of the absorbing medium (Z) is described by the following:

$$\frac{\sigma_{coh}}{\rho} \propto \frac{Z^2}{E} \tag{3.25}$$

where σ_{coh}/ρ is the mass attenuation coefficient for the process, with σ_{coh} representing the linear attenuation coefficient and ρ the density of the material. It can be deduced that for low-atomic-number materials, such as biological tissues and for high-energy photons, such as gamma-rays, this process will be insignificant.

If the interaction takes place with the atom as a whole, that is, with the combined action of all of the electrons, the processes is known as *Rayleigh* scattering. As in the above case, there is no significant loss of energy and the photon changes direction only slightly. Like Thomson scattering, this process is also of little significance in the field of environmental radioactivity.

3.2.5 Photonuclear Reactions

A sufficiently energetic photon can be absorbed by an atomic nucleus, thus resulting in the emission of a nucleon (proton or neutron) in a process called *photodisintegration*. An example of such a photonuclear reaction is the gamma-ray capture by the $^{206}_{82}Pb$ nucleus which then releases a neutron and becomes $^{205}_{82}Pb$. This reaction can be abbreviated by the nomenclature $^{206}_{82}Pb(\gamma, n)^{205}_{82}Pb$. Such a reaction is only possible if the energy of the incoming photon is greater than that of the binding energy of the emitted nucleon (see, for example, Turner (1995)). As reactions such as these can produce neutrons, this process is of importance in the field of radiation protection. In the area of environmental radioactivity, however, it has little significance as the photons have to have energies of several MeVs for such reactions to take place. Thus, like elastic scattering, photonuclear reactions have only been included here for the sake of completeness.

3.2.6 Total Mass Attenuation Coefficient

The mass attenuation coefficient (μ/ρ) given by Equation (3.8) represents the total for all of the attenuating processes that can occur for a given medium and photon energy. Thus, μ/ρ represents the sum of the individual photoelectric, Compton and pair production mass attenuation coefficients, as follows:

$$\frac{\mu}{\rho} = \frac{\tau}{\rho} + \frac{\sigma}{\rho} + \frac{\kappa}{\rho} \tag{3.26}$$

The relative importance of the individual mass attenuation processes for gamma-rays is illustrated in Figure 3.6. From this figure, it is clear that at low energies the

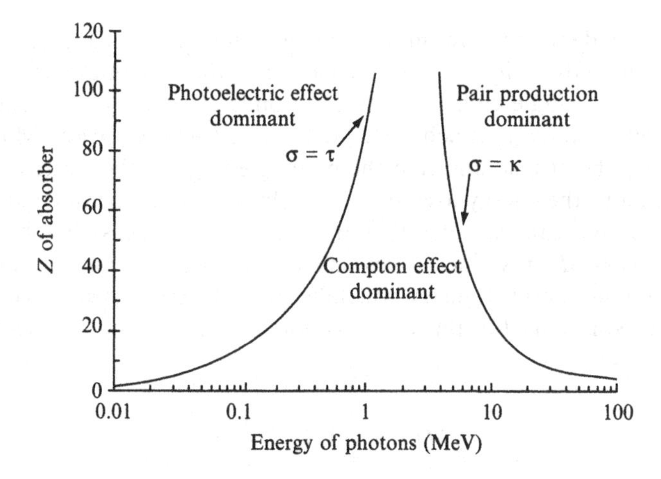

Figure 3.6 The relative importance of the individual mass attenuation processes for gamma-rays (adapted from Knoll, 1989)

photoelectric effect dominates, whereas for medium-energy photons (0.5–5 MeV) Compton scattering predominates. Pair production is only important for heavy elements and only when the energy of the photons is high.

3.3 INTERACTION OF CHARGED PARTICLES WITH MATTER

In this section we will examine how charged particles, such as alpha- and beta-particles, interact with matter. As explained earlier in Section 3.2, charged particles travel shorter distances than photons of similar energies and thus lose energy over smaller areas. This can result in severe damage to the material in which they are traversing. This can be particularly serious if radionuclides emitting α- and β-particles are inhaled. The effects of radiation damage on biological tissues are discussed further in Section 3.5 and Chapter 4. Some radiation detectors also rely on such interaction mechanisms and these are discussed in detail in Chapter 8. The section below discusses how particles interact and deposit their energies as they travel through matter.

3.3.1 Interaction of Alpha-Particles with Matter

As soon as an α-particle enters an absorber it undergoes Coulombic interactions with the atomic electrons. Depending on the proximity of the collision between the charged particle and the atomic electron, the latter is raised to a higher electronic shell of the absorber atom (*excitation*) or if the encounter is very close then the electron can be removed completely, thus leading to the *ionization* of the atom. During such encounters, the charged particle loses energy to the electron and suffers a deceleration in the absorber until it gives up all its energy through multiple interactions and stops in the absorber.

In order to understand how much energy is lost as a heavy charged particle travels through matter, it is useful to calculate the maximum energy that can be transferred in a single collision with an atomic electron. If we assume that the charged particle is moving much faster than the atomic electron, which can then be considered to be at rest, and that the binding energy of the electron is negligible when compared to the energy loss, then the collision can be considered to be *elastic* (Turner, 1995). We can consider that the interaction process involves a charged particle with mass M, travelling at velocity U colliding with an electron of mass m at rest. Maximum energy transfer will take place if the collision occurs head-on, and then the equations for the conservation of energy and momentum can be written as follows:

$$\frac{1}{2}MU^2 = \frac{1}{2}MV^2 + \frac{1}{2}mv^2 \tag{3.27}$$

and:

$$MU = MV + mv \tag{3.28}$$

where V and v are, respectively, the velocities of the charged particle and the electron after the collision. Equation (3.27) describes the conservation of energy, while Equation (3.28) describes the conservation of momentum relationships. The maximum energy transferred (E_{max}) during the collision will be the initial energy of the charged particle minus its energy after collision, as follows:

$$E_{max} = \frac{1}{2}MU^2 - \frac{1}{2}MV^2 \tag{3.29}$$

In order to calculate this difference, we first have to derive an expression for V from Equations (3.27) and (3.28) and then substitute into Equation (3.29) to determine E_{max}. The first step is to solve equation 3.28 for v, and thus:

$$v = \frac{M(U - V)}{m} \tag{3.30}$$

Now substitute for v into the energy conservation relationship (Equation (3.27)) to solve for V, as follows:

$$V = \frac{(M - m)U}{M + m} \tag{3.31}$$

We can now substitute for V into Equation (3.28) to arrive at an expression for E_{max}, as follows:

$$E_{max} = \frac{1}{2}MU^2 + \frac{1}{2}M\left[\frac{(M - m)U}{M + m}\right]^2 \tag{3.32}$$

Thus:

$$E_{\text{max}} = \frac{4mME}{(M+m)^2} \tag{3.33}$$

where:

$$E = \frac{1}{2}MU^2$$

in which E is the initial kinetic energy of the charged particle.

In the case of a heavy-charged particle where $m \ll M$, Equation (3.33) can be simplified to the following:

$$E_{\text{max}} = \frac{4mE}{M} \tag{3.34}$$

As the ratio of m/M for an electron and a proton is approximately 2000, E_{max} can be written as follows:

$$E_{\text{max}} \approx \frac{E}{500A} \tag{3.35}$$

where A is the atomic mass (total number of nucleons). Thus, as a charged particle travels through the absorber it looses energy in small steps (i.e. as 1/500 of the particle energy per nucleon) and it therefore has to lose energy through numerous such steps before it comes to a stop.

During the energy-loss process, the charged particle travels in a near straight-line path and thus has a definite *range* for a given energy and mass, and absorber. In addition to the charged particle causing ionization and excitation of atoms, if the proximity of the collision is small, then the ejected electron can also be sufficiently energetic to cause further ionization and excitation. These energetic electrons, called *delta-rays*, are an indirect mechanism by which the particle can transfer its energy to the absorbing medium. We thus have two mechanisms by which energy is imparted to the material, i.e. *directly* by the charged particle itself losing energy and secondly, *indirectly*, by the delta-rays losing energy.

3.3.2 Stopping Power

The *Linear Energy Transfer* (LET) and the *Stopping Power* of a substance for a given radiation are closely related quantities and fundamental when calculating doses. The stopping power (S) is defined as follows:

$$S = -\mathrm{d}E/\mathrm{d}x \tag{3.36}$$

where E is the energy of the particle and x is the distance travelled by the particle into the material (the units for S are usually MeV cm^{-1}). The stopping power is also referred to as linear energy transfer (LET), a quantity which is usually expressed as keV μm^{-1}.

There are several ways of calculating the stopping power for different media. The classical method of determining the value for stopping powers is the Bethe Formula which is given by the following:

$$S = -\frac{dE}{dx} = \left(\frac{4\pi e^4 z^2}{m_0 v^2}\right) NB \tag{3.37}$$

where:

$$B = Z\left[\ln\left(\frac{2m_0 v^2}{I}\right) - \ln\left(1 - \frac{v^2}{c^2}\right) - \frac{v^2}{c^2}\right]$$

where e is the electronic charge, z is the atomic number of the primary particle, N is the number density of the absorbing atoms, Z is the atomic number of the absorber, m_0 is the rest-mass of the electron, v is the velocity of the primary particle and c is the velocity of light.

The mean excitation and ionization potential (I) in Equation (3.37) is usually determined experimentally for the medium. In addition, note that for non-relativistic collisions where $v << c$, only the first term in B is important. In order for this equation to be valid, the velocity of the charged particles must be large when compared to the orbital electrons of absorbing atoms (Knoll, 1989). From Equation (3.37), we can infer that the energy loss varies as $1/v^2$, as the term B only varies slowly with particle energy. Thus, for a given type of particle the energy loss is greatest at low velocities or low energies. In other words, a slowly moving particle will be in the vicinity of an absorber atom for a longer time, thus increasing the likelihood of energy transfer. According to the Bethe formula energy loss also varies as z^2 and hence alpha-particles (charge of +2) will lose energy at a higher rate than protons (charge of +1), provided that they have the same velocity. Similarly, as the energy loss also varies as NZ, this implies that absorbers consisting of heavier elements will exhibit higher stopping powers than lighter materials. A major limitation of this equation is at low particle velocities where the incident particle can acquire electrons from the medium and lose its charge, so reducing z and hence reducing S. Figure 3.7 shows a comparison of the stopping power of water for different particles. At low energies, the first term in the Bethe formula increases (as it depends on $1/v^2$) and this causes the curve for particles, such as alpha-particles, to rise but then the logarithmic term falls and causes a maximum, known as the *Bragg peak*.

The range (R) of charged particles can be expressed in terms of the inverse of the stopping power, as follows:

$$R(T) = \int_0^T -\left(\frac{dE}{dx}\right)^{-1} dE \tag{3.38}$$

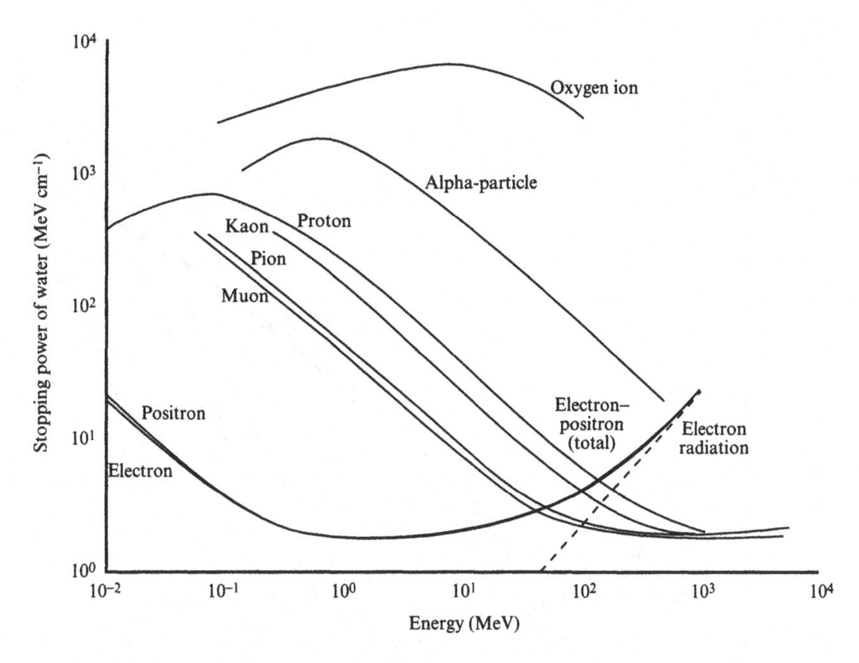

Figure 3.7 Stopping power of water for a range of charged particles and elementary particles (muon, pion, and kaon). From Turner, J. E., *Atoms, Radiation and Radiation Protection*, © Wiley, 1995. Reprinted by permission of John Wiley & Sons, Inc.

where T is the kinetic energy of the particle. Thus, the *range* is the distance travelled by the particle until it comes to rest. Databases of stopping powers and ranges for particles, as well as further discussion on the subject, can be found in various sources, such as ICRU (1984), ICRU (1993), Ziegler (1999), Seltzer *et al.* (2001) and Berger *et al.* (1999).

For particles such as protons and alpha particles, various empirical relationships have been developed (see, for example, Turner, 1995). For example, the following relationships can be used to determine the range of alpha-particles in air at 15°C and 1 atmosphere pressure:

$$R = 0.56E \text{ (for } E < 4) \tag{3.39}$$

$$R = 1.24E - 2.62 \text{ (for } 4 < E < 8) \tag{3.40}$$

where E is the energy of the alpha-particle in MeV. The ranges of some simple particles are shown in Figure 3.8, where the ranges are expressed as the linear range (in cm) multiplied by the density of the medium. The units thus become g cm^{-2}. An advantage of this type of unit is that one obtains similar values for similar materials. This also applies to the stopping power, which can be expressed as mass stopping power ($-\mathrm{d}E/\mathrm{d}\rho x$, where ρ is the density of the absorber material), in units of MeV cm^2 g^{-1}.

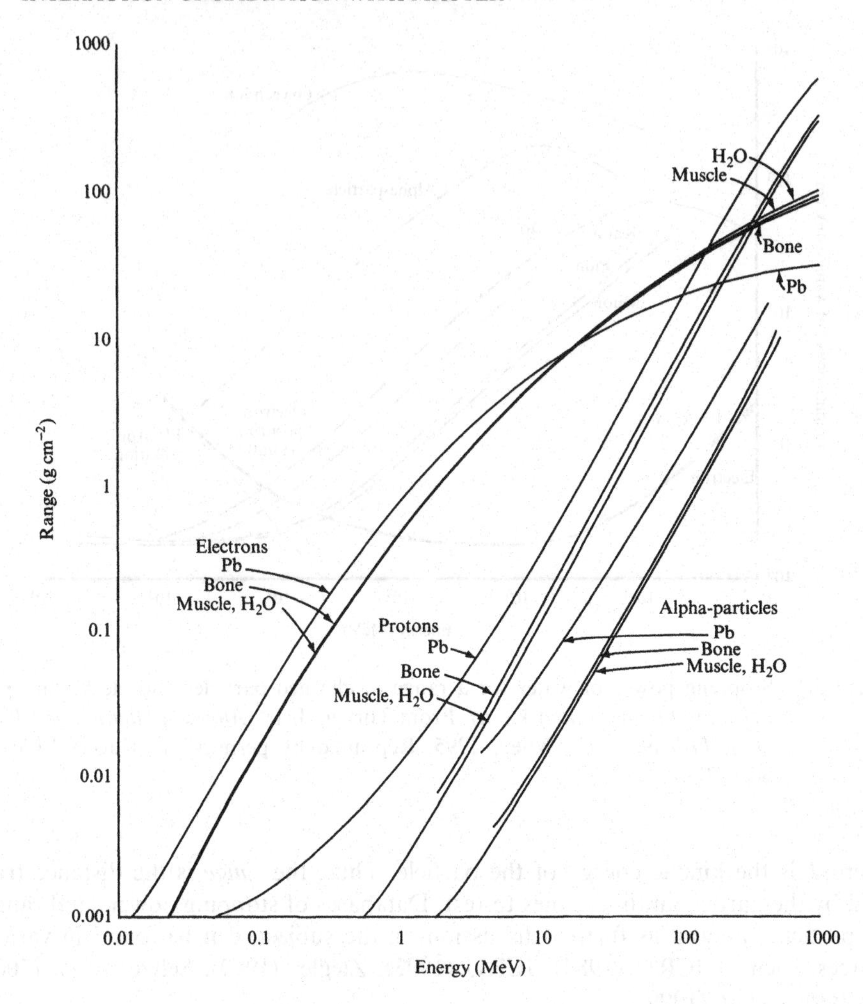

Figure 3.8 Ranges of protons, alpha-particles and electrons in water, muscle tissue, bone and lead. From Turner, J. E., *Atoms, Radiation and Radiation Protection*, © Wiley, 1995. Reprinted by permission of John Wiley & Sons, Inc.

3.4 INTERACTION OF BETA-PARTICLES WITH MATTER

The energy-loss mechanisms for electrons (such as beta-particles) and positrons differ considerably when compared to heavy particles such as protons and alpha-particles. Some of the key differences include the following:

1. The path of the electrons as it traverses a medium is not linear. As a result of numerous collisions, the trajectory is rather 'tortuous'.
2. As the incoming electron interacts with other electrons (hence both of equal mass), larger fractions of the incident energy are transferred.

3. *Bremsstrahlung* (electromagnetic radiation caused by the slowing down of charged particles) is much more important than for heavier particles.
4. Other interactions, such as elastic collisions, will also occur as a beta-particle moves through a medium, which again will affect its penetration.

The total stopping power of electrons can be considered to be the sum of collisional (col) and radiative (rad) energy losses, as follows:

$$-\frac{\mathrm{d}E}{\mathrm{d}x} = \left(-\frac{\mathrm{d}E}{\mathrm{d}x}\right)_{\mathrm{col}} + \left(-\frac{\mathrm{d}E}{\mathrm{d}x}\right)_{\mathrm{rad}} \tag{3.41}$$

It is interesting to compare these two energy-loss processes:

$$\frac{(\mathrm{d}E/\mathrm{d}x)_{\mathrm{rad}}}{(\mathrm{d}E/\mathrm{d}x)_{\mathrm{col}}} \cong \frac{EZ}{700} \quad \text{(Knoll, 1989)} \tag{3.42}$$

and:

$$\frac{(\mathrm{d}E/\mathrm{d}x)_{\mathrm{rad}}}{(\mathrm{d}E/\mathrm{d}x)_{\mathrm{col}}} \cong \frac{EZ}{800} \quad \text{(Turner, 1995)} \tag{3.43}$$

where E is in MeV. As the energies of beta-particles are typically less than a few MeV, radiative losses form a small fraction of the total energy loss due to ionization and excitation. In addition, radiative losses are only important for heavy absorbers and the ratio varies with Z. Energy loss in water, for example, is dominated by collisional processes.

As electrons can lose most of their energy in a single collision, one normally considers the average pathlength of the incoming electrons as approximately equal to its range. Some ranges for electrons in different materials are also shown in Figure 3.8. As with heavy particles, empirical relationships can be obtained for calculating the range of electrons in terms of their kinetic energy (T), as follows (Turner, 1995):

$$R = 0.412T^{1.27-0.0954 \ln T} \tag{3.44}$$

This equation is valid for $0.01 \leq T \leq 2.5$ (with T in MeV) and is used to calculate ranges (units of g cm^{-2}) in low-Z materials. For $T > 2.5$ MeV, the following relationship applies:

$$R = 0.530T - 0.106 \tag{3.45}$$

3.5 CHEMICAL AND BIOLOGICAL INTERACTIONS OF CHARGED PARTICLES

When considering interactions of nuclear radiation with cells, it is the interaction with water that is the most important. This is not surprising as human cells are mostly water (70–80%). Ionizing radiation will produce secondary electrons and in water these have energies < 100 eV. As a consequence of these interactions, the water molecule can change and dissociate into radicals that can cause damage to the cells. This damage can lead to symptoms such as radiation sickness, cataracts and cancer. The timescales over which such changes take place can range from $< 10^{-15}$ s for excitation of water molecules to occur, to years before cancer manifests itself.

In the initial stages, as the radiation enters the cells water molecules can become ionized to produce H_2O^+ and an electron or be excited (H_2O^*). This stages takes place very quickly ($< 10^{-15}$ s) and hence these changes occur in local track regions. During this physical stage, the ionized water molecule can react with another water molecule to form a hydronium ion (H_3O^+) and an hydroxyl radical (OH^-). The positive water molecule can also dissociate into an hydrogen ion (H^+) and (OH). Similarly, a negative water molecule ion (H_2O^-), formed by an interaction with a secondary electron, can also dissociate into a hydrogen atom (H) and OH^- radical.

The excited water molecule can undergo the following reactions to lose its excess energy:

$$H_2O^* \longrightarrow H_2O^+ + e^-$$

or:

$$H_2O^* \longrightarrow H + OH$$

The products H and OH, which have unpaired electrons and hence are very reactive, represent the damaging *free radicals*. In addition to these radicals, OH can undergo the following reaction to form a powerful oxidant, hydrogen peroxide, H_2O_2:

$$OH + OH \longrightarrow H_2O_2$$

This physiochemical stage lasts about 10^{-6} s. During the last chemical stages, such radicals can interact with the complex organic molecules which form chromosomes and cause genetic damage.

Radicals such as OH can lead to the breaking of DNA strands. This can take place directly or at a later stage (indirect effect). Some of the biochemical interactions take place in less than one second and most radicals have reacted within a millisecond. Cell divisions can result within minutes and death from acute exposure can result in less than a few weeks. Damage such as lung fibrosis can take months to develop and cancer and cataracts may appear years after exposure. Such biological effects will be discussed in more detail in Chapter 4.

3.6 KERMA

For indirectly ionizing radiation (photons and neutrons), an important quantity for calculating doses is *kerma*, which stands for Kinetic Energy Released per unit Mass. Kerma is the *initial* kinetic energy of all charged particles released by the incident radiation per unit mass and includes energy that may appear as Bremsstrahlung and

Auger electron energies. As secondary electrons can travel beyond the site of attenuation in a material, this site can be very different from the site of absorption. The International Commission on Radiation Units and Measurements (ICRU) (ICRU, 1998) defines kerma as follows: 'The kerma, K, is the quotient of $\mathrm{d}E_{\mathrm{tr}}$ by $\mathrm{d}m$, where $\mathrm{d}E_{\mathrm{tr}}$ is the sum of the initial kinetic energies of all charged particles liberated by uncharged particles in a mass $\mathrm{d}m$ of material'. Hence kerma can be expressed as follows:

$$K = \frac{\mathrm{d}E_{\mathrm{tr}}}{\mathrm{d}m} \tag{3.46}$$

Kerma has the same units as absorbed dose (grays) but is a measure of attenuation in a small volume and hence can differ from absorbed dose. Whereas absorbed dose increases with depth during irradiation, kerma, on the other hand, decreases with depth because of the attenuation of the primary radiation. Kerma can also be defined for a given reference material such as air – in this case it is called 'air kerma'.

4

Biological Effects of Ionizing Radiation and Radiological Protection

Ionizing radiation is a form of energy and, as such, can damage tissue. The interaction of ionizing radiation with matter has been described in Chapter 3. When ionizing radiation passes through tissue, the structure of molecules may be changed if component atoms are ionized or excited. If the molecule is part of the cell structure, the cell may be damaged. If the genetic material in the cell, deoxyribonucleic acid (DNA), is affected then the behaviour of the cell may be changed (see Sections 4.2 and 4.3). These effects on tissue are considered to be proportional to the energy deposited in the tissue. Thus, the fundamental dosimetric quantity is the absorbed dose, D – this is the energy absorbed per unit mass. In the SI system the unit is joule per kilogram and this is given the name gray (Gy) after a British scientist. The effects of radiation on tissue will depend not only on the average radiation dose in a tissue but also on how the energy is deposited within a cell. Alpha-particles are relatively large charged particles and transfer energy efficiently to tissue. In contrast, beta-particles and gamma-rays transfer energy less efficiently. Thus, when an alpha-particle passes through chromatin (the mixture of DNA and protein in the cell nucleus) it will deposit more of its energy than will a beta-particle or a gamma-ray. Thus, an alpha-particle will cause the most damage to the chromatin. The average energy deposited along the track of a particle is called the *linear energy transfer* (LET); alpha-particles (together with some other less common forms of ionizing radiation) are referred to as 'high-LET' radiation, whereas beta-particles and gamma-radiation are referred to as 'low-LET' radiation. The greater efficiency of 'high-LET' radiation in causing damage to cells is commonly expressed in terms of the *relative biological effectiveness* (RBE). The latter is the ratio of the dose of low-LET radiation to the dose of high-LET radiation that gives the same biological effect. In this chapter, dose to particular organs or tissues will be discussed, as well as the effects arising when the whole body has been exposed uniformly.

Currently, two broad categories of radiation-induced injury are recognized: *deterministic* effects and *stochastic* effects. Deterministic effects are characterized by a threshold dose below which they do not occur. This dose threshold is very much higher than the doses that are received in everyday life. Above the appropriate threshold, the severity of damage increases with dose. Stochastic effects can occur at lower doses than those that give rise to deterministic effects. In radiological protection, as we shall see later, the assumption is that there is not a threshold for stochastic effects and that the incidence of stochastic effects is proportional to the radiation dose – the greater the dose, then the greater the chance of incurring a stochastic effect. That is, until the dose is high enough to cause deterministic effects. The main stochastic effect of concern is cancer.

In this chapter, the quantities used to measure radiation dose to humans are described, together with the current system of radiological protection – the systematic approach to the protection of people from radiation.

4.1 RADIATION DOSE

In order to protect human beings from ionizing radiation, a way is needed of calculating a dose of radiation which takes into account that there are three distinct types of radiation, that radionuclides may enter the body and that different tissues or organs may be irradiated. The hierarchy of dose quantities now in use is described below. It is taken from the recommendations of the International Commission on Radiological Protection (ICRP). The scheme is illustrated in Figure 4.1.

4.1.1 Absorbed Dose

Absorbed dose is the amount of energy transferred by the radiation to the material being irradiated. The unit of absorbed dose is the gray, symbol Gy, named after a British scientist. One gray is one joule of absorbed energy per kilogram of material. It is a relatively large unit and sub-multiples such as a milligray (mGy), which is one thousandth of a gray, and a microgray, μGy, which is one millionth of a gray, are often used.

4.1.2 Equivalent Dose

A problem with using absorbed dose as an indication of the harm that radiation can do is that equal absorbed doses do not necessarily have equal biological effects. In particular, at low doses an absorbed dose of, say, 0.1 Gy of alpha-radiation is more harmful than an absorbed dose of 0.1 Gy of either beta- or gamma-radiation. Therefore, absorbed dose is weighted depending on the type of radiation in order to put all radiation on a common basis. The radiation weighting factors currently in use are 20 for alpha-radiation and one for beta- and gamma-radiation. The resulting quantity, the *equivalent dose*, is expressed in a unit called the sievert, symbol Sv, after a Swedish scientist. Like the gray, it is a large unit and sub-multiples are commonly used.

Therefore, for beta- and gamma-radiation, the equivalent dose is numerically equal to the absorbed dose, whereas for alpha-radiation, the equivalent dose is twenty times larger than the absorbed dose. Radiation weighting factors are given in Table 4.1.

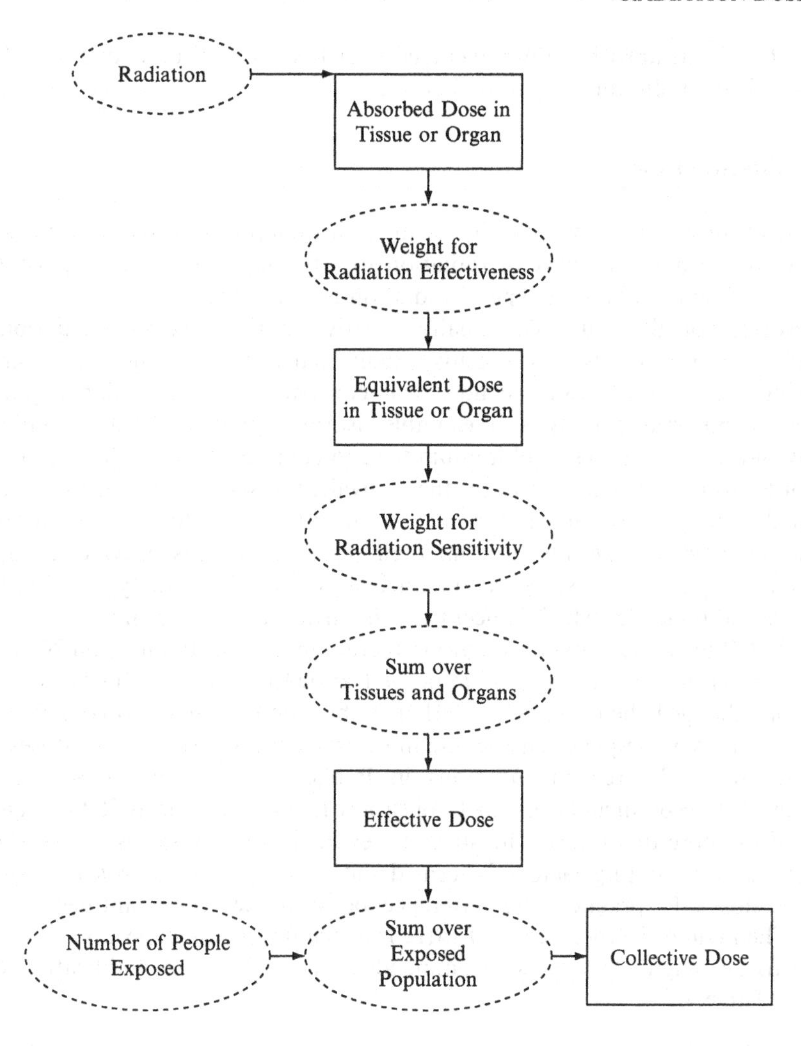

Figure 4.1 Hierarchy of dose quantities

Table 4.1 The ICRP radiation weighting factors (taken from ICRP Publication 60 (ICRP, 1991a)).

Type and energy range	Radiation weighting factor, w_r
Photons, all energies	1.00
Electrons and muons, all energies	1.00
Neutrons (depends on energy)	5–20
Protons, other than recoil protons, energy > 2 MeV	5.00
Alpha-particles, fission fragments, heavy nuclei	20.00

The equivalent dose in a tissue or organ T is given by the following equation:

$$H_T = \sum_R w_r D_{T,R} \quad (4.1)$$

where $D_{T,R}$ is the absorbed dose averaged over tissue or organ T, due to radiation R and w_r is the radiation weighting factor.

4.1.3 Effective Dose

Equivalent dose is a measure of the harm from radiation to a particular tissue. In other words, a dose of 1 mSv of radiation to, say, the liver will give a rise to the same risk of cancer whatever type of radiation is concerned.

However, not all tissues are equally sensitive to the effects of radiation. For example, the thyroid is less sensitive than many other tissues. Furthermore, following intake, some radionuclides will concentrate in particular organs and irradiate them preferentially. For example, isotopes of iodine will concentrate in the thyroid, while isotopes of plutonium tend to concentrate in the liver and bone.

In order to take this into account, the equivalent doses to different tissues can be weighted. The *effective dose* is the sum of the equivalent doses to the individual organs weighted to take into account their sensitivities. This provides a common basis for comparing exposures of human beings from different types of radiation and from radionuclides which concentrate in particular body organs.

The ICRP introduced tissue weighting factors in 1977 in Publication Number 26 (ICRP, 1977) and, as a result of improved knowledge of the effects of ionizing radiation, changed their values in 1991 in Publication Number 60 (ICRP, 1991a). This introduces a complication because in order to distinguish between doses calculated following the recommendations in Publication 26 from those calculated following the recommendations in Publication number 60, the ICRP changed the name of the dose quantities. The sum of weighted organ doses using the Publication 26 tissue weighting factors is termed the *effective dose equivalent* (instead of effective dose). In practice, the difference between the two quantities is usually small. The main difference occurs if there is a significant skin exposure.

The tissue weighting factors recommended by the ICRP in Publication 60 are given in Table 4.2.

Table 4.2 The ICRP tissue weighting factors (taken from ICRP Publication 60 (ICRP, 1991a)).

Tissue or organ	Tissue weighting factor
Gonads	0.20
Bone marrow (red)	0.12
Colon	0.12
Lung	0.12
Stomach	0.12
Bladder	0.05
Breast	0.05
Liver	0.05
Oesophagus	0.05
Thyroid	0.05
Skin	0.01
Bone surface	0.01
Remainder	0.05

The effective dose, E, is given by the following equation:

$$E = \sum_{T} w_T H_T \tag{4.2}$$

where H_T is the equivalent dose in tissue or organ T and w_T is the weighting factor for tissue T.

4.1.4 Committed Effective Dose

If a person is irradiated by a source of, say, gamma-radiation outside the body, he or she will receive a dose for as long as they are in the vicinity of the source. However, following an intake by ingestion or inhalation, some radionuclides persist in the body, thus irradiating tissue for many years. For these radionuclides, the total radiation dose will depend upon the half-life of the radionuclide, its distribution in the body and the rate at which it is lost from the body. Detailed mathematical models are available which enable the dose to be calculated for each year following an intake. The total effective dose delivered over the lifetime, taken to be 70 years for infants and 50 years for adults, is termed the *committed effective dose*. It is so called because once the radionuclide has been taken up into the body, the person is *committed* to receiving the radiation dose. One example of a radionuclide that persists in the body delivering a radiation dose over a lifetime is ^{239}Pu.

The ICRP has published values for the committed doses following intakes of 1 Bq of different radionuclides via ingestion or inhalation. These values are called dose coefficients and have been calculated for intakes by members of the public at six standard ages (ICRP, 1996) and for intakes by adult workers (ICRP, 1994a). Dose coefficients are calculated using complex models of the behaviour of radionuclides in the body. Of necessity, a number of general assumptions are made in their calculation. If doses calculated using the ICRP dose coefficients in a specific situation are so high as to cause concern, the appropriateness of the general assumptions should be checked by using the situation-specific information. For example, in the case of dose coefficients for inhalation of radionuclides by members of the public, it is assumed for non-gaseous radionuclides that the inhaled material is in the form of 1 μm AMAD (*Activity Mean Aerodynamic Diameter*). If, in reality, the inhaled material has a larger or smaller AMAD, the corresponding value for the dose coefficient could be different. One further point about dose coefficients is that many radionuclides have radioactive decay products. In growth of these decay products in the body following intake, the unit activity of the parent radionuclide is taken into account in the dose coefficients. However, the dose coefficients do not take account of any activity of the decay products in the initial intake.

The ICRP has published a compendium of dose coefficients on CD-Rom. This extends the results given in the publications referred to above by providing inhalation dose coefficients for ten particle sizes (ICRP, 1999a). A consistent but simpler set of dose coefficients has been published by the International Atomic Energy Authority (IAEA) (IAEA, 1996a).

In radiological protection, it is conventional to discuss the dose that an individual receives in a year. Indeed, dose limits are normally specified in terms of effective dose for such a period. In this context, effective dose means the sum of the effective doses from radiation sources external to the body, together with the sum of the effective doses from intakes of radionuclides into the body in the year. This sum is often simply referred to as the 'dose'.

4.1.5 Collective Effective Dose

If radiation effects are directly proportional to radiation dose without a threshold, then the sum of all of the doses to all of the individuals in a given population will be an indicator of the number of radiation-induced health effects. This sum of doses is termed the *collective dose*; it has the units manSv. For example, if in a population of 10 000 individuals each receives a dose of 0.1 mSv, the collective dose will be:

$$10\,000 \times 0.0001 = 1 \text{ manSv}$$

This formulation is only valid for dose levels below those that may cause deterministic effects.

Collective dose can be used as one input into the optimization of radiological protection (see below). The calculation and use of collective dose is complicated by the fact that releases of long-lived radionuclides to the environment may cause exposures of populations over many thousands of years into the future. The relevance of collective doses delivered at long times into the future to decision making today has been questioned.

Mathematically, the collective dose, S, can be defined as follows:

$$S = \int_0^{\infty} E \frac{\mathrm{d}N}{\mathrm{d}E} \mathrm{d}E \tag{4.3}$$

where $(\mathrm{d}N/\mathrm{d}E)\mathrm{d}E$ is the number of people receiving an effective dose of between E and $E + \mathrm{d}E$. The collective dose, as defined by the ICRP, includes all of the doses from a given source to the population under consideration, integrated to infinity. The equation can be recast as follows to take account of the time period of exposure:

$$S_j = \int_{t1}^{t2} \int_{E1}^{E2} \frac{E\mathrm{d}N_j}{\mathrm{d}E}(t)\mathrm{d}E\mathrm{d}t \tag{4.4}$$

where S_j is the collective dose to population N_j. If $t2$ and $E2$ are infinite, S_j is the collective dose, but if $t2$ is finite, the resultant quantity is known as the *truncated collective dose*. It is common in radiological protection to estimate truncated collective doses; for example, collective doses truncated at 500 years can be used as indicators of health detriment over the next 10 or so generations.

4.1.6 Dose Commitment

Unlike the other dose quantities mentioned above which can be related directly to exposures of individuals or populations, *dose commitment* is a calculational tool.

This is defined as the infinite time integral of the *per head dose rate* due to a particular event, such as discharges from a nuclear installation. For example, if a nuclear installation discharges a long-lived radionuclide to the environment, say ^{14}C, in a particular year, in each year afterwards the average dose to an individual in the World's population will be the collective dose in that year divided by the World's population. The total of these average individual doses summed over every year until the ^{14}C has decayed away is the corresponding dose commitment. Clearly, as ^{14}C takes many thousands of years to decay away, no single individual could receive this dose over their lifetime. However, the key point is that such a dose commitment is numerically equal to the maximum annual average individual dose rate if the discharges continued indefinitely at the same level. Similarly, if a particular practice is anticipated to continue for 50 years, the calculated dose commitment up to 50 years from one year's radionuclide releases would be numerically equal to the maximum annual individual dose, provided that releases continued at the same level. Mathematically, the dose commitment can be expressed as follows:

$$H_{c,T} = \int_0^\infty Ha_t(t)\mathrm{d}t \qquad (4.5)$$

where $H_{c,T}$ is the dose commitment and Ha_t is the per head dose rate.

It is important not to confuse dose commitment with committed dose (Section 4.1.4). In the example given above, the dose commitment up to 50 years would include the sum of the annual committed doses in each of the 50 years.

4.2 DETERMINISTIC EFFECTS

Deterministic effects usually only occur following exposures greater than about 1 Gy delivered over relatively short periods of time. They arise when the dose of radiation has been large enough to kill or otherwise prevent enough cells from functioning, to impair the function of an organ. This impairment of organ function may be transitory; an increase in the rate of cell loss due to the radiation exposure may be compensated for by an increase in the rate of replacement. However, if a sufficiently large number of the body's cells are damaged, repair ability may be overwhelmed and there will be a clinically detectable impairment of function in the irradiated tissue organ. Therefore, there will be a threshold below which the loss of cells is too small to impair tissue function to any detectable extent. Above this threshold, the severity of the effect will increase as more and more cells are damaged. Direct cell killing may not be the only process leading to deterministic effects. In many cases, cells will not be killed immediately but will only die when they attempt to divide, probably as a result of damage to the chromosomes. Thus, in general, the most sensitive tissues as far as deterministic effects are concerned will be those that rely on rapid cell division to function, e.g. bone marrow, intestine and skin. The occurrence of deterministic effects depends not only on the total dose received by the tissue but also on the rate at which the dose is received. If doses are delivered over a sufficiently protracted time, damage to the cells will be spread out in time and the tissue will have a greater chance of recovery due to cellular repair or repopulation from cell division.

Deterministic effects can be divided into fatal and non-fatal ones. Death may occur as a result of radiation damage to the red bone marrow, the gastrointestinal tract, the central nervous system, the lungs or the skin. Non-fatal deterministic effects include prodromal vomiting, prodromal diarrhoea, impaired lung function, acute radiation thyroiditis, hypothyroidism, lens opacities, sterility and, if the fetus is exposed, small head size (microcephaly).

Most deterministic effects occur within days or weeks of exposure although hypothyroidism and lens opacity may take some years to develop.

Although deterministic effects have a threshold, for any one effect the threshold dose will not be the same for everyone in the population. For example, the age, genetic predisposition and health of an individual may all affect radiosensitivity. Thus, deterministic effects commonly exhibit sigmoidal dose–response curves (Figure 4.2). Thus, it is usual to discuss deterministic effects in terms of their D_{50} dose – the dose where 50 % of the population will experience the particular effect. Where the effect results in death, the term is called the LD_{50}, and where it is some other effect, it is called the ED_{50}. In mathematical terms, the risk, R, of incurring a deterministic effect from doses of radiation, D, can be described by a Weibull-type function as follows (NRPB, 1996a):

$$R = 1 - e^{-H}$$
$$H = \ln 2\left(\frac{D}{D_{50}}\right)^{v} \qquad (4.6)$$

where H is known as the hazard function and D_{50} is the dose where one half of the exposed population exhibits the particular deterministic effect; V is the shape factor which determines the steepness of the risk function. This mathematical formulation does not have a threshold and so it is necessary to impose one artificially; commonly chosen values lie between 0.2 to 0.5 of the D_{50}.

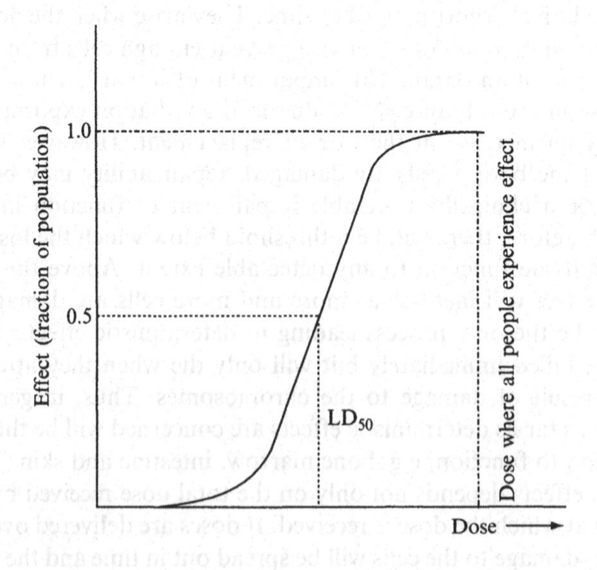

Figure 4.2 Typical dose–response curve for deterministic effects

The rate at which the dose is delivered also influences the likelihood of incurring a deterministic effect. Put simply, this is because if the dose is being delivered over an extended period of time, the irradiated tissue has the opportunity to repair itself. A proposed relationship between D_{50} and the dose rate $D\cdot$ in Gy h^{-1} is as follows:

$$D_{50}(D\cdot) = \theta_{\infty} + \frac{\theta_1}{D\cdot} \tag{4.7}$$

where θ_{∞} is the value of D_{50} at high dose rates and θ_1 (in $Gy^2\ h^{-1}$) describes the increase in D_{50} with decrease in dose rate (Scott, 1993).

Data on deterministic effects on humans comes from studies on the survivors of the atomic bombings of Japan and on the survivors of severe accidents, including the Chernobyl reactor accident in the former Soviet Union. Further information has been obtained from observations of the side effects of radiotherapy and on the effects on early radiologists.

In the following section on deterministic effects, the doses discussed are doses to the whole body from, say, penetrating gamma-radiation, or doses to particular specified tissues from gamma- or beta-radiation. The dose quantities, equivalent dose and effective dose, described in Section 4.1, may not be useful in estimating the risk of deterministic effects, particularly in the case of exposures to radiations with radiation weighting factors greater than unity. This is because radiation weighting factors are derived from considering stochastic effects. There is evidence that for types of radiation with weighting factors larger than unity, the relative effectiveness in causing deterministic effects is less than that for causing stochastic effects (ICRP, 1991a). Therefore, absorbed doses are used in this section.

The information in this section is taken largely from the ICRP (ICRP, 1991a) plus a review from the UK National Radiological Protection Board (NRPB) (NRPB, 1996a).

4.2.1 Fatal Deterministic Effects

Acute doses in the range 2–10 Gy are likely to cause death by damaging the bone marrow. This tissue produces the red blood cells which carry oxygen around the body, and some types of lymphocytes, or white blood cells, which fight infection. Thus, impairment of bone marrow function will have a variety of detrimental effects and death due to bone marrow failure will typically occur between 30 and 60 days following exposure. The severity of the damage to the bone marrow will depend on both the total dose and the rate at which it is delivered.

The most sensitive indicator of bone marrow damage is the level of lymphocytes in the blood: an acute radiation dose of around 1–2 Gy will reduce the blood lymphocyte level by around a factor of two within about 48 h following exposure. Blood concentrations of neutrophils and platelets also show dose-related decreases. Information on concentration of blood cells following an acute exposure can be used to predict the likelihood of survival and the necessity for treatment.

Data from a number of cases where individuals have been exposed to high doses of radiation, including Chernobyl accident victims, indicates that for healthy individuals receiving conventional medical care following exposure, the LD_{50} for bone marrow syndrome may be a whole body dose of around 5 Gy. In circumstances of limited medical care, the LD_{50} was lower at 3–5 Gy. These LD_{50}s refer to doses received over a few hours at most. If the dose is received in a more protracted manner, then the LD_{50} will increase. Clinically significant effects on the bone marrow occur with a threshold dose to the bone marrow of about 0.5 Gy delivered in a short period of time.

Higher acute doses can cause damage to the intestine. The membranes lining the intestine are constantly turning over with cells being lost to the lumen of the intestine. Impairment of the body's ability to replace these cells causes diarrhoea, loss of body fluids, infection, impairment of absorption of food and water, and ultimately to death. Limited human data suggest an LD_{50} of about 15 Gy for an acute exposure for death from radiation-induced intestinal damage. Animal studies indicate an LD_{50} of about 15 Gy for an acute exposure.

For predictive purposes, death from gastrointestinal syndrome is only likely to be important in situations involving doses from ingested radioactivity; if the exposure is from external irradiation, then death would occur in any case from bone marrow syndrome.

Inhalation of large quantities of radionuclides can cause damage to the lungs. Damage occurs in two phases: pneumonitis lasting from a few weeks after exposure for several months and longer-term changes including replacement of alveoli by collagen and connective tissue and fibrosis. Studies on the effects of radiation on lung function suggest that death may follow a dose to the lung of around 10 Gy from γ-emitters if delivered over a period of time of less than a day whereas larger doses of around 50 Gy are required if exposure is protracted over two weeks.

Contamination of the skin with large amounts of radionuclides can lead to skin damage. If the area of skin affected is large enough, then death can result. The first response of the skin to irradiation is a transient reddening, or erythema, resulting from capillary dilation. This effect may occur at doses to the skin in excess of 2 Gy. Depending on the radiation dose, this can be followed by dry desquamation (scaling of the skin), epilation (loss of hair) and moist desquamation (blistering). These effects arise from failure of cells in the basal layer to replace surface cells. The higher the dose, then the more severe is the reaction. If healing occurs, the resultant skin may be paler and thinner than normal skin.

The effects of radiation on the skin depend upon the dose and the area of skin irradiated. Erythema occurs in waves following a dose to the skin of greater than around 10 Gy – it appears quickly after exposure, reaching a peak of intensity at about one day. This is followed by a second wave of erythema between one and four weeks. Desquamation, or loss of skin tissue, follows erythema after an acute radiation dose of greater than 12 Gy. The severity of the response increases with both with the dose of radiation received and the area of skin exposed.

Whole-body doses in excess of about 50 Gy damage the central nervous system. Death usually occurs within two days of exposure. Death is caused by a combination of vascular damage, meningitis and encephalitis. Fluid also enters the brain, causing oedema.

A summary of the causes of death from deterministic effects following increasing whole-body doses is given in Table 4.3.

Table 4.3 Range of doses causing death from deterministic effects (taken from ICRP Publication 60 (ICRP, 1991a) and NRPB, 1996a)).

Whole-body dose (Gy)	Principal effects contributing to death	Time after exposure (days)
3–10	Damage to bone marrow	30–60
10–50	Damage to gastrointestinal tract and lungs	10–20
> 50	Damage to nervous system	1–5

4.2.2 Non-Fatal Deterministic Effects

Some non-fatal effects have been discussed above in connection with the response of the skin to irradiation. However, there are a range of other non-fatal effects. Such effects may, of course, be followed by fatal effects, depending on the dose and degree of medical care, etc.

Within the first 48 h of acute exposure of about 1 Gy, symptoms of gastrointestinal and neuromuscular effects arise. These are referred to collectively as the *prodromal syndrome* and include diarrhoea, nausea, vomiting, fatigue, intestinal cramps, sweating, fever and hypotension. These symptoms can occur without necessarily being followed by radiation-induced death or severe illness. They are thought to be caused by chemicals that are released from damaged tissue, triggering autonomic nervous centres in the brain stem.

There are human data on prodromal syndrome from observations on irradiated patients. Such studies indicate that the ED_{50} (the dose where the effect is observed in 50% of irradiated individuals) for prodromal diarrhoea is about 2.0 Gy for an acute exposure (in less than one day). For protracted exposures over one week, ED_{50} values for both prodromal diarrhoea and vomiting are both increased by a factor of about 2.

Acute radiation exposure can affect the thyroid gland – symptoms include thyroiditis (inflammation of the thyroid) and hypothyroidism (decreased activity of the thyroid). The thyroid concentrates iodine because the principle hormone it secretes, thyroxin, contains this element. Radioiodine may be released into the environment following a nuclear reactor accident and so, in such circumstances, there is the potential for significant doses to the thyroid. There are human data on the response of the thyroid to acute radiation exposure from studies on individuals receiving the radioisotope ^{131}I for medical purposes. These data suggest that acute radiation thyroiditis is unlikely to occur at doses to the thyroid from radioiodine of less than about 200 Gy. Hypothyroidism has a lower threshold of the order of a few tens of grays although it is not clearly defined.

The lens of the eye is particularly sensitive to deterministic effects as damaged cells are not readily repaired or removed. The initial site of radiation-induced damage is the continually dividing epithelial cells on the anterior surface of the lens. The clinical manifestation of radiation damage is lens opacity (the clouding of the lens or part of the lens). At low radiation doses, these may take years to develop, remain small and cause no significant impairment of vision; however, at high doses they may develop quickly in the form of cataracts. Human data come from various sources, including studies on patients given radiotherapy and the survivors of the Japanese atomic bombings. These studies indicate that the latent period between exposure and the appearance of lens opacities is a function of dose and varies between 0.5 and 35 years. For acute exposures, the lowest dose producing a cataract was about 5 Gy but clinically detectable lens opacities could be observed at acute lens doses of 2 Gy. These thresholds increase with dose protraction.

Acute doses of radiation can cause sterility in both males and females. In males, sperm is produced continuously after puberty and the process appears to be readily affected by radiation exposure. Doses to the human testes as low as 0.15 Gy may cause a significant, if transient, depression in sperm count. The duration of aspermia increases with the magnitude of the dose. Permanent male sterility is likely to occur after an acute dose to the testes larger than about 3.5 Gy to 6 Gy. Continuing (chronic) exposures of 2 Gy per year or greater are also likely to cause permanent sterility. Female fertility may be impaired temporarily following acute doses to the ovary as low as 0.6 to 1.5 Gy. The dose causing permanent sterility depends on age at exposure; the threshold decreases with increasing age. For example, a study an radiotherapy patients receiving doses of between 2.5 and 5 Gy showed that about 60 % of patients under 40 years of age became sterile, whereas about 80 % of older women became sterile.

Exposure during pregnancy may cause small head size (microcephaly) in the offspring. This has been studied in mothers receiving X-ray therapy treatment during pregnancy and also in the 'A' bomb survivors. There are some inconsistencies in the data but roughly the ED_{50} for microcephaly is an acute exposure of around 1 Gy.

4.3 STOCHASTIC EFFECTS

Stochastic effects are those where their probability of occurrence, but not their severity, depends upon the radiation dose. The main types of stochastic effects following exposure to ionizing radiation are the incidence of cancer in those exposed and the incidence of hereditary disease in their descendants.

4.3.1 Studies on Exposed Populations

The fact that exposure to ionizing radiation can cause cancer has been known for many decades but it is important to understand that exposure to ionizing radiation increases the likelihood of contracting cancer – no level of radiation dose is certain to cause cancer. Analogies are often made with cigarette smoking; smoking cigarettes increases the probability of getting lung cancer and this probability increases with the number of cigarettes smoked but not all individuals who smoke will get lung cancer.

The main body of data on cancer risk from ionizing radiation comes from the Life Span Study (LSS) of the Japanese Atomic Bomb survivors. An overview of these studies is provided by Schull (1996). This information is supplemented by studies on other irradiated populations such as patients irradiated to treat ankylosing spondylitis (a medical condition where some joints ossify) and other medical conditions. Data have also been obtained from individuals exposed to radionuclides in the past at work, in particular uranium miners and radium dial painters. Studies on uranium miners from various countries have shown an increased mortality from lung cancer that has been attributed to exposure to ^{222}Rn and its daughters (BEIR IV, 1988). An increased incidence of head sinus carcinoma and of bone cancer has been observed in individuals, primarily radium dial painters, who have been occupationally exposed to long-lived isotopes of radium (Rowland, 1994). Further information is increasingly being obtained from follow-up studies on individuals who have been exposed to radiation during work in the nuclear industry (e.g. Muirhead *et al.*, 1999).

There are many problems in establishing a risk factor for ionizing radiation, including the following:

1. There is a relatively high natural incidence of cancer in the population and so establishing the increase due to an incremental exposure to ionizing radiation can be difficult.
2. There is a minimum time period between exposure to ionizing radiation and the appearance of a radiation-induced cancer. This minimum period varies with age at exposure and with the type of cancer – this is referred to as the *latent period.* Such periods range from a few years for leukaemia and bone cancer to over ten years for some solid tumours. The time-dependence for cancer incidence following the latent period is not known for all types of cancer. The risk of leukaemia and bone cancer is largely expressed within twenty five years of exposure to ionizing radiation but for other tumours it is not clear whether incidence reaches a peak and then declines, or whether some other pattern of incidence is followed. Thus, unless an exposed population has been followed until all of the cancer risk has been expressed, estimates of excess cancers will be uncertain.

There are 90 000 or so individuals in the Life Span Study (LSS) of Hiroshima and Nagasaki survivors. By far the major dose to these individuals came from the acute gamma- and neutron-radiation as the bombs detonated. One of the first requirements for this study was an estimate of the radiation dose received by each survivor. This estimate had to take account of the position of each survivor relative to ground zero, as well as factors such as the shielding afforded by buildings. There were particular difficulties in estimating doses; for example, it emerged in the 1980s that the relatively high humidity of the air at the time the bombs detonated significantly reduced the neutron dose at Hiroshima to below that which may otherwise have been expected. The number of excess cancers had then to be established and this requires an estimate of the normal background incidence rate in the exposed population – in principle, this can be deduced from observations on an otherwise identical but unexposed population. The numbers of excess cancer deaths attributable to the 'A' bomb radiation is not large. In the period from 1950, when the LSS started, to 1990 there were a total of 7827 cancer deaths of which it is estimated, from models such as the ones described below, that there are 85 excess deaths from leukaemia and 335 from solid cancers (Pierce *et al.*, 1996). Most of the excess deaths from leukaemia occurred during the first fifteen years following exposure. For solid tumours, however, it appears that the pattern of excess risk resembles a lifelong elevation of the natural cancer rate which, in itself, generally increases with age. The Life Span Study is continuing. At the time of the most recent estimate of radiation risks (see ICRP, 1991a), around two thirds of the exposed population were still alive and so mathematical models had to be used to estimate the total numbers of excess cancers over the lifetime of the exposed population.

Broadly, two types of models have been used to project over time data from only a limited period in the lives of individuals, as follows:

1. The *additive risk model* in which radiation induces cancer at a rate independent of the spontaneous rate after the latent period.
2. The *multiplicative risk model* in which, after a latent period, the excess cancers are given by a factor applied to the age-dependent incidence of natural cancers in the population. The factor may vary with time following the exposure.

One important difference between these two models is that as the spontaneous incidence of cancer increases with age, a multiplicative risk model with a constant relative risk will predict an increasing incidence of radiation-induced cancer with age. A simple additive risk model would predict a constant incidence of radiation-induced cancer with age (see Figure 4.3).

The Life Span Study indicates, so far, that a multiplicative model provides a better basis for estimating lifetime risks, at least for the common types of solid cancer. Such models cannot, however, be applied indiscriminately as there is evidence that the risk of cancer may decline many years following exposure. This may be the case for leukemia, in particular, and possibly bone cancers, but it could be true for all types of cancer many years after exposure. Thus, multiplicative risk models applied over a lifetime could significantly overestimate the risks. Nevertheless, these models are used, in one form or another, to estimate the total of cancers in the exposed population that are attributable to the excess radiation exposure.

The Japanese A-bomb survivors include individuals who received relatively high doses of radiation – of the order of hundreds of milligray. Individuals in many of the

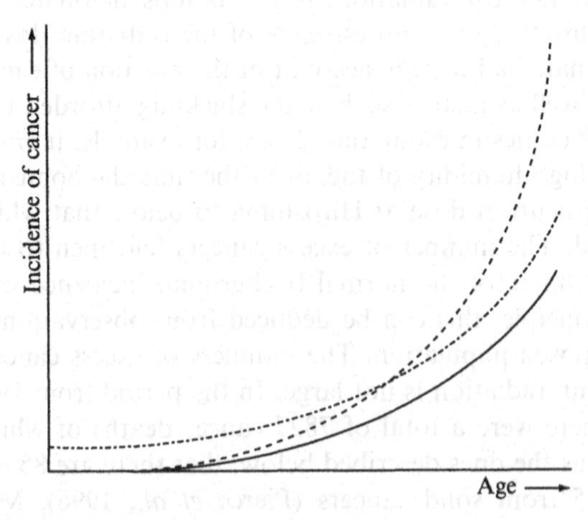

Figure 4.3 Modelling cancer risks: (——), spontaneous incidence; (-·-·-), additive risk model; (- - - -), relative risk model

other groups studied also received high doses. It has proved difficult to show effects below doses of around 100 mGy. One reason is that the risk factor is small, and, at low doses, a very large population would have to be irradiated in order to show a statistically significant increase over the high spontaneous cancer rate. However, some evidence for cancer induction at lower doses has been obtained from epidemiological studies on children who have been irradiated *in utero* during diagnostic X-ray examinations during pregnancy. These studies have shown an excess of cancer from incremental doses in the range 10–20 mGy. One of the reasons that this was observed is the relatively low spontaneous incidence of childhood cancer.

Most of the quantitative information on radiation-induced cancer in humans comes from groups, predominantly the A-bomb survivors, that were exposed to relatively high levels of radiation. Evidence from various experimental studies on animals indicates that, at least for low-LET radiation, the risk per unit dose at such high dose rates is often greater than at low dose rates. Thus a dose and dose rate effectiveness factor (DDREF) is used in extrapolating risks obtained from studies involving high doses and dose rates to the low doses and dose rates relevant to the general environment. Various animal and cellular studies involving dose fractionation, as well as analysis of pooled data on various cancers in the A-bomb survivors, indicates a DDREF of around 2 to 9 (ICRP, 1991a).

The Chernobyl accident (see Section 6.4.3) resulted in significant contamination of areas of the former Soviet Union. Epidemiological studies are being conducted in an attempt to establish whether cancer rates are elevated in these areas. So far, it appears reasonably certain that rates of childhood thyroid cancer are elevated, perhaps 5- to 10-fold, in the heavily contaminated regions (UNSCEAR, 2000b)). This increase is probably linked to exposure during childhood to radioactive iodine released during the Chernobyl accident. There are a number of problems in interpreting the data and in establishing the relationship between risk of thyroid cancer and dose to the thyroid. Issues include uncertainty concerning the underlying rate of thyroid cancer (before the accident), difficulties in establishing the magnitude of the thyroid doses received, the influence of screening in possibly finding more thyroid cancers than would otherwise have been found and the relatively short follow-up time in that more excess thyroid cancers may appear in the future. However, it appears at the moment that the numbers of thyroid cancers are broadly within the range that would have been expected on the basis of studies not involving the Chernobyl accident. Attention has also been focused on leukaemia, as the disease is one of the early carcinogenic effects of ionizing radiation, with a latency period of 2–3 years. However, the current consensus is that there is no increased level of leukaemia in residents of contaminated areas (UNSCEAR, 2000b). Similarly no elevation in the incidence of solid tumours (other than thyroid cancer) has been found, although the longer latency period for such cancers raises the possibility that an excess will be found in the future.

Data on risks following long-term exposure to relatively low-levels of radiation have recently become available from studies on occupationally exposed individuals. An international study involving combining data on nearly 96 000 radiation workers from Canada, the UK and the USA has shown a statistically significant increase in deaths from leukaemia (excluding chronic lymphocytic leukaemia – a disease that is not thought to be induced by radiation) and a trend of increasing risk with increasing dose (IARC, 1994). The excess relative risk of death from leukaemia was similar to that obtained from the Japanese A bomb survivor study with some indications for a DDREF of around 1.5. The results for other cancers did not show a statistically significant excess of deaths. Overall, the study concluded that the risk estimates that form the basis for current radiological standards are not seriously in error.

If DNA in male or female germ cells is damaged by ionizing radiation, heritable effects may, in principle, be experienced in the irradiated population. Evidence for such radiation effects is scarce – no statistically significant excesses have so far been observed in exposed human populations. In the absence of any direct evidence, experimental studies on the incidence of genetic effects in irradiated animals, mainly mice, have been used to project estimates of such effects to exposed human populations. There are obvious uncertainties associated with this approach and it is generally thought to result in an upper estimate for the risk of genetic effects.

Overall, epidemiological studies on populations of exposed individuals show a statistically significant increase in mortality from some cancers at doses of around 100 mSv upwards. Doses in the range 10 to 20 mSv when received *in utero* may also increase the risk of childhood cancer (NRPB, 1995). However, so far epidemiological studies lack the power to provide convincing evidence for the radiation induction of cancer at the dose rates typically encountered in the environment (typically around a few millisieverts a year). Thus, interest has turned to molecular studies on the mechanism of cancer induction by ionizing radiation in an attempt to provide an insight into effects at low dose rates.

4.3.2 Molecular Studies

Experimental data on the cellular and molecular mechanisms of radiation tumorigenesis provide a scientific framework within which stochastic effects on humans can be viewed. These studies may not be able to provide direct quantitative information on radiation risks to man at low doses and dose rates but a knowledge of molecular mechanisms may at least establish whether the same processes are occurring at various different dose rates.

The first stage in radiation-induced tumorigenesis is almost certainly damage to chromosomal DNA – the hereditary material. In many cases, this damage will be repaired by the cell's own surveillance and repair mechanisms. In the normal course of events, each human cell will sustain many hundreds of spontaneous DNA damage events per hour. This damage is mainly caused by thermodynamic instability and attack by reactive chemical radicals produced during normal biochemical reactions; the damage probably contributes to the background natural cancer risk. However, the cell's capacity to repair damage is such that the rate of spontaneous single gene mutation per cell generation is very low, at around one in a million or less. If ionizing radiation caused the same type of DNA damage as arises spontaneously, it would be expected to be repaired efficiently and would thus have an insignificant effect on the incidence of cancer. However, as is described below, there is some evidence to suggest that the damage caused to DNA by ionizing radiation is, in some instances, not identical to the spontaneous damage.

DNA comprises two complementary interacting molecular chains. The vast majority of spontaneous DNA damaging events involve only one of the chains and so a repair enzyme can use information from the complementary chain to ensure error-free repair (Ames, 1989). In contrast, when ionizing radiation interacts with a DNA molecule, it could cause coincident damage to both DNA strands; such damage would be inherently more difficult to repair correctly as the repair enzymes could no longer use the complementary strand to provide the necessary information. This type of damage could easily lead to gene mutations (see NRPB, 1995 and Cox, 1996). If the gene affected plays a role in certain key cellular activities, such as cell growth, reproduction or differentiation, the result can be a cell with a predisposition to undergo aberrant cellular growth and behaviour; in other words, it will have the potential to turn into a cancer cell.

Currently, two overall categories of tumour initiating gene mutations are recognized, as follows:

(i) activation of proto-oncogenes, leading to increased activity of genes involved in growth stimulation;

(ii) inactivation of tumour suppression genes, i.e. loss of activity of a gene involved in growth inhibition.

Molecular studies on human cancers suggest that the second mechanism is possibly the most important, with one reason being that it may require rather more specific damage to activate a gene than to inactivate one. A number of tumour-suppressor genes, together with their associated cancers, have been identified and the list is growing. One particular suppressor gene, p53, is changed in a number of solid human cancers (see Weinberg, 1991); this gene may have a role in the control of the cell cycle (Lane, 1992).

There is a considerable body of evidence that most tumours originate from a mutation in a single cell (Woodruff, 1988). However, the initial damage to a cell, termed *initiation*, may not in itself be enough to result in clinically identifiable cancer. Subsequent events, known generally as *promotion*, may help stimulate the growth of a mutated cell, thus causing it to begin uncontrolled cell division. Tumour promoting agents include phorbol esters, bile acids, cellular growth factors and hormones (UNSCEAR, 1993).

The next stage in the process leading to a full malignancy is the accumulation of further gene mutations in the growing clone of pre-cancerous cells. Ultimately, these mutations result in the reproductive and invasive properties characteristic of malignant cells. This is a complex process involving changes to many aspects of cellular behaviour, including cell adhesion, mobility, etc. (Hart and Saini, 1992). There are two likely explanations for these further mutations: they may merely be a consequence of the natural rate of spontaneous mutation or the initial gene mutation may cause some instability in the mechanism of cell division. There is some evidence for the latter explanation (see NRPB, 1995).

The molecular biology studies, particularly the ideas on double-strand DNA damage, provide general support for the principle that any radiation dose, no matter how low, carries with it some risk of a stochastic effect. A generalized overall scheme for radiation-induced carcinogenesis is shown in Figure 4.4.

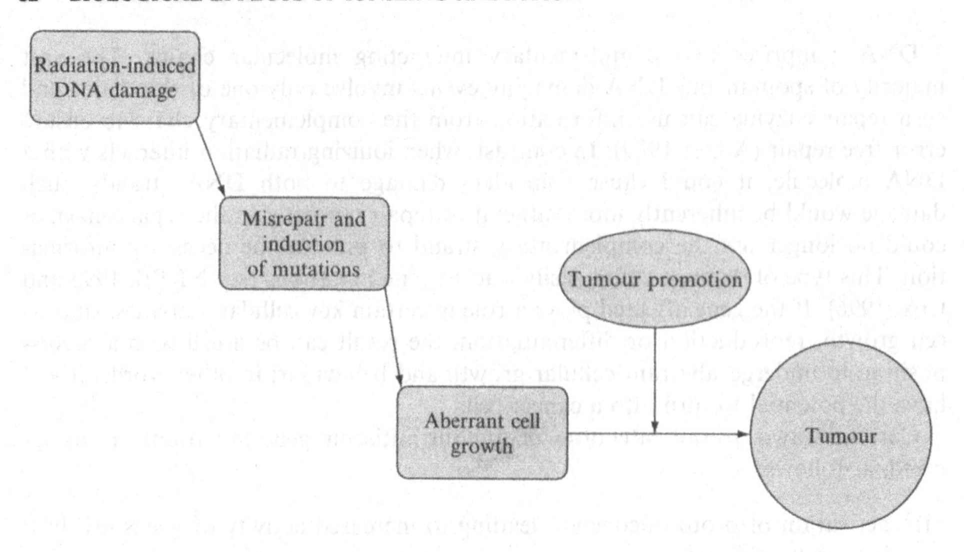

Figure 4.4 A general scheme for tumour initiation

One outstanding topic in the general area of radiation-induced cancer is whether there are individuals in the population who, due to their genetic make-up, are more susceptible to such effects than are others. Around 20% of cancer in the general population may have a significant genetic component (Bodmer, 1991); furthermore, there are a number of recognized hereditary defects in DNA repair or tumour-suppressor genes that can lead to an increased susceptibility to spontaneous cancer (see UNSCEAR, 1993). Are such individuals at higher risk of radiation-induced cancer? There is evidence that individuals suffering from the rare genetic DNA repair deficient condition, ataxia-telangiectasia, have an increased susceptibility to deterministic radiation effects during radiotherapy, although there were too few patients to establish any increased risk to stochastic effects (see Arlett, 1992). Currently, it is considered likely from the available human and animal data that most cancer-prone humans carrying tumour-suppressor gene mutations will have an increased susceptibility to radiation-induced cancer but such individuals are rare (Cox, 1996). It should be pointed out, however, that the increased risk to such individuals due to additional radiation doses at levels commonly found in the environment would not be detectable because of their high risk of developing cancer spontaneously. Overall, this is a developing area of work and the ICRP has published a recent review (ICRP, 1998b).

4.3.3 Conclusions – the Linear No-Threshold Hypothesis

The evidence described in this section favours a linear no-threshold dose response relationship between dose and the likelihood of incurring a stochastic effect. For radiological protection purposes, as described in the following sections, this assumption is assumed to hold true down to essentially zero dose. In other words, a radiation dose, no matter how small, is assumed to confer a proportional risk to the health of the individual receiving the dose.

The relationship between radiation dose and the likelihood of incurring a stochastic effect can be expressed as a risk factor, which is the probability of incurring a stochastic effect per sievert. Thus:

$$\text{Likelihood of incurring a stochastic effect} = \text{Dose (Sv)} \times \text{Risk factor (Sv}^{-1}\text{)} \quad (4.8)$$

Data on risks from ionizing radiation have been reviewed by the ICRP, taking as a basis published studies by UNSCEAR and BEIR (Biological Effects of Ionizing Radiation, a committee of the US National Academy of Sciences). The ICRP risk factors (see Equation (4.1)) for general use in radiological protection are given in Table 4.4. These risk factors are derived for a representative world population and to some extent depend upon the assumptions concerning underlying spontaneous cancer rates. This should be borne in mind if estimates are required of the numbers of radiation-induced health effects in a particular exposed population.

A no-threshold dose–response relationship means that it is not possible to specify a dose below which no risk to health will occur. This is very important for radiological protection. Since it is not possible to ensure the individuals are never exposed to ionizing radiations, it has to be recognized that some risk will always be involved in activities involving exposure to ionizing radiation. The question is, therefore: 'how much risk should individuals be exposed to?'. There cannot be a single, absolute answer to this question, since the level of risk, which could be tolerated, must be related to the level of the anticipated benefits from the activity giving rise to the exposure. In other words, the assumption of a no-threshold dose–response relationship means that radiation protection is based on the concept of balancing risks against benefits. Moreover, it leads to the conclusion that the level of risk that can be tolerated is dependent upon the source of the exposure, since the anticipated benefits arise from different practices or intervention measures. The linear no-threshold dose–response relationship also means that (over the range of dose that the assumption holds) incremental risks resulting from different sources of exposure may be considered independently. An increment of dose of 100 μSv carries with it the same incremental risk to an individual regardless of whether that individual's total exposure from all sources is 0.5 mSv or 50 mSv. It therefore follows that the balancing of risks against benefits can be meaningfully carried out for individual operations, which may add or subtract an increment of exposure total for different people, as well as for the total for dose received by particular individuals.

There is, however, some evidence suggesting a threshold in the dose–response relationship and even some suggestion that small doses may be 'good for you' (see, for example, Luckey, 1982). At the moment, such views are in the minority but data on radiation effects are accruing continually, particularly from molecular

Table 4.4 Risk factors for stochastic effects[a] (taken from ICRP Publication 60 (ICRP, 1991a)).

Exposed population	Fatal cancer	Non-fatal cancer	Severe hereditary effects	Total
Adult workers	0.04	0.01	0.01	0.06
Whole population	0.05	0.01	0.01	0.07

[a] In Sv^{-1}.

studies, and opinions may change in the future. Discussion of the possibility of a threshold in the dose–response relationship and criticism of the assumption of a linear no-threshold dose–response are the subjects of a recent review to which the interested reader is referred (Wade Patterson, 1997).

4.4 RADIOLOGICAL PROTECTION

Many beneficial activities involve ionizing radiation, either directly or indirectly. These activities include medical research and industrial uses of radionuclides or ionizing radiation. Such activities can cause exposure of workers and incidental exposure of members of the public. People may also be exposed to elevated levels of natural radioactivity.

The International Commission of Radiological Protection (ICRP) has developed a system for protecting mankind from the hazards of ionizing radiation. The latest overall recommendations were published in 1991 (ICRP, 1991a). The International Atomic Energy Agency (IAEA), a UN agency, promulgates and interprets the ICRP recommendations, particularly for developing countries. Thus, the ICRP recommendations form the basis for radiological protection in most countries throughout the world.

4.4.1 The ICRP System of Radiological Protection

In Publication 60, the ICRP recommended a system of radiological protection which is intended to provide an appropriate standard of protection for mankind without unduly limiting the beneficial practices giving rise to the exposure. Central to this system of protection is the linear no-threshold assumption, i.e. that any dose of radiation is assumed to carry with it a risk to the health of the individual and this risk is assumed to be proportional to the dose (see Section 4.3.3). The ICRP system is divided into two, namely *practices* and *interventions*. A practice is a deliberate activity which, as a by-product, results in the increased exposure of individuals. Examples of practices are the generation of electricity by nuclear power and the production of radiopharmaceuticals. The objective of radiological protection in these situations is to control this increase in exposure to acceptable levels. The system of protection for practices is described in the next sub-section. Its primary application is during the development and design of practices that could cause exposure. For example, during the design of, say, a nuclear medicine laboratory, the ways that people could be exposed to the radioactivity being used would have to be taken into account. The facility would have to be designed to meet the principles of protection for practices, including complying with dose limits. However, the vast majority of people's exposures are not from practices but come instead from natural sources. Mostly, these do not cause concern but in some situations they can rise to levels where the health implications may warrant action. In circumstances where the source of exposure already exists in the environment, or is not directly under control, the ICRP system of protection for intervention is used. This system is similar to the system of protection for practices being based on the same underlying biological assumptions. Similarities and differences between the two systems are discussed in the final sub-section. Schematic representation of practices and interventions are given in Figures 4.5 and 4.6, respectively, and examples of circumstances where the principles for practices or interventions would be applied are given in Table 4.5.

Practices

The ICRP defines practices as 'those human activities that increase overall exposure to radiation [by] introducing whole new blocks of sources, pathways and

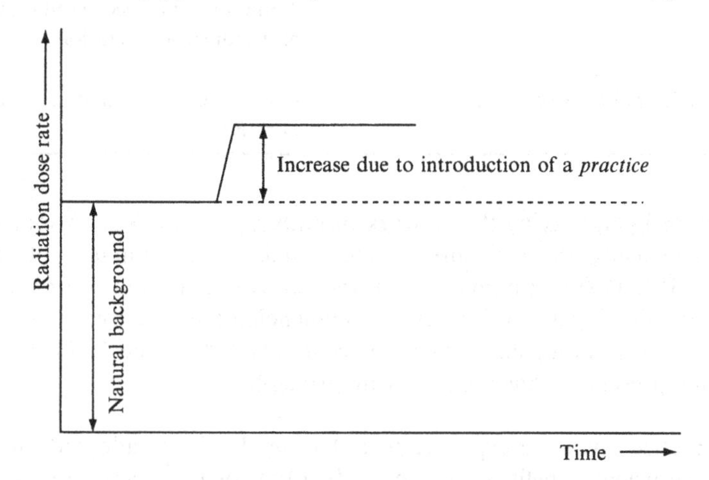

Figure 4.5 Schematic (not to scale) showing that a practice adds radiation doses over and above those from the natural background: In the ICRP System of Protection, such an increase is limited to acceptable levels and the source of the additional exposure has to be worthwhile

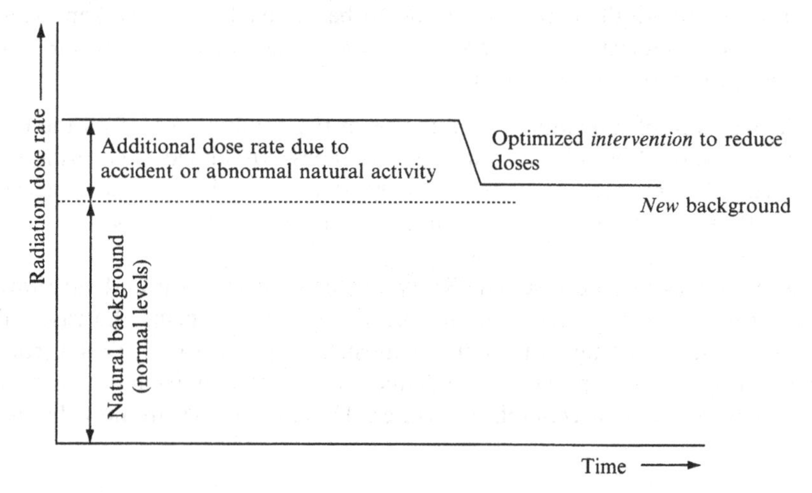

Figure 4.6 Schematic (not to scale) showing that an intervention reduces radiation doses. In the ICRP System of Protection, the actions taken to reduce doses have to be worthwhile

Table 4.5 Example situations where the principles of protection for (a) practices, and (b) interventions would apply.

Appropriate principles for protection	
Practices	Interventions
Public exposure from normal operation of nuclear power stations	Public exposure from accidental releases from nuclear power stations
Release of land following decommissioning nuclear sites	Land which is already in the public domain but has been contaminated by past industrial operations
Disposal of solid radioactive waste	Exposures to high natural concentration of ^{222}Rn

individuals, or by modifying the network of pathways from existing sources to man and thus increasing the exposure of individuals or the number of individuals exposed' (ICRP, 1991a). Emphasis is on the control of the source of exposure and this can generally be planned for before commencing the practice.

The system of radiological protection recommended by the ICRP for proposed and continuing practices has the following principles:

(a) No practice involving exposure to radiation should be adopted unless it produces sufficient benefit to exposed individuals or to society to offset the radiation detriment it causes *(Justification of a practice)*.

(b) In relation to any particular source within a practice, the magnitude of individual doses the number of people exposed, and the likelihood of incurring exposures where these are not certain to be received should be kept as low as reasonably achievable, with economic and social factors taken into account *(The optimization of protection)*.

(c) The exposure of individuals resulting from the combination of all the relevant practices should be subject to dose limits. These are aimed at ensuring that no individual is exposed to radiation risks that are judged to be unacceptable from those practices under any normal circumstances *(Individual dose limits)*.

These principles can be restated. Radiation can cause harm and therefore any intended use should be worthwhile (justification) and, this being the case, all *reasonable* steps should be taken to reduce exposures (optimization). Doses from uses of radiation should be kept within pre-defined levels (dose limits and constraints) in order to prevent any unacceptable exposures. These principles are now discussed in turn.

Justification of practices The ICRP definition of the justification of a practice requires only that the net benefit of a practice, including the waste management, be positive. In other words, that overall the practice does more good than harm. The radiation detriment associated with the practice must be taken into account but so must other detriments and costs; it may be the case that the detriment from radiation is only a small part of the total. Thus, the justification of a practice goes far beyond the scope of radiological protection.

The process of justification is required not only when a new practice is being introduced, but also when an existing practice is being reviewed in the light of new information about its efficiency or consequences. If such a review indicates that a practice no longer produces sufficient benefit to offset the detriment, withdrawal of the practice should be considered. However, withdrawal of a practice may not result in the withdrawal of all of the associated sources of exposure, and the implications of replacing one practice with another may have to be considered.

Optimization of protection This is a fundamental principle of radiological protection and is also one of the most difficult to apply in the area of public exposure. It arises from the assumption that no radiation dose is entirely free from risk. Therefore, merely complying with a dose limit is not sufficient and doses below the limit should be reduced when this can be reasonably carried out. This does not mean that everything technically feasible should be done to reduce doses as there will come a point when further reductions become unreasonable in terms of their cost and other disadvantages.

Generally stated, it is the balancing of the benefit of radiation dose reduction against the resources expended in achieving that reduction. This process should take into account any additional harm involved in achieving that reduction, together with any relevant social and economic factors. The ICRP has clarified the application of the principle as follows: 'If the next step in reducing detriment can be achieved only with a deployment of resources that is seriously out of line with the consequent reduction, it is not in society's interest to take that step...'.

In optimizing radiological protection for a particular practice or source within a practice, account should be taken of exposures of both members of the public and of the workforce. For example, the introduction of effluent treatment plant, e.g. holding tanks, at a facility might reduce the level of public exposure, but might also lead to an increase in occupational exposure.

The difference between the principles of justification and of optimization sometimes causes confusion: justification relates to the practice itself – it should do more good than harm, while optimization is the process of establishing the appropriate level of protection for that practice.

In order to avoid inequities that may arise during the optimization of protection and in order to take account of the fact that some individuals may be exposed to more than one controlled source, the ICRP recommends that the process of optimization is constrained by restrictions on the doses to individuals, termed *dose constraints*. Any option that implies critical group doses higher than the selected value of the appropriate dose constraint should be rejected.

Dose limitation The ICRP recommends limits on the doses that are received by individuals from the operation of practices. These limits are set such that continued exposure at a dose above the limit would, in the ICRP's view, be unacceptable on any reasonable basis. The ICRP's principal limit for members of the public is

1 mSv per annum. Higher doses up to 5 mSv in a year are allowable as long as the five-year average is below 1 mSv. Limits are also recommended for the lens of the eye and localized areas of the skin as these tissues may not be protected against deterministic effects by the limit on effective dose. It is important to recognize that this ICRP dose limit is directed at doses from controlled sources of ionizing radiation. Doses incurred in circumstances where the only available course of action is intervention must not be included in comparisons with the dose limit. The ICRP states, among other things, that exposure from artificial radioactivity already in the environment is one example of a situation that can only be influenced by intervention.

In setting the dose limit for members of the public, the ICRP took account of the biological data on the health risks from radiation. The lifetime fatal cancer risk at an exposure rate of 1 mSv per year is estimated by the ICRP at 0.4% which represents an increase of about 1.5% of the natural likelihood of dying of cancer, which is currently 20–25% in most Western countries. The average annual risk from continuous lifetime exposure at a rate of 1 mSv per year is 4×10^{-5}. Consideration of risks was, however, not the only factor in the ICRP's decision on the value for the dose limit. Account was also taken of doses from natural background and their variation. The ICRP concluded: 'This natural background may not be harmless, but it makes only a small contribution to the health detriment which society experiences. It may not be welcome, but variations from place to place (excluding the large variations in the dose from radon in dwellings) can hardly be called unacceptable' (ICRP, 1991a).

It is generally not possible to estimate doses received by particular individual members of the public. Furthermore, the dose will depend upon the habits of the individual and, as these may change over a period of time, any dose estimate will show corresponding changes. To help overcome these problems, the ICRP recommends that dose limits and dose constraints are applied to the mean dose in the appropriate *critical group*. A critical group is defined as the group of members of the public expected to receive the highest dose from a given radiation source or sources. This group is usually relatively small in size, comprising of a few to a few tens of individuals. The average dose in the critical group is a more robust quantity than the dose to any particular individual because although individual habits may vary, the average habits in the most exposed group are likely to be fairly constant.

The ICRP recommends higher dose limits for workers in industries using radioactivity. The rationale for this is that workers are incurring risks from exposure to ionizing radiation on a voluntary basis whereas any exposure to members of the public will be involuntary. The ICRP's dose limit for workers is 20 mSv in a year to be averaged over a period of five years with no more than 50 mSv in a single year. The ICRP also recommends limits on some organ doses for both workers and members of the public. The ICRP dose limits are given in Table 4.6.

Dose limits do not mark any abrupt change in biological risk. They reflect a judgement on how acceptable doses are from controlled uses of radioactivity.

Dose limits are used in circumstances where there is reasonable certainty that doses will occur and their magnitudes can be estimated with some certainty. However, following disposal of long-lived solid radioactive waste, doses may only arise

Table 4.6 Recommended dose limits of the ICRP (taken from ICRP Publication 60 (ICRP, 1991a)).

Parameter	Workers	Public
Effective dose	20 mSv in a year[a]	1 mSv in a year[b]
Annual equivalent dose in:		
lens of eye	150 mSv	15 mSv
skin	500 mSv	50 mSv
hands and feet	500 mSv	—

[a]Averaged over 5 years with no more than 50 mSv in a year.
[b]In a special circumstance, higher doses can be allowed in a year provided that the five-year average is no higher than 1 mSv per year.

many hundreds or thousands of years into the future; their magnitude will depend upon events and processes that have probabilities associated with them. In this situation, the use of dose limits may have to be adapted; this is discussed further in Chapter 5.

Derived environmental Criteria

Dose limits refer to radiation doses to individuals. Doses arise from radionuclides in the environment and it is possible to calculate the concentration of a particular radionuclide in a specified environmental material that could cause a certain dose to an individual. To carry this out, information is required on the habits of the person and on the routes by which doses are received. These routes are called *exposure pathways* and they depend upon the environmental material and the particular radionuclide. For some materials, the exposure pathways are simple – the exposure pathway for radionuclides in milk is the consumption of milk and milk products. For other materials, a number of exposure pathways may become involved, including ones with complex chains linking the radionuclide to man. An example of a material that can lead to several exposure pathways is soil. A radionuclide in soil can cause exposure in a number of different ways. If the radionuclide is a gamma-emitter it can cause exposure of people standing on the soil without the radionuclide entering the body – this is an example of *external exposure.* The radionuclide could be incorporated into foodstuffs, hence resulting in *internal exposure* when radionuclides are taken into the body. Soil can be re-suspended and subsequently inhaled, also resulting in internal exposure.

Thus, it is possible to derive radionuclide concentrations in environmental media that are related to dose limits or some other dose criterion by a defined model that includes assumptions about the habits of the exposed individuals. In order that the calculated concentrations are generally applicable, it is usual to assume habits typical of critical groups. The guidelines of the World Health Organization (WHO) for drinking water quality and the UK NRPB's Generalized Derived Limits are examples of derived environmental criteria.

The WHO's 'Guidelines for Drinking Water Quality' (WHO, 1993) includes a section on radionuclides. These guidelines are intended to allow drinking water suppliers to determine whether action may be required to reduce radionuclide levels in drinking water. They apply to both natural and artificial radionuclides in drinking water. The guidelines adopt a two-stage approach. First, there are screening levels of $0.1\ Bq\ l^{-1}$ and $1\ Bq\ l^{-1}$ for gross alpha- and gross beta-activity, respectively. If measured concentrations are below these levels, no further action is required. These screening levels are based on the levels of radiologically significant radionuclides that could give rise to a dose of about 0.1 mSv per year. Secondly, if either of the screening levels is exceeded, further analysis is required to establish the radionuclides involved and the dose implied by drinking the water for a year. If a dose in excess of 0.1 mSv per year is indicated, the situation should be investigated more closely and remedial action should be investigated.

The UK NRPB publishes *Generalized Derived Limits* (GDLs). The term *Derived Limits* refers to measurable quantities that are related to the dose limit by defined assumptions. The UK NRPB has calculated GDLs for various radionuclides in a range of environmental media, including soil, water, grass and basic foods (NRPB, 1996b). GDLs are levels of radionuclides in the environmental media that would give rise to a dose of 1 mSv per year to a member of a critical group interacting with that material. They are calculated using cautious assumptions concerning the location of the critical group, its dietary and other habits, and the source of food, such that compliance with the GDL ensures virtual certainty of compliance with the dose limit. A detailed example of how a GDL is calculated is given in Chapter 16. Briefly, the GDL for a radionuclide in soil takes account of the following pathways:

- External exposure from standing on soil
- Inadvertent ingestion of soil
- Inhalation of re-suspended soil
- Ingestion of plant products grown on soil
- Ingestion of animal products from animals (cows, sheep, pigs, etc.) – reared on the soil

The concentration of radionuclide in soil that would give rise to an annual dose of 1 mSv is calculated taking account of doses delivered via all of these exposure pathways. The importance of each exposure pathway depends upon the radionuclide's parent element. For example, ingestion of vegetables and cereals is important for ^{106}Ru, ^{134}Cs and ^{137}Cs, whereas inhalation of re-suspended soil is important for actinides such as ^{238}Pu, ^{239}Pu and ^{241}Am.

Interventions

The principles of protection for intervention are used when the source of exposure is not under control and the exposure is usually already occurring. The main features are as follows:

(a) The proposed intervention should do more good than harm, i.e. the reduction in harm resulting from the reduction in dose should be sufficient to justify the disruption and costs, etc. involved in the intervention *(The justification of intervention)*.

(b) The form, scale, and duration of intervention should be optimized so that the benefit of the reduction in dose should be maximized *(The optimization of intervention)*.

These principles say that one should not intervene to reduce people's doses unless it is worthwhile (Principle (a)) and, if it is worthwhile, then it should be done in the most effective manner (Principle (b)). This is because intervening to reduce will usually have additional consequences, particularly disruption and economic consequences. For example, if a house is found to have relatively high levels of radon, measures can be taken to reduce these, such as increasing the ventilation rate or sealing the floors. These will inevitably involve financial costs and disruption to the occupants. It is therefore necessary to balance these undesirable consequences against the benefit gained from reducing the health risk, in order to determine whether the action is worth undertaking – in other words, is justified. Thus, in the case of intervention, the balancing of harm against benefits is undertaken, but it is the intervention measures that should be justified and optimized, rather than anything relating to the source of the exposure, as is the case for practices.

This difference between the justification and optimization of practices and intervention is significant, because it can result in very different risks being considered tolerable in the two situations. In the case of practices, an individual's existing exposure from radiation will be increased by operation of the practice, and so it is right that significant benefits are achieved from that practice to offset the increased risk. Moreover, where those who are exposed only indirectly experience the benefits, it is reasonable that the magnitude of the increased risk should be strictly limited. Therefore, the dose limit for members of the public from the operation of practices is relatively small. In the case of intervention, however, the exposure exists already; any intervention measures undertaken to reduce this exposure will result in some harmful consequences to the individual. It is therefore necessary to ensure that any benefit achieved by the intervention outweighs these harmful consequences. Where the disruption and financial cost involved would be very significant, a correspondingly high reduction in exposure would be required to make the action justified. This point may be illustrated by comparing the ICRP's dose limit for members of the public of 1 mSv per year with the ICRP's recommended range for action levels for radon in houses, which corresponds to 3–10 mSv per year. In the case of intervention to reduce radon levels, the expenditure and disruption incurred by the individual is relatively low. This means that the action level for radon is also relatively low (although still higher than the public dose limit). For intervention after an accident, the undesirable consequences can be much larger, particularly if it is necessary to evacuate a large number of people. Therefore, it is not surprising that criteria for initiating intervention after accidents are higher, and often significantly higher, than the dose limits for members of the public for practices.

Therefore, the use of dose limits for a practice (see Table 4.6) as a basis for deciding on the level at which intervention to reduce doses should be invoked might involve measures which are out of all proportion to the benefit obtained. Accordingly, the ICRP recommends that dose limits for a practice do not apply to intervention situations. Instead of dose limits or constraints, it is the level of *avertable dose* that is important in intervention situations. Doses are averted by taking particular actions or *countermeasures*. The avertable dose is the dose saved by the introduction of a particular countermeasure and it should be compared with the disadvantages of the countermeasure (cost, disruption, etc.) in order to see if the countermeasure is worthwhile, i.e. optimization of radiological protection. However, it may also be necessary to specify in advance a dose level above which action should be taken. This leads to the following terminology:

- *intervention level* – the level of avertable dose above which a particular countermeasure should be taken;
- *action level* – the level of dose or directly measurable quantity above which action should be taken.

In intervention situations, actions are taken to *save* or *prevent* doses from occurring. For example, the ICRP has issued a guidance on when to implement actions to stop people receiving doses from accidental release of radionuclides.

A significant accident at a Western nuclear power plant is considered to be very unlikely but if one occurred, it could lead to a release of radionuclides to the atmosphere. Such a release could cause people in the neighbourhood to receive significant doses from, for example, external irradiation from radionuclides in the plume or from inhalation of material in the plume. Individuals further afield would be less at risk because the normal atmospheric dispersion, together with radioactive decay, would reduce airborne concentration. If an accidental release is considered a possibility, there are broadly two actions that can be taken in the environment to reduce the dose that may be received by people in the neighbourhood. First, people could be evacuated and taken to a place where doses from the accident are likely to be low or non-existent. However, it will significantly disrupt people's lives, causing distress, etc. Secondly, people can be instructed to shelter in buildings, keeping the doors and windows closed. The protection afforded by the building will help to keep doses down but this will not be as effective as evacuation. However, sheltering is not as disruptive as evacuation. Evacuation and sheltering are termed countermeasures. The ICRP principles for intervention require that the benefit obtained by a countermeasure in terms of the dose it saves should at least offset the harm the countermeasure causes by, say, disrupting people's lives. There are problems in presenting these benefits and harm on a common basis but the ICRP has developed guidance on when to implement countermeasures (ICRP, 1991b). The ICRP's guidance on intervention levels for sheltering and evacuation are given in Table 4.7.

The guidance on implementation of countermeasures is presented in the form of ranges. This is because in any given accident situation the disadvantages of taking the countermeasure will be different. Looking at sheltering, the idea is that if the dose saved by sheltering is 50 mSv or more, then the countermeasure should almost

Table 4.7 The ICRP intervention levels (taken from ICRP Publication 63 (ICRP, 1991b)).

Type of intervention	Intervention level (averted dose)	
	Almost always justified	Range of optimized values
Sheltering	50	Not more than a factor of ten lower than justified value
Evacuation	500[a]	—

[a]In under a week.

always be taken; if the dose saved is above 5 mSv, then the countermeasure should be considered. Similar arguments apply to evacuation.

The action taken to reduce excessive exposures to radon (^{222}Rn, see Chapter 1) is another example of intervention. However, in this case the intervention criterion is specified in terms of an action level. Radon in the home is the main cause of human exposure to ionizing radiation in most parts of the world. It comes from ^{226}Ra, a member of the ^{238}U decay chain, in the soil and builds up indoors. It is not the radon gas itself that irradiates people, but some of its short-lived decay products. The indoor levels of radon are given in Chapter 1 – the range is large, with values ranging to over 2000 Bq m^{-3}, resulting in doses up to 100 mSv per year. The ICRP recommends that national authorities should set an action level for radon intervention measures in the range 200 to 600 Bq m^{-3} (ICRP, 1994b). The ICRP stresses that different countries will need to set different action levels, depending upon the scale of the problem and the availability of resources. The ICRP states: *the best choice of action level may well be that level which defines a significant, but not unmanageable, number of houses in need of remedial work.* Worldwide, the majority of countries that have set such action levels have specified levels in the range 200 to 400 Bq m^{-3}. The USA has the lowest action level, at 150 Bq m^{-3}.

Practices and interventions: differences and similarities

The two ICRP systems are based on the same general principles of justification, optimization and limitation but they are applied in different ways, as summarized in Table 4.8. One difference lies in the application of the principle of limitation and this has led to confusion and misunderstanding. People misunderstand the nature of the dose limit for members of the public for practices; this does not mark an abrupt change in the nature or magnitude of biological risk. Rather, it represents a judgement on the acceptability of the risks incurred from the incidental exposure of people from the deliberate operation of practices. Therefore, it is not directly relevant to other situations where people are exposed. This aspect needs to be remembered.

One last point is that in both practices and interventions, decisions are usually taken on the basis of increments in exposure or decrements in exposure from particular sources or groups of sources; the total dose received by an individual from all sources is not taken into account. Recently, however, the ICRP has issued guidance that in circumstances where the long-term exposure of members of the public from radioactivity in the environment exceeds in total around 100 mSv per

Table 4.8 Application of radiological protection principles to practices and interventions.

Principle	Practice	Intervention
Justification	Overall, the practice should do more good than harm. This decision is taken by society and radiological factors are only one input. Doses and risks are increased	The intervention should do more good than harm. The reduction in doses and risks is a benefit and may be the only one
Optimization	Radiological protection is optimized within dose or risk constraints. The benefit is received by society as a whole; the detriment may only be received by a particular population group	Protection is optimized. The benefit is experienced by the population group who are having their exposure, or potential for exposure, reduced. The costs may be spread over society
Limitation	Total individual doses from all sources subject to control do not exceed dose limits or some control on risk in the case of potential exposures	The ICRP recommends against the application of dose limits for deciding on the need for, or scope of, intervention. However, every effort should be made to avoid the incidence of serious deterministic injuries

year, intervention will nearly always be required to reduce this dose. In radiological protection terminology, the intervention would be justified on the basis of the magnitude of the health risk to the exposed individuals. Conversely, the ICRP notes that in circumstances where this long-term dose is less than about 10 mSv per year, intervention is not likely to be justified on the basis of the health risks alone. These figures of 100 mSv and 10 mSv per year serve as guidelines for applying the system of protection for intervention in the case of long-term exposures of members of the public. These types of situations could include exposures arising from practices that have long since closed down. Examples include contamination from past radium luminizing and nuclear weapons production or testing.

4.5 REGULATION AND RADIOACTIVITY

In many, if not all, developed countries the ICRP Recommendations form the basis for regulatory systems for controlling the use and disposal of radionuclides. Furthermore, there are many emergent and developing countries that would benefit from such a regulatory infrastructure. To aid such countries, the International Atomic Energy Agency (IAEA) publishes 'Basic Safety Standards for the Protection against Ionizing Radiation Sources', the latest version of which was published in 1996 (IAEA, 1996a). These basic safety standards develop the ICRP recommendations into a basis for a regulatory system and give practical guidance in this respect.

It is useful at this point to briefly describe the background of the IAEA, which came into existence in 1957. Its Statute was agreed at an international conference held at the United Nations (UN), with its role broadly being to promote the peaceful use of atomic energy. It is specifically authorized under the terms of its Statute to establish basic safety standards. The IAEA has its headquarters in Vienna.

The IAEA's basic safety standards closely reflect the ICRP recommendations as outlined above. However, it is worthwhile to elaborate on two specific features, i.e. exclusion and exemption. As described in Chapter 1, almost all materials are radioactive to some extent; natural radionuclides are ubiquitous and radiation is pervasive but it would be impractical to attempt to regulate such situations. Thus, the basic safety standards *exclude* exposures that are not amenable to control. The examples given in the standards are exposure from ^{40}K in the body, from cosmic radiation at ground level and from unmodified concentrations of radionuclides in most raw materials.

There is a further aspect to regulatory control. It has been recognized for a number of years that there may be circumstances where application of the full regulatory process for ensuring radiological safety is burdensome and unwarranted because the risks associated with the particular use of radionuclides are trivial. In these circumstances, the radiation sources (and the organizations using them) may be exempted from some or all regulatory control. According to the basic safety standards, such *exemption* may be granted if the associated dose is less than or equal to 10 μSv per year and radiological protection is optimized. The basic safety standards go on to say that if the collective dose from one year of the unregulated practice is one manSv or below, then it can be assumed that radiological protection is optimized. A dose of 10 μSv corresponds to a risk of death from fatal cancer of between one in a million (10^{-6}) and one in ten million (10^{-7}). Such levels of risk are commonly regarded as trivial.

The IAEA's basic safety standards contains radionuclide-specific levels that allow automatic exemption (*unconditional exemption*) from the requirements of the standards except that the use of the radionuclides must be *justified*. These levels are expressed in terms of total activities and activity concentrations. Values for over 300 radionuclides are listed. If a user of radionuclides holds total activities less than the specified activity limits or the radionuclide concentrations in the relevant materials are less than the concentration limits, the user may be exempted. The radionuclide specific levels were calculated by applying the dose criteria noted above using a set of exposure scenarios representing situations where radionuclides are being used. Exempt activity concentrations range from 1 Bq g^{-1} for most α-emitting actinides to 1×10^6 Bq g^{-1} for tritium and ^{37}Ar. The total activity levels range from 1×10^3 Bq for most α-emitting actinides to 1×10^{12} Bq for ^{83}Kr.

The European Union also issues basic safety standards that implement the ICRP recommendations – these are legally binding on the Member States. These standards contain the same radionuclide-specific exemption levels as are in the IAEA's standards.

The concept of exemption is similar in many respects to the concept of *clearance*, which is the release of materials without further restriction from an industrial operation (a practice) that is subject to regulation. This is discussed in Section 5.5, where some comparisons between the two concepts are made.

4.6 CONCLUSIONS

There are two main types of health effects caused by ionizing radiation, namely deterministic effects and stochastic effects. A deterministic effect has a threshold dose below which it does not occur. Thresholds for deterministic effects are higher than the dose levels generally experienced in the environment. Stochastic effects do not have a threshold but the probability of a stochastic effect occurring increases with increasing dose. Stochastic effects include cancer. For radiological protection purposes, it is assumed that there is a linear relationship between dose and the probability of occurrence of stochastic effects. The ICRP's principles of radiological protection are derived on the basis of this assumption. An objective of radiological protection is to reduce the incidence of stochastic effects to acceptable levels without unduly limiting the beneficial uses of radioactivity.

5

Management of Radioactive Waste

5.1 BACKGROUND

Radioactive wastes are produced in gaseous, liquid or solid forms by many industrial processes, notably the generation of electricity by nuclear power. The processes leading to the generation of radioactive wastes and the quantities so produced are described in Chapters 6 and 7. The safe disposal of radioactive waste is an issue that attracts considerable public interest and concern and is likely to become more acute in the near future as many nuclear power stations close down and ways are sought for disposal of decommissioning wastes. The disposal of radioactive wastes is covered by the ICRP's System of Radiological Protection but there are also wider issues that are not completely addressed by the ICRP. An example is the question of discharges of radionuclides from installations in one country irradiating the inhabitants of another (this is an issue for all pollutants, of course – not just radioactive ones). In an attempt to address such issues and to provide a framework for the safe management of radioactive waste, the International Atomic Energy Agency (IAEA) has elaborated nine principles of radioactive waste management (IAEA, 1995a). These principles are derived from the IAEA's basic objective of radioactive waste management, i.e. that radioactive waste is dealt with in a manner that protects human health and the environment now and in the future without imposing undue burdens on future generations. The nine principles are are as follows:

(1) Radioactive waste shall be managed in a way that secures an acceptable level of protection for human health and the environment.

(2) Radioactive waste shall be managed in such a way as to provide an acceptable level of protection for the environment.

(3) Radioactive waste shall be managed in such a way as to assure that possible effects on human health beyond national borders will be taken into account.

(4) Radioactive waste shall be managed in such a way that predicted impacts on the health of future generations will not be greater than relevant levels of impact that are acceptable today.

(5) Radioactive waste shall be managed in such a way that will not impose undue burdens on future generations.

(6) Radioactive waste shall be managed within an appropriate national legal framework including clear allocation of responsibilities and provision for independent regulatory functions.

(7) Generation of radioactive waste shall be kept to the minimum practicable.

(8) Interdependencies among all steps in radioactive waste generation and management shall be taken into account.

(9) The safety of facilities for radioactive waste management shall be appropriately assured during their lifetime.

In any one country, application of radiological protection principles and of waste management principles cannot be left as a matter of chance; a *regulatory authority* should exist. The functions of the regulatory authority that are relevant to waste management include the following: preparation of regulations, review of applications to dispose radioactive materials, the approval or rejection of these applications and the granting of authorizations, the conduct of periodic inspections to verify compliance, and the enforcement against any violations of regulations, standards and license conditions.

There are two broad approaches to the management of radioactive waste, i.e. the 'dilute-and-disperse' approach and the 'concentrate-and-contain' approach. In both cases, both the ICRP's principles of protection and the principles for radioactive waste management developed by the IAEA are relevant, although the manner in which they are applied may differ, as described in this chapter. The first of the ICRP's principles, i.e. *justification*, is applied, however, in the same way to any practice, no matter what type of waste is produced. The principle of justification requires that overall the practice should do more good than harm. Waste management and disposal operations are an integral part of the practice generating the waste and therefore should be considered in the assessment of the justification of that practice (ICRP, 1997). It is wrong to regard them as a free-standing practice, needing its own justification.

Returning to the disposal options, in the *dilute-and-disperse approach* wastes containing very low concentrations of radionuclides are discharged directly to the environment. The wastes are normally in a liquid or gaseous form. They are commonly referred to as *effluents*. The advantages of this approach are that some form of verification and control is usually possible. Environmental monitoring can be undertaken and doses assessed in a retrospective assessment (see Section 5.2). In the *concentrate-and-contain approach*, the waste is usually in a solid form and is isolated from man's environment in order to minimize the possibility of exposures. In the case of relatively short-lived radionuclides (half-lives of less than around a few years), it may be possible to isolate the waste in secure stores until radioactive decay reduces the amount of activity to harmless levels. It is also possible to treat liquid and gaseous wastes containing radionuclides of half-lives of a few days or weeks, or possibly longer, in this manner. However, it is generally considered that wastes containing significant amounts of longer-lived radionuclides will have to be isolated in a repository. The fact that the waste is in a concentrated form leads to its own specific problems; if the integrity of the repository is breached, individuals could receive large exposures. This obviously becomes more of an issue for wastes containing very long-lived radionuclides, due to the long timescales involved and the fact that knowledge of the repository may be lost over such time periods. This is dealt with in detail in the section on solid waste. In general, the concentrate-and-contain approach relies on isolation to protect individuals in current and near-future generations. In the longer term, radioactive decay will reduce activity levels but it is important to note that it is impossible to guarantee isolation for the periods of time required for very long-lived radionuclides to decay to insignificant levels. Over such time periods, waste forms are likely to breakdown and radionuclides could disperse in, say, groundwater, ultimately leading to the possibility of doses to future individuals.

For some types of waste, there is a choice whether to use the dilute-and-disperse approach or the concentrate-and-contain approach. Examples are contaminated liquids or gases where it is possible to remove radionuclides using chemical or physical means and isolate them as, say, solid waste. In such cases, optimization studies can be undertaken to determine the radiologically optimum choice. The various options should be identified and their features examined as far as possible, including capital, operating and maintenance costs, the implications for waste management, and the effect on individual and collective doses for both the public and workers. In the past, a technique called *cost–benefit analysis* has been used with varying degrees of success as an aid to taking such decisions. This technique involves assigning monetary costs to each of the relevant factors and then calculating the cost associated with each option; the one with the lowest is likely to be the optimum option. One important input into cost – benefit analysis is collective dose (see Section 4.1.5). The collective dose arising from each option can be estimated and the corresponding monetary cost can be calculated on the basis of values for cost per unit collective dose (NRPB, 1993). The ICRP has criticized the use of cost – benefit analysis in taking waste management decisions, stating, in its policy for radioactive waste disposal (ICRP, 1997) that ICRP's policy is more judgmental, being summarized, in essence, by the following statement: 'Have I done all that I reasonably can to reduce these radiation doses?'. Guidance has also been provided by the ICRP on how the quantity 'collective dose' should be used. Collective dose is an aggregated quantity and problems arise by use of collective doses from very small doses to large populations and from doses occurring over very long periods of time. In the case of very small doses, the following argument has been advanced in some quarters: Why should the collective dose be significant when the constituent individual doses are not? However, the current view, as articulated by the ICRP, is that the component of collective dose due to small individual doses cannot be ignored solely because the individual doses are small. Where possible, however, it would provide more information for decision-making purposes to separately identify that component of the total collective dose which is delivered at very small individual doses; decision-makers could, for example, place lower weight in the decision-making process on this component versus the components or collective doses that are delivered at higher individual doses. The weight attached to collective doses delivered over long time periods into the future is also addressed by the ICRP. Uncertainty is the main problem. Both the magnitude of the individual doses and the size of the exposed population become increasingly uncertain as the time increases. There is also the issue that current judgements about the relationship between dose and health effects (or detriment) may not be valid for future generations. In the light of such arguments, the ICRP suggests that forecasts of collective doses over periods longer than several thousand years and forecasts of health detriment over periods longer than several hundreds of years 'should be examined critically'. Overall, the ICRP argues that when collective doses are being considered in decisions on waste management options, it is more informative to present the doses in blocks of individual dose and time intervals.

There are a number of techniques available for optimizing radiological protection. An example study which illustrates the factors that have to be taken into account and the general approach that can be followed is given at the end of this chapter in Section 5.7. In general, optimization can involve a number of complex trade-offs, including:

- trade-off between doses to the public and doses to the workers involved in waste treatment and disposal operations;
- trade-off between present doses resulting from effluent discharges and future doses associated with disposal of solid waste resulting from solidification of those effluents;
- choice between options whose characteristics are known with different degrees of certainty.

Increasingly, however, social and political factors are dictating that the concentrate-and-contain approach is strongly preferred. The two options are illustrated in Figure 5.1 and are now described in more detail.

5.2 DILUTE-AND-DISPERSE

Historically, this option has been used for slightly contaminated liquids and gases such as treated effluents from the spent fuel storage ponds at nuclear power stations and reprocessing facilities where spent fuel is stored.

Discharges occur in two ways, i.e. gases and fine particulate material may be discharged to atmosphere, while liquids, soluble substances and suspended solids can be discharged in water to either the marine environment or the freshwater environment.

Discharges to atmosphere cause radiation doses to individuals who are in the plume of activity as it disperses downwind. People can receive doses from gamma- or beta-emitting radionuclides in the plume, while the radionuclides themselves remain outside the body – this is an example of external irradiation. Individuals may also inhale radionuclides and this causes internal irradiation.

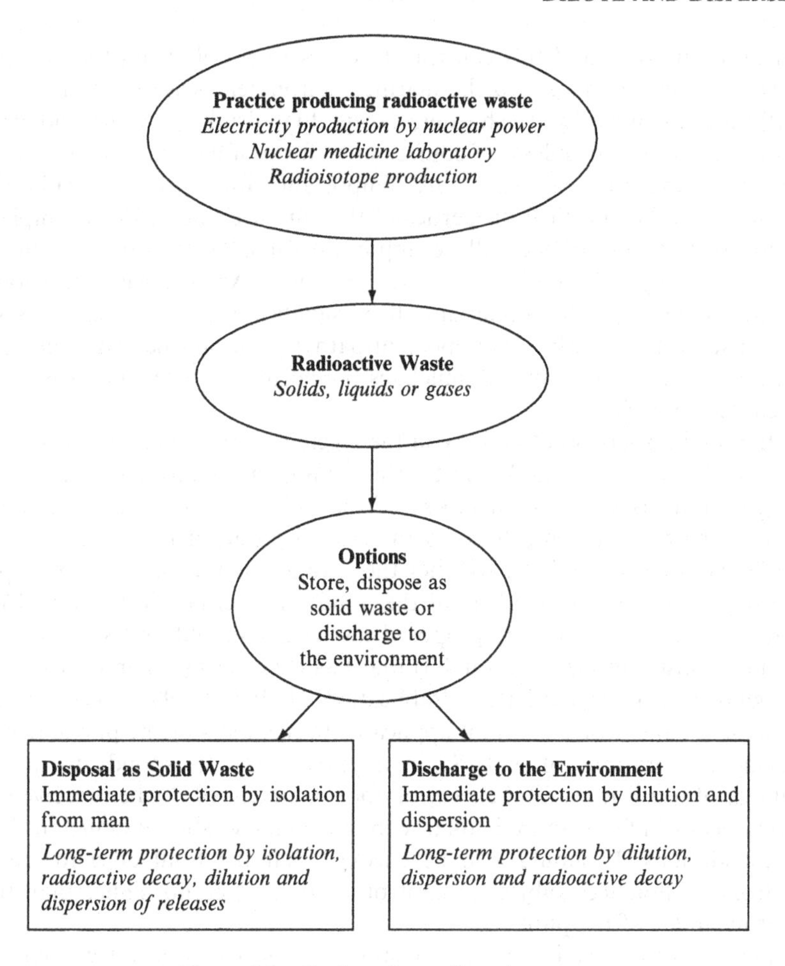

Figure 5.1 Options for radioactive waste

Radionuclides in discharged particulate material (dusts, etc.) may eventually deposit on the ground by, among other mechanisms, gravitational settling, thus leading to radiation exposure after the airborne plume of activity has passed. The exposure pathways arising from deposited radioactive material include external exposure from gamma- and beta-emitters and internal exposure from consumption of radionuclides incorporated into foodstuffs. Material deposited on the ground may be re-suspended due to the action of winds, etc., and subsequently inhaled. Radionuclides that are isotopes of biologically available elements, i.e., carbon, sulfur and tritium, may also become directly incorporated into foodstuffs, even when discharged as gases or vapours.

Discharges to the freshwater environment may result in internal doses from the consumption of any drinking water abstracted from the water body and from consumption of freshwater fish that have incorporated radionuclides. Some radionuclides readily bind to sediments due to their chemical properties; thus, these radionuclides will become bound to river beaches and banks, and cause doses by external exposure or inhalation of re-suspended material. Freshwater may be used for irrigation of crops and this can lead to doses to people from food-chain pathways. Doses from discharges to the marine environment arise from similar exposure pathways to those for freshwater, except that drinking water and irrigation pathways will not occur unless, of course, some form of desalination is involved.

Different exposure pathways are important for different radionuclides, depending upon the chemical properties of the parent element. For example, external exposure from the plume will be important for beta- or gamma-emitting isotopes of chemically inert noble gases, e.g., ^{85}Kr and ^{41}Ar, whereas inhalation of the plume and of re-suspended material will be significant for alpha-emitters such as ^{239}Pu. Consumption of fish is an important pathway for marine discharges of ^{137}Cs, while external exposure from radionuclides bound to beach sediments is important for discharges of ^{95}Zr.

Dilution and dispersion of activity in the receiving environmental medium assists in keeping doses at acceptable levels. Restricting the quantities of radionuclides discharged controls doses to members of the public. Prior to discharging radionuclides to the environment, doses to the critical group of members of the public should be assessed to establish whether they are acceptable. The general approach to assessing exposures of members of the public is described in Chapter 16. For a new facility, the doses from the proposed discharges should be less than the value of the dose constraint (see Section 4.3) that has been set by the regulatory authority. As guidance, the ICRP has, in Publication 77 (ICRP, 1997), recommended an upper value for the dose constraint applicable to members of the public from waste management operations of 300 μSv in a year. This corresponds to an average annual risk of death from fatal cancer of approximately 10^{-5} when received in each and every year of life by an individual with the characteristics assumed in the ICRP Recommendations. It should also be checked that the total dose to the critical group from all practices subject to control is likely to be no higher than the dose limit for members of the public.

Compliance with dose limits and constraints is on its own, not enough; it must also be established that the proposed discharges are the radiologically optimum solution. This is an important stage in the process of deciding whether the form or scale of a proposed discharge to the environment is acceptable. As mentioned earlier, this involves making an evaluation of the costs and efficacies of available control options and of the possibility of changing the process or activity under consideration such that radioactive waste is not generated or, at least, its generation is reduced to the minimum practicable. For routine discharges of radioactive materials into the environment, the main types of control options are to provide either storage facilities for gaseous and liquid effluents, so that short-lived radionuclides can decay before release, or treatment facilities that remove radionuclides from the effluent stream for disposal by other means. Relatively simple factors such as changing the height of planned discharge stacks may also deserve consideration.

If the regulatory authority is satisfied that the proposed discharges are acceptable, it is normal to set legal limits on the quantities of radionuclides that can be released in given periods of time. These are referred to as *discharge limits* and are usually specified in a formal document called an *authorization*. The limits should satisfy the requirements of the optimization of protection and the condition that doses to the critical group shall not exceed the appropriate dose constraints. They should also reflect the requirements of a well-designed and well-managed practice and, should provide a margin for operational flexibility and variability. Limits may be set in a number of different formats, e.g. limits on individual nuclides, limits on groups of nuclides, etc.

In order to demonstrate that discharges are in compliance with the limits, monitoring of the radionuclide levels in the discharged effluent is normally required. This often involves taking samples of effluent for radiochemical analysis. Similarly, in order to check the assumptions used to evaluate critical group doses, monitoring of radionuclide levels in the sector of the environment receiving the discharge is also often required. The calculation of critical group doses from the results of environmental monitoring programmes is described in Chapter 16. If the result of such a *retrospective* dose assessment using the results from environmental monitoring suggests higher than expected doses, there may be the possibility of revising the discharge authorization. Environmental monitoring also provides an additional means, besides effluent monitoring, of checking for unexpected releases. The requirements for monitoring are usually specified in the discharge authorization. The IAEA has published guidance on the authorization of effluent releases that links specific requirements, such as the scale of environmental monitoring, to assessed critical group doses, to which the interested reader is referred (IAEA, 2000).

5.3 CONCENTRATE-AND-CONTAIN

Solid radioactive wastes are generated both by the nuclear and non-nuclear industries (see Chapters 6 and 7). The general manner of disposal is some form of burial. Wastes containing very low levels of radionuclides may, in some cases, be disposed of to landfill sites with ordinary refuse. Wastes containing higher levels of radionuclides require greater isolation from man's environment (the biosphere). Spent nuclear fuel and the highly concentrated fission product wastes arising from the reprocessing of spent fuel require the highest degree of isolation; currently, deep geological disposal at depths of several hundreds of metres is being considered. Deep geological disposal is also an option for wastes with lower concentrations of radionuclides.

Disposal options for radioactive waste, other than very low level waste, depend upon the so-called *multi-barrier approach*. Here each feature of the disposal option, e.g. the waste form, the containers, the lining of the disposal facility, the geological formation in which the facility is placed, etc., inhibits the migration of radionuclides from the waste to the biosphere. Thus, if one barrier fails to behave as expected, the other barriers should be sufficient to ensure safety.

One important aspect that has to be taken into account in developing an acceptable strategy for dealing with solid radioactive waste is that doses from disposals of long-lived wastes may only occur many years into the future. Following burial of the wastes, the degradation of the waste and the slow leaching and migration of the radionuclides may mean that radionuclides only appear in man's environment after extended periods of time. Nevertheless, future populations deserve protection from our actions today. The levels of current doses are regulated under the ICRP's system of radiological protection (see Section 4.3) using dose limits and dose constraints. However, it is difficult to calculate a unique dose to future individuals from disposal of solid waste because the calculated dose will depend on what assumptions are made about future human habits, e.g. what foods are produced, how much is consumed and at what rate, etc. The further into the future one looks, then the more uncertain one becomes about human habits. Furthermore, in the far future there may be significant changes to the environment that can influence projected dose levels; for example, global warming may cause sea-level rise and, conversely, the long-term weather cycles could result in glaciation with a concomitant drop in sea level. Ice pressure from glaciation could also cause faulting of rocks and thus increase the rate of migration of radionuclides from a waste disposal facility built in the rock. There is also the possibility that future individuals could receive doses by inadvertently drilling or digging into a waste site – this is referred to as *human intrusion*. The intruders may receive doses from handling the waste (not knowing exactly what it is) and the act of digging or drilling into the waste could increase the flux of radionuclides to the biosphere. Thus, there are large uncertainties surrounding any estimate of doses to future individuals from disposals of solid radioactive waste, particularly if long-lived radionuclides are involved, which introduces difficulties in using dose limits and dose constraints as criteria for acceptability. This has led to a search for other radiological protection criteria against which to assess the suitability of disposal options for solid radioactive waste. However, as this chapter shows, uncertainty remains an important factor when considering disposal options for long-lived radioactive wastes.

The problems of ensuring safety in the future are less acute for disposals of low-level solid short-lived waste ($t_{1/2} <\sim 30$ years). In such cases, it is reasonable to assume that the disposal facility will remain under the control of the operator for a period of up to, say, a few hundreds of years, during which time the radionuclides will have decayed to negligible levels. This period of control is referred to as the *institutional control period*: inadvertent intrusion is prevented and the performance of the disposal facility can be monitored so that corrective actions can be taken if necessary. An issue regarding long-lived wastes is: can we assume that institutional control will continue in perpetuity? The current view is no.

The timescale that can be involved in solid waste disposal assessments are put into perspective in Table 5.1 where the half-lives of some important radionuclides are set beside the times of historical and predicted future human and geological events. It is thought provoking that the half-life of iodine-129 is comparable with the time since the evolutionary line of man branched from the apes.

5.4 RADIOLOGICAL PROTECTION CRITERIA FOR SOLID WASTE DISPOSAL

The fundamental principles governing the disposal of solid radioactive waste have been developed over several years. The IAEA principles for radioactive waste management, described in Section 5.1, apply, for example. One overriding principle that has dominated discussions about disposal of radioactive waste, in particular long-lived waste, is that future generations should be afforded the same degree of protection as is the current population. Furthermore, we do not want to place the responsibility for the management of our wastes on future generations. One problem

Table 5.1 Timescales and waste disposal (taken from NRPB, 1992).

Years	Historical	Future	Nuclide
100	Discovery of radioactivity	Greenhouse effects	
1000	Norman conquest	Large ecological changes, e.g. lakes fill with weeds	
	Egyptian pyramids		
		Mineral and energy resources exhausted?	^{14}C
10 000	Discovery of agriculture		^{239}Pu
	Last glaciation of northern Europe	Next glaciation	
	Use of fire and tools by humans		
100 000	Emergence of the Neanderthal man	Time between major glaciations	^{99}Tc
1 000 000	Emergence of *Homo sapiens*	Stable geological formations remain relatively unchanged	^{237}Np
	Evolutionary branching between humans and apes		
10 000 000		Spontaneous appearance of new families of species	^{129}I
100 000 000	Dinosaurs populated the earth	Large-scale movements of continents (thousands of kilometres)	
1 000 000 000	Appearance of multicellular organisms	Significant probability by this time of nearby supernova, or meteorite impacts	
		Increase in solar intensity sufficient to erase life on earth	^{238}U
		Age of earth	
		Sun becomes red giant	

is how to set radiological protection criteria which reflect these principles in order to be able to evaluate the performance of a proposed disposal facility. One solution is to set criteria in terms of individual risk. This can be viewed as an extension of the use of dose limits, as a dose limit essentially acts as a limit on the chance of contracting a fatal radiation-induced cancer. Thus, we have:

$$\text{Dose } (H) \times \text{Risk} = \text{Chance of contracting fatal cancer} \tag{5.1}$$

The above equation assumes that an individual will receive the dose. However, if it is not certain that the dose will be received, then the chance of contracting fatal cancer must be less. If the probability of the dose being received is taken into account, the equation becomes:

$$\begin{aligned}&\text{Probability of receiving the dose } H \times \text{Dose } H \times \text{Risk factor} =\\&\text{Chance of contracting fatal cancer}\end{aligned} \tag{5.2}$$

Thus, a limit on the chance of contracting a fatal radiation-induced cancer can achieve the same purpose as the dose limit, and it can, in principle, be applied to a much wider range of situations. Thus, the term 'risk' can be broadly defined as follows:

Probability of receiving exposure $H \times$ Probability that the exposure will give rise to a serious health effect

Following this approach, the risk to individuals in the future from a waste disposal facility could be assessed by calculating the dose that could be received by individuals in a particular set of future conditions, multiplying the dose by the perceived likelihood that the particular set of circumstances will occur, and finally multiplying by the risk factor per unit dose (see Section 4.2.1). In this way, risks to future individuals could be assessed and compared with a risk criterion. Sets of future conditions are often called *exposure scenarios*. For example, one exposure scenario could represent the normal leaching of radionuclides from the waste in a repository (this would be the so-called normal evolution scenario) and another scenario could represent direct inadvertent human intrusion into a repository by, say, exploratory drilling.

Therefore, the 'risk', R, becomes:

$$R = \gamma \sum_{i} P_i E_i \tag{5.3}$$

where γ is the risk (or probability) of fatal cancer per unit dose, and P_i is the probability of scenario i which if it occurs, gives rise to an effective dose E_i (assuming that this is less than about 0.5 Sv and R refers to the same individual).

Using this approach, it is important that the set of exposure scenarios chosen covers the range of plausible future conditions for the particular waste disposal facility. The sum of the scenario probabilities should be unity. Each scenario can represent a series of possible futures that have very similar radiological consequences and so can be treated together as one possible future evolution.

It can require considerable effort to identify an adequate set of scenarios. Expert judgement, computer models of possible future conditions and natural analogues may assist in the process. Assigning probabilities to scenarios is equally difficult and can appear arbitrary. This overall approach is discussed in several publications (NRPB, 1992; Cooper et al., 1992). The concept of 'risk', as described in this section is used in some national regulations as a basis for regulating solid radioactive waste disposal facilities. However, due to the apparent arbitrariness of using risk in this way, other countries use a more judgmental approach. The ICRP has recently issued guidance on this subject which contains elements of both approaches.

5.4.1 ICRP Recommendations

The ICRP's system of radiological protection described in Chapter 4 does not contain any numerical criteria for assessing the long-term safety of solid waste disposal facilities. The ICRP has, however, recently issued guidance on the radiological criteria for the disposal of solid long-lived radioactive waste (ICRP, 1998b). This guidance further develops the 1990 Recommendations and the policy for the disposal of radioactive waste (ICRP, 1997) specifically in the context of the disposal of solid long-lived radioactive waste. The disposal options considered are surface or near-surface disposal and geological disposal. All types of solid long-lived radioactive wastes are covered, including solid high-level waste and large-volume low-level wastes that can arise from the processing of mineral ores. The issues addressed include the radiological protection criteria that should be applied in establishing whether future populations are being adequately protected, and guidance on how to apply the optimization principle in waste disposal. Particular consideration is given to evaluating the significance of exposures arising from direct inadvertent human intrusion into a waste disposal facility.

The ICRP points out that protection of future generations to the same extent as current generations implies the use of the current dose and risk criteria derived from considering associated health detriment. Therefore, in principle, protection of future generations should be achieved by applying these dose and risk criteria to the estimated future doses and risks in appropriately defined critical groups. These criteria are dose limits and dose and risk constraints. However, as it is impossible to know in advance what the total dose to future individuals will be from practices because we do not know what other practices will be in use in the future, dose limits cannot be used and so the appropriate criteria are the constraints: constrained optimization is therefore the central approach to evaluating the radiological acceptability of a waste disposal system.

Knowledge of the disposal facility may be lost in the future and, consequently, it cannot be assumed that any mitigation measures would be carried out to reduce doses should these reach unacceptable levels. An effective disposal facility will, however, retain the waste during the period of greatest potential hazard with only residual radionuclides entering man's environment in the distant future. In this future time period, two broad categories of exposure situation are distinguished, i.e. natural processes and human intrusion. The term 'human intrusion' covers inadvertent human actions affecting repository integrity and potentially having radiological consequences. It is more likely to occur after knowledge of the repository has been lost, i.e. in the far future. The term, 'natural processes', includes all of the processes that lead to the exposure of individuals other than human intrusion. Application of radiological protection criteria to these two categories of exposure situations is different.

Natural processes

These processes include the foreseen gradual degradation of the repository together with other, less likely, natural processes that may disrupt the performance of the repository. Therefore, the objective of protecting the public in such circumstances would have to consider both the probability of occurrence and the magnitude of the corresponding exposures. This can be achieved by either aggregating the probabilities and corresponding doses (or rather the risk equivalent of the dose) in an overall evaluation of risk or by separate consideration of the dose and associated likelihood of occurrence of the exposure in a disaggregated approach. The key criterion is the dose constraint for members of the public. An upper numerical value of 300 μSv per year has been recommended by the ICRP (see Section 5.1); this corresponds to a risk constraint of the order of 10^{-5} per year. In the aggregated approach, the total risk to a representative critical group is compared with the risk constraint as described earlier in this section. This is conceptually satisfying but as has been noted requires a comprehensive evaluation of all relevant exposure situations and their associated probabilities of occurrence – a process that can be difficult to achieve in a transparent and convincing manner. In the disaggregated approach, likely or representative release scenarios are identified and the calculated doses from these scenarios are compared with the dose constraint. The radiological significance of other, less likely, scenarios can be evaluated from a separate consideration of the resultant doses and their probabilities of occurrence. The ICRP considers that although a similar level of protection can be achieved by these approaches, more information can be obtained for decision-making purposes from the disaggregated approach.

All of these approaches require assessments of doses or risks to critical groups. Due to the long timescales under consideration, the habits and characteristics of the critical group, as well as those of the environment in which it is located, can only be assumed. Thus, the critical group is hypothetical. It should be chosen on the basis of reasonably plausible assumptions, taking account of current lifestyles and site- or region-specific information. In many cases, different exposure scenarios with their associated critical groups will have different probabilities of occurrence and therefore the highest dose may not correspond to the highest risk. Because of this, it is important to clearly present the different exposure scenarios and their associated probabilities of occurrence.

The long timescales under consideration impact on the importance attached to the assessed doses and risks. These estimates, according to the ICRP, should not be regarded as measures of health detriment beyond times of around several hundreds of years into the future. In the case of longer time periods, they represent indicators of the protection afforded by the disposal system.

Human intrusion

The possibility of elevated exposures from human intrusion is an inescapable consequence of the decision to concentrate waste in a discrete disposal facility. An intrusion event could result in radioactive material being brought to the surface, thus resulting in the exposure of nearby populations to significant radiation doses. How should such events be taken into account in evaluating the radiological acceptability as a waste disposal option?

In principle, a risk-based approach, considering both the probability and consequences of human intrusion, could be used to evaluate the radiological significance of human intrusion. The ICRP, however, cautions against this approach as there is no scientific basis for predicting the nature or probability of the corresponding future human actions. Instead, it is suggested that the radiological consequences of intrusion should be considered. These consequences should not, however, be compared with the dose constraint of 0.3 mSv per year for members of the public; this constraint applies during the process of optimization of protection and, by definition, intrusion will have bypassed all of the barriers which were considered during the optimization process. So, what should the consequences be compared with? The ICRP considers that in circumstances where human intrusion could lead to doses to those living around the site sufficiently high that intervention on current criteria would almost always be justified, reasonable efforts should be made to reduce the probability of intrusion or to limit its consequences. The general criteria for intervention in the case of long-term exposures have been established in ICRP Publication 82 (ICRP, 1999b); the ICRP suggests that these could be used to evaluate the significance of human intrusion. Briefly, these criteria are that an existing annual dose of 10 mSv may be used as a generic reference level, below which intervention is not likely to be justifiable, whereas at 100 mSv per year and above, intervention is considered to be almost always justifiable. The term 'existing annual dose' refers to the sum of the existing and persisting annual doses to individuals at a given location. The exposure that may occur from a repository is one component of this. The doses should be assessed by using plausible stylized exposure scenarios representing human intrusion events. Examples show how this approach can be applied. High-level waste would cause doses from intrusion higher than 100 mSv per year and thereafter reasonable steps should be taken to reduce the probability or the consequences of intrusion. Not much can be done about the consequences, but placing the waste in a deep repository away from natural resources that might be mined in the future could reduce the probability of intrusion. Conversely, wastes with low activity levels where intrusion doses are less than 10 mSv per year might be suitable for disposal in shallow trenches.

This guidance from the ICRP addresses the exposures to individuals in local population groups but there is the issue of the exposure of the intruder. This is not directly addressed by the ICRP.

Technical and managerial aspects

The importance of these other factors is pointed out in the ICRP guidance. One of the key principles is the concept of defence in depth which provides for successive passive safety measures which enhance confidence that the disposal system is robust and has an adequate safety margin. This is referred to earlier as the *multi-barrier concept*. In addition to the technical principles, an essential managerial principle for all individuals and organizations involved in the repository development process is to establish a consistent and pervading approach to safety which governs their actions.

Compliance with radiological criteria

This is a difficult topic because of the inherent uncertainties in our estimates of radiological impacts on future generations. For this reason, demonstration that radiological protection criteria will be met in the future is not as simple as a straightforward comparison of estimated doses and risks with the appropriate constraints. Judgement may be required. The ICRP points out that the dose and risk constraints should be considered as reference values for time periods farther into the future, and additional arguments, such as those described in Section 5.4.2, should be brought to bear when judging compliance. A decision on the acceptability of a disposal system should be based on reasonable assurance rather than on an absolute demonstration of compliance with numerical criteria.

Overall, the ICRP's view on compliance is that provided the appropriate constraint for natural processes has been appropriately satisfied, that reasonable measures have been taken to reduce the probability of human intrusion and that sound engineering, technical and managerial principles have been followed, the radiological protection requirements can be considered to be satisfied.

5.4.2 Other Criteria

The uncertainties inherent in estimating risks to future individuals have prompted a search for other indicators of the safety of radioactive waste repositories. In assessing the long-term safety of a repository, mathematical models are used to estimate leaching from the waste, the flux through the geosphere and the behaviour of radionuclides in the biosphere. The processes are shown in Figure 5.2. At various stages in the process, intermediate quantities are calculated, such as flux from the repository to the geosphere. Estimates of such quantities may have less uncertainty associated with them than do estimates of risk or dose; thus, they could provide a measure of the safety of the repository if they could be compared with some known data. These other indicators include flux through the repository barriers (i.e. backfill, waste canisters, etc.), fluxes to the biosphere, estimated environmental concentrations and radiotoxicity (IAEA, 1994a). The last of these, radiotoxicity, is an interesting indicator – it is based on one of several radiological parameters, such as activity per unit volume and the number of annual limits of intake by ingestion of inhalation contained in the waste.[1] The radiotoxicity of the waste can be compared with the radiotoxicity of, say, uranium ore. As the radioactive content of the waste falls due to radioactive decay, the radiotoxicity of the waste will fall until at some point in the future it becomes equal to that of the parent ore – the crossover time. Depending on the particular assumptions, most radiotoxicity curves in the literature show crossover times of the order of a few thousand to a few hundred thousand years. There are limitations to the usefulness of radiotoxicity indices as they only reflect hazard potential and not the actual hazard posed by the waste in the repository.

The flux of radionuclides from a repository facility to the biosphere has also been used as an indicator as it can be compared with the flux of natural radionuclides from the geosphere to the biosphere. This approach avoids making assumptions about the future biosphere but the connection with health risks is not clear.

The time until the peak risks occur or until the flux of radionuclides to the biosphere peaks can also be a useful indicator of safety – the longer this time period, then the greater the isolation potential of the repository. Overall, however, the utility of these other safety indicators suffers from the absence of definitive comparators.

[1] An annual limit of intake (ALI) is conventionally described as 'the quantity of a specified radionuclide that has to be ingested or inhaled in a year by a worker in order to receive the dose limit for workers' (see Section 4.4.1).

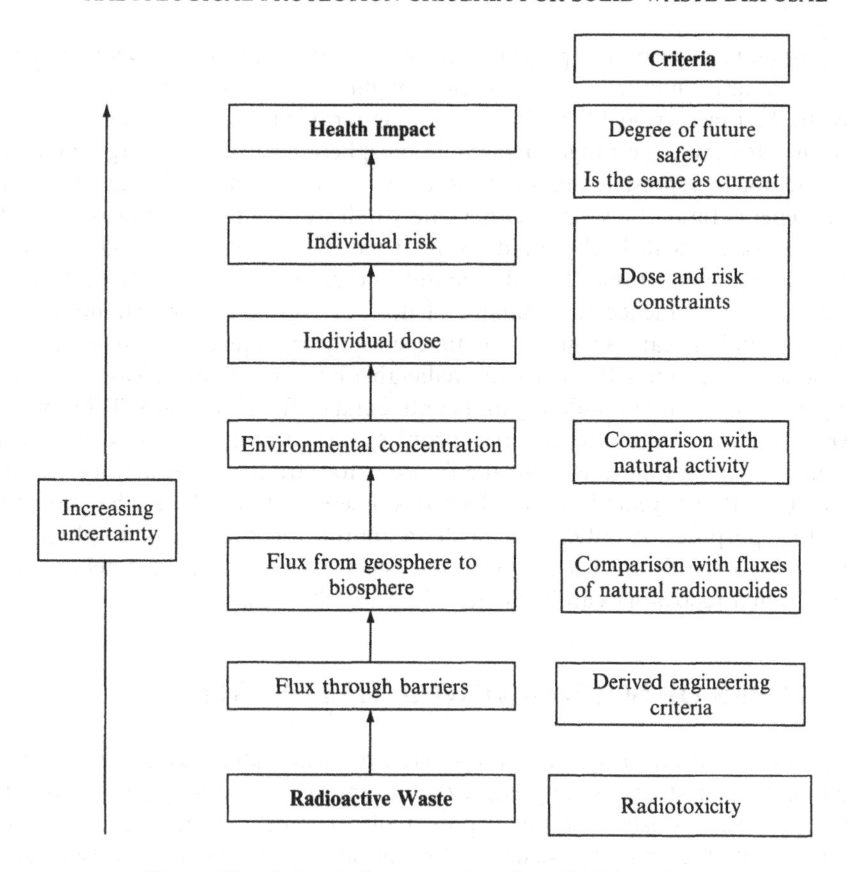

Figure 5.2 Safety indicators (taken from IAEA, 1994a)

5.4.3 Time Frames

Peak radionuclide releases from a repository for long-lived waste may occur many hundreds of thousands of years into the future. However, the uncertainties associated with estimates of risks and with estimates of other safety indicators increases with the time into the future being considered. Although uncertainty increases continuously with time, it has been suggested that, as an aid to estimating repository safety, the future is divided into the following time frames (IAEA, 1994a; NRPB, 1992):

- closure to 10^4 years
- 10^4 to 10^6 years
- beyond 10^6 years

The way in which a safety assessment is undertaken would be different within each time frame in a manner that reflects the reliability of the data on which the assessment is based. In particular, detailed assumptions about environmental conditions and human habits in the far future are unlikely to be justified.

Assessments covering the period from repository closure to 10^4 years could reasonably assume climatic conditions, etc. roughly equivalent to present day conditions. In the time period 10^4 to 10^6 years, long-term climatic changes are very likely to occur. However, the range of possible biosphere conditions is large and, rather than attempt to represent this range in assessments, it may be better to assume a single 'generic' future biosphere. Other safety indicators such as radiotoxicity indices and comparisons with background levels of natural radionuclides will become important in this time frame. For the period beyond 10^6 years into the future, little credence can be attached to estimates of dose or risk as the uncertainties become great: one million years is roughly the time since the emergence of *Homo sapiens* and over periods of around 10^7 years unpredictable large scale geological changes may take place, e.g. mountain building and continental drift, etc. (see also Table 5.1).

An alternative approach to the problem of carrying out risk assessments covering very long timescales into the future is to introduce a regulatory cut-off at, say, 1000 or 10 000 years beyond which risk assessments would not be required for regulatory purposes. A criticism of such an approach is that it could fail to protect individuals in the future because the time when the first activity reaches the biosphere from a repository can be greater than 10 000 years.

5.5 TYPES OF SOLID RADIOACTIVE WASTES

Most countries have a system of radioactive waste categorization and, broadly speaking, these systems usually reflect the hazard associated with particular wastes. Categorization of waste at source is potentially extremely useful as it can avoid the possibility of disposing of waste of low hazard in expensive disposal facilities intended for waste of high hazard.

In general terms, radioactive wastes are often divided into three overall categories, i.e. 'low', 'medium' and 'high', depending upon the concentrations of radionuclides in the waste. The precise definition of these categories differs between countries but can generally be described as follows.

Low-level waste (LLW) is waste that contains or is suspected of containing low concentrations of radioactive material. Low-level waste can be handled by using simple protection measures such as rubber gloves and no special shielding is required. Typically, it consists of paper towels, rubber gloves, overshoes and other rubbish. Significant amounts of low-level wastes, as lightly contaminated building materials, could be generated by reactor decommissioning. In the UK, low-level waste is defined as having less than 4×10^9 Bq alpha-emitting radionuclides per tonne and less than 1.2×10^{10} Bq beta/gamma-emitting radionuclides per tonne. One particularly important type of low-level waste is the large volume of waste produced by the mining and extraction of minerals.

Medium-level waste (*intermediate-level waste* (ILW)) is waste that contains concentrations of radionuclides between low- and high-level waste. Handling generally requires shielding and remote devices. These wastes come from nuclear power plants and reprocessing plants and typically consist of fuel cladding, reactor components and effluent treatment residues such as spent ion-exchange resins.

High-level waste (HLW) is waste in which the concentrations of radionuclides are so high that the material is hot and will remain so for decades. It requires cooling, heavy shielding and remote-handling devices. It arises from reprocessing, initially in a liquid form and contains nearly all of the non-volatile fission products in the spent fuel. This waste is subsequently vitrified to form hard blocks of glass. Spent nuclear fuel that is not reprocessed may also be classified as high-level waste.

A fourth category of waste is sometimes recognized, namely *alpha waste*, which is waste containing long-lived alpha-emitters such as ^{239}Pu. It is handled in the same general way as low- or medium-level waste but with special protection to isolate it from humans: inhalation is a particularly important exposure pathway for material contaminated with alpha-emitting radionuclides.

The IAEA issued guidance on the categorization of radioactive waste in 1994 (IAEA, 1994b). The objective of this guidance is to recommend a method of deriving a classification system and to suggest a general method of classifying radioactive wastes. The classification system proposed by IAEA has three major classes of radioactive waste, as follows:

- *Exempt Waste*. This is waste containing such low concentrations of radionuclides that it can be released or cleared from nuclear regulatory control as the associated radiological hazards are negligible. Radionuclide limits for this category of waste are termed *clearance levels*. These levels should be set by national authorities, taking into account criteria established for exemption from regulatory control (see below). In other words, the doses implied from disposing of these wastes should not be greater than around 10 μSv per year. The term 'cleared waste' is now considered preferable to 'exempt waste'.
- *Low- and Intermediate-Level Waste*. These are wastes containing such amounts of radionuclides that action is required in order to ensure the protection of workers and members of the public. This class covers a very wide range of radioactive waste, from wastes that are just above exempt wastes to wastes requiring shielding and cooling because of their high content of radionuclides (high-level waste, see below). A range of disposal options may be available for low-and intermediate-level wastes.
- *High-Level Waste*. This is waste containing such high levels of radionuclides that a high degree of isolation from the biosphere is required over long time periods. Such waste normally requires both shielding and cooling.

The proposed boundary levels for these waste categories are shown in Table 5.2 . These levels are given in terms of the parameters most relevant to the safe handling and disposal of the waste. The low- and intermediate-waste category (LILW) is divided into short-lived LILW and long-lived LILW catagories. Long-lived LILW contains radionuclides with half-lives greater than about 30 years. This distinction is useful because the radiological hazard associated with short-lived radionuclides can be significantly reduced by radioactive decay over a relatively short time period. Thus, near-surface disposal may be a suitable option for short-lived LILW, whereas deep geological disposal may be required for long-lived LILW.

Table 5.2 Waste categorization scheme (from IAEA, 1994b).

Waste classes	Typical characteristics	Disposal options
Exempt waste	Activity levels < clearance levels	No radiological restrictions
Low- and intermediate level waste (LILW)	Not exempt or high-level waste	—
Short-lived waste (LILW-SL)	Restricted long-lived radionuclide concentration	Near-surface or geological disposal facility
Long-lived waste (LILW-LL)	—	Geological disposal facility
High-level waste	Thermal power above 2 kW m^{-1}	Geological disposal facility

Clearance is an important concept. This refers to the release of material from regulatory control. The underlying idea is that there are substances including wastes, etc. that are produced by the industrial use of radionuclides that could be released from control and treated as if they are essentially not radioactive. One example might be the release from regulatory control of slightly contaminated material from the decommissioning of nuclear power stations: the material might be steel, etc. and have some economic value if it is recycled and, furthermore, it might not be radiologically justified to dispose of it as waste.

Candidate materials for clearance can include waste materials for recycling from within the nuclear industry and wastes from the use of radioisotopes in hospitals, research laboratories or general industry. In some cases, particularly the recycling of materials from the nuclear industry, the volumes of materials can be large.

The issue is to define radionuclide levels below which the material can be released, or cleared from regulatory control. It is now generally accepted that the same radiological criteria should be applied to clearance as apply to exemption (see Section 4.5), i.e. the doses and risks should be 'trivial'. This leads to the following terminology:

- *Unconditional clearance*, where doses are trivial no matter what, within reason, happens to the cleared material.
- *Conditional clearance*, where doses are only trivial if certain conditions are complied with, e.g. the material is consigned to specified recycling facilities. This term is however falling into disuse because the attachment of conditions means that the process is identical to an authorized release of material.

A number of studies have been conducted in several countries to establish clearance levels for various materials in various situations. The results from several of these studies have been used by the IAEA to provisionally propose unconditional clearance levels for solid radioactive materials (IAEA, 1996b). The clearance levels derived by the IAEA range from 0.3 Bq g^{-1} for the actinides, ^{60}Co, ^{137}Cs and other radiologically significant radionuclides, to 3000 Bq g^{-1} for beta-emitters such as tritium and ^{35}S. Thus, material with lower activity levels could be cleared from regulatory control. As this book is being written, further work is being undertaken by the IAEA to define clearance levels for radionuclides that are applicable to any solid material. Similar work has been carried out within the European Community where guidance on 'general' clearance levels has been issued (EC, 2000a); these apply to materials of any origin or source.

A specific study into clearance levels for recycling of metals from the nuclear industry has been conducted under the auspices of the European Commission (EC, 1998). Significant quantities of metals, up to 10 000 tonnes a year, may be produced in the European Community from the anticipated programme for decommissioning nuclear facilities. The project investigated the possible doses that could arise from the recycling of metals and derived clearance levels in terms of Bq of radionuclide per gram of metal on the basis that the maximum dose should be 10 μSv per year. The exposure situations considered for the recycling of steel are illustrated in Figure 5.3. The situations include the handling and cutting of scrap, disposals of the slags and dusts following smelting, and uses of the final steel product. Consideration of this wide range of situations is necessary because different radionuclides behave differently during the processing of the scrap steel. For example, cobalt isotopes tend to stay with the steel during smelting, whereas plutonium isotopes concentrate in the slag and caesium isotopes concentrate in the dusts filtered from furnace 'off-gases'.

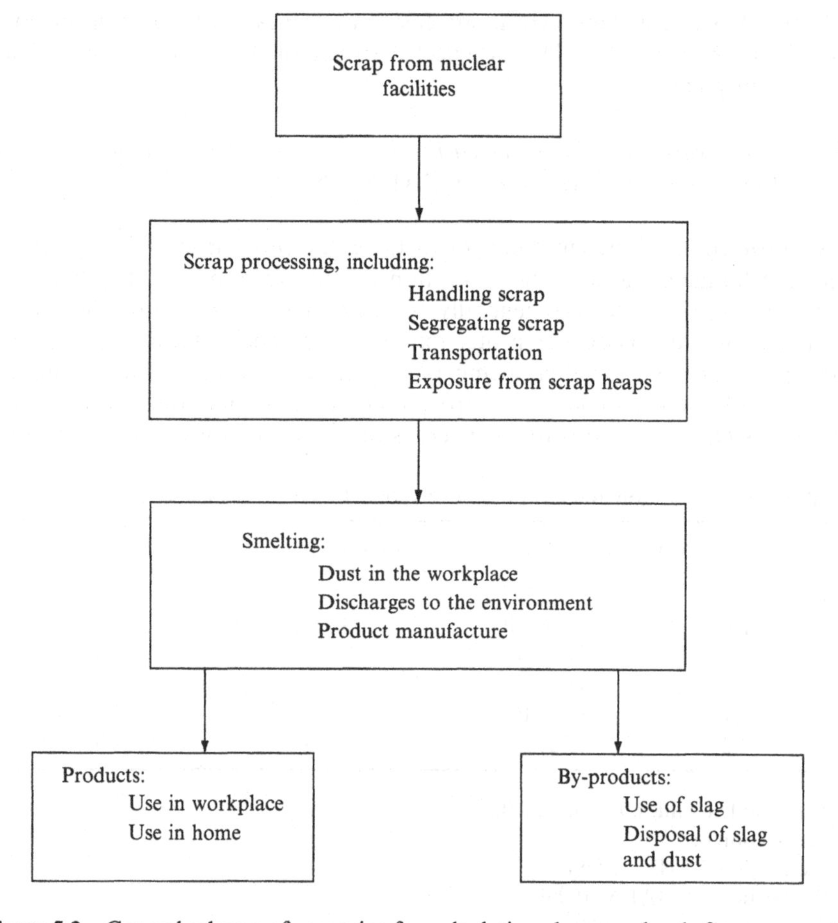

Figure 5.3 General scheme of scenarios for calculating clearance levels for scrap metal

Clearance is an important idea because it provides a means for potentially economically valuable material to be recovered while avoiding wasteful use of disposal capacity and maintaining radiological protection standards. The concept is related to exemption (see Section 4.4) and there is an important interaction between the two from a regulatory point of view: radionuclide-specific clearance levels should be set at or below the radionuclide-specific exemption levels to avoid the possibility of material being released from regulatory control at one location only to enter it again at another. A comparison of some published clearance levels and exemption levels are given in Table 5.3. Currently, there are moves within the international community to merge the concepts of exemption and clearance so as to produce a single set of radionuclide-specific values that serve both purposes; in effect, such a set would define a radioactive material for regulatory purposes (see Cooper, 2000). The world awaits!

5.6 ENVIRONMENTAL ASPECTS

The previous sections of this chapter have been almost entirely concerned with the protection of humans as in the ICRP system of radiological protection as described in Chapter 4. There is, however, an increasing awareness that the environment as a whole should be protected. This is reflected in one of the principles of radioactive waste management:

> *Radioactive waste shall be managed in such a way as to provide an acceptable level of protection of the environment* (IAEA, 1995a).

Such protection of the environment includes the protection of living organisms other than humans and also the protection of natural resources, e.g. land, forest, water, raw materials, etc. The generally accepted view has been that the protection of humans should ensure the protection of other species (ICRP, 1977). In this regard, protection of individual members of other species is probably an impossible objective – the aim should be to protect populations and thus ensure that the ecological balance is not disturbed. There is now a search for criteria against which

Table 5.3 Exemption and clearance levels for selected radionuclides[a].

	Exemption[b]	Metal clearance[c]	Unconditional clearance[d]	General clearance[e]
^{3}H	5.6×10^{5}	1.4×10^{3}	3×10^{3}	8.6×10^{2}
^{14}C	1.8×10^{4}	7.6×10^{1}	3×10^{2}	6.3×10^{1}
^{60}Co	6.6	5.8×10^{-1}	3×10^{-1}	9.9×10^{-2}
^{90}Sr	1.6×10^{2}	8.9	3	1.1
^{137}Cs	2.9×10^{1}	5.7×10^{-1}	3×10^{-1}	3.8×10^{-1}
^{239}Pu	2.2	2.5×10^{-1}	3×10^{-1}	1.4×10^{-1}

[a]Un-rounded values, in units of Bq g^{-1}.
[b]See Section 4.5.
[c]See Section 5.5 (EC, 1998).
[d]See Section 5.5 (IAEA, 1996b).
[e]See Section 5.5 (EC, 2000a).

to assess the requirement to protect the environment (IAEA, 1999a) and work is proceeding on how to combine protection of the environment with protection of people in order to produce a common protection system (Pentreath, 2002).

5.7 WASTE MANAGEMENT SYSTEMS

Radioactive wastes can arise in various physical forms, i.e. liquids, gases or solids. There will sometimes be a choice to be made as to whether to release radioactive wastes directly to the environment in liquid or gaseous waste streams or to extract radionuclides from the waste streams and concentrate them for disposal as solid radioactive waste. Such a decision has to take account of many factors and is one example of optimization of protection.

One of the most important factors to be taken into account is compliance with dose limits and constraints (see Chapter 4); if projected doses from discharges to the environment are too high, the levels of discharges will have to be reduced. One way of doing this is treating the effluent to remove radionuclides. For example, ion-exchange resins can be effective in removing radioactive ionic species. Other examples are the use of carbon filters to remove radioiodine and the evaporation of liquid waste streams. If doses from discharges are below dose limits and constraints, the principle of optimization of protection still applies and doses should be reduced to as low as reasonably achievable. The question is: by how much should discharges be reduced? Unless the resultant solid wastes comprise particularly short-lived radionuclides, they will require disposal and many of the issues discussed in this chapter will become relevant, e.g. doses and risks to future generations and their associated uncertainties.

There are choices to be made at many stages of the nuclear fuel cycle and it is important that such decisions are made in an informed manner. There have been a number of studies investigating ways of comparing options. One, in particular, is described here – a European Community project that compared, on the basis of costs and the radiological impact of management and disposal of all associated wastes, the reprocessing of light-water reactor spent fuel with its direct disposal by emplacement in a deep geological formation on land (Mobbs *et al.*, 1991; Schaller *et al.*, 1991). The study considered the spent fuel arising from a 20 GW(e)[1] reactor park.

The waste characteristics and management methods were based on current and proposed practices in the UK, France and Germany.

The steps considered for the direct disposal of spent fuel were as follows:

- storage of spent fuel
- preconditioning by disassembling the fuel elements and encapsulation in disposable gas-tight containers (DSCs)
- storage of these containers
- final conditioning by placing the DSCs into thick-walled final disposal containers

[1] GW(e), one gigawatt (10^9 watts) of electricity.

- transport to the disposal site
- disposal in a granite formation

The steps considered for the reprocessing of spent nuclear fuel were as follows:

- storage of spent fuel
- reprocessing to produce fission-product-concentrate liquor (HLW) and ILW and LLW streams, plus gaseous and liquid effluents that are discharged
- ILW and LLW solidified in cement, resin or bitumen
- vitrified HLW stored for at least 30 years
- transport of wastes to disposal site
- HLW disposal in granite, ILW in clay, and LLW in shallow land burial

The comparison took account of costs, occupational exposures and public exposures. Costs and levels of occupational exposure were based on the best available information from current experience. The radiological impact on members of the public was estimated by using computer models. In the solid waste assessments, the two general types of exposure scenario were considered: migration of radionuclides to man's environment and intrusion. The results show that for disposal of vitrified high-level waste and of spent fuel, the highest risks come from the migration scenarios. In contrast, for disposals of both low-level and intermediate-level wastes, intrusion scenarios dominate.

One interesting feature of the results is that for the migration scenario, the risks from spent nuclear fuel disposal are two orders of magnitude higher than for vitrified high-level waste disposal and yet both arise largely from ^{99}Tc. The main reason is the difference in the release from the waste form. In the case of spent fuel, the container is assumed to last for 100 years and a proportion (5 %) of the ^{99}Tc is taken to be present in the cladding/fuel gap and on the fuel grain boundaries – this technetium would be released rapidly, over a period of a few years, once the fuel container has corroded away. The remainder of the technetium is released over 100 000 years or so. For vitrified high-level waste, the container is assumed to last for 1000 years and leaching of all of the waste is assumed to take place at the slow rate, over approximately 100 000 years. The combination of a rapid initial release from the spent fuel, together with a geosphere transit time that is of the same order as the half-life of the radionuclide, means that a significant proportion of the spent fuel inventory of ^{99}Tc reaches the biosphere, whereas this is not the case for the vitrified high-level waste. A longer container lifetime or a more strongly sorbing geosphere would result in the calculation of similar risks for both options. The assumptions about container lifetime, etc. were based on the best available information but the results illustrate how sensitive answers can be to values chosen for key parameters.

In order to compare the two options, i.e. reprocessing and direct disposal, the costs and various radiological impacts (referred to as *attributes*) were assigned weights and combined in a multi-attribute hierarchy. The weights, or importance, assigned to each attribute reflect their perceived significance in the decision-making process. The numerical value of each attribute, e.g. critical group dose, is then multiplied by the weighting factor and then each is summed to yield an overall value for the option. In this way, two or more options can be compared. The method has considerable attractions but assigning values to weighting factors can be problematic.

It was not possible to compare reprocessing with direct disposal in the EC study because it emerged that some important aspects had not been considered. These aspects were as follows:

(a) Reprocessing is a current commercial practice, whereas conditioning for direct disposal was still at the planning stage.

(b) Reprocessing leads to the recovery of potentially economically valuable uranium and plutonium.

(c) National policies may favour the re-use of plutonium in fast breeders or light-water reactors (LWRs).

(d) National programmes may already include reprocessing facilities for other reactor types.

Nevertheless, the following general points arose from the study: there is a cost advantage for the direct disposal option but this should be weighed against the other factors such as long-term radiological impact and the recovery of useful material during reprocessing, cask storage is the most expensive option for spent nuclear fuel, deep geological disposal costs are dominated by the high costs of site selection and pilot facility operation, and the radiological impact on members of the public from spent fuel disposal is estimated to be higher than that from disposal of reprocessing wastes.

This study has been described in some detail in order to illustrate some of the issues that can arise in optimizing radioactive waste management. More generally, decisions have to be taken about the handling of most types of radioactive wastes. Following generation, wastes can be sorted and segregated according to radionuclide content, chemical properties and physical properties, etc. It may be of considerable benefit to segregate wastes suitable for near-surface disposal from those requiring deep disposal – in this way, the relatively expensive deep capacity can be used most effectively. Volume-reduction techniques may also be used in order to maximise the lifetime of a disposal facility. Incineration and mechanical compaction are examples of volume-reduction techniques.

Wastes may require conditioning in order to be transported, handled or disposed of. For low- and intermediate-level wastes, solidification in bitumen or concrete is often used, followed by encapsulation in steel drums. The drums are stored prior to disposal. At each stage in the handling of the waste, up to its final emplacement in a repository, radiation exposures to the workforce will have to be as low as reasonably achievable and kept within accepted limits. The possibility of accidental releases of radionuclides will also have to be taken into account.

6

Nuclear Power

Nuclear power is an important source of energy – in 1995, nuclear power generated around 17% of the world's electricity (see Section 1.6 above). The proportion of electricity generated by nuclear means in individual countries is given in Table 6.1. Civil uses other than electricity generation have been proposed for nuclear reactors. These other uses include district heating, desalination and coal gasification. They are largely still to be developed.

Nuclear reactors depend upon a reaction between neutrons and atomic nuclei of the fuel. Uranium, which is the common fuel for reactors, consists principally of the nuclei of two isotopes, i.e. ^{238}U and ^{235}U: the proportions by weight in natural uranium metal are 99.3 and 0.7%, respectively. In a reactor, neutrons produced by fission of ^{235}U are slowed down by a moderator so they cause further ^{235}U to undergo fission – thus, a chain reaction is propagated. During fission, ^{235}U atoms

Table 6.1 Nuclear power in 2001.

Country[a]	Reactors in operation		Proportion of total electricity (%)
	Number of units	MW(e)	
Canada	14	9998	11.8
France	59	63 073	76.4
Germany	19	21 122	30.6
Japan	53	43 491	33.8
Russia	29	19 843	14.9
Republic of Korea	16	12 990	40.7
Ukraine	13	11 207	47.3
UK	35	12 968	21.9
USA	104	97 411	19.8
Other	96	59 224	—

[a]Countries with more than about 10 000 MW(e) installed capacity are separately identified (IAEA, 2001a). In addition, a further 31 reactors are under construction of 27 756 net MW(e) capacity, mainly in the Far East and former Eastern Bloc countries.

are split into isotopes of various elements (called fission products) with the release of several high-energy neutrons, some gamma-radiation and heat that is taken away by coolant and used to generate electricity by powering steam turbines. Some fission products absorb neutrons efficiently, thus inhibiting the chain reaction, and so the fuel requires replacement before all of the ^{235}U is used up. The spent nuclear fuel is highly radioactive and requires remote handling, together with shielding to prevent excessive human exposure.

There are two main options for dealing with spent nuclear fuel: direct disposal or processing to recover spent ^{235}U, followed by disposal of the separated fission products. The entire scheme from isolation of natural uranium from uranium-bearing ores to recovery of unused ^{235}U from spent nuclear fuel is termed the 'nuclear fuel cycle'.

Figure 6.1 illustrates the general nuclear fuel cycle. The word 'cycle' refers largely to the fact that unused ^{235}U can be recovered from the spent fuel by reprocessing and used in fabricating new fuel and returned or 'cycled' again through a reactor. There are releases of radionuclides to the environment at various points in the cycle and these are shown in Figure 6.1. The ICRP's system of radiological protection and the IAEA's principles of radioactive waste management are both applicable to the control of such releases (see Chapter 4).

The nuclear fuel cycle is described in Section 6.2, but first the basic physics of nuclear reactors is described, together with various types of nuclear reactors.

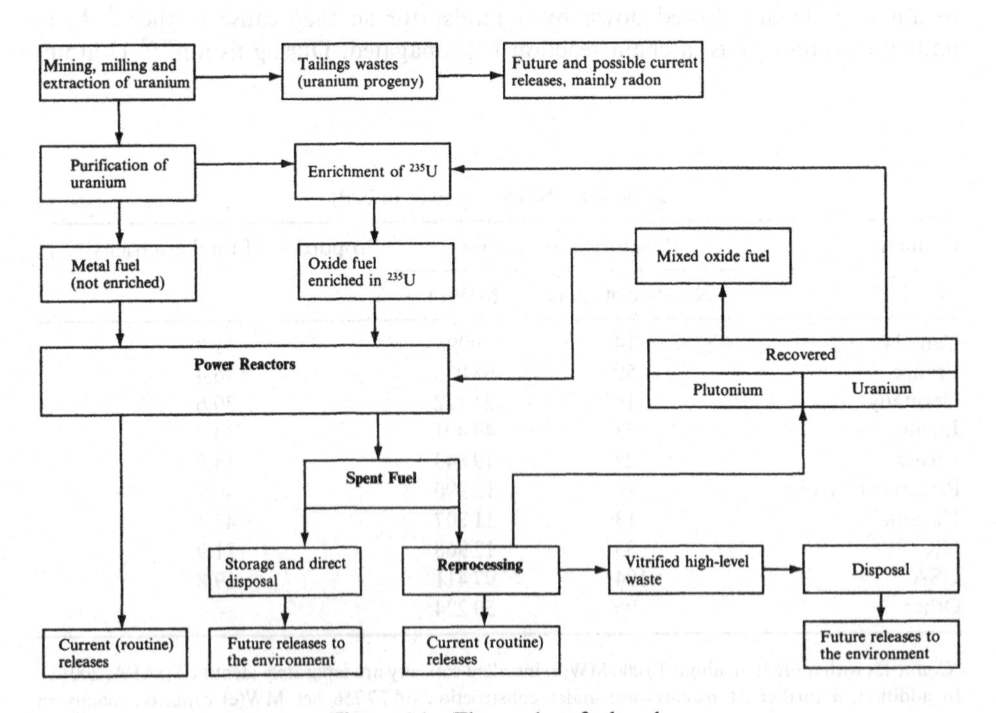

Figure 6.1 The nuclear fuel cycle

6.1 NUCLEAR REACTORS

6.1.1 Basic Physics of Nuclear Reactors

This section gives an elementary description of nuclear fission; advanced textbooks should be consulted for a more comprehensive description (e.g. Burcham and Jobes, 1995, and Kaplan, 1963).

In nuclear fission, a neutron is absorbed by a nucleus which then splits into two approximately equal fragments, called *fission products*. Neutrons, usually around two or three, are released, together with around 200 MeV of energy per fission. The energy is distributed roughly as follows: 170 MeV as kinetic energy of fission products, 5 MeV as neutron kinetic energy, and 25 MeV as gamma-rays, beta-particles and antineutrinos. Energy released during the fission of ^{235}U is about a million times more than is released when the same mass of coal is burnt. The release of energy can be explained as resulting from the fact that the nucleons in nuclei of mass number from around 50 to 150 are bound together more tightly than those in nuclei of higher or lower mass number (see Section 2.2). Thus, the fission of a nucleus of atomic mass of around 240 into two roughly equal fragments will release energy.

The neutrons released during fission can induce further fissions and thus cause a chain reaction. The vast majority of neutrons are emitted promptly but a small fraction (of about 0.006% for ^{235}U) is released over a period of a minute or so. These so-called *delayed neutrons* are associated with the radioactive decay of the fission fragments. They are a key feature in the safety of a nuclear reactor in that they are important for the propagation of the chain reaction. Thus, any buildup or reduction in the flux of neutrons in the reactor, and the corresponding changes in power levels, will take place relatively slowly. If the flux of neutrons is so large that *prompt* neutrons alone can sustain the chain reaction, a situation called a *prompt criticality* occurs. This cannot be controlled and power buildup is rapid. This was a factor in the Chernobyl accident (see Section 6.4.3).

The likelihood of an interaction between a nucleus and a neutron that leads to the fission of the nucleus can be represented by the microscopic cross-section of the nucleus. This depends on the neutron energy and the characteristics of the particular type of nucleus (isotope) involved.

Nuclei capable of undergoing fission can be divided into those that are *fissile* and those that are *fissionable*. Fissile nuclei undergo fission with neutrons of all energies and their cross-section increases as the energy of the neutron decreases. Generally, fissile nuclides have either an even number of protons and an odd number of neutrons, or odd numbers of both. The fissile nuclides of main importance are ^{233}U, ^{235}U and ^{239}Pu.

Fissionable nuclei only undergo fission with neutrons above a minimum, or threshold, energy of about 1 MeV and have either an odd number of protons and an even number of neutrons, or even numbers of both.

The yield of nuclei, or fission products, produced during fission shows a characteristic mass distribution with two peaks. The light peak is generally centred on mass number 90, while the heavy peak is centred on mass number 140. Because the neutron-to-proton ratio increases with mass number, the fission products will have a neutron-to-proton ratio above that required for stability. In other words, they are 'neutron-rich' and will adjust their neutron-to-proton ratio by beta-emission. This radioactive decay produces heat termed *decay-product* or *fission-product* heating. It causes long-term heat generation that has to be taken into account in handling spent nuclear fuel.

Fission of nuclei is not the only process caused by interaction with neutrons. Another process is absorption, followed by release of binding energy as gamma-rays – this is called *radiative capture*. One important example is the production of the fissile ^{239}Pu from ^{238}U, as follows:

uranium-238 + a neutron ⟶ uranium-239 ⟶ neptunium-239 ⟶ plutonium-239

β-decay β-decay

This forms a basis for fast reactors (see below). Furthermore, neutron capture by fission products is important because it takes neutrons out of the chain reaction and ultimately would 'poison' the reactor. Two particular fission products have large cross-sections for the absorption of neutrons, i.e. ^{135}Xe and ^{149}Sm. Xenon-135 is formed by decay of ^{135}I and is lost by beta-decay to ^{135}Cs and via neutron capture. One interesting point is that if an operating reactor is rapidly shut down or 'tripped', ^{135}Xe will not be lost by neutron capture and will build up for a few hours by decay of its precursor ^{135}I($t_{1/2}$, 6.6 h). This can have important implications for the possibility of a reactor being brought back to power following a 'trip'.

The common types of nuclear reactor all use ^{235}U as fuel and it is possible to estimate the amount that is consumed in producing power. Each fission produces about 190 MeV of useful energy, which corresponds to 3.04×10^{-11} joules (190 MeV $\times$ 1.6×10^{-13} joules/MeV). Thus, given that 1 joule is equivalent to 1 watt second, the number of fissions required to produce 1 joule of energy is 3.3×10^{10}. Given that 86% of neutron absorptions lead to fission, the number of ^{235}U nuclei consumed during one day of operation of a 1 MW thermal reactor is as follows:

$$\frac{10^6 \text{W} \times 3.3 \times 10^{10} \text{ fissions J}^{-1} \times 86\,400 \text{ s}}{0.86}$$

$$= 3.32 \times 10^{21} \text{ atoms per day}$$

Since 235 g of ^{235}U contains 6.02×10^{23} atoms (the Avogadro constant), the mass of uranium consumed at 1 MW power is therefore:

$$\frac{3.32 \times 10^{21} \times 235}{6.02 \times 10^{23}} = 1.3 \text{ g d}^{-1}$$

6.1.2 Types of Nuclear Reactors

Nuclear power reactors can be divided broadly into *thermal* reactors and *fast* reactors. The vast majority of power reactors are thermal reactors.

Thermal reactors use ^{235}U as fuel, either in the proportion in natural uranium (0.72% ^{235}U by mass) or slightly enriched up to say 3% ^{235}U. Because of the low fissile content of the fuel, the neutrons have to be slowed down to promote fission. The material that slows the neutrons in a reactor is called a *moderator*. Those in common use are D_2O (so-called heavy water), graphite and H_2O. Natural uranium can be used in reactors moderated with D_2O or graphite but reactors moderated with H_2O have to use fuel enriched in ^{235}U.

Fast reactors gain their name from the fact that the neutrons are not deliberately slowed down by a moderator. In these reactors, fissile ^{239}Pu is produced from ^{238}U by neutron capture (see Section 6.1.1). A mixture of ^{238}U and ^{239}Pu fuels these reactors. The core is usually surrounded by a blanket of ^{238}U in which more ^{239}Pu is generated. The ^{239}Pu produced in the core is consumed as fuel, while that produced in the blanket can be chemically separated to provide material for further cores. The essential requirement for ^{239}Pu to be produced is that after losses due to absorption in fuel cladding, etc., there must be at least two neutrons produced from one fission: one neutron to sustain the chain reaction and one neutron for production of ^{239}Pu via neutron capture by ^{238}U. Such conditions arise in fast reactors but not in thermal reactors. The production of fissile material is often represented by a conversion ratio – this is the ratio of the rate of production of new fissile material to the rate of consumption of original fissile material in the core. Conversion ratios are about 1.2 for fast reactors and about 0.76 for thermal reactors. There are only a few fast breeder reactors in operation – two in France and one in the former Soviet Union. A fast breeder reactor also operated in the UK.

The features that largely characterize the various types of power reactors are the type of fuel and its enrichment, the type of coolant, and the type of moderator. The common types of power reactors are characterized in this way in Table 6.2.

The quantities of electricity generated by various reactor types worldwide are given in Table 6.3.

6.2 NUCLEAR FUEL CYCLE

Nuclear reactors require fuel and an adequate means of dealing with spent fuel. The nuclear fuel cycle comprises the mining and milling of uranium, the manufacture of fuel elements, the production of electricity using the fuel and the treatment of spent nuclear fuel, which may include reprocessing to recover unused ^{235}U and other fissile material. The complete nuclear fuel cycle is illustrated above in Figure 6.1. The main stages in the nuclear fuel cycle are described below.

6.2.1 Mining, Milling and Extraction of Uranium Ore

This occurs in several areas of the world, including Africa, Canada, the United States of America, Australia and parts of Eastern Europe. Uranium-bearing ore that is economically worth mining contains at least around 0.2% U_3O_8. Uranium-238 is at the head of a long chain of radionuclides (see Appendix 3). Over geological time periods, the decay chain will come to equilibrium and daughter radionuclides will be present with the uranium ore. This has important implications: first, workers mining

Table 6.2 Common types of nuclear power reactor.

Name	Acronym	Fuel	Enrichment (%U-235)	Coolant	Moderator	Comments
Pressurized-water reactor	PWR	Uranium oxide	Yes (2–3%)	Light water	Light water	The same light water acts as moderator and coolant. It is under pressure to prevent boiling and takes heat away from the reactor core to heat exchangers
Boiling-water reactor	BWR	Uranium oxide	Yes (2–3%)	Light water	Light water	The same light water acts as moderator and coolant. It is allowed to boil and generates steam directly to power electricity- generating turbines
Canadian deuterium reactor	CANDU	Uranium metal	No (0.7%)	Heavy water	Heavy water	Separate heavy water is used for moderator and coolant. The heavy water moderator is cooled separately
Magnox	—	Uranium metal	No (0.7%)	Carbon dioxide gas	Graphite	An early type of reactor used mainly in the UK. Carbon dioxide gas takes heat away to heat exchangers
Advanced gas cooled	AGR	Uranium oxide	Yes (2–3%)	Carbon dioxide gas	Graphite	A development of the Magnox reactor, used solely in the UK. Operates at higher temperatures and is more efficient
RBMK	—	Uranium oxide	Yes (~ 2%)	Light water	Graphite	A hybrid reactor type used solely in the former Eastern Bloc countries

Table 6.3 Reactor types and net electrical power in 2000 (from Nuclear Engineering International, 2001).

Reactor type	Operating commercial reactors			Commercial reactors under construction	
	Number of units	Total MW(e)[a]	% of Total MW(e)	Number of units	MW(e)
PWR	258	241 055	65	24	22 811
BWR	91	82 002	22	4	4663
GCR[b]	18	3288	0.9	—	—
AGR	14	9164	2.5	—	—
CANDU	34	17 957	4.8	5	3145
RBMK	13	13 600	3.7	1	1000
FBR[c]	4	1280	0.3	—	—
Other	5	213	0.05	—	—
Total	437	368 559	—	34	31 619

[a]Total installed capacity. The electricity delivered to the grid will be lower. The thermal energy generated is normally about three times the electricity capacity and depends upon the efficiency of the reactor type.
[b]GCR, Graphite-moderated, gas-cooled reactors – mainly UK Magnox.
[c]FBR, Fast breeder reactor.

the uranium ore will be exposed to a radiation hazard, and secondly, the waste products from the extraction of uranium will be radioactive.

The radiation hazard to uranium miners arises mainly from internal irradiation from inhalation of ^{222}Rn and its daughters (commonly considered together as 'radon'). The hazard from radon is generally greater in deep mines than in open mines where the radon can disperse. Ores containing ^{226}Ra, the immediate precursor of radon, have been mined for centuries, but usually to recover other elements. A notable example is silver mining in the Schneeberg district of Saxony in Germany. This commenced around 1470. Mining has been very extensive in this area with some shafts reaching depths of about 400 m. The silver-bearing ore also contained uranium, which was discarded as an unwanted by-product called 'pitchblende' (see Chapter 1).

As early as the 16th century, an unusually high increase of lung disease was observed in mine workers. The incidence of the lung disease, by now known as 'Schneeberger Lung Disease', increased in the 17th and 18th centuries. In 1879, the disease was recognized as cancer. Following the discoveries of ^{226}Ra and ^{210}Po in pitchblende, attention focused on radioactivity as a possible cause of the disease in miners. By the late 1930s, it was generally accepted that inhalation of radon was a possible cause of the high incidence of lung cancer in uranium miners. It was not until the 1950s, however, that the importance of short-lived decay products in air was recognized. Epidemiological studies on uranium miners have helped to quantify the risk from radon exposure. The average concentration of radon in most mines in the Schneeberg area in the early part of this century was between 70–120 kBq m^{-3} – this is between 3000 and 6000 times higher than the average radon concentration levels in homes in the UK. Today, radon exposure in mines is addressed by the

ICRP's system of radiological protection (see Chapter 4) and risks are controlled to acceptable levels.

In 1995, about 33 500 tonnes of uranium were mined. The main producer was Canada, with 31 % of the total. Niger, the Russian Federation, Australia and Kazakhstan each produced between 7 and 9 % of the 1995 total (IAEA, 1995b).

After the uranium-bearing ore has been mined, it is transported to nearby mills where it is crushed and ground with water to form a slurry. The U_3O_8 is separated from the slurry using either acid or alkali leaching, depending on the type of ore. The most commonly used process involves leaching with sulfuric acid, together with an oxidizing agent, such as hydrogen peroxide. Alkaline leaching is used on ores that would otherwise require processing with large amounts of acid due to their lime content; leaching is carried out with a solution of sodium carbonate containing an oxidizing agent, which is usually sodium chlorate. The uranium goes into solution together with traces of impurities such as iron, molybdenum and vanadium. The solid residue is nearly equal in dry mass to that of the original ore and contains many of the daughter radionuclides of ^{238}U, including ^{226}Ra. This residue is called *mill tailings*. It is a slurry typically comprising 70 to 80 % of sand-sized particles and 20 to 30 % finer-sized particles and is commonly stored in a tailings pond. The tailings are kept underwater to prevent drying out and subsequent dispersal of radioactive dust. The covering of water also helps to limit the release of radon to the atmosphere. The tailings contain nearly all of the ^{230}Th (half-life of 80 000 years) and ^{226}Ra (half-life of 1600 years) that was in the original ore and thus will remain radioactive for many thousands of years.

Uranium mill tailings are a significant environmental problem in some parts of the world. For example, the need for uranium for atomic bomb production led the Soviet government to initiate an intensive uranium mining programme in the former East Germany in 1946; only limited consideration was given to environmental and health problems associated with the tailings, which were often simply dumped on land. Contamination of watercourses occurred and sometimes houses were built close to the tailings, hence leading to human exposure.

Clearly, areas contaminated by mill tailings will eventually require remediation to ensure that risks to health are acceptable. Furthermore, there are large quantities of tailings in storage in tailings ponds in various parts of the world. These ponds will remain radioactive for many thousands of years and an acceptable means of treating these wastes in order to limit health risks to future generations will eventually have to be found.

Following separation from the tailings, uranium is often precipitated from solution by using ammonia. The resultant solid is separated and heated to yield 'yellowcake', which is partially purified U_3O_8.

6.2.2 Chemical Refining

The next stage in the process is purification of the uranium. The yellowcake is dissolved in nitric acid and the uranyl nitrate is extracted by using tributyl phosphate (TBP) in kerosene or dodecane solvent. Complexed uranium in organic solvent is separated from the impurities, which remain in the aqueous phase, and the uranium product is then back-extracted into a dilute nitric acid solution. The resulting solution is then concentrated by evaporation and calcined to produce UO_2. The UO_2 is usually reacted with hydrogen fluoride at 200 °C to 300 °C to produce UF_4. The solid UF_4 is then reacted with fluorine gas at 500 °C to yield UF_6 gas which is condensed at −10 °C and placed in cylinders for shipping to enrichment or fuel-fabrication facilities.

Discharges of radionuclides associated with this stage of the fuel cycle are small, comprising mainly low concentrations of uranium daughters.

6.2.3 Enrichment

The major reactor types require fuel enriched in ^{235}U. There are two principal methods for enrichment, i.e. (a) gaseous diffusion and (b) gas centrifugation. The latter is superseding the former worldwide.

(a) *Gaseous diffusion* depends upon the fact that gases of different mass will diffuse through porous membranes at slightly different rates. Lighter gases will diffuse faster. However, because there are only slight differences in mass between $^{235}UF_6$ and $^{238}UF_6$, several hundred diffusion stages are required and such plants are large and consume considerable amounts of power.

(b) *Gas-centrifuge enrichment* uses centrifugal forces to separate $^{235}UF_6$ from $^{238}UF_6$. The $^{238}UF_6$ molecules will move preferentially outwards from the axis. The degree of enrichment per stage is higher than that in the diffusion process and this method also uses less energy.

The major enrichment plants are in France, the UK, the Russian Federation and the USA, with France and the USA using the gaseous diffusion process and the UK using centrifugation.

The process of enriching in ^{235}U also increases activity levels of the natural α-emitting radionuclide ^{234}U. This is not a particular problem for handling enriched uranium produced directly from natural uranium. However, as ^{234}U is not entirely destroyed in a nuclear power reactor, enriching uranium that has been recovered from spent nuclear fuel by reprocessing (see Section 6.2.5) will result in a steady buildup of ^{234}U levels in the enriched uranium product. There are other problems associated with using recycled uranium: an artificial radionuclide ^{232}U is present in recycled uranium and this radionuclide's decay chain includes a number of gamma-emitters that grow in over a period of about ten years to a total gamma-activity much greater than that of natural uranium.

The main waste product from enrichment is uranium depleted in ^{235}U to 0.2–0.3 % and is also depleted in ^{234}U. This so-called 'depleted uranium' can be used as 'breeder' material in Fast Breeder Reactors (FBRs) and also finds some uses in situations where high-density materials are required, e.g. balancing weights in aircraft. One particular use that has attracted attention recently is in armour-piercing munitions. These were first used in the Gulf War in 1991 and have also been used more recently in conflicts involving NATO troops in the former Yugoslavia. Concerns have been raised following these conflicts over the health implications of residues of depleted uranium remaining in the environment. Due to the reduced amounts of ^{235}U and ^{234}U, depleted uranium is about 60% less radioactive than natural uranium. Nevertheless, the primary constituent, ^{238}U, being an α-emitter can present a radiological hazard if inhaled in sufficient quantities. Therefore, problems are only likely to occur in situations where depleted uranium has been dispersed in large quantities in areas where it can be re-suspended and then inhaled. It should also be borne in mind that in many situations the chemical toxicity of depleted uranium is likely to be more important than the radiological risks.

Laser-based enrichment technologies are being developed in some countries but are unlikely to be operational before the year 2005. Possibly the most likely to be successful is a process called *atomic vapour laser isotope separation*. This process relies on the selective ionization of ^{235}U vapour by a laser tuned to one of the wavelengths of ^{235}U. An electromagnetic field is then used to separate the ionized ^{235}U.

6.2.4 Fuel Fabrication

The fuel used in nuclear power reactors, depending on the type of reactor, can be natural uranium metal, uranium oxide enriched or not in ^{235}U, or mixed uranium and plutonium oxides (MOX).

Natural uranium fuel is used in early types of gas-cooled reactors (e.g., Magnox reactors – see Table 6.1). This fuel is currently produced in Western counties at a rate of about 2400 tonnes of heavy metal (t HM) per year but as these reactors reach the end of their life, production of this type of fuel will be phased out. The uranium metal fuel is produced from UF_4 by heating with magnesium or calcium metal:

$$UF_4 + 2Mg \text{ (or } 2Ca) \longrightarrow U + 2MgF_2 \text{ (or } 2CaF_2)$$

The major types of nuclear power reactor use uranium oxide fuel. Enriched uranium oxide fuel is used in PWRs and BWRs, whereas CANDU reactors use uranium oxide with the uranium in natural isotopic composition. Enriched uranium is produced as UF_6 and three processes have been used to convert this into UO_2, as follows:

(i) Treatment with CO_2 and NH_3 to produce ammonium uranyl carbonate, which is then heated to produce UO_2.

(ii) Treatment with ammonia to produce ammonium diuranate, followed by reduction with hydrogen to UO_2.

(iii) Reduction of UF_6 to UF_4 with hydrogen, and then treatment with steam to produce UO_2.

6.2.5 Power Generation

The various reactor types have been described above in Section 6.1.2. Discharges of radionuclides occur during the normal operations of nuclear power stations. Gaseous and volatile radionuclides, e.g. isotopes of krypton, xenon and iodine, may leak from the fuel. Spent fuel is often stored at the reactor site in water-filled ponds and this can result in very low levels of liquid releases when the ponds are purged.

Table 6.4 summarizes the discharges from the different types of power reactor. The data are taken from the 2000 report of the United Nations Scientific Committee on the Effects of Atomic Radiation (UNSCEAR) and are normalized to unit-energy production. It should be emphasized that the numbers represent worldwide averages (UNSCEAR, 2000a).

Some reactor designs have characteristic releases associated with them. For example older UK Magnox reactors (see Table 6.2) with steel pressure vessels discharge the noble gas argon-41, which is produced by neutron activation of natural argon-40 in the shield cooling air.

Power generation also produces solid radioactive wastes. Fuel is eventually used up, requiring removal from the reactor and replacement. Wastes arise from other activities such as treatment of effluents and replacement of worn-out components. Generally, the annual arisings of solid waste from a light water reactor (PWR or BWR) are up to 100 m^3 of low- and intermediate-level waste and 30 tonnes of spent nuclear fuel (source, EU Website – see Bibliography). Reactors do not have an indefinite lifespan and eventually have to be permanently shut down and decommissioned. This process also produces wastes – both effluents and solid waste. Decommissioning of a typical light-water reactor can produce around 10 000 to 15 000 tonnes of solid waste, much of which will have little or no additional radioactivity (see Section 6.3.4).

6.2.6 Spent-Fuel Treatment

Management of spent nuclear fuel is one of the crucial parts of the nuclear fuel cycle. Currently, there are essentially three methods for dealing with spent nuclear

Table 6.4 Normalized releases of radionuclides from nuclear power stations[a].

Nuclear power station type	TBq[b] (GW(e) year)$^{-1}$						
	Noble gases	Tritium (gaseous)	Carbon-14[c]	Iodine-131	Gaseous particulate	Tritium (liquid)	Other liquid
PWR	13	2.4	0.22	2×10^{-4}	1×10^{-4}	19	8×10^{-3}
BWR	180	0.86	0.51	3×10^{-4}	0.35	0.87	1×10^{-2}
GCR[c]	1100	5.7	4.5	—	3.7×10^{-4}	6.6	1.42
AGR	26.4	1.9	1.25	1×10^{-5}	2.4×10^{-5}	360	0.6
CANDU	250	330	1.6	1×10^{-4}	5×10^{-5}	340	4.4×10^{-2}
RBMK	460	26	1.3	7×10^{-3}	8×10^{-3}	11	6×10^{-3}
FBR	210	49	0.12	2×10^{-4}	1×10^{-3}	1.7	2.3×10^{-2}

[a]Data for PWR, BWR, CANDU, RBMK and FBR are taken from UNSCEAR (2000a) and are averages for the years 1995 to 1997, except for ^{14}C which are averages for the years 1990 to 1994. Data for GCR are for 1996, while data for AGR are for 1998/1999. Information was taken from the Annual Reports of British Energy (AGR) and Magnox Electric (GCR).
[b]1 TBq = 10^{12} Bq.
[c]UK Magnox.

fuel, namely direct disposal after suitable conditioning – the 'once-through' fuel cycle, reprocessing to recover useable fissile material, e.g. ^{235}U and ^{239}Pu, and disposal of the remaining wastes, and long-term above-ground storage until the best option becomes clear.

Countries following the once-through fuel cycle include Canada, the USA and Sweden. However, as yet no disposal facilities are available for the spent fuel and this is being stored as an interim measure in above-ground facilities. Countries adopting the reprocessing option include France, the UK, Japan and Russia. Some countries, e.g. Mexico and the Republic of Korea, are adopting a 'wait-and-see' philosophy (Semenov *et al.*, 1995).

The total arisings of spent nuclear fuel worldwide by the end of 1995 were 165 000 t HM and this is more than twenty times the current annual reprocessing capacity (IAEA, 1996c). Annual arisings of spent nuclear fuel are around 10 000 t HM. Thus, the direct disposal route is likely to be the main option for spent nuclear fuel. However, the first repositories for spent fuel are unlikely to be in operation before 2010, by which time about 215 000 t HM of fuel will be in storage (Semenov *et al.*, 1995).

Commercial civil nuclear reprocessing occurs at two major locations, i.e. Sellafield in the UK and Cap de la Hague in France. Minor amounts of fuel are reprocessed in India, Russia and Japan. The projected reprocessing capacities for 1995 and 2010 are given in Table 6.5. The reprocessing process starts with the dissolution of spent fuel in nitric acid. The tributyl phosphate (TBP) process (see also Section 6.2.2) is then used to extract usable uranium and plutonium. In this process, the highly radioactive acidic solution of dissolved spent fuel is mixed with TBP in solvent. The resultant two-phase mixture is agitated in pulsed mixing columns. Plutonium and uranium, together with small amounts of other radionuclides, pass into the TBP solvent phase, leaving the vast majority of the fission products in the acidic aqueous phase. The uranium is purified to remove the contaminants, which are mainly isotopes of ruthenium and neptunium, and the resultant uranyl nitrate is concentrated by evaporation. Uranyl nitrate is then converted into uranium trioxide by thermal denitrification. This is sent to an enrichment plant for conversion to uranium hexafluoride and subsequent enrichment and re-use as fuel. The plutonium is purified by solvent extraction in extractors that are specially designed to avoid any possibility of a criticality. Purified plutonium nitrate

Table 6.5 Projected reprocessing capacities[a] (adapted from IAEA, 1996c).

Country	1995	2010
France	2205	1605
India	150	250
Japan	90	800
Russia	400	400
UK	2710	2710
Total	5555	5765

[a]In tonnes of heavy metal per year.

is converted into plutonium oxide via calcination of the oxalate. The plutonium may be used in the production of mixed oxide fuels (MOXs) which are being used increasingly in some countries as a fuel for PWRs. Using plutonium in this way can help to preserve uranium resources. A fuel cycle involving energy production using enriched uranium fuel, followed by reprocessing of the spent fuel, re-use of the recovered plutonium in MOX fuel and then disposal of the spent MOX fuel, uses about 20% less uranium per unit quantity of electricity than the alternative of not utilizing the recovered plutonium.

The liquid high-level waste arising from reprocessing contains over 97% of the original radioactivity present in spent fuel. These wastes are concentrated by reduced-pressure evaporation and stored in purpose-built stainless steel tanks. The stored liquid is cooled continuously to remove the heat generated by radioactive decay and agitated to avoid settling of suspended solids. The concentration of dissolved salts in the tanks is monitored carefully because self-evaporation could lead to precipitation of dissolved solids. Reprocessing wastes have been stored safely in this manner for over 20 years, although it is a costly process. The preferred method of dealing with these wastes is vitrification, followed by storage before disposal in a deep geological repository. The vitrification process involves incorporating the high-level wastes into borosilicate glass. This is a stable matrix and is much easier to handle than the liquid high-level waste. The latter waste is taken from the storage tanks and converted into a powdered solid by a specially designed evaporator and calciner. The powder is then mixed with the appropriate chemicals and heated to convert it into glass. The glass blocks are then welded into steel containers, which in the UK will be stored in air-cooled vaults for at least 50 years. After this time, the decay heat will have reduced to levels that make disposal easier.

Discharges of radionuclides occur during reprocessing. Gaseous fission products can be released during the initial handling and dissolution of the fuel, and the various storage, extraction and purification processes can lead to liquid wastes containing very low levels of radionuclides. Current commercial reprocessing plants have complex and efficient waste treatment processes to minimize routine liquid and gaseous discharges to the environment. For example, the reprocessing plant operated by British Nuclear Fuels Limited (BNFL) at Sellafield has an ion-exchange plant (known as SIXEP, the Site Ion-Exchange Plant) to remove traces of caesium and strontium from liquid wastes before discharge. The plant uses the ion-exchange mineral 'Clinoptilolite' which, following use, will be disposed of as intermediate-level waste.

Discharge data normalized to electricity production are presented in Table 6.6 and are again taken from the 2000 UNSCEAR report. These data are averages for the period 1995–1997 for the major reprocessing plants at Sellafield and Cap de la Hague, together with the Japanese pilot plant at Tokai-Mura. All of the reprocessing plants discharge significant quantities of the gaseous fission product ^{85}Kr. However, there are some noticeable differences in discharge levels for other radionuclides hidden in the averaging; discharges of ^{90}Sr and ^{106}Ru in liquid effluent are higher from Cap de la Hague than from Sellafield, whereas the reverse is the case for ^{137}Cs. It should be pointed out, however, that discharges of ^{137}Cs in liquid effluents from Sellafield dropped dramatically in the mid 1980s from 325 TBq in 1985 to 17.9 TBq in 1986.

Table 6.6 Discharges from reprocessing plants[a] (from UNSCEAR, 2000a).

Discharge mode	TBq[b] (GW(e) year)$^{-1}$							
	^{3}H	^{14}C	^{85}Kr	^{90}Sr	^{106}Ru	^{129}I	^{131}I	^{137}Cs
Liquid	255	0.4	—	0.8	0.5	0.04	—	0.2
Gaseous	9.6	0.3	6900	—	—	0.001	5×10^{-5}	1×10^{-5}

[a]These data are averages for the years 1995–1997 for the major reprocessing plants at Sellafield, UK, and Cap de la Hague, France, together with the Japanese pilot reprocessing plant at Tokai-Mura. During this time period, the fuel reprocessed corresponded to 160 GW(e).
[b]1 TBq = 10^{12} Bq.

As mentioned above, some countries are considering direct disposal of spent nuclear fuel. Clearly, there are few opportunities for the release of radionuclides during the storage and direct disposal of spent fuel. However, the long-term radiological impact from spent nuclear fuel disposal may be somewhat higher than that from the corresponding amount of vitrified high-level waste (HLW) arising from reprocessing due to the presence of radionuclides in the spent fuel that would have been released to the environment during reprocessing.

Interest in whether reprocessing is a viable option or not has promoted a number of comparative studies on reprocessing versus direct disposal. One of these was described in Section 5.7 above. A more recent one has been undertaken by the Nuclear Energy Agency (NEA) of the Organization for Economic Co-operation and Development (OECD), where the reprocessing fuel cycle, with use of the recovered plutonium in MOX fuel, is compared with the once-through cycle where spent fuel is sent for disposal (NEA, 2000). The two cycles are shown in Figure 6.2 Essentially, using recovered plutonium in MOX fuel reduces the requirement for mining of uranium by about 20 % for generation of the same amount of electricity. Taking this into account, a series of generic calculations, based upon current 'best practice', showed that the radiological impact of the two fuel cycles is likely to be similar (Table 6.7). Furthermore, there is considerable uncertainty surrounding the assessed doses from mining and milling as, depending upon the disposal conditions for the waste, collective doses could be over an order of magnitude greater than those shown in the table. Doses from disposal of spent fuel or of vitrified high-level waste from reprocessing were not taken into account as they were considered to be broadly similar. Generally speaking, this study shows that factors other than radiological impact are likely to be important determinants in deciding whether or not to use the reprocessing option. These factors include the availability of reprocessing plant, security of energy sources, and national security considerations.

6.3 THE RADIOLOGICAL IMPACT OF ROUTINE RELEASES OF EFFLUENTS AND SOLID WASTES FROM THE NUCLEAR POWER INDUSTRY

The generation of electricity by nuclear power involves the processing of materials that are naturally radioactive and, as a by-product, produces larger amounts of

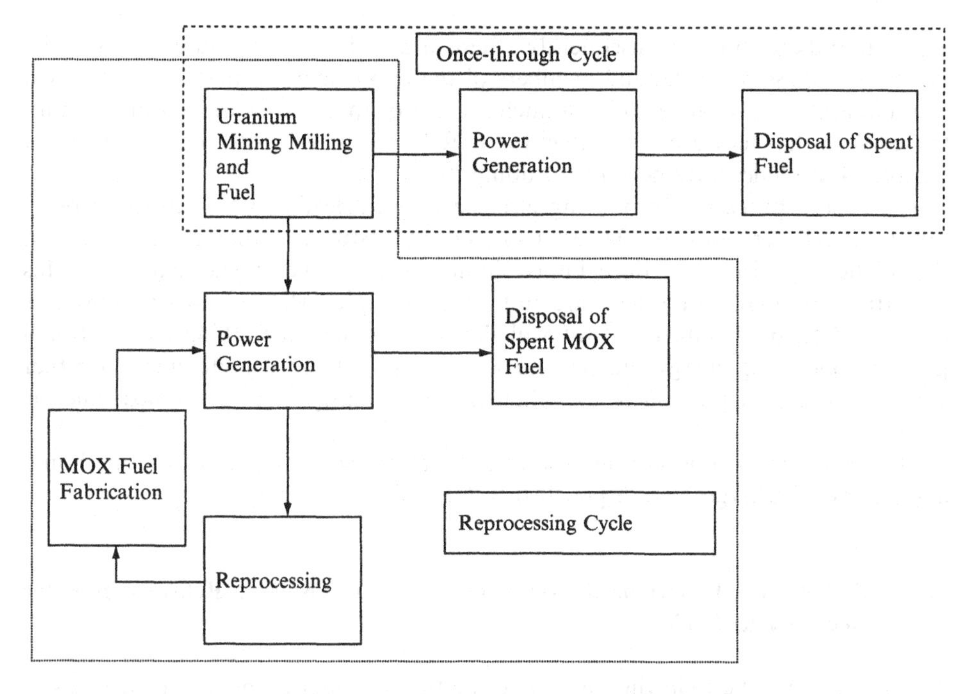

Figure 6.2 Schemes for the Once-through and reprocessing cycles

Table 6.7 Summary of doses to the public and workers from major fuel cycle stages of each option[a] (from NEA, 2000).

Stage of fuel cycle	Public (generic calculations)			Workers (operational data)	
	Collective dose truncated at 500 years (manSv (GW year)$^{-1}$)		Critical group (mSv per year)	Annual collective dose (man Sv (GW year)$^{-1}$)	
	Once-through	Reprocessing		Once-through	Reprocessing
Mining and milling	1.0	0.8	0.3–0.5	0.02–0.18	0.016–0.14
Fuel conversion and enrichment	0.0009	0.0009	0.02	0.008–0.02	0.006–0.016
Fuel fabrication	0.0009	0.0009	0.02	0.007	0.094
Power generation	0.6	0.6	~0.0006	1.0–2.7	1.0–2.7
Reprocessing and vitrification[a]	NA	1.2	0.4	NA	0.014
Transportation	Trivial	Trivial	Trivial	0.005–0.02	0.005–0.03
Disposal	—	—	—	—	—
Total	1.6	2.6	NA	1.04–2.93	1.14–2.99

[a]NA, not applicable.

artificial radionuclides. Consequently, there are radiation exposures of workers involved in these processes and members of the public may be incidentally exposed to radiation via discharges of radionuclides. All of these exposures should be controlled such that they are acceptable; the ICRP's system of protection, described in Chapter 4, provides a framework for doing this.

The overall process of producing electricity by nuclear power is clearly a practice in ICRP terminology. Thus, it should be justified, radiological protection should be optimized and dose limits should be respected. Some aspects of this are straightforward but others are not. For example, whether nuclear power is considered to be 'justified' or not will depend, among other things, on national policies concerning energy sources and will also reflect social preferences, together with historical factors such as whether the country has developed nuclear weapons.

The remainder of this section describes the radiological impact of routine liquid and gaseous discharges and disposals of solid radioactive waste.

6.3.1 Radiological Impact of Routine Liquid and Atmospheric Releases from the Nuclear Fuel Cycle

This section and the following ones describe the radiological impacts of releases of radionuclides to the environment. The reader is advised to refer back to Chapter 4 for information on radiological protection and to Chapter 16 below for information on the way in which doses are calculated.

UNSCEAR has published estimates of the collective doses arising from releases from the nuclear fuel cycle (Table 6.8). The results were calculated by using mathematical models described in the UNSCEAR reports. Collective doses to local and regional populations are presented separately from the collective doses to the world's population. The collective dose to the local and regional population will, with the exception of that arising from ^{222}Rn emanation from mill tailings, generally be delivered over a few years following release. The collective doses to the world's population from power production and reprocessing arise largely from ^{14}C, which delivers exposures over protracted periods of time at very low individual dose rates to a very large population – hence the relatively large value. The results are based on worldwide average discharges; nevertheless, they show that the radiological impact on the regional population from mining and milling is considerably higher than that from power production and reprocessing. When the world's population is considered, both reprocessing and power production give rise to larger collective doses than mining and milling due to the release of ^{14}C. Considerable uncertainties, however, surround the estimated collective doses from mining tailings as their estimated magnitude depends upon the assumptions made about permeability to radon of the cover over the tailings; completely exposed tailings dumps with no covering to prevent radon emanation could give rise to collective doses at least an order of magnitude larger than those presented here.

The total collective dose over the next 10 000 years to the world's population from the entire period of nuclear power production to 1997 is about 250 000 manSv. Around 80 % of this total arises from discharges of ^{14}C from power stations and

Table 6.8 Collective doses from routine releases of the nuclear fuel cycle (based on data for the years 1995–1997 from UNSCEAR, 2000a).

Stage of cycle	Normalized collective effective dose truncated at 10 000 years (manSv (GW(e) year)$^{-1}$)	
	Local and regional population	World population
Mining and milling of uranium[a]	7.5	7.5
Power production[b]	0.4	30
Reprocessing[c]	0.1	65

[a]Mainly from ^{222}Rn, although there is large uncertainty surrounding this value.
[b]Based on a weighted mean of reactor types (see Table 6.1).
[c]Normalized to fuel reprocessed which is currently around 0.04 of that produced.

reprocessing plants. The collective dose over all time would be about a factor of two higher than the 10 000-year figure.

To place these numbers in perspective, the annual collective dose to the world's population from inescapable natural background radiation is roughly 20×10^6 manSv. Natural ^{14}C produced in the upper atmosphere by the interaction of cosmic rays with nitrogen-14 contributes about 120 000 manSv per year.

UNSCEAR also provide information on some annual critical group doses, which are estimated to be 1 μSv for PWRs, 7 μSv for BWRs, 10 μSv for CANDU, 10 μSv for GCRs, 20 μSv for RBMKs and 0.1 μSv for FBRs. Current reported critical group doses for the major reprocessing plants are around 200 μSv per year or less. These doses are all lower than the ICRP's dose limit for members of the public.

A number of assessments of regional or national nuclear power programmes have been published. A comprehensive evaluation of critical group doses around civil nuclear sites in England and Wales was published in 1994 (Robinson *et al.*, 1994). This considered the situation in the early 1990s and covered a range of nuclear fuel cycle facilities, including fuel fabrication, uranium enrichment, power generation and reprocessing. The habits of people living near the facilities were taken into account; the doses that members of the public receive from discharges of radionuclides depend upon their particular habits. For example, if there is a house where people are growing their own foodstuff near to a particular site, the maximum doses from atmospheric discharges may be higher than if people with those habits did not live in the proximity.

The study on sites in England and Wales summed doses to individuals from all discharges of radionuclides and other routes of exposure in order to compare them with the dose limit. This is important because a critical group near one site may receive doses from discharges from other sites. The critical group doses calculated for the reprocessing plant operated by BNFL at Sellafield illustrates this point. A group of seafood consumers was identified as a likely critical group for Sellafield discharges, and doses to this critical group were estimated from environmental measurements. It was shown that doses were received not only from the Sellafield discharges but also from discharges of natural radionuclides from a factory at Whitehaven, a few miles away, that was processing phosphate rock (industrial processes releasing natural radionuclides to the environment are described in

Chapter 7). The total annual dose to this critical group of about 250 μSv is divided roughly equally between the contributions from Sellafield and the phosphate-rock-processing factory. The critical group for the Sellafield operations only was a group of individuals living near to the site. This critical group receives doses from direct gamma-radiation from the Magnox reactors at Sellafield, together with doses from, largely atmospheric, discharges from the site as a whole (Sellafield is a complex site with reprocessing facilities, waste storage facilities and Magnox power stations). The total annual critical group dose is about 190 μSv and direct external irradiation contributes about 57 %. This analysis of critical group doses around Sellafield shows that, for a particular industrial facility using radionuclides, the critical group for comparison with the dose limit may not be same one that receives the highest dose from that facility (it must be remembered that the dose limit is applied to the sum total of doses from all sources subject to control).

A number of assessments of the overall radiological impact of nuclear power have been undertaken, in particular under the auspices of the European Commission. These assessments use models of environmental transfer (see Chapter 16). In 1986, the European Commission set up the Project MARINA Committee to investigate the radiological impact of radionuclides in northern European marine waters (EC, 1990a). The area under study comprised the North Sea and Atlantic waters out to the Mid-Atlantic Ridge and as far south as the Straits of Gibraltar.

The Committee concluded that the most significant anthropogenic source up to the mid-1980s was liquid discharges from civil nuclear sites. The maximum collective dose rate from this source, i.e. 330 manSv, occurred in 1979 and the main contributor was Sellafield, which accounted for 90 %, followed by Cap de la Hague, the other main European reprocessing plant, which accounted for 9 %. The dominant exposure pathway was fish consumption (~ 50 %) and the most significant radionuclide was ^{137}Cs, which contributed over 50 % to this collective dose. However, discharges from Sellafield declined considerably in the early 1980s, resulting in a reduction in both the total exposure of EC population from nuclear fuel cycle sources and in the relative contribution from Sellafield. Indeed, the corresponding collective dose to the EC population in 1986, of approximately 130 manSv, is dominated by pre-1985 discharges, with Sellafield contributing about 50 % of this dose. The corresponding contribution from Cap de la Hague discharges rises to about 20 %. It is worth pointing out that even the peak collective dose rate, i.e. 330 manSv per year, is an order of magnitude lower than the collective doses arising from natural radionuclides in the marine environment (Table 6.9).

The methodology used in the Project MARINA study formed the basis for two other studies undertaken under the auspices of the European Commission, namely Project MARINA-MED, a study of the radiological impact on the population of the European Community of radioactivity in the Mediterranean Sea (EC, 1994), and Project MARINA-BALT, a corresponding study on the Baltic Sea (EC, 2000b).

Most of the activity in controlled discharges from nuclear plants into the Mediterranean comes from the French plant, Marcoule, which discharges into the River Rhone. Over the study period 1980 to 1991, this site accounted for about 77 % of the total beta-activity and almost all of the alpha-activity discharged from nuclear

Table 6.9 Collective exposure of the EC population from radionuclides in northern european marine waters (from EC, 1990a).

Source of radionuclides	Collective dose rate (manSv per year)	
	Peak (year)	1986
Civil nuclear site discharges	330 (1979)	130
Weapons test fall-out	~100 (1964)	~30
Natural radionuclides	3400 (1986)	3400
Chernobyl fall-out	~140 (1986)	~140

sites to the Mediterranean Sea. Discharges also occur from biomedical facilities, which are mainly of short-lived radionuclides, and industrial sites processing phosphate ores to produce fertilizer which release natural radionuclides of the ^{238}U chain, of which the most important is ^{210}Po. Figures for releases from phosphate plants were not available and estimates of doses were made by using the limited number of measurements on environmental materials. Estimates of typical critical group doses suggested annual doses of about 0.5 μSv from ^{137}Cs (a nuclide related to nuclear industry discharges) and 500 μSv from ^{210}Po (a naturally occurring nuclide that is also discharged from phosphate processing plants). Annual doses from ^{137}Cs were about an order of magnitude higher in the late 1980s due to contamination from the Chernobyl accident. Calculations of collective doses showed that the collective dose over the next 500 years from liquid discharges from nuclear plants between 1980 and 1991 is about 2 manSv, with a maximum annual collective dose rate of about 0.2 manSv in 1984. The reprocessing plant at Marcoule is the most important contributor. These values are around two to three orders of magnitude lower than the corresponding figures for radionuclides in northern European waters (see Project MARINA results), reflecting both the lower discharges to the Mediterranean and the lower seafood productivity. The collective dose from Chernobyl contamination (see Section 6.4.3) of the Mediterranean over the next 500 years was assessed at about 4 manSv, i.e. a factor or two higher than that from controlled discharges from the nuclear fuel cycle.

Contamination from the Chernobyl accident was also the dominant man-made contributor to doses from radioactivity in the Baltic Sea. Peak annual doses from this source were around 200 μSv, whereas doses from other man-made sources were, at most, an order of magnitude lower. The Chernobyl fall-out also dominates the collective doses to the EU population from man-made sources in the Baltic. This study took account of releases of activity to the Baltic over the period 1950 to 1996. Collective doses over the next 500 years, i.e. up to about the year 2400, to the EU population were estimated. The contributions to the collective dose from man-made sources, 2560 man Sv, was 1700 manSv from the Chernobyl fall-out, 650 manSv from nuclear weapons test fall-out, 200 manSv from discharges from European reprocessing facilities (mainly Sellafield), 1 manSv from nuclear power stations and 0.5 manSv from nuclear research facilities. For comparison, the annual collective dose rate from natural radionuclides in the Baltic Sea was estimated at about 400 manSv, which gives a total for the approximately 50-year study period (1950 to 1996) of about 20 000 manSv.

An overall summary of the three MARINA studies is given in Table 6.10. This shows that the relative importance of the different man-made sources varies depending upon the particular sea area, but overall reprocessing discharges to northern European waters dominate due to the high discharges in the 1970s and 1980s. Chernobyl is the second most important contributor dominating for the Baltic Sea, which is the closest sea area to the source investigated.

The three MARINA projects were interested only in exposures arising from the marine environment and did not, for example, take account of doses via freshwater pathways from many continental reactors which are situated on rivers. However, other work has considered all exposures of the EC population from routine atmospheric and liquid discharges from EC nuclear sites (EC, 1995). Collective doses were calculated for discharges in two particular years, i.e. 1977 and 1986. (These years were chosen to bound a decade of nuclear power.) Table 6.11 shows the collective doses, truncated at 500 years, to the EC population from discharges in each of the two years. In overall terms, the collective exposure of the EC population

Table 6.10 Collective doses to the EU population up to year 2400 from radioactivity in marine waters.

Source	Northern[a] European waters	Mediterranean[b] Sea	Baltic Sea[c]	Total
Routine discharges from the nuclear fuel cycle	5300	2.0	200	5500
Chernobyl fall-out	1000	—	1700	2700
Weapons test fall-out	1600	4.0	650	2250
Natural activity[d]	3400	NR[e]	400	—

[a]Data taken from EC, 1990a.
[b]Data taken from EC, 1994.
[c]Data taken from EC, 2000b.
[d]The figures in this row are annual collective dose rates and are included for comparative purposes.
[e]NR, not reported.

Table 6.11 Collective doses over the next 500 years to the EC population from discharges in selected years (in manSv) (from EC, 1995).

Site	Discharge year	
	1977	1986
Nuclear power stations	40	60
Reprocessing plants:		
Cap de la Hague	30	45
Dounreay	1	2
Karlsruhe	1	1
Marcoule	7	20
Sellafield	300	16
Total	380	140

is less for 1986 discharges than for discharges in 1977. The reason is a reduction in the levels of liquid discharges from Sellafield. Conversely, discharges from the French reprocessing plants at Cap de la Hague and Marcoule (a largely military reprocessing plant) are higher in the later year and this is reflected in an increase in the collective exposure from these two sources. The increase in the collective exposure from power stations between 1977 and 1986 is largely due to an increase in the number of power units from 60 to 125.

Although perhaps outside the main scope of this book, it is informative to compare the radiological impact on workers in the nuclear power industry with the impact on members of the public. A study of the impact of the UK nuclear power industry showed that in the late 1970s, the annual collective dose to the UK population from routine discharges was higher than the annual collective dose to the workers from their operation of the nuclear power facilities. However, with the reduction in discharges referred to above, by the mid-1980s the annual collective dose to the workers was higher than that to the public (Cooper, 1992).

There have been other studies on the radiological impact of nuclear power in other regions of the world. China has an expanding nuclear power industry and assessments of its radiological impact have been published (Pan *et al.*, 1996). The annual collective dose to the Chinese population from current releases of radionuclides from the Chinese nuclear power industry is estimated to be about 20 manSv, of which the vast majority arises from the mining and milling of uranium. This is less than 0.01 % of the annual collective dose to the Chinese population from natural background radiation (the average individual dose from natural background radiation in China is 2.3 mSv (Pan *et al.*, 1996)).

6.3.2 Solid Radioactive Wastes

The total quantities of solid radioactive wastes generated by the nuclear fuel cycle are not large. Worldwide, nuclear power generation facilities each year produce around 200 000 m^3 of low- and intermediate-level waste and 10 000 m^3 of high-level waste (IAEA Managing Radioactive Waste Fact Sheet). The total annual radioactive waste arisings in the EU are estimated to be 50 000 m^3, of which about one third is long-lived low- and intermediate-level waste and two thirds is short-lived low- and intermediate-level waste. Only 1 % is high-level waste (source, EU Website – see Bibliography). These volumes can be compared with the total arising of industrial wastes in the EU of 1×10^9 m^3, of which 1×10^7 m^3, is categorized as toxic industrial waste.

The various categories of waste may require different disposal options. Low-level wastes may be disposed of in shallow land facilities which are typically 20 m deep trenches lined or otherwise and may be capped with clay or some other semi-impermeable cover. Some types of medium-level wastes have been disposed of in shallow burial facilities but deep geological disposal is generally considered the most appropriate means of disposal. High-level wastes require isolation from the biosphere in deep repositories.

A total of about 2×10^6 m^3 of low- and medium-level waste has been disposed of in EU countries in shallow-land and deep repositories up until the year 2000. In

EU countries, disposals in shallow-land facilities have been carried out in Spain, France, Finland, Sweden and the UK. Deep disposals have been carried out by Germany; around 42 000 m^3 of low-level waste and 260 m^3 of intermediate-level waste were disposed of to a repository in the Asse Salt mine between 1967 and 1978. Assessments of all repositories have been made for authorization purposes. In the UK, risks are assessed against a risk target of 10^{-6} per year.

On an international scale, over 90 repositories for low- and intermediate-level waste have been constructed, of which about 70% are engineered near surface facilities, about 25% are simple near-surface facilities, with the remainder being mined cavities and deep-geological repositories. Around 80 repositories have also been constructed in Belarus to deal with wastes from the Chernobyl accident (IAEA, 1997).

Historically, some radioactive wastes have been disposed of to sea (see IAEA, 1999b). Until 1983, some western countries carried out sea dumping of some types of radioactive wastes. The dumping at sea of all types of wastes, including radioactive waste, is regulated under the terms of the 1972 *Convention of the Prevention of Marine Pollution by Wastes and other Matter*, known as the *London Convention*. In 1985, a resolution was passed at the London Convention which imposed a non-binding moratorium on sea-dumping. This was subsequently (1993) converted into an indefinite ban. A total of about 112 000 tonnes of low- and intermediate-level waste were disposed of in the north-east Atlantic from 1949 to 1982. The nations involved were the UK, Belgium, France, Switzerland and The Netherlands. The quantities of radionuclides involved are given in Table 6.12. The 42 PBq dumped at the north-east Atlantic dumpsite represents 92% of the total dumped in the world's oceans by western countries.

Sea dumping from 1971 onwards in the north-east Atlantic was undertaken under the auspices of the Nuclear Energy Agency (NEA) of the Organization for Economic Co-operation and Development (OECD). The NEA established a 'Co-ordinated Research and Environment Research Programme' (CRESP) in 1981 to improve understanding of the environmental processes relating to the dumping operation at the north-east Atlantic site. The work of CRESP was taken into account in NEA's 1985 review of the radiological implications of dumping at the north-east Atlantic dumpsite (NEA, 1985). During the review, a new model of dispersion in Atlantic waters was developed, which was coupled with models representing various exposure pathways in order to estimate doses. The maximum individual dose from all dumping up to 1982, was estimated to be 0.02 μSv per year to a hypothetical critical group of mollusc consumers in the Atlantic. The dose was calculated to arise in, roughly, 200 years time, with the dominant radionuclide being ^{239}Pu.

Smaller quantities of radioactive wastes were dumped by western countries at other locations: in total, about 3.5 PBq were dumped off the east and west coasts of the US with another 0.02 PBq being dumped in the west Pacific.

Radioactive wastes were also dumped at sea by the former Soviet Union. The main dumping site was the Kara and Barents Seas surrounding the Arctic Island of Novaya Zemlya – an island where the Soviet Union had conducted around 80 atmospheric nuclear weapons tests. Low-level liquid wastes, solid low- and intermediate-level wastes, and entire nuclear reactors were dumped. The reactors, some complete with spent fuel, came from the nuclear icebreaker 'Lenin' and from nuclear submarines. The total activity in the seven reactors containing spent nuclear

Table 6.12 Quantities of radioactive waste dumped in the north-east Atlantic from 1949 to 1982 (from NEA, 1985).

Gross mass (tonnes)	Total activity (TBq)[a]		
	Alpha	Beta/Gamma	Tritium
112 000	680	11 600	15 200

[a] $1\,TBq = 10^{12}\,Bq$.

fuel was estimated to be around 37 PBq at the time of dumping. The corresponding figure for the reactors without spent fuel is less than 3.7 PBq. The total amount of solid, low- and intermediate-level waste was estimated to be around 600 TBq at the time of dumping, with an additional 880 TBq in liquid waste (Sjoblom *et al.*, 1999). Long-lived radionuclides in the solid wastes will be released following gradual corrosion of the waste form. The IAEA undertook an assessment of the radiological implications of this dumping with a view to establishing whether some form of remediation is justified in radiological protection terms (Sjoblom *et al.*, 1999). Remedial actions could include retrieving some of the more active waste objects for disposal elsewhere or enclosing some of the objects *in situ* in an impermeable coating. The assessment showed that collective doses up the year 3000 from the dumped waste were only about 10 manSv. Future doses to individuals in populations groups inhabiting areas bordering the Kara and Barents seas were estimated to be around 1 μSv per year. Higher annual doses, however, of up to 4000 μSv could be received by military personnel patrolling the island of Novaya Zemlya. On the basis of these results and taking account of the hazards involving in undertaking any remedial actions, it was concluded that remediation is not warranted on radiological protection grounds.

6.3.3 Development of Disposal Options for Radioactive Wastes

Low-level radioactive wastes, and to a lesser extent intermediate-level wastes, have been disposed of in an authorized manner in several countries. However, disposal sites for high-level wastes have still to be developed (some high-level wastes were disposed of in the former Soviet Union but in a manner that would not be acceptable under current radiological protection criteria). In this section, two studies on high-level waste disposal are briefly described to illustrate some of the issues involved. The studies are the US Yucca Mountain Project (NRC, 1995a) and the European Commission's PAGIS Project (EC, 1990b).

The Yucca Mountain Project

Spent nuclear fuel is not reprocessed in the USA. The fuel from commercial nuclear power reactors is stored in water pools and in dry stores at over 70 sites in 43 states. There are also high-level wastes that have been produced by the defence programme. Ultimately, these two types of high-level waste will be disposed of to a deep-geological repository and the process of selecting a suitable site for such a repository has been underway since around 1975. In 1987, Congress directed the DoE's Office of Civilian Radioactive Waste Management to concentrate only on the Yucca Mountain Site in Nevada. Since that time, a site-characterization study has been ongoing to assess the suitability of the site.

The Yucca Mountain region comprises largely of layers of consolidated volcanic ash. The intention is to sit the repository about 300 m below ground in unsaturated rack about 300 to 500 m above the water table. By law, the repository should eventually hold about 63 000 tonnes of spent nuclear fuel from commercial power reactors and about 7 000 tonnes of high-level wastes from the defence programme. Up to 100 years after disposal operations begin, the repository would be sealed by back-filling the tunnels and sealing the entrance ramps and shafts.

There has been some debate in the US concerning the criteria that should be applied in assessing the suitability of the proposed repository in the Yucca Mountain region. The Energy Policy Act of 1992 requires the Environmental Protection Agency (EPA) to develop standards specifically for the Yucca Mountain site.

Under the direction of Congress, the US National Academy of Sciences (NAS) formed a committee to review the scientific basis for the standards for protecting the public from radionuclide releases from the proposed repository. Its findings were published in 1995 and illustrate many of the issues surrounding the evaluation of the safety of deep-geological repositories. The committee was asked to address a number of specific questions.

The committee concluded that the standard for protecting the public should be set in terms of the risk to individuals of adverse health effects (see Chapter 5) and, furthermore, that risks to critical groups should be calculated. It was recognized that releases of radionuclides may occur over many thousands of years into the future and the current USA standard contains a time limit of 10 000 years for the purpose of assessing compliance. The committee saw no scientific basis for imposing a time limit in this manner and recommended that 'the compliance assessment be conducted for the time period when the greatest risk occurs, within the time limits imposed by long-term stability of the geological environment'.

On the issue of human intrusion, the committee considered two questions, namely:

(1) Is it reasonable to assume that a system of active controls, e.g. permanent institutional control, can be developed that will prevent an unreasonable risk of human intrusion?

(2) Is it reasonable to make scientifically supportable predictions of the probability that a repository's engineered or geologic barriers will be breached as a result of human intrusion over a period of 10 000 years?

In answer to the first question, the committee concluded that although it was reasonable to assume that a system of active control would be in place for an initial period after the repository has been closed and sealed, there is no defensible basis for assuming that such a system can be relied upon for times far into the future. Furthermore, a system of passive controls, such as markers over the repository, is unlikely to be effective as they may require renewal, and languages, etc. may change, thus making interpretation difficult.

Similarly, the committee's answer to the second question was also 'no'. The reasons were as follows. Inadvertent human intrusion is usually considered to result from a search for resources by, say, exploratory drilling. However, it was argued, we do not know what a valuable resource will be in the future and neither do we know the nature of future sub-surface exploration techniques. Both of these issues have been covered in other assessments (see, for example, the PAGIS project below) by assuming current levels of technology.

The committee qualified its answers to the questions on human intrusion with the following observations. First, active and passive controls must help to reduce the risk of human intrusion, at least in the short term. Secondly, assessments of the consequences of human intrusion in the form of, say, drilling into the repository are informative in that they provide an insight into the degree to which the ability of a repository to protect human health would be degraded by such events.

The NAS committee's views were considered, together with opinions voiced at other public consultations, in formulating the EPA's final 'rule', issued as 40CFR, Part 197 in June 2001. This specifies an individual dose standard of 150 μSv per year to be applied up to 10 000 years (the regulatory period) after repository closure from the undisturbed repository; over this time period, releases from the repository are not expected to occur. A standard is also specified for inadvertent human intrusion. This is based on a stylized drilling scenario in which drilling through the waste into the underlying aquifer leads to the release of contaminated groundwater. The standard requires an estimate to be made of the time period after which an exploratory driller would not recognize that drilling into a man-made artifact had occurred. Doses from the use of water would be calculated for this time. If the exposure occurred before the end of the regulatory period, the dose would be compared with the 150 μSv standard; if it occurred afterwards, it would be included, as information, in the repository environmental impact assessment.

PAGIS – the Performance Assessment of Geological Isolation Systems for Radioactive Waste

The European Commission's PAGIS project started in 1982. Broadly, its objectives were as follows:

(1) To set up a common methodology for undertaking performance assessments on underground repositories.

(2) To draw conclusions about the acceptability of various geological formations and engineered barriers to protect present and future generations from the consequences of high-level waste disposal facilities.

Three candidate host rock formations were investigated, namely clay, granite and salt deposits. As an alternative, emplacement beneath deep-ocean seabed sediments was also considered. The project concluded with a seminar in 1989 (EC, 1990b).

The project emphasized the multi-barrier approach. The repository system was considered to be a series of barriers to the migration of radionuclides back to man's environment. The essential feature of the concept is that if one or more barriers fail, the others would be sufficient to keep man's exposure to a very low level. The barriers would be both man-made, such as the waste form, canister and back-fill, and natural, such as the host rock and overlying strata. The importance of each individual barrier will vary between different disposal options.

To make the analysis as realistic as possible, the data used were taken from actual sites. For example, the site for the repository situated in a salt formation was taken to be in the Gorleben salt dome. For each disposal option, one or more variant sites were also chosen to show the influence of changing local characteristics.

The assessment procedure was, for each site, to develop a model covering the most likely future evolution of the site – this was termed the *normal evolution scenario*. Typically, if water is present, this scenario would represent corrosion of the canisters containing the waste, slow dissolution of the glass matrix, and migration of radionuclides through the engineered barriers and through the geosphere to man's environment. A series of different scenarios was also modelled that represented low probability perturbations such as human intrusion events – these were termed *altered evolution scenarios*. The overall risk from the repository was then estimated by summing the risks to a critical group from all mutually exclusive scenarios.

The project also investigated the *uncertainties* in the estimates of doses and risks, together with their *sensitivity* to changes in parameter values. Both of these were carried out by making repeated calculations while changing the values of the parameters used in the model across their likely ranges.

Some of the main features of the results from the performance assessment were as follows:

1. For the clay-option normal evolution scenario, there is no release of radionuclides to the biosphere within the first 100 000 years. The peak doses occur around one million years and around ten million years into the future, with the most important radionuclides being ^{99}Tc and ^{237}Np, respectively. The peak annual doses are calculated to be less than a microsievert.

2. For the granite-option normal evolution scenario, no activity is released to the biosphere before at least 10 000 years and peak annual doses do not occur before one million years into the future. The doses are all less than a few microsieverts.

3. Salt is a somewhat different disposal option to granite or clay in that it is essentially dry and no releases of radionuclides to the biosphere would be expected for several millions of years until erosion processes expose the repository. The altered evolution scenarios considered include human intrusion due to solution mining of the salt. Annual doses in this scenario are still only, at most, several tens of microsieverts and the probability of such intrusion occurring would also have to be taken into account in considering the results.

4. Doses from the normal evolution scenario for sub-seabed disposal are less than a nanosievert a year and are predicted to peak in around 100 000 years after

disposal. The main contributors to the dose are selenium-79 and technetium-99. The highest dose from an altered evolution scenario is about one microsievert.

6.3.4 Decommissioning

The first generation of nuclear power reactors are now ending their useful life and are being taken out of operational service. By 2000, around 90 commercial reactors and 250 research reactors of various types were permanently shut down. There will be increasing numbers of reactors requiring dismantling in the near future, as those built from the mid 1960s onwards cease operation. The number reaching 40 years of age, approaching the maximum operating lifetime, rises rapidly from 2005 to 2010, reaching a peak of around 30 reactors a year in 2015 and staying high for around a decade.

The decommissioning of reactors is commonly viewed in three stages:

- *Stage 1 decommissioning* – the removal of the spent fuel for reprocessing, storage or direct disposal;
- *Stage 2 decommissioning* – the removal of equipment and material outside the bioshield (the screening preventing harmful radiation levels);
- *Stage 3 decommissioning* – the removal of the bioshield and all of the material within it, leaving a 'green-field' site, i.e. one suitable for any further purpose.

The decommissioning of reactors involves the removal of quantities of mildly radioactive material such as steel, copper, aluminium and concrete. The quantities of radioactive waste depend upon the type of reactor and the time elapsed after cessation of power generation: increasing the time will reduce the activity levels due to radioactive decay.

Table 6.13 shows the approximate amounts of active material which will have to be dealt with during Stage 2 and Stage 3 decommissioning of PWRs and GCRs. The data in this table, together with the information that follows on decommissioning PWRs and GCRs, are taken largely from Davis *et al.* (1991). Material removed during Stage 2 decommissioning may include fuel storage ponds, heat exchangers, effluent treatment plant, etc; the total quantities of material are similar for the two reactor types although the breakdown between steel and concrete is different. However, the large mass of graphite moderator in the core of a GCR and the much larger mass of concrete surrounding it means that there is a much greater amount of material to be handled during Stage 3 decommissioning of a GCR than for a PWR.

Table 6.13 Wastes arising during decommissioning (in tonnes) (from Davis *et al.*, 1991).

	PWR		GCR[a]		
	Steel	Concrete	Steel	Concrete	Graphite
Stage 2	3770	7600	10 830	1010	—
Stage 3	610	600	5040	30 320	4400

[a]Based on figures for the UK Magnox reactor.

For a PWR, the majority of the radionuclides left after the defuelling, around 10^{18} Bq, will be concentrated in a relatively small mass of stainless-steel components inside the pressure vessel, close to the reactor core. This comprises a mass of about 200 tonnes and has an average specific activity exceeding 1×10^{15} Bq per tonne at shut-down. The mild steel pressure vessel itself weighs about 400 tonnes and is less active: roughly, 1×10^{12} Bq per tonne at shut-down. For this reason, several countries are developing techniques to remove the whole reactor vessel intact, including the components within it, thereby making use of the vessel as a primary radiation shield. The activity of these material falls steeply for the first few years with the decay of ^{55}Fe and then ^{60}Co, and then it begins to level off at the activity of the remaining longer-lived radionuclides, ^{63}Ni, and then ^{94}Nb, ^{93}Mo and ^{59}Ni. These innermost steel components will clearly remain too active to be disposed of as low-level waste for the foreseeable future. The activity of the pressure vessel itself will fall to roughly 4×10^{9} Bq per tonne after 100 years.

The total residual activity in a typical twin-Magnox-reactor station is somewhat lower than for a PWR and will be spread out over a larger mass, consisting of graphite and a few hundred tonnes of steel pressure and vessel components. The graphite core amounts to 3100 tonnes for a twin-reactor station. Much of its initial activity is ^{60}Co, deriving from the cobalt impurity in the graphite, but after 10 years or so the activity of the graphite is dominated by ^{14}C, which persists for thousands of years, hence rendering the graphite too active to be disposed of as low-level waste. However, over 1000 tonnes of reflector graphite around the core may fall into the low-level waste category within 100 years.

The mass of concrete surrounding the core is much larger than for the PWR. At 10 years after shut-down, a substantial mass of this will be classed as non-active, but the majority will probably still be radioactive waste. At 100 years, the mass that is predicted to be radioactive waste lies between a few hundred and about 11 500 tonnes. The mass that is declared to be non-active will depend on the practicality of separating layers of concrete and on what is deemed to be acceptable for the mass over which measurements may be averaged.

The large amounts of lightly contaminated metals (steel, copper from wiring, and aluminium) that may arise during the decommissioning, has prompted studies on whether these materials can be released as scrap metal. Work has been undertaken within the EU and elsewhere on defining levels of contamination below which the material can be released to the general scrap metal pool. This release of material would be referred to in radiological protection terms as 'clearance' (see Chapter 5).

There are three broad strategies for decommissioning after the fuel has been removed (after Stage 1 has been completed), as follows:

- All of the components are dismantled, decontaminated where possible and removed for disposal. This could take five to ten years.
- The plant is kept intact for an extended period of time under surveillance and control. Projected time periods can range up to 150 years. During this time period, much of the activity will decay away. The plant is then dismantled.
- The reactor is entombed in concrete and covered in soil, for example. It is left in that state with no intention of dismantling. Surveillance will be required for the

time period that the reactor components present a hazard, which could be around 100 years.

The land that remains after the reactor components have been removed may be contaminated to very low levels with artificial radionuclide, but there is, at the moment, no international advice on the activity levels at which clean-up is required before the land can be released for general use. The process of releasing such land would, however, almost certainly be categorized as a practice in ICRP terminology (see Section 4.3.2). In other words, the doses to members of the public making use of the land in the future would have to be below the dose limit but it may not be necessary to clean up the land to the extent that all detectable artificial radionuclides have been removed.

There is now a considerable body of practical experience in dismantling nuclear facilities and the interested reader should consult Wampach *et al.* (1995) among other publications.

6.4 NUCLEAR ACCIDENTS

Modern nuclear installations have sophisticated systems to prevent accidents but, in the past, there have been at least three significant incidents resulting in releases of radionuclides to the environment. This section describes the three most well-known accidents, namely the Windscale fire, the Three Mile Island accident, and the Chernobyl accident. When reading these descriptions of the accidents, it is worth recalling that the principles of protection for intervention are applied to the protection of people from the consequences of accidents (see Chapter 4).

6.4.1 The Windscale Accident

On 10 and 11 October 1957, a fire in the Number 1 pile at the Windscale establishment in Cumbria, UK, led to an uncontrolled release of activity to the atmosphere. The resultant plume of activity travelled over much of England and over parts of northern Europe. The following description of the accident and its consequences is largely taken from Clarke, (1989).

The reactor, one of two identical ones on the site, was graphite-moderated and used natural uranium fuel. The coolant, i.e. air, was drawn through the reactor by large fans and was then discharged to atmosphere via 125 m high stacks. Each stack had a large array of filters at the top – known as 'Cockcroft's Folly' after their originator.

The normal operating temperature of the reactor was fairly low, i.e. 150°C, and this allowed so-called 'Wigner energy' to build up in the graphite. Wigner energy is energy stored as deformation of the graphite structure. This energy could be released spontaneously in an uncontrolled manner as had happened earlier in 1952, thus alerting the operators to the problem. Consequently, from that time the Wigner energy was released in a controlled way by periodic heating of the core. The heating was achieved by reducing the flow of cooling air.

In the 1957 incident, the release of Wigner energy did not take place as expected; one part of the core overheated, the cladding of some fuel elements melted, and both the graphite and uranium caught fire. Restoring the flow of cooling air at what was thought to be the end of the Wigner energy release only made matters worse.

During the fire, the fission products released from the oxidizing fuel were carried through the core and up the stack with the coolant. The filters reduced the amount of particulate material released but were not effective in preventing the release of volatile fission products, such as ^{131}I. Besides fission products, other nuclides were released, with the most notable being ^{210}Po which was being produced by neutron irradiation of bismuth ingots in the core. The major releases of activity occurred around 2400 hours on 10 October when CO_2 injection was used and between 0900 and 1100 hours on 11 October when water was initially introduced. The releases to the atmosphere are given in Table 6.14.

At the time of the accident, the radionuclide identified as of principal concern was ^{131}I; its route to the population was through consumption of cow's milk. The prompt imposition of a ban on milk supplies reduced intakes of radioiodine via the pasture–cow milk pathway. Nevertheless, the highest organ irradiated was the thyroid and the expected health effect would be an increase in the risk of thyroid cancer – the majority being non-fatal, although still requiring treatment. The appearance of any cancers would be over a few decades after irradiation. The lifetime risk of radiation – induced thyroid cancer in the most exposed individuals is estimated at about 4×10^{-4}. These risks are for the few infants measured who were estimated to receive thyroid doses of up to 160 mSv; for adults with the highest doses, the risks were about a factor of five less. The highest *annual* risk thus corresponds to about 8×10^{-6} per year. The average natural risk of thyroid cancer in the UK is at about 9×10^{-6} per year for males and 22×10^{-6} per year for females – lower at younger ages and increasing throughout life. Taking into account all nuclides and pathways, the total fatal risk to the most exposed individual is about 3×10^{-4} or 6×10^{-6} per year.

Table 6.15 shows the collective doses from the accident, received in the local area, the whole of the UK and Europe (including Scandinavia). The inhalation route is the most important, contributing 50 % of the collective dose to the UK

Table 6.14 Radionuclides released during the 1957 Windscale accident (from Crick and Linsley, 1984).

Radionuclide	Amount released (TBq)[a]
^{89}Sr	3
^{90}Sr	0.07
^{106}Ru	3
^{131}I	740
^{132}Te	440
^{133}Xe	12 000
^{137}Cs	22
^{210}Po	8.8
^{239}Pu	0.0016

[a] 1 TBq = 10^{12} Bq.

Table 6.15 Collective effective doses due to the Windscale 1957 fire (in manSv) (from Clarke, 1989).

Pathway	Cumbria	UK	UK and Continental Europe
Inhalation	35	900	900
Ingestion:			
milk	88	570	590
other foods	12	170	190
External:			
ground deposit	12	190	210
plume	5	54	57
Total[a]	150	1900	2000

[a]Rounded values.

population, the milk-ingestion pathway accounts for 30%, and the remaining 20% is shared between the other ingestion pathways and external irradiation. The most important radionuclides contributing to the collective dose are ^{131}I (37%) and ^{210}Po (37%), followed by ^{137}Cs (15%).

6.4.2 The Three Mile Island Accident

Three Mile Island is in Middletown, Pennsylvania, USA. The site's Number 2 reactor was involved in the accident. This was a PWR built by Babcox and Wilcox which started operating in 1978. The accident happened shortly afterwards on 28 March 1979 and so the inventory of fission products in the core was relatively small.

The basis of this accident was straightforward, i.e. the failure of a pressure-release valve to close, although the sequence of events surrounding the accident is complex. Briefly, the reactor was running at nearly full power when there was a failure in some ancillary equipment which prevented the main feedwater pumps from operating. This led to an increase in temperature and pressure in the primary cooling circuit. After a few seconds, a pressure-relief valve opened to relieve the pressure. About ten seconds later, the reactor control rods went into the core to shut down the reactor – this happened automatically because the coolant pressure had exceeded a predefined level. After a few seconds, as expected, the temperature and pressure of the coolant began to fall but the pressure-relief valve failed to close. This valve should have closed when the pressure had dropped to below a predetermined level. Signals available to the operator failed to show that the valve was still open. Coolant poured through the open valve, and as the pressure in the primary coolant system continued to decease, voids formed in portions of the system, thus causing a redistribution of the remaining water. This resulted in a level indicator erroneously signalling that the system was full of water. Therefore, the operator stopped adding cooling water. Eventually, the coolant level dropped to below the top of the fuel rods. The cladding began to melt and chemical reactions between the zirconium cladding and the steam from the overheated coolant produced hydrogen. The presence of hydrogen made it difficult to refill the pressure vessel with coolant. Adequate cooling was established after about 16 hours into the accident but, by then, about one third of the core had melted. The reactor is now unserviceable and the cost of dismantling the reactor and associated buildings that are contaminated with fission products is estimated at one billion dollars (1988 prices).

Most of the radionuclides released from the core were contained within the containment building enclosing the reactor. The quantity of ^{131}I released was nearly 1000 times less than in the Windscale accident, although that of ^{135}Xe was 25 times greater. In comparison with the noble gases, far less caesium, strontium and iodine were released from the Three Mile Island accident than from the Windscale accident because of the tortuous pathway from the core to the atmosphere and, as the containment system was intact, the air inside the buildings was released through the normal filtration system (Clarke, 1989).

The principal pathway of exposure of humans was external radiation from radioactive decay of the noble gases in the plume as it dispersed. Monitoring off-site confirmed this result; the maximum individual off-site dose from external gamma-radiation has been estimated at 0.83 mSv from thermoluminescent (TLD) detectors at 0.5 mile distances from the site. Based on this dose, the risk to such an individual is 3.6×10^{-6} over the next few decades, or perhaps 10^{-6} per year. The average dose to the 2 000 000 people who live within 50 miles was about 15 μSv (NRC, 1995b) and the number of health effects is estimated to be one fatal cancer among 325 000 fatal cancers that will occur naturally in that population. If the whole population of the USA is considered, the number of theoretical cancers from the Three Mile Island accident could double but the expected naturally occurring numbers increase by a factor of about 100 (Clarke, 1989).

6.4.3 The Chernobyl Accident

On 26 April 1986, Unit 4 of the Chernobyl nuclear power station, in the Ukraine, suffered a major accident – prolonged release of radionuclides followed. The four reactors at Chernobyl were of the RBMK type, with Units 1 and 2 being completed in 1977 and Units 3 and 4 by 1983. Two further units were being constructed on the site at the time of the accident. A comprehensive account of the radiological implications of the Chernobyl accident has been published by the Nuclear Energy Agency (NEA, 1995), from which the following description is taken. The events leading up to the accident are outlined in Table 6.16.

Table 6.16 Sequence of events leading to the Chernobyl accident.

Date and/or time	Event
Early 1986	Proposed experiment to test the time period that the steam turbines would generate electricity after steam cut-off. The experiment was originally intended to run with the reactor at zero power at a time when the reactor was being shut down for routine maintenance. Generation of steam prior to cut-off was to be achieved using fission product decay heat

April 1986	Plan changed – in case the experiment failed, the operators decided to run the experiment with the reactor at low power so that the experiment could be repeated. This meant that the emergency core-cooling systems had to be disconnected as they would have shut down the reactor during the first experiment
April 25	The reactor was being brought out of service for routine maintenance and it was intended to conduct the experiment in the afternoon. By noon, the reactor was running at half-power. At 14:00, the emergency core cooling system was disconnected – this, however, probably did not contribute to the accident
April 25, 14:00	The local electricity grid control centre demanded that the reactor be kept on-line until 23:10 as there was high electricity demand and there were delays in bringing a coal-fired station on-line. The experiment was then rescheduled for 01:00 on April 26. The reactor was operated at half-power until 23:10 and this allowed a buildup of ^{135}Xe – a reactor 'poison'
April 25, 23:10	The operators started to reduce power to conduct the experiment. An operator error meant that the power dropped by far more than was wanted. More ^{135}Xe was produced and it became very difficult to raise power to the 700–1000 MW(th) intended for the experiment – the normal operating power was 3200 MW(th)
April 26, 00:30	The operators removed almost all of the control rods in an attempt to increase the reactor power, but this rose to only about 200 MW(th) by 01:00. Very little steam was produced
April 26, 01:03 to 01:07	Additional coolant pumps were switched in as part of the test procedure. This resulted in more heat being removed from the core and to problems with the water level in the steam separator. Cold water passed through the core, resulting in even lower steam production and less power as more neutrons were absorbed. To maintain power, more control rods were removed from the core
April 26, 01:21:50	To stabilize the water levels in the steam separator, the operators sharply reduced the supply of cold water to the core. Steam was produced and power increased due to the increase in available neutrons. Some control rods began to automatically enter the core to compensate for the reduced absorption by the coolant
April 26, 01:23:04	The experiment was started by closing the main steam valve to the generators. The consequent reduced flow of coolant resulted in more steam being produced and the reactor power increased due to the positive void effect. Some control rods began to enter the core automatically but the automatic trip systems had been disconnected in case a repeat was required (see above)
April 26, 01:23:40	The shift manager tried to shut down the reactor as there was a rapid increase in power. Control rods were inserted manually but this took several seconds. Within three seconds, a prompt criticality occurred and there were two explosions. *The accident happened*

The accident occurred during a test and not during normal operation. The purpose of the test was to see how long the electricity-generating turbines would continue to generate electricity after the reactor had been shut down. This information was thought to be needed because reactors not only generate electricity but they also consume small amounts to power, for example, the pumps that circulate cooling water. In the event of a reactor shut-down, this electricity would normally be provided by the electricity grid. If, however, this failed, diesel generators would be used but there is a time delay before they can be started. Thus, the test was to see in there was sufficient momentum in the spinning turbines to generate sufficient electricity in the period before the diesel generators started. The test had to be conducted when the reactor was being shut down. Such a shut down was scheduled for 25/26 April 1986.

The origins of the accident lie in an inherently poor reactor design, together with major errors by the operators. The RBMK design is unique to the Soviet Union. This is a graphite-moderated, boiling light-water reactor using slightly enriched (around 2%) fuel. Water passes through the core in pressure tubes surrounding each fuel element. The water boils in the upper part of the vertical tubes to produce steam which is fed directly to turbines to generate electricity. One problem with the design, which played a large part in the accident, is the 'positive void coefficient'. This can be explained as follows. In the operating reactor, some neutrons are absorbed by water in the cooling tubes. If these tubes contain steam, fewer neutrons are absorbed and, as they would be moderated by the graphite moderator, an increased number of fissions would be caused in the fuel – the result is that the reactor produces more power, so exacerbating the problem as more water would be converted to steam. This effect is not particularly important at normal operating levels because at relatively high temperatures ^{238}U is a good absorber of neutrons and any increase in power caused by the positive void effect would be compensated for by the increased absorption of neutrons by ^{238}U. However, at low-power levels, less than around 20% of the maximum, the positive void effect is dominant and, as a consequence, the reactor can become unstable in this operating region and prone to power surges. Even when operating in the low-power region, the positive void effect will not be important provided that sufficient control rods are present in the core (control rods absorb neutrons).

The human factors in the accident were that the operators failed to follow operating procedures during the test and allowed the reactor to enter the low-power zone with insufficient control rods present. The test went disastrously wrong; in a fraction of a second, the power increased one hundred-fold. At 0123 hours on Saturday, 26 April 1986, two explosions destroyed the core of Unit 4 and shifted the 1000 tonne reactor cover plate. Hot pieces of fuel were ejected directly from the destroyed reactor building and the graphite, now open to the air, caught fire.

The heat generated by the burning graphite dispersed radionuclides high into the atmosphere. The fire lasted for 10 days, with the most significant radionuclide releases occurring during this period. Some much smaller releases occurred over the following 10 days. The total release during the accident is estimated to be of the order of 1.2×10^{19} Bq – details are given in Table 6.17.

All the noble gases in the core were released, together with between 20% and 60% of the volatile radionuclides, e.g. ^{137}Cs and ^{131}I. However, only up to 3.5% of the refractory radionuclides were released. Some 3–4% of the core is believed to have been ejected to the atmosphere, of which some 0.3–0.5% was deposited on the site and some 1.5–2% was deposited within 20 km, leaving about 1–1.5% of the core dispersed beyond 20 km.

Table 6.17 Estimates of the radionuclides released from the 1986 Chernobyl accident[a].

Nuclide	Half-life (years)[b]	Quantity released (TBq)[c]
^{85}Kr	10.76	33 000
^{89}Sr	0.14	80 000
^{90}Sr	28.8	8000
^{99}Mo	0.01	210 000
^{95}Zr	0.18	140 000–196 000
^{103}Ru	0.11	120 000–170 000
^{106}Ru	1.02	30 000
^{131}I	0.02	1 760 000
^{133}I	0.002	910 000
^{132}Te	0.01	1 000 000
^{133}Xe	0.01	6 500 000
^{134}Cs	2.06	44 000–50 000
^{137}Cs	30.07	74 000–85 000
^{140}Ba	0.04	170 000
^{141}Ce	0.09	120 000–200 000
^{144}Ce	0.78	90 000–140 000
^{239}Np	0.001	1 700 000
^{238}Pu	88	30–35
^{239}Pu	24 100	30–33
^{240}Pu	6567	42–53
^{241}Pu	14.35	5900–6300
^{242}Cm	0.45	900–11 000

[a]Data taken from UNSCEAR (2000a,b). Estimates are approximate with some ranges given.
[b]To two decimal places.
[c]1 TBq = 10^{12} Bq.

The weather conditions changed frequently while the major releases occurred (the first 10 days) causing radionuclides to be dispersed in several directions. Patterns of ground deposit were strongly influenced by rainfall levels while the plume was overhead. There are several areas of significant contamination in what is now Belarus, Ukraine and Russia, which are characterized by ^{137}Cs depositions of 500 kBq m^{-2} or greater. As expected, the most highly contaminated area is close to the reactor: ^{137}Cs deposits generally exceed 1500 kBq m^{-2} within 30 km of the site. The total deposit of ^{137}Cs in contaminated areas (defined as having a deposit of $>$ 37 kBq m^{-2}) of the former Soviet Union is estimated at 29 000 TBq, which represents about one third of the total release (see Table 6.17); for comparison, the corresponding residual activity from weapons testing is about 500 TBq (UNSCEAR, 2000a). The contaminated areas represent approximately 50 000 km^{-2}. Outside the former Soviet Union, a number of areas of northern Europe, in total about 45 000 km^{-2}, suffered ^{137}Cs contamination in the range 37 to 200 kBq m^{-2}.

The first indications of an accident were detected in Sweden when workers entering a nuclear power station were found to be contaminated: this contamination had originated from Chernobyl and had contaminated the workers outside the power station! The plume activity was tracked as it passed over Europe.

Countries outside Europe received very little deposition, although activity was detected throughout the northern hemisphere.

At Chernobyl, around 200 000 individuals were involved in 1986 and 1987 in the initial response to the accident. It was in this time period that the highest radiation exposures were received. These people were drawn from the site operators, firemen from the nearby towns and settlements, and the military – the latter are referred to as the 'liquidators'. In total, up to 1996, between 600 000 and 800 000 individuals have been involved in the post-accident clean-up. This total does, however, include people who worked on contaminated areas away from the immediate vicinity of the reactor.

The Soviet authorities have reported that acute radiation syndrome of varying clinical severity was diagnosed in 237 subjects who were all workers at the site or were brought in to fight fires or for the immediate clean-up. The doses, of between 2 and 16 Gy, were principally due to external irradiation by γ- and β-rays. A complication in the treatment of these patients was the development in 48 individuals of severe skin burns covering up to 90% of the body from aerosols deposited on the surface of the skin and clothing. A total of 28 people died of the immediate radiation effects of the accident (deterministic effects, see Section 4.1). On average, the liquidators received doses of around 100 mSv; some, approximately 10%, received doses of about 250 mSv, with one or two percent receiving doses in excess of 500 mSv. A few tens of individuals may have received potentially lethal doses of a few sieverts.

An 'exclusion' zone of 30 km radius was established around the plant and the population contained therein, around 90 000 people, were evacuated. Evacuation started on 27 April 1986. Within a few weeks, a further 26 000 people were evacuated from the most contaminated areas of Ukraine and Belarus. Doses from external irradiation, prior to evacuation, to the individuals from within the 30 km zone ranged from 1 to 660 mSv (Vargo, 2000) although only about 0.1% received doses greater than 250 mSv. Overall, less than 10% of the individuals received doses from external exposure greater than 50 mSv. These evacuated individuals also had significant intakes of ^{131}I, which gives rise to doses to the thyroid. These doses showed a strong age-dependence. The average thyroid dose to children less than one year old was 2.5 Gy, whereas thyroid doses to adults were about an order of magnitude lower.

In the other areas of the former Soviet Union with significant contamination, relatively high doses were also received. In the first few weeks following the accident, consumption of ^{131}I in cow's milk was an important route of exposure. People living in the Gomel Region of Belarus appear to be among the most affected, with children receiving reported thyroid doses of up to 40 Gy; the average thyroid dose to children aged 0 to 7 in this area being about 1 Gy.

An excess of thyroid cancer has now been reported in some areas of the former Soviet Union, in particular Belarus, which is probably related to the thyroid doses received in the first few months following the accident. Prior to the accident, the annual incidence of thyroid cancer in children in Belarus was about one per million or less. However, by 1994 the annual incidence had increased to around 36 per million children. Smaller increases have been reported in affected areas of the Ukraine and the Russian Federation (IAEA, 1996c). The excess does not appear to be exhibited in children who were born after six months following the accident. This suggests that the increased risk may be related to the relatively high doses received shortly after the accident and not to the much lower doses received since.

However, problems in detecting statistically any raised incidence of thyroid cancer against the normal background incidence could be responsible for this finding. Recent work has shown that there is a statistically significant increased risk at thyroid doses of around 50 mGy and above, and that the increased risk is a linear function of dose up to around 1 Gy (Jacob *et al.*, 1999). Thyroid cancer is treatable and by 1995 there had been only three attributable deaths from this cause in the exposed population. The extent of the future incidence of thyroid cancer is uncertain but the disease is likely to remain at its present level for some decades.

Controls were placed on the consumption of foodstuffs in affected areas and because of this action most of the exposure since the summer of 1986 is now due to external irradiation from ^{137}Cs. Doses in these areas in the time period 1986 to 1989 ranged from 5 to 250 mSv, with an average of 40 mSv. The total collective dose in these areas in this time period is estimated to be 9 700 manSv.

Doses to individuals outside the former Soviet Union were relatively low. The average dose to a UK inhabitant in the year following the accident was about 40 μSv, although somewhat higher doses were received in parts of Western Europe where it rained during passage of the plume. Reported average effective doses during the year following the accident were around 660 μSv in Austria, 760 μSv in Bulgaria, 570 μSv in Rumania, 500 μSv in Finland, 490 μSv in Italy, 370 μSv in Greece, and 300 μSv in Germany. The corresponding doses in France, Belgium and Denmark were all below 40 μSv (UNSCEAR 1988; NEA 1987). Important pathways were external irradiation from deposited activity and ingestion of foodstuffs, with the latter tending to dominate. Doses in subsequent years are usually much smaller.

The collective doses in various regions of Europe and the former Soviet Union are shown in Table 6.18. Within the European Community, about 1000 theoretical non-fatal cancers would be predicted over the next few decades. The number of theoretical fatalities from all cancers in the European Community due to Chernobyl is predicted to be in the region of 4000, spread out over a few decades following the accident. Over the same period, about 20 million or so people are expected to die in the European Community countries from cancer of one type or another. The effects of Chernobyl seem unlikely to be seen in any epidemiological study undertaken in Europe. The collective dose over all time to the world's population from the Chernobyl accident was calculated by UNSCEAR (1988) to be 600 000 manSv. This estimate is slightly lower than those presented in Table 6.18 and only considers doses from ^{137}Cs.

The destroyed reactor is now contained within a concrete 'sarcophagus'. This was completed in November 1986 and contains highly contaminated debris together with the remains of the reactor's fuel. The total inventory of long-lived radionuclides in the sarcophagus is estimated to be about 7×10^{17} Bq. Concerns have been expressed about the stability of the sarcophagus; international studies are currently underway to investigate this topic, together with other issues related to the remediation of the Chernobyl site.

6.4.4 Accident Response

The Chernobyl accident woke the international community up to the fact that serious nuclear accidents can have effects on neighbouring countries and also at

Table 6.18 Collective doses over all time from the Chernobyl reactor accident.

Area	Population	Collective effective dose (manSv)
Evacuated	116 000	4000[b,c]
European Soviet Union	75 000 000	200 000–2 000 000[d]
European Community[a]	300 000 000	78 000[d]
UK	55 000 000	3 000[d]

[a]Includes the following countries: Belgium, Denmark, France, Federal Republic of Germany, Greece, Ireland, Italy, Lumembourg, The Netherlands, Portugal, Spain and the UK.
[b]Data from UNSCEAR (2000a,b).
[c]About 8000 manSV from external exposure was averted by the evacuation (from UNSCEAR, 2000a,b).
[d]Data from Clarke (1989).

considerable distances. Questions were asked about basic nuclear safety standards, the early notification of accidents, and the international response to accidents.

Attention focused on the IAEA as a forum for obtaining international agreements on nuclear safety and the international response to nuclear accidents. From its inception, the IAEA had had a role in co-ordinating emergency planning among member states but this role was a relatively minor one. The Three Mile Island accident (see Section 6.4.2) prompted the IAEA to re-evaluate its system; guidelines on information exchange were issued. However, these were only guidelines and the international community now wanted something more tangible.

In the aftermath of the Chernobyl accident, three international conventions were developed under the auspices of IAEA, as follows:

- *The Convention on Early Notification of a Nuclear Accident*. This was adopted in September 1986 and requires Member States to report accidents at nuclear sites to potentially affected Member States, either directly or via the IAEA, and to the IAEA itself. Data essential to an assessment of the situation must also be transmitted.
- *Convention on Assistance in the Case of Nuclear Accident or Radiological Emergency*. This was also adopted in 1986 and requires Member States to notify the IAEA of the expertise, etc. that they could provide to another Member State in the event of an accident. In the event of a request for assistance, each Member State decides on what it can provide, with the IAEA acting as a clearing house.
- *Convention on Nuclear Safety*. This was adopted in 1994 (see IAEA, 1994c). Its objective is to commit participating Member States to a high level of nuclear safety by setting international benchmarks to which these States would subscribe. It is somewhat unusual as it is an 'incentive convention'; there are no legal sanctions for breaking its terms – instead, States are required to submit reports to regular meetings of the contracting parties where the reports are peer reviewed. In effect, the consequences of failure to comply with the terms of the Convention would be to be regarded as 'a bad boy'!

In another post-Chernobyl development, an 'International Nuclear Event Scale' (INES) has been proposed as a means of helping to communicate the gravity of a reported accident. This scale has been developed by a group of experts convened jointly by the IAEA and NEA. The scale classifies accidents into seven levels. Level 1 is the classification for the least serious accidents, representing a deviation from the authorized operating regime with no off-site consequences. At the other end of the scale, Level 7 represents a major accident with a large release of radionuclides leading to widespread health and environmental effects. The levels between 1 and 7 represent increasingly serious accidents; those which lead to off-site consequences start at Level 3, representing a serious incident.

On the INES, the Windscale accident and the Three Mile Island accident are both categorized as Level 5, i.e. a serious incident. The Chernobyl accident is a Level 7 – a *major accident*.

The Chernobyl accident has also led many countries to have an increased interest in systems and procedures for dealing with the consequences of reactor accidents. For example, in the UK an automatic monitoring network has been set up with the purpose of detecting and monitoring a plume of activity from a nuclear accident in another country.

Guidance on Contaminated Foodstuffs

The Chernobyl accident also raised concerns about contaminated foodstuffs, particularly in international trade. Extensive areas of agricultural land had been contaminated, both inside and outside the former Soviet Union. Food produced in these areas may be consumed by the local population and could also, potentially, be bought and sold on the international market. The question was asked: what standards should apply in this situation? Different countries tended to adopt their own levels for controlling the sale of contaminated foodstuffs but there was the issue of international trade in foodstuffs. In 1986, following the accident, the Commission of the European Communities established provisional levels for controlling the import and export of foodstuffs to and from Member States, as well as between them. The levels adopted for ^{137}Cs were 370 Bq l^{-1} for milk and 600 Bq kg^{-1} in other foodstuffs. Such levels are referred to as either *intervention levels* or *action levels* because in the case of foodstuffs above the level, action or intervention should be undertaken. These values have been adopted in further recommendations governing imports of agricultural products originating in third countries (i.e. not EU Member States) which are contaminated following the Chernobyl accident.

The EU has also issued intervention levels for marketed foods and animal feeds that would apply in EU Member States in the case of a future accident. The values for foodstuffs for radionuclides of half-life greater than 10 days (which would usually be mainly ^{134}Cs and ^{137}Cs) range from 400 Bq kg^{-1} for baby foods to 12 500 Bq kg^{-1} for 'minor' foods.

Guidance from the World Health Organization and the Codex Alimentarius Commission As one of the many post-Chernobyl actions by international organizations, in 1988 the World Health Organization (WHO) issued guidelines for radionuclides in foods (WHO, 1988). These are intended to apply after widespread contamination resulting from a major accident and in countries at some distance from the site of the accident.

The guideline values were derived on the following basis:

(i) an annual effective dose criterion of 5 mSv (or 50 mSv to the thyroid);

(ii) a normalized hypothetical diet based on worldwide representative values of higher than average consumption rates for different categories of foods;

(iii) two dosimetric classes of radionuclides, i.e. one for actinides (using a representative high dose coefficient of 10^{-6} Sv Bq^{-1}), and one for other nuclides (using a representative low dose coefficient of 10^{-8} Sv Bq^{-1}).

The guideline values (Table 6.19) were calculated for each radionuclide group and each foodstuff in isolation. Guideline values were also calculated for infants for milk and drinking water using infant-specific consumption rates and dose coefficients. These are presented in Table 6.20. It was found that the values for drinking water provided for general application (see Table 6.19) adequately protected the infant from this pathway, except for strontium-90, where a separate value is required.

In all cases, if several radionuclides have been released during an accident leading to the contamination of several food components, the addition of doses from the nuclides and foods will have to be allowed for in any decision taken to control doses from the consumption of foodstuffs.

Codex Alimentarius Guideline levels for radionuclides in foods for use in international trade following contamination by nuclear accidents were issued in 1989 by the Codex Alimentarius Commission (Table 6.21) (CODEX, 1989). This Commission is a body set up jointly by the Food and Agriculture Organization (FAO) of the United Nations and WHO. Their guideline values are largely based on the WHO guidance described above, but have been greatly simplified. Guideline levels are expressed as radionuclide concentrations in food, and there is no distinction

Table 6.19 WHO guideline values for derived intervention levels for foods (Bq kg^{-1}).

Class of radionuclide	Food							
	Cereals	Roots and tubers	Vegetables	Fruit	Meat	Milk	Fish	Drinking water
High-dose coefficient (actinides)	35	50	80	70	100	45	350	7
Low-dose coefficient (others)	3500	5000	8000	7000	10 000	4500	35 000	700

Table 6.20 WHO guideline values for derived intervention levels in foods for infants.

Radionuclide	Derived intervention Level (Bq l^{-1})
^{90}Sr	160
^{131}I	1600
^{137}Cs	1800
^{239}Pu	7

Table 6.21 Codex Alimentarius guideline levels for radionuclides in foods following accidental nuclear contamination for use in international trade.

Dose coefficient (Sv Bq^{-1})	Foods destined for general consumption[a]	
	Representative radionuclides	Level (Bq kg^{-1})
10^{-6}	^{239}Pu, ^{241}Am	10
10^{-7}	^{90}Sr	100
10^{-8}	^{131}I, ^{134}Cs, ^{137}Cs	1000

Dose coefficient (Sv Bq^{-1})	Milk and infant foods[a]	
	Representative radionuclides	Level (Bq kg^{-1})
10^{-5}	^{241}Am, ^{239}Pu	1
10^{-7}	^{131}I, ^{90}Sr	100
10^{-8}	^{134}Cs, ^{137}Cs	1000

[a]These levels apply to food as prepared for consumption and could be unnecessarily restrictive if applied to dried or concentrated foods prior to dilution or reconstitution.

except for baby food. The levels apply only to radionuclides contaminating food moving in international trade following an accident and not to naturally occurring radionuclides that have always been present in diet. The Commission defines their guideline levels as follows:

Guideline Levels are intended for use in regulating foods moving in international trade. When the Guideline Levels are exceeded, governments should decide whether and under what circumstances, the food should be distributed within their territory or jurisdiction.

In calculating the guideline levels, the Commission made the following assumptions:

(i) The dose criterion is 5 mSv per year.

(ii) 550 kg of food is consumed per year, all of which is contaminated.

(iii) Dose coefficients can be divided into three groups, as follows:
- those with a dose coefficient of 10^{-6} Sv Bq^{-1} (examples are plutonium-239 and other actinides);
- those with a dose coefficient of 10^{-7} Sv Bq^{-1} (examples are strontium-90 and some other beta-emitters);
- those with a dose coefficient of 10^{-8} Sv Bq^{-1} (examples are caesium-134, caesium-135 and iodine-131).

For infant foods and milk, a dose coefficient of 10^{-5} Sv Bq^{-1} was used instead of the 10^{-6} Sv Bq^{-1} value and iodine-131 was assigned to the 10^{-7} Sv Bq^{-1} category.

The general formula for calculating these guideline levels is as follows:

$$\text{Guideline level} = \frac{\text{Dose criterion } (5 \times 10^{-3}\ \text{Sv})}{m \times d}$$

where m is the mass of food consumed per year (kg), and d is the dose coefficient (Sv Bq^{-1}).

Thus, for actinides, the guideline level for the general population would be given by the following:

$$\frac{5 \times 10^{-3}}{550 \times 10^{-6}} = 9.1\ \text{Bq kg}^{-1}$$

This is rounded to 10 Bq kg^{-1}. The level for strontium-90 would be ten times greater, while the level for caesium-134 would be one hundred times greater.

Due to the conservative assumptions in the calculations, such as the fact that all foods are contaminated to the same degree, the Commission considered that it was very unlikely that application of their levels would result in anybody receiving a dose greater than a fraction of a millisievert. The guideline levels apply in the year following the accident.

Basic safety standards The Codex Alimentarius guideline levels provide the basis for the generic action levels for foodstuffs given in Schedule 5 of the Basic Safety Standards. These Standards add, however, that if food is scarce or there are other serious social or economic considerations, higher optimized action levels for food and drinking water would be expected to be used.

6.5 NUCLEAR WEAPONS

Mass for mass, nuclear reactions release amounts of energy far in excess of that produced during chemical reactions. In nuclear power stations, this energy is released over extended time periods at relatively low levels: in nuclear weapons, the release occurs in a fraction of a second, thus leading to a considerable increase in temperature and pressure. The resulting shock wave can cause destruction over wide areas.

There are two basic types of nuclear weapon, namely fission weapons, where the energy is released during the splitting of heavy atoms, and fusion weapons, where the energy is released during the joining together or fusion of light atoms.

6.5.1 Fission Weapons

Nuclear fission has been discussed above in Section 6.1.1. The fissile material used in weapons is ^{235}U or ^{239}Pu. In order to achieve a rapid release of energy via fission, a large number of nuclear fissions must happen over a very short time period. A single nuclear fission produces two or more neutrons and in order to get a very rapid chain reaction, the likelihood of these neutrons interacting rapidly with other fissile nuclei has to be maximized. One way of doing this is to bring together enough fissile material so that the so-called *critical mass* is exceeded. This can be explained as follows. In a mass of fissile material, the neutrons produced by fission may interact with other nuclei, thus causing further fission, or may be lost from the surface of the material. All other things being equal, the proportion of neutrons lost from the surface will reflect the surface-to-volume ratio. Therefore, as the mass of material is increased and the surface-to-volume ratio falls, there will come a point when a self-sustaining chain reaction occurs in the mass of material – this is the critical mass. For a nuclear explosion to take place, the critical mass must be exceeded so that a rapid multiplicative chain reaction of fission occurs. When this occurs, it is referred to as a *super-critical mass*. Clearly, the amount of material forming a critical or super-critical mass will depend upon its shape.

There are two general ways to produce a super-critical mass in a nuclear weapon. In the first method, two or more pieces of fissile material are brought together very rapidly. One way of doing this is to 'fire' one piece of fissile material at another piece using a device resembling a gun barrel. In the second method, a sub-critical mass of fissile material is made super-critical by compressing it using a conventional explosive. The compression increases the density of fissile nuclei in the material and increases the likelihood of interaction with neutrons.

The nuclear weapons used at Hiroshima and Nagasaki were fission bombs. The Hiroshima bomb used ^{235}U, while the Nagasaki bomb used ^{239}Pu. Fission bombs have now become superseded by fusion bombs which can generate more explosive force.

6.5.2 Fusion Weapons

Fusion weapons are also known as 'thermonuclear weapons' or 'hydrogen' bombs. In these weapons, energy is produced by fusion of hydrogen isotopes into heavier nuclei. Typically, this produces about three times as much energy as the fission of an equal amount of ^{235}U.

Nuclear fusion requires temperature of several millions of degrees centigrade; the only practical way of reaching such temperatures in a weapon is by nuclear fission. Thus, the fusion reaction is essentially initiated by a fission weapon.

As well as energy, the nuclear fusion reactions produce neutrons fast enough to cause fission of ^{238}U. Thus, in order to increase the explosive power, the fusion bomb can be surrounded by a blanket of ^{238}U. The high-energy neutrons interact with the ^{238}U, hence causing fission and consequent release of energy. The entire sequence can be represented as follows:

$$\text{fission} \longrightarrow \text{fusion} \longrightarrow \text{fission}$$

In a typical weapon, energy is produced in roughly equal amounts by fission and fusion processes.

6.5.3 Weapons Fall-Out

Detonation of a nuclear weapon will produce fission products and activation products. If the detonation is on or just above the Earth's surface, radioactive debris will be swept upwards with the fireball and some will lead eventually to worldwide fall-out. The majority of atmospheric nuclear weapons tests have been conducted at two test sites, i.e. the Arctic island of Novaya Zemlya (by the Soviet Union) and the Bikini and Eniwetok Islands in the Pacific Ocean, (by the US). Other test sites include the Chinese test site at Lopnor in Western China, Christmas Island in the Pacific Ocean and the Australian desert (where the UK conducted tests), the French test site at Mururoa Atoll in the Pacific Ocean, and the US test site in the Nevada Desert. Three types of weapons test fall-out are generally recognized. First, there is the relatively heavy debris that deposits close to the test site – this is referred to as *local fall-out*. Secondly, there is the finer material that is injected into the atmosphere but is heavy enough to deposit moderately quickly with respect to large-scale atmospheric mixing processes – this is *tropospheric fall-out* and occurs in the latitude band around that of the test site. Finally, there is the very fine material and gases which stay in the atmosphere long enough to mix in the hemispheric air mass – this is referred to as *global fall-out*. The relative amounts of these types of fall-out are reported to be 12, 10 and 78 %, respectively (Aarkrog, 1996). General environmental levels of weapons-test-derived radionuclides are described above in Section 1.1.2.

In terms of explosive yield, 1962 was the peak year for atmospheric weapons testing. Tests carried out in 1962 accounted for 62 % of the total explosive yield from all atmospheric tests (545 Mt of trinitrotoluene (TNT)). The levels of weapons testing in the early 1960s were such that the resultant annual doses in the northern hemisphere were tens of microsieverts. The peak annual dose from weapons test fall-out to an average individual in the UK is estimated to be about 140 μSv in 1963 – a level which is higher than most critical group doses for nuclear installations nowadays. The current annual average doses from weapons test fall-out in the UK 3–4 μSv (Hughes and O'Riordan, 1994).

Collective doses from weapons test fall-out have been estimated by UNSCEAR (UNSCEAR, 1993). Over all time, the world's population will receive an estimated 3×10^7 manSv – the main contributor being ^{14}C, which accounts for 2.6×10^7 manSv. Of the remainder, about 50 % results from ^{137}Cs. Around 20 % of the total collective dose will be delivered up to the year 2200 and this includes the majority of the doses from all radionuclides, with the exclusion of ^{14}C.

6.5.4 Nuclear Test Sites and Defence Industry Facilities

Nuclear weapons testing has taken place in several areas of the world. These regions are necessarily fairly remote from human habitation. Prompted, perhaps, by the cessation of the 'Cold War,' there has been increasing interest in the residual radioactive contamination in some of these areas and, in particular, on whether certain of them are, or will be, suitable for human habitation. There has also been concern at some locations over the doses individuals may have received from local fall-out from the tests; this concern is often caused by suggestions of increased cancer incidence in the exposed population. Similar concerns have been expressed over past releases from some nuclear weapons installations. Dose assessments covering these types of situations can be broadly divided into those that attempt to estimate doses that individuals may have received from past releases of radionuclides – these are often termed *dose reconstruction studies* – and those that look at doses people may receive in the future from re-occupying areas where there is now residual contamination-these are termed *prospective studies*. This section gives examples of both types of assessment and concludes with a description of contaminated areas arising from defence industry activities in the former Soviet Union.

Dose reconstruction studies

Dose reconstruction studies have been undertaken in various parts of the world to estimate doses that particular populations have received from specified releases of radionuclides to the environment. There are several reasons for conducting dose reconstruction studies (see Miller and Smith, 1996). One is public concern about past releases from nuclear tests and weapons production facilities. This has been a reason in the US, in particular, where some groups have held the view that they may have been deliberately misled about past releases from nuclear weapons facilities. A second reason is as an input into epidemiological studies on the effects of radiation doses. Typically, in this type of study a population that has been exposed to a radiation dose additional to normal background levels is followed for many years in order to establish whether there is any excess incidence of cancer and, if so, its magnitude. The additional radiation doses that individuals in the population have received from the particular radionuclide release are reconstructed from available data. Comparing any excess cancer incidence with the additional radiation dose can yield information on the dose response for ionizing radiation (see Chapter 4).

The general principles underlying dose reconstruction studies have been described in a number of publications (NRC, 1995b; Miller and Denham, 1994). Briefly, when investigating a particular release to the environment, the extent of the radionuclide contamination of environmental materials, including foodstuffs, has to be quantified for the time period and geographical area of interest. This information is combined with data on human habits to calculate radiation doses. Dose reconstruction studies are usually concerned with radionuclide releases that occurred many years ago – at a time when environmental monitoring and record keeping were not as extensive as they are today. Thus, the levels of the radionuclide releases often have to be reconstructed from the available data, together with a knowledge of the process that gave rise to the release. Mathematical models of radionuclide behaviour in the environment are then used on the release estimate to calculate levels in environmental materials. At this stage, any relevant historical data on environmental measurements can be used to check the model predictions or, indeed, replace them for some exposure pathways if sufficient are available. The final stage is to estimate the doses to humans by using information on human habits to derive intakes of radionuclides, etc. In this section, two recent dose reconstruction studies are briefly described. The first investigates doses received by local populations from weapons testings in the Nevada Desert in the US and the other study assesses the doses received locally from discharges from the Hanford weapons facility, also in the US.

Over 100 above-ground nuclear weapons tests were conducted in the Nevada Desert between 1951 and 1958. Over half of these tests deposited measurable amounts of radioactive fall-out in parts of Nevada and Utah. Washington and Kane counties in Utah received relatively high levels of contamination in 1953 when fall-out was carried from the tests more rapidly than expected.

Since the late 1970s, there have been various reports of an increased incidence of leukaemia and other cancers in south western Utah and to answer these concerns, a project was started with the objective of calculating the radiation doses the population in this area has received from the fall-out. In common with other projects of this type, in order to make the assessment as realistic as possible, it was based upon environmental measurements taken at the time of the tests. One of the problems faced by the project was now to relate the many thousands of gamma-dose rate measurements made on environmental materials to the deposition of the complex mixture of radionuclides released during the tests. A dynamic food chain model was also developed to estimate doses to individuals via food chain pathways (Kirchner *et al.*, 1996).

Results have been published for bone marrow doses and for thyroid doses. Bone marrow doses were estimated for comparison with leukaemia rates (Stevens *et al.*, 1990). Average bone marrow doses from the fall-out in Washington and Kane counties were estimated to be 19 and 9.1 mGy, respectively, for individuals who remained in the specified county for the period 1952 to 1958. The bone marrow doses in the other counties in Utah varied from 4.5 to 1.5 mGy. The main contributor to the bone marrow doses was external exposure from radionuclides deposited on the ground. Analysis revealed a weak association between bone marrow doses from the fall-out and all types of leukaemia but this association was not statistically significant (Stevens *et al.*, 1990). The weapons testing produced significant quantities of ^{131}I and ^{133}I. Isotopes of iodine concentrate in the thyroid gland and could cause thyroid cancer. Thyroid doses to individuals in two of the counties receiving the largest radioactive deposits, Washington County, Utah and Lincoln County, Nevada, have been estimated (Till *et al.*, 1995). The thyroid dose to a typical individual living in Washington County from the weapons testing was estimated to be 360 mGy; the corresponding dose in Lincoln County was 49 mGy. A typical individual is one who had lived in the particular county for the entire fall-out period, drank cow's milk and had been born before 1 January 1951. Drinking cow's milk is important because this is the main exposure pathway, contributing around 73% of the thyroid dose.

As a result of the prevailing weather conditions, particular tests dominate the doses received by individuals living close to the test site. For example, test 'Harry', which took place on 19 May 1953, resulted in estimated thyroid doses at St George, Utah, of up to around 0.8 Gy to infants fed on milk from cows grazing on fresh pasture (Whicker *et al.*, 1996). Estimates of collective thyroid doses suggest that the largest number of fall-out-induced thyroid cancers might have developed in Salt Lake County (Whicker *et al.*, 1996), but, so far, no excess has been reported.

Discharges from the Hanford site in Southeastern Washington State, US, are the subject of a second dose reconstruction study (Shipler *et al.*, 1996). Hanford was one of the sites chosen for the atomic bomb project in 1942 (the Manhattan Project). One of the site's functions was to produce the plutonium for bombs and construction of the first plutonium producing reactor commenced in 1943. The first three reactors started operating in 1944 and 1945. The chemical plant to separate plutonium from the irradiated fuel started operation in 1944. The Cold War prompted an expansion of the site, with six new reactors commencing operation by 1963. Further chemical separation plants were also constructed. All but one of the reactors had a 'once-through' cooling water system and discharged the water directly to the Columbia river. This resulted in radionuclide releases to the river – one mechanism was induction of radioactivity in impurities in the water as the cooling water passed by the core. From the mid 1960s onwards, the 'once-through' reactors were shut down at the rate of one a year until 1971 when all had ceased operation. Another reactor using a different cooling system operated until 1987.

Public concerns over operations at Hanford led to the initiation, in 1986, of a project to evaluate, among all things, the doses that local inhabitants had received. The project considered both discharges to atmosphere and discharges to the Columbia river.

The main radioisotope of interest for atmospheric releases was ^{131}I. This was released when the spent fuel rods were dissolved in acid prior to extraction of plutonium. The major releases occurred between 1944 and 1951. Iodine-131 releases totalled about 27 000 TBq between these two years. The major release of about 20 000 TBq occurred in 1945. Releases after 1951 were considerably lower because effluent treatment systems were added. Other radiologically significant radionuclides released to atmosphere were ^{90}Sr, ^{103}Ru and ^{106}Ru, ^{144}Ce and ^{239}Pu.

The significant radionuclides released to the Columbia river were, ^{24}Na, ^{32}P, ^{65}Zn, ^{76}As and ^{239}Np. The peak discharge years were the late 1950s and early 1960s. For example, ^{24}Na discharges peaked in 1960 at 51 000 TBq per year and ^{32}P discharges peaked in 1961 at around 800 TBq per year (Farris *et al.*, 1996).

The calculations of doses from atmospheric releases made use of mathematical models of radionuclide transfer in the environment (see Chapter 13), together with measurement data on levels of radionuclides in environmental materials. However, it was not until the mid-1950s that the importance of the cow's milk pathway for radioiodine was recognized at Hanford; consequently, radionuclide levels in milk were not monitored during the period of high radioiodine releases (the mid-1940s) and the dose estimates rely exclusively on mathematical models. The dose calculations were undertaken for a number of categories of individuals in order to cover the range of possible local habits. The results of the calculations are described in detail in Farris *et al.* (1996). The calculations took account of uncertainties in the source term and in environmental conditions – the median values from the calculations are reported here.

Radioiodine concentrates in the thyroid. The highest doses to this organ were estimated to have been received by infants consuming milk from a family cow fed on fresh pasture supplemented on alfalfa and grain. The thyroid dose in the year of maximum ^{131}I release, i.e. 1945, is estimated to be about 1.92 Gy at the geographical area of maximum radioiodine concentration. The average cumulative thyroid dose to a child residing in this area and consuming milk from a cow fed on fresh pasture during the period of high radioiodine release (1945 to 1951) is about 2.4 Gy. The assumption that the cow is fed on fresh pasture is important; ^{131}I has a half-life of about eight days and will decay if feed is stored, thus resulting in lower ^{131}I concentrations. The calculated doses for milk produced by cows fed on stored feed were about five times lower than those for cows fed on fresh pasture.

The maximum annual effective dose to an adult over the period 1945 to 1972 (the time the plutonium producing reactors were operating) is estimated to be about 10 mSv in 1945. The dose drops by around three orders of magnitude by 1972. The dominant radionuclide up to 1960 is ^{131}I. This radionuclide accounts for 99.8% of the annual dose in 1945. The dominant radionuclide from 1960 onwards is ^{144}Ce, which accounts for about 80% of the effective dose to an adult in 1965. The change in radionuclide contribution is due to increased fuel cooling time in the later years, during which the short-lived radionuclides such as ^{131}I will decay, and emission controls at the separation plant.

Dose from releases to the Columbia River were also estimated from a combination of environmental monitoring data and computer modelling. Doses were calculated for three types of individual, characterized on the basis of their use of the river for food, drink, recreation or as a place of work. The annual effective dose to a representative maximally exposed individual peaked at about 1.4 mSv in around 1960 – the main contributions coming from ingestion of fish containing ^{65}Zn and ^{32}P. Annual doses to this representative individual were above 0.1 mSv for the entire period of reactor operation (1945 to 1971).

The cumulative effective dose to the representative maximally exposed individuals for atmospheric releases and for releases to the Columbia River over the period 1944 to 1992 are estimated to be 12 and 15 mSv, respectively. The results from this study are, among other things, an input into the 'Hanford Thyroid Disease Study'.

Prospective assessments

Studies are also being undertaken into the doses people could receive from reinhabiting areas that have been used for nuclear weapons testing. Such dose estimates are important inputs into decisions on whether the areas should be decontaminated or remediated in some other way before human access is allowed. An example of such a study described here is an assessment of possible future doses to Aborigines occupying the former UK nuclear weapons test site at Maralinga in Australia. Other similar studies are being undertaken, under the auspices of the IAEA, on former test sites on certain Pacific Islands. These sites are the former US site in the Marshall Islands and the French sites at Mururoa and Fangataufa Atolls.

The UK conducted nuclear weapons tests at Maralinga and Emu in South Australia between 1953 and 1963. The programme involved nine nuclear explosions, together with 'minor trials' involving dispersion of radioactive material. Much of the radioactivity was short-lived and has decayed away but contamination by long-lived radionuclides remains. These radionuclides include isotopes of caesium, plutonium, americium, cobalt, strontium and uranium. Activity concentrations are such that, even in the 1990s, access has to be restricted. The major areas of contamination resulted from the 'minor trials' and distinct plumes of activity remain on the ground.

The contamination was of concern, particularly to Aborigine populations who wished access to the area for historical reasons. In order to assess the radiological significance of the contamination, a Technical Assessment Group (TAG) was set up in 1986 by the Australian Government. A programme of work was instigated, involving scientists from Australia, the US and the UK, to evaluate the doses likely to be received by future occupants of the test sites. The area is semi-arid and inhalation of re-suspended material can be a major exposure pathway under such circumstances. Intakes of radionuclides via contamination of cuts and abrasions can also be important. A programme of experimental work was undertaken to establish the bioavailability of radionuclides following intakes of the contaminated dusts present at the test sites (see Stradling *et al.*, 1992). Environmental monitoring established the extent of the contamination by individual radionuclides.

The possible future doses to individuals in an Aboriginal population with a semi-traditional life-style were calculated (Haywood and Smith, 1990 and 1992). The objective was to calculate typical doses rather the higher doses that could be received by small sub-groups with extreme habits. The following exposure pathways were taken into account:

(i) inhalation of dust re-suspended from the ground;

(ii) ingestion of foodstuffs from the area and any associated soil or dust;

(iii) external exposure from gamma-emitting radionuclides in the ground;

(iv) external exposure from β-emitting radionuclides in the ground, or contaminating the skin or clothing;

(v) contamination of sores and wounds.

Information on the habits of Aborigine populations was combined with measurements and bioavailability data, if appropriate, to calculate doses. The latter were calculated for adults, children, one-year-old infants and three-month-old infants. The calculated doses range over three orders of magnitude, thus reflecting the heterogeneous nature of the contamination. Projected effective doses from permanent inhabitation of particular areas range from a few millisieverts at the edge of 'minor trials' plumes to up to 0.5 Sv in small areas immediately around the test sites. The most important exposure pathway is inhalation; the most important radionuclide is ^{239}Pu.

The estimates of possible future doses are one input into a decision as to whether the contaminated land requires restrictions on its future usage or clean-up and, if so, to what extent. The TAG concluded that the annual risk of fatal cancer as a result of exposure to ionizing radiation at the test sites should be less than 10^{-4} (one in ten thousand) for each year in the individual's life-span (TAG, 1990). Decontamination of areas where this risk level could be exceeded was estimated to cost some hundreds of millions of dollars; the cost of fencing off the areas was estimated to be substantially less. At the time of the study there were no internationally agreed criteria for deciding on the need to decontaminate areas such as the weapons test sites in Australia. However, the ICRP has recently issued guidance for situations where there is prolonged exposure of members of the public. Broadly speaking, this is that in circumstances where total annual doses are below 10 mSv it will seldom be justified to intervene to reduce doses or potential doses. The desirability for intervention will increase with the value of the assessed dose until at annual dose levels above 100 mSv intervention will almost always be justified (ICRP, 1999b).

The Australian and UK Governments reached agreement in 1993 on financial matters concerning remediation. It is intended that there will be unrestricted access to the area, with the exception of a small zone of a few square kilometres in area where camping and full-time occupancy will not be allowed. The project involves removal of contaminated soil and further treatment of the pits containing previously disposed-of plutonium and other material.

Environmental contamination in the former Soviet Union

The nuclear weapons industry has caused significant radioactive contamination in some parts of the former Soviet Union (Tsaturov and Anisimova, 1994). This contamination is considered separately to that from the Chernobyl reactor accident as that was a civil nuclear accident (see Section 6.4.3). Areas suffering the most significant levels of contamination from the Soviet nuclear weapons industry are the Southern Urals, where weapons materials were processed, together with the weapons test sites at Novaya Zemlya and Semipalatinsk.

The 'MAYAK' production association was set up in the former Soviet Union in 1949 for the production of plutonium for the military. This is situated in the Chelyabinsk Region of the Southern Urals. Possibly due to the urgency attached to the production of nuclear weapons, MAYAK operations have led to significant contamination of areas of the Southern Urals, either from accidents or as the result of deliberate actions The most serious of the accidents was a chemical explosion, in September 1957, in a tank of high-level waste at Kyshtym, a part of the MAYAK complex. The explosive yield was equivalent to about 70 to 100 tonnes of TNT (Aarkrog, 1996). This accident dispersed about 74 PBq of activity, mostly in the form of short-lived fission products such as ^{95}Zr and ^{144}Ce. Smaller amounts of longer-lived radionuclides were released, of which the most significant was 2 PBq of ^{90}Sr (Aarkrog *et al.*, 1992). The contaminated area, of about 20 000 km^2, extends in a narrow band for a few hundred kilometres to the north-east of Kyshtym – about 300 000 individuals were affected (Romanov and Drozhko, 1996). The Soviet authorities adopted a safety limit for members of the population of 74 GBq of ^{90}Sr per km^2, which corresponded to an area of about 1000 km^2. These restrictions were subsequently partially lifted and by 1982 farming, etc. had been resumed in 82 % of the affected area.

Deliberate actions at the MAYAK complex have caused considerable contamination of the upper reaches of the Techa River. This river rises in the Southern Urals (not far from Ekaterinburg) and drains into the Arctic Ocean via a number of other rivers, including, finally the River Ob. Between 1949 and 1951, about 100 PBq of medium- and high-level wastes were released directly into the Techa River from MAYAK reprocessing activities (Trapeznikov *et al.*, 1994). Strontium-

90 contributed about 6 PBq and ^{137}Cs contributed about 12 PBq (Romanov and Drozhko, 1996). Most of the activity was released in 1950 and 1951. In attempts to remedy the situation, a number of dams and artificial reservoirs were constructed between 1951 and 1964 in the upper reaches of the river. Around 7500 inhabitants were also moved away from 22 settlements situated on the river system (another estimate puts the number of individuals evacuated at 28 000 (Romanov and Drozhko, 1996)). The cumulative doses to these individuals from the contamination prior to their relocation, ranged up to 1.4 Sv (Tsaturov and Anisimova, 1994). A recent study of the Techa River from 50 km from the discharge point to its confluence with the Iset River, 240 km downstream, has shown that this section of the river now contains a very small fraction, less than around 0.1 %, of the ^{90}Sr and ^{137}Cs that was discharged into it (Trapeznikov *et al.*, 1993). The study concluded that most of the ^{90}Sr and ^{137}Cs is still retained in the reservoir complex but more than 1 PBq may be present downstream of the Techa River or even in the Arctic Ocean. Caesium-137 and strontium-90 contamination of the flood plains of the upper reaches of the river is such that the use of an area of 80 km^2 is restricted.

Since the early 1950s, radioactive wastes have been stored or disposed of in open reservoirs at MAYAK. One of these reservoirs, Lake Karachay, has been used for the disposal of about 4500 PBq of radioactive waste (Aarkrog *et al.*, 1992; Tsaturov and Anisimova, 1994). In 1967, during a hot spring that followed a dry winter, the lake partially dried out and wind-blown radioactive dust contaminated an area of about 1800 km^2. A total of 22 TBq, including 11 TBq of ^{137}Cs and 4 TBq of ^{90}Sr, were dispersed by the wind (Romanov and Drozhko, 1996). The level of contamination was much lower than from the 1957 accident.

Estimates of doses to individuals living near the Techa River have been published (Tsaturov and Anisimova, 1994). The highest cumulative doses were received by the individuals mentioned above who were relocated from settlements close to the MAYAK facilities. Doses to individuals in other areas are lower and generally decrease with distance downstream. Currently, around 20 350 PBq of high-level waste are stored in tanks at MAYAK, with another 4440 PBq of waste in Lake Karachay and 7.4 PBq in the Techa River reservoir system (Fetisov *et al.*, 1994). The high-level wastes and the wastes in Lake Karachay will require treatment to convert them into a form that can stored and/or disposed of safely. Since the accident in 1967, the lake has been progressively filled in, with the intention being to fill it in completely, but, as the lake bed and sides are not sealed, problems are arising from the contamination of groundwater. About 3.5×10^6 m^3 of contaminated water are estimated to have filtered down through the bedrock underlying the lake to form a lens of contaminated water extending over an area of about 10 km^2 (Romanov and Drozhko, 1996). The front of the lens may be moving south at a rate of up to 80 m per year and could eventually contaminate sources of potable water.

Other installations in the Soviet Union were involved in the production of nuclear weapons and have caused environmental contamination to a greater or lesser extent. These facilities include Tomsk-7, situated on the River Ob, and the plutonium-producing facility Krasnoyask-46, which is located on the Yenisei – a river that discharges into the Arctic Ocean close to the estuary of the Ob. There is information that suggests that discharges from both of these installations have caused significant environmental contamination (Bridges and Bridges, 1995).

The Soviet Union, like the Western nuclear powers, has conducted nuclear weapons tests underground, underwater and in the atmosphere. Tests conducted in the atmosphere are of most significance from the viewpoint of impact on the world's population as radionuclides are dispersed globally. Tests conducted either on or just above the ground, such that the fireball touches the ground, are likely to cause the greatest local contamination due to activation of materials in the soil. Between 1949 and 1962, 124 nuclear weapons tests were conducted at various heights in the atmosphere at the Soviet test site at Semipalatinsk in Kazakhstan, with another 81 being conducted on the Arctic island of Novaya Zemlya. A number of sub-surface tests were also conducted at both sites (see Basabikov *et al.*, 1994, for data on the types and numbers of Soviet tests). The civilian population of Novaya Zemlya was evacuated before the tests were conducted and, consequently, the radiological impact of local fall-out was minimal. However, at least some of the tests at Semipalatinsk did have implications for the local population. The most significant of these from this viewpoint was the first Soviet test, which was conducted in August 1949: doses to individuals in neighbouring populated areas may have exceeded 50 mSv (Basabikov *et al.*, 1994). Current levels of ^{137}Cs contamination at the test site have been investigated (Dubasov *et al.*, 1994). Levels in some areas of the test site ranged up to 180 GBq km^{-2}, with around 0.5% of the area having levels above 37 GBq km^{-2}. Levels in adjacent populated areas can reach 11 GBq km^{-2}. A comparison of levels of ^{137}Cs contamination in the former Soviet Union and in some other areas of the world is given in Table 6.22.

Table 6.22 Caesium-137 levels in the former Soviet Union and comparison with global weapons fall-out levels.

Area	Caesium-137 (GBq km^{-2})[a]	Reference
< 30 km from Chernobyl	1500	NEA, 1995
Areas of Belarus, Ukraine and Russia contaminated by the Chernobyl accident	500	NEA, 1995
Semipalatinsk test site	Up to 180	Dubasov *et al.*, 1994
Semipalatinsk test site: local populated areas	Up to 11	Dubasov *et al.*, 1994
Russia: global and tropospheric weapons test fall-out[b]	1.8 to 18	Dubasov *et al.*, 1994
Average weapons test fall-out, 40° to 50°N	5.2	Aarkrog, 1996
Weapons test fall-out levels in the UK[b]	2 to 9	Cawse and Baker, 1990

[a]Note that these activity levels can be converted into doses by using factors, derived by models, that relate the concentration of ^{137}Cs in soil to the dose from external radiation to an individual standing on the soil. If the deposit is a fairly recent one and the activity has not 'mixed into' the soil, the dose to an individual standing on soil that has received a deposit of 10 GBq km^{-2} of ^{137}Cs will be 1.35×10^{-8} Sv h^{-1}. It should be noted that the dose rate indoors will be less due to shielding by building materials. Furthermore, as the deposit ages and moves down through the soil, the dose rate will fall due to shielding by the soil itself.

[b]Levels vary depending upon rainfall – rain washes activity out of the atmosphere.

6.6 CONCLUSIONS

Nuclear power has two manifestations, namely nuclear weapons and nuclear power for generation of electricity. The testing of nuclear weapons has injected significant quantities of fission and activation products into the atmosphere. It has had a far higher radiological impact on the world's population to date than has the generation of electricity by nuclear means: the collective dose from weapons test fall-out to the year 2200 is estimated to be 6×10^6 manSv, with a value over all time of 3×10^7 manSv. Electricity generation by nuclear power has had a much lower radiological impact; the collective dose from the entire nuclear power production up to the year 1989 is around 4×10^5 manSv. The collective dose from the Chernobyl reactor accident lies between the two, with estimates ranging from 6×10^5 manSv to 2×10^6 manSv. It is a sobering thought that one, admittedly very severe, accident to a single reactor resulted in a significant radiological impact greater than that of the routine discharges from the world's total nuclear power production industry. In most countries, individual doses from routine discharges from the nuclear power industry are controlled within the ICRP's system of radiological protection and are, by and large, low. There are, however, particular areas of the world which have suffered significant levels of contamination either from nuclear weapons testing or from accidents – the restoration of such areas is an issue that remains to be fully addressed.

7

Releases of Radionuclides from Non-Nuclear Power Industries

Chapter 6 describes releases of radionuclides to the environment from the nuclear power industry. However, there are a variety of medical and industrial uses for radionuclides which can result in releases to the environment and the processing of ores and other naturally radioactive materials can produce low-level radioactive wastes. These are the subjects of this chapter.

7.1 NATURALLY RADIOACTIVE RAW MATERIALS

Ores and other raw materials are often naturally radioactive, usually due to the presence of radionuclides in the thorium and uranium series (see Appendix 3). Some ores and minerals have concentrations of radionuclides many times the average for soils and rocks. The minerals are usually phosphates, oxides or silicates of transition elements or rare earths. Industries processing such materials will often produce wastes containing elevated amounts of the natural radionuclides. It is important to remember that the presence of radioactivity is usually incidental; in other words, the materials are not being processed because they are radioactive but for their other properties. In ores, due to the long geological timescales since the materials were laid down in the Earth's crust, the decay chains have usually come into equilibrium. The equilibrium will be disturbed during the separation of products (e.g. metals, etc.); the activity in separated materials containing short-lived radionuclides unsupported by their parents, may be initially high but will rapidly decay away, whereas separated ^{238}U or ^{232}Th will take millennia to come into equilibrium with their complete decay chains. Thus, the reader will find the discussion of radioactive equilibria in Chapter 3 useful when studying this chapter. One further point deserves mention: the decay chains of both ^{238}U and ^{232}Th contain an isotope of the inert gas radon, which could escape from the ore body. The isotope in the ^{232}Th decay chain is very short-lived (56 s) and thus has little opportunity to escape, but ^{222}Rn, the isotope in the ^{238}U series, has a half-life of 3.8 d and could escape. If a fraction of the ^{222}Rn escapes, the concentration in the ore body of its daughters, which include isotopes of bismuth, polonium and lead, will be similarly reduced. Loss of ^{222}Rn from an ore body will depend upon many factors, such as the porosity and degree of fracturing.

There are a number of industries that use materials that are naturally radioactive. These industries range from production of phosphates for fertilizers and detergents from phosphate rock to the production of paint pigments from particular metal ores. A list of the ores and minerals used, together with an indication of the activity concentration of the natural radionuclides, is given in Table 7.1. Examples of the concentrations of radionuclides in wastes and residues from these processes are given in Table 7.2. The industries are described in the following sections.

7.1.1 Phosphate industry

Phosphate rock is processed at several locations worldwide to produce phosphates, largely for fertilizers but also for use in detergents. The major deposits are sedimentary in origin and contain natural uranium which was incorporated by ionic substitution during the sedimentation process. Concentrations of radionuclides of the ^{232}Th chain are relatively low. Major deposits of phosphate rock occur in Morocco, Florida and the Kola region of Russia. Rock is also mined in other regions of the world including Israel, Jordan, China and Tunisia (see Baetsle, 1991). On geological timescales, the ^{238}U chain (see Appendix 3) will come into equilibrium. The concentration of radionuclides in the rock depends to some extent upon its geographical origin but is typically around 1700 Bq kg^{-1} of ^{238}U, in equilibrium with its daughters down to ^{226}Ra (see Appendix 3), together with variable activities of the ^{226}Ra daughters depending upon whether there is a significant loss of ^{222}Rn from the rock. About 15 % of known phosphate deposits are of

Table 7.1 Activity concentrations in ores and mineral sands (in units of Bq kg^{-1}).

Ore/mineral sand	Industrial use	^{238}U	^{226}Ra	^{232}Th	^{40}K
Phosphate ore	Phosphates	30–5000	30–5000	25–2000	3–200
Monazite sand	Rare earths	370	—	1800	160
Monazite	Rare earths	6000–40 000	—	8000–900 000	—
Bastnaesite	Rare earths	—	—	400	—
Xenotime	Rare earths	3500–500 000	—	180 000	—
Thorianite	Thorium	—	—	2 500 000–5 500 000	—
Tin ores	Tin	1000	—	300	—
Pyrochlore	Niobium	10 000	—	80 000	—
Titanium ores	Titanium	70–9000	—	70–9000	—
Ilmenite	Titanium	2000	—	1000	—
Zircon sands (baddeleyite)	Refractory materials	10 000	3000–4000	10 000	—
Bauxite	Aluminum	400–600	—	400–600	—
Coal	Various	20	—	20	40
Iron ore	Iron and steel	15	—	—	—

Table 7.2 Activity concentrations in industrial residues and wastes (in units of Bq kg^{-1}).

Material	^{238}U	^{226}Ra	^{232}Th	^{40}K
Tin slag	1000–4000	1000–4000	230–340	—
Oil scale (old process)	—	Up to 4000	—	—
Oil scale (new process - scale inhibition techniques)	—	40–100	—	—
Rare earth extraction by-products	—	3000–450 000	—	—
T_iO_2 production residues from ilmenite	—	Up to 400 000	—	—
Monazite processing residues	—	Up to 450 000	—	—
Zircon processing				
residues	—	2000–50 000	—	—
sludge	—	200–7000	—	—
Copper slag	—	500–2000	—	—
Aluminium processing sludge	260–540	150–330	—	—
Fly ash	400	—	—	—
Blast furnace slag from steel production	150	—	150	—

igneous or metamorphic origin. In these rocks, the concentrations of ^{232}Th chain radionuclides can be higher than those of the ^{238}U chain.

One of the main ways of processing phosphate rock is treatment with sulfuric acid or, less commonly, hydrochloric acid. Raw phosphate ore is dissolved in acid, producing phosphoric acid and, if sulfuric acid is used, a calcium sulfate precipitates. The general equation is as follows:

$$Ca_{10}(PO_4)_6F_2 + 10H_2SO_4 + 20H_2O \longrightarrow 10CaSO_4.2H_2O + 6H_3PO_4 + 2HF$$

The precipitate ($CaSO_4.2H_2O$), called phosphogypsum, contains most of the ^{226}Ra that was present in the original ore, together with smaller proportions of the preceding members of the ^{238}U chain. Demand for phosphate fertilizer means that significant quantities of phosphogypsum by-product are produced; annual worldwide production of phosphogypsum was estimated to be between 2.2×10^8 t and 2.8×10^8 t in the year 2000. The phosphogypsum is often discharged to the aquatic environment as a sludge or, simply dumped near the plant. A study in the early 1980s suggested that around 14 % of phosphogypsum is reprocessed, 58 % is stored and 28 % is disposed of to the aquatic environment (Carmichael, 1988). The preferred means of disposal varies between countries. For example, the phosphate industry in the United States produces around 3×10^7 t of phosphogypsum per year which is stored in stacks or mine cuts. It has been estimated that the total inventory of phosphogypsum stored in Florida will reach 10^9 t by the year 2000 (Burnett *et al.*, 1996). In contrast, the phosphogypsum by-product from the large Dutch phosphate processing industry in Rotterdam is discharged to the aquatic environment. The two plants on Rotterdam Harbour process about 1.2×10^6 t of rock per year and produce about 2×10^6 t of phosphogypsum which is discharged to the harbour. Thus, around 1300 GBq each of ^{226}Ra, ^{210}Pb and ^{210}Po are discharged. About 90 to 95% of these radionuclides dissolve and are transported to the open sea. The average annual dose due to these discharges from the consumption of marine produce caught in Dutch coastal waters is estimated to be 1–2 μSv. A small amount of activity remains bound to sediments in the harbour and results in a local enhancement of ^{226}Ra of between 10 and 70 Bq kg^{-1} of sediment (Timmermans and van Der Steen, 1996). For over 40 years, the by-products were also discharged to the marine environment from a phosphate rock processing plant at Whitehaven in the UK. In 1992, plant operations changed from the processing of phosphate rock to processing crude phosphoric acid and discharges essentially ceased. Prior to this time, the discharges led to measurable increases in ^{210}Po concentrations in local shellfish. Coincidentally, the plant is situated a few kilometres from the nuclear fuel cycle complex at Sellafield and estimates of doses to individuals in the early 1990s showed that the critical group for Sellafield liquid discharges also received significant exposures from ^{210}Po from the phosphate plant's discharges: the annual dose to the critical group from ^{210}Po was about 125 μSv, which is approximately equal to the total annual dose from Sellafield's liquid discharges (Robinson *et al.*, 1994). Similarly, the phosphate rock industry in Morocco disposes of about 25 000 tonnes of phosphogypsum per day to the Atlantic Ocean when operating at nominal capacity (Becker, 1989). Other industrial discharges of natural radionuclides to marine waters include releases to the Seine Estuary in France and to the Tinto and Odiel Rivers in Spain (see McDonald *et al.*, 1996).

In the past, the precipitate has been dehydrated for use as wall plaster in the building industry. This use has led to concerns over human exposure to ^{222}Rn produced from the ^{226}Ra in the gypsum. Other uses for phosphogypsum have included land treatment to provide plant nutrients or to reclaim acidic or sodic soils. Although the radionuclides in the phosphogypsum may not pose a problem in an agricultural context, there is the possibility that land so treated could be unsuitable for future housing development because of ^{222}Rn ingress from the added ^{226}Ra. Such considerations led the US Environmental Protection Agency (EPA) to rule that phosphogypsum could only be used for land treatment if the average ^{226}Ra concentration did not exceed 370 Bq kg^{-1}. The phosphoric acid can be treated in various ways to produce fertilizer which contains most of the uranium and thorium present in the original ore. Use of this fertilizer will also lead to human exposure.

Phosphogypsum waste has been used as a building material in some parts of the world, notably Belgium, with internal walls being typically finished with a layer 1–2 cm in depth. The presence of ^{238}U and ^{232}Th and their decay products means that the material will produce radon (^{222}Rn) and thoron (^{220}Th) gases, thus causing exposure on the occupants. Thoron is very short-lived (half-life, 55.6 s) and covering the phosphogypsum surface with latex paint will reduce transfer to the air by between a factor of 10 to 20, hence significantly reducing the exposure. The additional annual effective doses from occupancy of houses with phosphogypsum-covered internal walls is estimated to be 0.1 mSv from radon exposure with an equal contribution from thoron if the phosphogypsum covering is painted, rising to about 1 mSv if not (Vanmarcke, 1996).

The chemistry of the hydrochloric acid process is different and fertilizers that are essentially free of ^{226}Ra and ^{238}U are produced. The uranium and radium isotopes are solubilized and enter the waste stream. The uranium and thorium is subsequently precipitated as a sludge during the effluent treatment process but the radium remains in solution and will be discharged with the liquid wastes unless specific steps are taken to remove it.

Elemental phosphorous can be produced from phosphate rock from a thermal process; the phosphoric acid produced has a very high degree of purity. Phosphate rock is mixed with silica and coke and then treated electrically to a temperature of 1300 °C. The equation for this chemical reaction is as follows:

$$2Ca_3(PO_4)_2 + 6SiO_2 + 4C \longrightarrow 6CaSiO_4 + 4CO + 4P$$

The phosphate is chemically reduced by the coke and can be recovered as elemental phosphorous and phosphoric acid. Natural radionuclides concentrate predominantly in the dense metallic slag. Radium-226 concentrations of around 1500 Bq kg^{-1} in the slag have been reported (Baetsle, 1991). Exceptions are ^{210}Po and, to a lesser extent, isotopes of lead. These are relatively volatile and may enter the gas phase. It has been estimated that, in general, around 30% of the ^{210}Po present in the ore is released in gaseous effluents (Baetsle, 1991). Annual atmospheric discharges from a thermal phosphate rock processing plant in the Netherlands are around 45 GBq of ^{210}Pb and 540 GBq of ^{210}Po; the corresponding liquid releases are 25 GBq and 100 GBq, respectively. The highest critical group doses are from atmospheric discharges and are just less than 40 μSv per year (Timmermans and van der Steen, 1996).

Slags from the thermal processing of phosphate rock have been used in the past as infill in the road building and general construction industry. The presence and elevated levels of members of the ^{238}U series in these materials could lead to external exposure of individuals in the vicinity and thus has led to concerns being expressed about the desirability of such use of the slags.

7.1.2 Production of Special Metals from Ore or from By-Products of other Processes

Several special metals are extracted from ores or from by-products of other industrial processes that are enriched with natural radionuclides of the ^{238}U, ^{235}U or ^{232}Th series. During the extraction of metals, these natural radionuclides will eventually find their way to by-products or and to wastes such as slags, off-gas filter dusts and liquid or gaseous effluents. Exposure of the workforce can occur during the processing of the ores, with the main exposure pathway being inhalation of contaminated dust (Hipkin and Paynter, 1991). The main metal producing processes of interests are described below.

Tin production

Tin smelters use feedstocks comprising natural ores and tin-rich residues from other industrial processes. Tin ores typically contain about 1000 Bq kg^{-1} of ^{238}U and around 300 Bq kg^{-1} of ^{232}Th, both in equilibrium with their daughters (Harvey *et al.*, 1994). The principal sources of tin ores are found in Bolivia and South-East

Asia. The tin-rich residues from other industrial processes may have elevated concentrations, up to 10 000 Bq kg^{-1}, of ^{210}Pb and ^{210}Po (Hipkin and Paynter, 1991).

After some initial purification steps involving crushing and washing, the ore is mixed with charcoal as a reducing agent and heated to around 1000 °C in order for the chemical reaction to take place. The product is purified by re-smelting and, in some cases, electrolytic refining. The slag produced by the initial smelting step contains the non-volatile natural radionuclides (^{238}U, ^{226}Ra, etc.) at concentrations of around 1000 to 4000 Bq kg^{-1}. Polonium-210 volatilizes during the initial heating and precipitates from the furnace off-gases in the off-gas treatment processes.

Lead and bismuth can be produced as by-products of the tin smelting process. The lead will contain trace amounts of ^{210}Pb from the original ore, together with its decay products, i.e. ^{210}Bi and ^{210}Po. The ^{210}Bi in the initial feedstock will separate with the bismuth by-product. Bismuth-210 has a half-life of around 5 d and thus will rapidly decay away to its longer-lived daughter ^{210}Po whose activity in the product can reach around 100 000 Bq kg^{-1} (Hipkin and Paynter, 1991). This activity will, in turn, decay away with the half-life of ^{210}Po of 138 d (the decay product of ^{210}Po is stable ^{206}Pb).

One of the world's largest tin smelters operated in the UK on the Humber Estuary from 1937 to 1992 when it closed for commercial reasons. This refinery specialized in recovering tin from low-grade materials, including by-products from other industrial processes. At its peak production, it generated about 10 % of the world's tin. A number of studies have been conducted into releases of natural radionuclides from this plant because of public concern over a possible association with an abnormally high incidence of childhood leukaemia in the nearby town of Kingston-upon-Hull (situated 14 km to the east). The average annual intake of raw materials was 8.2×10^4 t. These materials were very heterogeneous but could contain on average up to between 10^3 and 10^4 Bq kg^{-1} of ^{238}U and ^{226}Ra. The natural decay series was not in equilibrium in a number of feedstocks; in particular, the ^{210}Po levels could be higher or lower than expected. The fate of ^{210}Po during tin production has been studied. The total annual intake was estimated to have been around 100 GBq and, of this, about 30 % was isolated in the tellurium dross produced during the production of a lead/bismuth alloy product, 48 % decayed within the refinery, 19 % went to waste slag and 2 % was discharged to atmosphere (Baxter *et al.*, 1996). The uranium and thorium present in the feedstock largely went with the slag which was occasionally used as road construction material; otherwise, it was generally disposed of to landfill sites.

Niobium production

As it is resistant to corrosion, the metal niobium is used in chemical plants and nuclear reactors. It also finds application in electronic components. The metal occurs in combination with other elements in a number of ores, including tantalite, columbite, furgusonite and samarskite. The most important niobium containing-containing ore is probably pyrochlore ($(Na, Ca, Ce)_2Nb_2O_6F$), which derives its name from the Greek as it turns a green colour upon heating. Niobium-bearing ores are found in Australia, Nigeria, Uganda, Kenya, Tanzania and Canada. Pyrochlore contains ^{238}U and ^{232}Th, together with their daughters; typical concentrations of ^{238}U and ^{232}Th are 10 000 and 80 000 Bq kg^{-1}, respectively.

Pyrochlore is mixed with aluminium powder, aluminium wire and iron chippings, and the mixture is heated in a furnace to produce ferroniobium alloy. The ^{238}U and ^{232}Th partition with the furnace slags, which are mainly disposed of to landfill sites (Harvey *et al.*, 1994). A second method of producing niobium is by melting the ore with sodium or potassium hydroxide, dissolution in hydrochloric acid and treatment with chlorine at 750–800 °C.

Rare earth production

Rare earths is an old term for the *lanthanides.* They are found in a number of minerals, but the main ore is monazite – a phosphate mineral containing thorium and the rare earths. As a result of natural weathering processes, monazite occurs in mineral sands in various parts of the World, e.g. India and Australia.

Following various pre-treatments, including gravimetric and electromagnetic sorting, the rare earths are separated by a furnace process and can be further purified by electrolysis at high temperatures. The main uses for rare earths were in glass polish and as cigarette lighter flints. Current usage includes alloying materials in metallurgy, in the manufacture of permanent magnets and in motor vehicle catalytic converters.

Monazite is rich in thorium (usually between 0.5 and 28 % by mass) and thus typically contains around 10 000 Bq kg^{-1} of ^{232}Th plus daughters. The thorium may be extracted separately (see below) but, if not, the radionuclides will partition with the slags.

Thorium production

Economically important minerals containing thorium are monazite (see above), thorite and thorianite. Thorium is obtained by concentrating the minerals and then treating with acid to produce thorium salts. The main use for thorium was in the production of gas mantles but this usage is declining and the element is now finding wider usage in, for example, the electronics industry. The metal is also used in welding electrodes where it is added to tungsten in order to facilitate ignition. Clearly, radioactive isotopes of thorium will remain with the parent element and appear in the final product. The activity concentrations of thorium isotopes in the various products have been reported by Dalheimer and Henricks (1994). Thorium-containing welding electrodes have activity concentrations for both ^{232}Th and ^{228}Th of about 100 Bq g^{-1}; concentrations of the two isotopes in gas mantles are about an order of magnitude higher and special alloys, such as are used in jet engines, have concentrations of around 70 Bq g^{-1}.

The presence of radioactive thorium isotopes in the final product can result in human exposure. For example, it has been estimated that use of a thorium gas mantle in a caravan for around 20 hours per year will result in a dose of about 100 μSv to a child occupant from inhalation of volatilized thorium isotopes; the corresponding dose to an adult is about 50 μSv (Hughes and O'Riordan, 1994).

Titanium oxide production

Titanium metal is used in the construction of aircraft, ships and chemical plant. The oxide is used extensively as a pigment in the paint industry and also in the rubber, ceramic and varnish industries.

The element occurs in the minerals ilmenite ($FeO.TiO_2$) and rutile (TiO_2). The metal is isolated by a process involving crushing the ore with coal, followed by heating and treatment with chlorine and separation of $TiCl_4$. Titanium 'sponge' is produced by reduction of the chloride and can be purified by further smelting.

Titanium ores contain ^{238}U and ^{232}Th series radionuclides at concentrations of between 70 and 9000 Bq kg^{-1}, depending upon the origin of the ore (Dalheimer and Henricks, 1994). Volatile radionuclides are released to atmosphere during the heating stages of the separation process.

Refractory materials

Silicates and oxides of the element zirconium have excellent refractory properties and find uses in components of, for example, high-temperature furnaces. The silicate exists naturally as zircon ($ZrSiO_4$) in the mineral sands which occur in Australia, India, and several other areas of the world. The oxide occurs naturally as the ore baddelyite, which is found in Brazil, Sweden, India and Italy. Both zircon deposits and baddelyite contain natural radionuclides in the thorium and uranium series. Zircon sands contains around 10 000 Bq kg^{-1} of ^{238}U and of ^{232}Th in secular equilibrium with their daughters (Scholten, 1996).

Refractory components are made from zircon sands by mixing the sand with alumina and sodium carbonate, followed by smelting at high temperature (National Group for Studying the Implications of Zircon Sand, 1985). In common with other processes involving high temperatures and ^{238}U series radionuclides, the radionuclides of volatile elements, such as polonium and lead, are released to atmosphere. Discharges from a typical plant may be around 150 MBq of ^{210}Po and 50 MBq of ^{210}Pb per year (Harvey *et al.*, 1994).

Doses from the production of special metals

Annual doses from the production of special metals have been estimated (Harvey *et al.*, 1994). Occupational doses range up to of the order of 1 mSv per year, being highest for production of refractory materials and for production of ferroniobium alloy from pyrochlore. Doses to members of the public from discharges are lower, being at most around a few μSv per year.

7.1.3 Steel Production

Steel production can result in discharges of small amounts of radioactive material to the environment. Modern steel production plants require the iron ore feedstock to be in the form of fairly large lumps to aid air flow through the furnace. Much of the raw iron ore is too fine to be used and has to be sintered prior to use. In the sintering process, a mixture of iron ore and coke is heated so that the ore partially melts and forms agglomerates which, after cooling, are broken up for use in the blast furnace. The raw materials contain trace amounts of uranium and thorium and their decay products. The average activity of ^{238}U and decay products in the iron ore used in the UK is about 15 Bq kg^{-1} (Harvey, 1998). During the sintering process, some of the ^{210}Pb and ^{210}Po is volatilized and a fraction may escape with the off-gases through a discharge stack. A UK study estimated that annual critical group doses from these discharges could be in the range of 2–20 μSv (Harvey, 1998). The major route of exposure was from ingestion of contaminated foodstuffs that had been grown in the vicinity.

7.2 ENERGY PRODUCTION

Fossil fuels contain primordial radionuclides to a greater or lesser degree and the combustion of these fuels can lead to radiation exposure from dispersion of the radionuclides into the environment. This is of most significance in the burning of coal, a process which accounts for around 40 % of the world's electricity.

7.2.1 Coal Combustion

Coal normally contains trace quantities of potassium-40, together with nuclides in the ^{238}U and ^{232}Th decay series. Typically, these are present at the same or lower concentrations than in soil. Activity concentrations of ^{40}K, ^{238}U and ^{232}Th are around 50, 20 and 20 Bq kg^{-1}, respectively, with the decay products of ^{238}U and ^{232}Th in equilibrium (UNSCEAR, 1988). However, some types of coal are reported to contain somewhat higher concentrations of natural radionuclides (Pan, 1993, cited in Sandor *et al.*, 1996) and coal used in Romanian power stations may have up to six times more ^{40}K and twice the UNSCEAR value of ^{238}U (Botezatu *et al.*, 1996). Another example is coal mined near the German town of Freital. Coal and uranium ore occur together in this area and the coal reportedly contains ^{238}U at concentrations up to 15 000 Bq kg^{-1} (Martin *et al.*, 1997).

Following combustion of the coal, radionuclides concentrate in the fly ash, some of which may escape from the stack, thus causing additional radiation exposure of the surrounding population. The fraction of fly ash escaping from the stack depends upon the effectiveness of the stack filtration system (modern plant will have this equipment while older plant may not). A US study suggests that around 3 % of the activity in the coal is released to atmosphere in plants of old design, with the proportion falling to around 0.5 % in modern plants (Roek *et al.*, 1987).

Concentrations of radionuclides in escaping fly ash are reported by UNSCEAR (1988) to be enriched around five times when compared with the original coal, although the enrichment of the relatively volatile radionuclides, ^{210}Pb and ^{210}Po, can be over an order of magnitude. The enrichment is lower in lignite-fuelled plants due to the lower furnace temperatures in plants which burn this type of coal (Hedvall and Erlandsson, 1996).

Annual critical group doses for discharges from modern coal-fired plant are reported by UNSCEAR (1988) to be around 1 μSv per GW(e) year, with the corresponding figure for old plant being 20 μSv. Modelling studies suggest that the critical group dose arising from discharges from a 3.5 GW(e) UK coal-fired power station is about 1.5 μSv per year, with the main contributing pathway being ingestion of locally produced foodstuffs and the major contributing radionuclide being ^{210}Po (Smith *et al*, 2001). Studies on one of the largest power stations in the world, Naticoke (Canada), suggest even lower doses at around 0.1 μSv per year (Tracy and Prantl, 1982).

Estimates of collective doses from discharges from coal-fired power stations range from a few manSv per GW(e) year to over 100 manSv per GW(e) year (see Hedvall and Erlandson, 1996). Such differences may reflect differences in the modelling approaches adopted, as well as variations in the discharges due to the age of the plant and the type of coal used. For comparison, the annual collective dose to the UK population in the mid 1990s from the annual production of electricity by coal-fired power stations is estimated to be 5 manSv (Hughes and O'Riordan, 1994). For comparison, the corresponding value for the nuclear energy production in the UK is estimated at about 14 manSv (Hughes, 1999)

Individuals working in coal-fired power stations may also receive additional radiation exposures, up to possibly between 100 and 200 μSv per year (UNSCEAR, 1988). Inhalation of dust can be an important exposure pathway; in some power stations residual dust and ash is stored in wet slurry ponds which can dry out at the edges, so leading to re-suspension of the dust. A recent study in the UK, however, estimated lower doses of around 10 μSv per year, with external exposure from radionuclides in the ^{238}U and ^{232}Th chains being the most significant exposure pathway (Smith *et al.*, 2001). Such estimates are based on assumptions and not directly on measurements and so different assumptions made in the studies are likely to be responsible for the differences.

The disposal of ashes and dusts will also give rise to doses. Significant quantities of these wastes are produced; the rate of production is estimated at 3×10^7 t per year in the European Union and 6×10^7 t per year in the USA (Martin *et al.*, 1997). Doses to members of the public can arise from storage of ash prior to disposal. Ash piles are present at some, but not all power stations. Dust can be re-suspended by wind action and radon will also be produced. Both of these processes could cause exposure of members of the public. Gamma-emitting radionuclides in the ash could also cause external exposure of nearby individuals. Estimates suggest that doses to members of the public from ash piles are not likely to be high; a UK study indicated doses of at most a microsievert with inhalation of radon and daughters being the most significant exposure pathway (Smith *et al.*, 2001). Some of the waste is disposed of to landfill or to sea and some is used as bulk construction material. Doses to operatives at disposal sites from handling dry ashes and dusts are estimated to lie in the range 100 to 200 μSv per year (Martin *et al.*, 1997) although a recent study has reported lower doses of around 5 μSv per year for landfill workers (Smith *et al.*, 2001).

Waste ash is also extensively used as bulk material in concrete manufacture and, for example, in road construction. Some ash is used in the production of lightweight building blocks; these are typically made by sintering a mixture of iron slag and fly ash. Since fly ash contains elevated levels of natural radionuclides, so will the construction materials. However, the resultant concentrations will depend upon the relative amounts of fly ash used. A study conducted in the Netherlands, where relatively large amounts of fly ash are used in cement production, suggests that ^{226}Ra and ^{232}Th concentrations in the cement are between 70 and 170 Bq kg^{-1} (Scholten *et al.*, 1993, quoted in Martin *et al.*, 1997). In general, however, the combined concentrations of ^{226}Ra and ^{232}Th in cements are reported to rarely exceed 100 Bq kg^{-1} (Martin *et al.*, 1997).

Doses to workers during the manufacture of construction materials are reported to lie in the range 100 to 200 μSv per year by Martin *et al.*, (1997), whereas slightly lower values of around 40 μSv per year were estimated by Smith *et al.* (2001). The dose does, of course, depend upon the working conditions and the activity concentrations in the materials being used. Members of the public occupying buildings constructed from these materials will be exposed to external radiation emitted from the natural radionuclides, predominantly in this case ^{226}Ra and daughters. Work in Germany (Ettenhauber and Lehmann, 1986), suggests that annual doses from external exposure from occupancy of buildings made from bricks or concrete are around 700 and 400 μSv, respectively. The estimates were made for bricks containing ^{226}Ra at 60 Bq kg^{-1} and concrete containing ^{226}Ra at 30 Bq kg^{-1}. These concentrations are of the same order as the typical values for concrete reported above. It is important to recall, however, that all building materials are radioactive to some extent, be it of natural or artificial origin; therefore, it is the incremental dose from building materials containing fly ash that is the important parameter. This incremental annual dose is around 200 μSv for building materials containing 30% ash; this corresponds to a concentration of ^{226}Ra of 30 Bq kg^{-1} and ^{40}K of 900 Bq kg^{-1} (Smith *et al.*, 2001). Radon contributes about 30%, with the remainder coming from external exposure.

Mining coal may also give rise to additional doses, due mainly to exposure to elevated levels of radon. Miners working in deep coalmines in the UK are estimated to receive an additional annual dose of about 600 μSv (Hughes and O'Riordan, 1994). A summary of doses from the coal-fired power station industry is provided in Table 7.3.

7.2.2 Other Forms of Energy Production

Other forms of energy production involving the combustion of fossil fuels can lead to the dispersion of natural radionuclides into the environment. Hedvall and Erlandsson (1996) have recently reviewed these processes.

Oil is a major fossil fuel. The concentration of natural radionuclides in mineral oils is lower than in coal and the amount of fly ash produced during combustion is also less (NCRP, 1987). However, oil-fired power stations tend not to have such extensive off-gas treatment systems as do modern coal-fired stations and so critical group doses from their discharges may not be too dissimilar to those from a modern coal-fired plant (~ 1 μSv per year (UNSCEAR, 1993)).

The main radiological problem associated with the oil industry, albeit a relatively minor one, is the presence of ^{226}Ra in scale that forms in pipework (Testa *et al.*,

Table 7.3 Annual doses from the coal-fired power industry.

Situation	Annual dose (μSv)[a]
Occupational exposure	
Coal mining	1000
Power station operation	10–100
Fly ash handling	10–100
Manufacturing building material from fly ash	10–100
Public exposure	
Discharges of fly ash to atmosphere	1
Storage of fly ash	1
Occupying a house made from materials containing fly ash	100

[a]Doses are representative of the maximums likely to be received in most situations and are rounded to orders of magnitude.

1994). This typically occurs with oil and gas fields formed in the Jurassic Period. It results from mineral impurities in the oil or gas which precipitate in pipework due to one or more of the following processes: injection of water into the well to enhance recovery of oil, pressure or temperature changes, and evaporation in the case of gas wells (Testa *et al.*, 1993). The scale is a precipitate of either barium/strontium sulfate or calcium carbonate with which radium salts have co-precipitated, so leading to the presence of ^{226}Ra in the scale. Radium-228 may also be present in a proportion reflecting the ratio of ^{238}U to ^{232}Th in the originating strata but higher members of the decay chains are usually present in only trace quantities in the scales. Thus, as ^{228}Ra has a fairly short half-life, i.e. 5.75 years, in comparison with ^{226}Ra (1600 years), its concentration will decline over a period of years as the scale ages. Concentrations of ^{226}Ra up to 4000 Bq kg^{-1} have been found in the past but with the introduction of scale-inhibition techniques current concentrations are typically in the region of 40–100 Bq kg^{-1}. The short-lived daughters of ^{226}Ra (^{222}Rn to ^{214}Po; see Appendix 3) rapidly grow into approximately the same activity as the parent in about one month. Actinium-228, the short-lived daughter of ^{228}Ra, grows in rapidly over a few days.

Natural gas is another fossil fuel that is used extensively in some parts of the world. Although the concentrations of uranium and thorium will obviously be very low, it can contain radon which has leached in from the surrounding strata; radon concentrations of up to 50 Bq l^{-1} have been reported (UNSCEAR, 1988). There appear to be no estimates available of the critical group doses from releases from gas-fired power stations but the additional annual doses to individuals using natural gas for cooking in the home have been estimated to be in the range 4–10 μSv. Slightly higher annual doses of up to 50 μSv may arise from the use of natural gas heaters in poorly ventilated situations (NCRP, 1987). These additional doses from the radon in natural gas are trivial when compared with the doses that arise from the seepage of this radionuclide into buildings from the underlying ground (see Chapter 1).

7.3 THE CONTROL OF EXPOSURES ARISING FROM THE INDUSTRIAL USE OF MATERIAL CONTAINING NATURAL RADIONUCLIDES

The controls placed on the releases of natural radionuclides from industries using materials that are naturally contaminated with significant amounts of natural radionuclides have typically not been as extensive as those placed on the nuclear industry, although there is no reason in principle why they should be treated any differently from one another.

Industries using these materials are often excluded or exempted from some or all regulatory requirements; this situation often arose because the industry was already in operation at the time when controls were being placed on releases from the emergent nuclear industry. In the UK, the accumulation and disposal of radioactive waste is controlled by the *Radioactive Substances Act*. However, certain naturally occurring elements up to specified concentrations are excluded from control. Furthermore, there are a number of exemption orders which exempt some naturally occurring radionuclides at higher concentrations from requirements of the Act. In the USA, the regulation of material containing natural radionuclides that is associated with the nuclear industry is regulated at a federal level, as is thorium-containing waste and waste from the mining of metals, while the decision whether to regulate other situations is left to the individual States.

Whether or not to regulate disposal of wastes containing elevated levels of natural radionuclides generated incidentally by industrial processes that are not involved in the nuclear industry or radionuclide industry is an issue currently under discussion at national and international levels. These industries, many of which have been operating for many tens of years, may not see the necessity for additional controls; a counter argument is that the health risks associated with exposure to radionuclides in the wastes may be little different to those associated with regulated wastes from the nuclear industry.

Some situations have caused particular concerns. For many years, phosphogypsum and ash from coal-fired power stations has been used as a component of building materials – no doubt saving natural resources in the process. The phosphogypsum (see Section 7.1.1) and ash (see Section 7.2.1) will, however, contain enhanced levels of natural radionuclides and thus the use of building materials containing such substances could cause exposure of members of the public. The issue is: what are the acceptable levels of artificially added natural radionuclides given that many building materials are naturally radioactive in their own right? Guidance on this situation has been issued in the European Union (EC, 1999). The guidance applies equally to natural building materials as well as to materials containing wastes from other processes. Essentially, building materials giving rise to annual doses less than 300 μSv from external irradiation would be exempted from any restrictions on radiological grounds. Restrictions would be placed on materials giving rise to greater doses, depending on the practicability of so doing, with the proviso that annual doses higher than 1000 μSv would be tolerated in only exceptional circumstances. This guidance has the benefit of practicability but it is not clear how the dose criteria were decided upon.

7.4 RESIDUES FROM PAST INDUSTRIAL ACTIVITIES

Processing of naturally occurring radioactive materials has taken place at many locations, world-wide. Over time, some of these industries have closed down while others have moved location. Processing residues and wastes can contaminate areas once occupied by these industries.

Extraction of uranium and other metals from uranium-rich ores has led to the accumulation of large quantities of naturally radioactive wastes at a number of locations. The extraction of uranium has been discussed in Chapter 6. Uranium-bearing ores are also a source of a number of other economically valuable metals – Radium is one. This was once used extensively in medicine and as a luminizing material (see Section 7.5.2). It has now largely been superseded in both of these roles by other materials and the industrial production plants have largely been closed down. Because the concentration of radium in ores is very low, between 10 and 400 t of ore was needed in order to produce 1 g of radium. Wastes were often simply dumped next to the processing plants. Most extraction of radium occurred at a plant at Olen in Belgium which accounted for about half of the 4.5 kg of radium produced in the world since 1898 (see Vandenhove, 2000). The Olen plant operated between 1922 and 1969. Liquid effluents from the factory were discharged into a nearby brook that, consequently, became contaminated with ^{226}Ra. Flooding of the brook caused contamination of several hectares of land. Solid wastes were dumped on an area of land of about 9 to 10 hectares within the boundary of the plant.

Contamination around the brook was very heterogeneous with some small areas exhibiting dose rates of 100 μSvh^{-1} (typical background dose rates above uncontaminated soils are around three orders of magnitude lower), having ^{226}Ra concentrations of up to 480 Bq kg^{-1} (the world-wide average for soils, etc. is 40 Bq kg^{-1}). The Belgian authorities established an action level of 0.15 μ Sv h^{-1} for remediation of these contaminated areas: soil from areas with higher dose rates would be removed and replaced. About 9700 m^3 of soil is involved (Vanmarcke and Zeevaert, 2000).

The on-site dump is fenced off and so the only routes for exposure of members of the public are from radon emanation. The highest annual dose from this source, at a dwelling near the dump, is around 4 mSv. However, ^{226}Ra has a half-life of 1600 years and so the dump will remain radioactive for thousands of years. If control over the dump is lost in the future, it is possible that future individuals could receive much higher doses if dwellings were constructed directly on top of it. Because of this possibility, options have been investigated for the treatment of the dump in order to minimize future exposures (Vanmarcke and Zeevaert, 2000). The most favourable option appears to be moving the 125 000 m^3 of waste to a nearby area and covering it with layers of sand and clay, together with a waterproof membrane. The total cost of restoration is estimated at 9 million Euros.

Other sites have been found to be contaminated with radium, including old watch factories and metal-processing plants (see Vandenhove, 2000). Unusually, perhaps, contamination has been found at some military and ex-military airfields in the UK. The reason is that in the 1940s and 1950s surplus aircraft and aircraft parts were often burnt as a means of disposal. In those days, aircraft dials were luminized with radium, which remains with the ash after burning. The dials contained between 6 and 880 kBq of radium. Ashes were either buried in shallow pits or were simply raked out over the surrounding ground, leading to radium contamination. This had occurred at RAF Carlisle in Cumbria, UK, where around 140 m^3 of material containing 350 MBq of activity had to be removed prior to release of the land for other uses (Gibbs, 1994).

The existence of sites contaminated by past human actions raises an important issue: how much clean-up should be undertaken? The most obvious answer is to clean up until the site has activity levels no higher than those in the surrounding uncontaminated environment, i.e. the ambient 'background' level. There are a number of problems with such an approach, as follows:

(i) It can be difficult to establish the 'background' level as this can vary naturally over quite small distances.

(ii) The extent of the implied clean-up may be large and could be disproportionate when the benefits in terms of risk reduction are taken into account.

(iii) The contaminated material that is removed has to be disposed of in a manner that would lead to lower doses to people, which usually means disposal in a purpose-built repository. Removal of material followed by dumping at another location merely transfers the problem!

In a radiological protection context, the problem can be viewed in the following way. If the contaminated land or property is not already in the public domain and the intention is to release it for redevelopment, the principles of protection for practices should be applied. In other words, future doses to members of the public should be assessed as being no higher than a dose constraint, which in turn should be a fraction of the dose limit for members of the public. Within this 'boundary' condition, radiological protection should be optimized: all reasonable steps should be taken to reduce doses to future individuals. If, however, contamination is found on land or property that is already in the pubic domain, the principles of protection for intervention should be applied. Again, all that is reasonable should be done to reduce doses to people but, in this case, the dose limits do not apply.

This radiological protection approach also has its drawbacks. People occupying property that is found to be contaminated could expect the same degree of protection as people who are going to occupy land that is to be released into the public domain. The fact that the dose limit applies in the latter situation, but not in the former, could result in individuals occupying land that is subsequently found to be contaminated being allowed to receive higher doses. Such equity considerations have led a number of authorities to apply dose limits and constraints in both types of situations.

7.5 RELEASES FROM RADIOISOTOPE PRODUCTION AND USE

Radionuclides and ionizing radiation have many uses in medicine and in general industry (see Longworth, 1998). The radionuclide sources have to be produced; a process that can result in discharges of radionuclides into the environment. The use of the sources, particularly in medicine, can also lead to some releases to the environment. The broad levels of discharges are described in Chapter 1; this present section discusses some of the particular uses of radionuclides in more detail, together with their associated discharges. First, the methods used to produce radionuclides for industrial and medical uses are briefly described. This is followed by a discussion of the industrial and medical uses of radioisotopes. Some of the industrial and medical uses of common radionuclides are given in Table 7.4.

7.5.1 Production of Radionuclides

Early applications made use of radionuclides isolated from natural sources. With the advent of accelerators and reactors, artificial radionuclides could be produced and these have found many and varied uses in industry and medicine. Over 10 000 radionuclide-based products are now in use.

The main means of producing these artificial radionuclides is by bombardment of target nuclei by neutrons or charged particles (protons, deuterons, alpha particles, etc.). Irradiation by neutrons is usually achieved by placing the target in the neutron flux generated by a nuclear reactor. The capture of the neutron is normally accompanied by emission of photons (gamma-rays). Examples of radionuclides produced in this way are ^{60}Co from ^{59}Co, and ^{24}Na from ^{23}Na. The product of a simple, single neutron capture is a radionuclide of the same element as the target and so it cannot be separated from the target by chemical means. This can limit the specific activity of the product. One way of producing radionuclides with high specific activities is to produce a radionuclide by neutron capture which then decays by beta-decay to yield a daughter that can be separated chemically. An example is the production of ^{125}I. This radionuclide has a half-life of 60 d and is used in clinical tests, and in the treatment of thyroid and prostate disorders. It is produced by the beta-decay of ^{125}Xe ($t_{1/2}$, 18 h), which in turn is produced by neutron capture of ^{124}Xe.

Many radionuclides of use in medical diagnosis and treatment are relatively short-lived. This feature is important in reducing unwanted exposures and for management of any wastes. However, it introduces problems if the radionuclide is to be used at places distant from nuclear reactors or accelerators. One way of overcoming these problems is to use a radioisotope *generator*. This comprises a longer-lived parent radionuclide which decays to the desired shorter-lived daughter. The parent is normally immobilized on a column, thus allowing the daughter to be eluted as required. A major example is the ^{99}Tc generator. Technetium-99m is a gamma-emitting radionuclide with a half-life of about six hours; it decays to ^{99}Tc ($t_{1/2}$, 213 000 years), and is used extensively in the radiopharmaceutical industry. The parent in the generator is ^{99}Mo ($t_{1/2}$, 2.75 d) which decays by beta-particle emission to ^{99m}Tc. The main production route for ^{99}Mo is fission of ^{235}U by thermal neutrons, followed by chemical separation from other fission products.

Particle accelerators provide a means of producing artificial radionuclides that are deficient in neutrons. Accelerators use potential difference to accelerate ions which then bombard a target material. The most common reaction is absorption of a proton followed by emission of a neutron. Examples are the production of ^{109}Cd,

Table 7.4 Radionuclides used in medicine, research and industry (based on information taken from IAEA, 1998).

Nuclide	Half-life[a]	Use
^{3}H	12.3 y	Luminous devices; research; clinical diagnosis
^{14}C	5.73×10^{3} y	Clinical diagnosis; research (particularly biomedical)
^{22}Na	2.6 y	Research; clinical diagnosis
^{24}Na	15 h	Clinical diagnosis; detecting leaks in pipes
^{32}P	14.3 d	Clinical therapy; biomedical research
^{35}S	87.4 d	Research
^{36}Cl	3.01×10^{5} y	Research
^{42}K	12.4 h	Biomedical research
^{45}Ca	163 d	Research
^{47}Ca	4.53 d	Study of cellular functions and bone formation
^{51}Cr	27.7 d	Research – mainly into blood cell survival
^{59}Fe	44.5 d	Clinical measurements; research
^{57}Co	271 d	Research; to diagnose pernicious anemia
^{58}Co	70.8 d	Clinical measurements; research
^{60}Co	5.3 y	In cancer treatment; to sterilize instruments, etc.
^{67}Ga	3.26 h	Medical diagnosis
^{75}Se	120 d	Biomedical research (protein studies); clinical measurements
^{85}Sr	64.8 d	To study bone formation
^{89}Sr	50.5 d	Clinical therapy (reduction of pain from bone cancer)
^{85}Kr	10.7 y	Leak testing; thickness gauges
^{90}Y	2.67 d	Clinical therapy
^{99m}Tc	6.02 h	In clinical diagnosis and research by binding to specific proteins and antibodies – produced from Mo-99 (half-life, 2.75 d)
^{111}In	2.83 d	Clinical measurements; research
^{123}I	13.2 h	In diagnosis of thyroid and other metabolic disorders
^{125}I	60.1 d	In clinical tests and in diagnosis of thyroid disorders
^{131}I	8.04 d	To treat thyroid disorders
^{192}Ir	73.8 d	To test pipes; in cancer treatment
^{127}Xe	36.4 d	Inert gas used in lung function studies
^{133}Xe	5.24 d	Inert gas used in lung function and blood flow studies
^{147}Pm	2.62 y	Luminous devices; thickness gauges
^{169}Er	9.4 d	Biological research
^{198}Au	2.69 d	Clinical therapy
^{197}Hg	2.67 d	Clinical measurements
^{203}Hg	46.6 d	Biological research
^{201}Tl	3.04 d	In cardiology and in cancer detection
^{204}Tl	3.8 y	Thickness gauges
^{232}Th	1.4×10^{10} y	Gas mantles; welding electrodes
^{238}Pu	87.7 y	Power sources for satellites
^{241}Am	432.7 y	Smoke detectors

[a]h, hour; d, day; y, year.

which is used as a source of soft X-rays, and ^{56}Co, which is used for measuring wear in steel components.

7.5.2 Industrial Uses of Radionuclides

There are many uses for radionuclides in industry (see Eyre, 1996). One of the first uses was in the production of radioluminescent dials and markers; the energy of the radiation is converted into light by a phosphor. The first radionuclide to be used in this way was ^{226}Ra although it has now been superseded by ^{147}Pm and tritium – both less radiotoxic. It is worth recalling that some of the earliest effects of ionizing radiation were seen in the individuals who painted dials with radium. Significant quantities of ^{226}Ra were used in luminizing aircraft dials during the Second World War and areas where surplus aircraft were disposed of may contain measurable quantities of this radionuclide. Furthermore, the history of many old radium luminizing factories is poorly documented and industrial sites contaminated with this isotope still turn up from time to time.

Typically, the quantities of ^{226}Ra in watches were around 3700–11 000 Bq. Aircraft dials could contain much larger amounts, up to around 70 000 Bq (Kathren, 1984). Radon-226 and its daughters emit gamma-radiation that can irradiate the wearer. The annual dose to the area of skin under a typical radium-luminized watch has been estimated as 24 mSv if the watch is worn continuously (Kathren, 1984); the corresponding dose from a pocket watch is higher at about 1 Sv (NCRP, 1987). The latter dose is high but below the level where effects on the skin would be seen. The fact that wearers were receiving doses, together with the increased availability of replacements due to the growth of the nuclear industry, led to the abandonment of ^{226}Ra for luminizing. Production of ^{226}Ra-luminized wristwatches ceased in most countries towards the end of the 1960s but production of clocks continued for a longer period. It has been estimated that around 9 million radium-luminized timepieces were sold annually in the USA between 1974 and 1977 (NCRP, 1987). The number of radium-luminized wristwatches in circulation prior to the end of the 1960s must have been considerably larger. Since the early 1970s, ^{226}Ra has been replaced as a luminizing agent by the weak beta-emitters ^{147}Pm and tritium. The watch casing stops the low-energy beta-particles emitted by these radionuclides. However, in the case of tritium, leakage can result in radiation exposure via inhalation and absorption through the skin; the corresponding annual dose is estimated to be about 1 μSv (Hughes and O'Riordan, 1994). Promethium-147 can give rise to exposures via 'bremsstrahlung' and from gamma-emitting impurities; the annual doses from a pocket watch luminized with this radionuclide and kept in the hip pocket for 16 hours a day is estimated to be about 0.35 mSv to the skin and less than 1 μSv to the whole body (Hughes and O'Riordan, 1994). As well as the benefits, tritium and ^{147}Pm have some disadvantages when compared with ^{226}Ra; their much shorter half-lives and weak emissions mean that larger amounts have to be used to produce luminescence.

The ionizing properties of radiation have been used in anti-static devices which reduce the buildup of electrical charge on certain materials; explosives and photographic film are examples. Polonium-210 has been used in this role in quantities up to

37 MBq per static eliminator (NCRP, 1987). The short half-life of this nuclide, i.e. 138 d, is a disadvantage and both ^{98}Sr and ^{241}Am have been used in the past.

Another use of radionuclides which makes use of ionization of air is in spark-gap indicators; the purpose of these is to facilitate formation of electric ignition sparks. Cobalt-60 has been proposed as the source in amounts up to about 37 000 Bq per spark generator (NCRP 1987). For similar reasons, some fluorescent lamp starters contain small amounts of thorium. The purpose of the thorium is to produce ionization within the starter; a typical starter may contain about 185 mBq of thorium (NCRP, 1987).

A major use of radionuclides is in thickness, level and density gauges. Two main techniques are used, i.e. transmission and back-scattering. Both beta-emitters and gamma-emitters are used, depending upon the type of application. However, the majority of applications involve beta-particle transmission or back-scattering.

In the transmission technique, the source is placed on one side of the sample with the detector on the opposite side. The attenuation of the beam can be correlated with, for example, the thickness of the material. The equations governing attenuation of radiation beams are described in Chapter 3. In the back-scattering technique, the detector is placed on the same side of the sample as the source to detect back-scattering; this technique is mainly used to determine the thickness of coatings. Radionuclides used as sources include, ^{63}Ni, ^{14}C and ^{204}Tl.

7.5.3 Medical Uses of Radionuclides

Radionuclides are used extensively in medicine for diagnosis and treatment. A survey carried out in the UK in 1988/89 showed that around 430 000 administrations of radiopharmaceuticals were carried out. Proportionally similar figures are likely to apply in other western countries. The most frequently used radionuclide reported in the UK study was ^{99m}Tc which accounted for about 70% of the total number of procedures. Some uses for individual radionuclides are given in Table 7.4.

Very broadly, the use of radionuclides in diagnosis falls into three groups, as follows:

(1) The radionuclide is an isotope of a naturally occurring element whose metabolism can be used to investigate a particular medical condition.

(2) The radionuclide is an isotope of an element that can mimic a particular element in the body.

(3) The radionuclide is linked to a biologically active molecule to 'label' it.

The radionuclides are typically gamma-emitters and their distribution in the body is followed by a gamma-camera. An example is given below but this is only one from a myriad of uses.

Technetium-99m can be complexed with a *lopophilic* (fat-loving) chemical, hexamethylpropytenamineoxine, which in turn will bind to leucocytes (white blood cells that help to fight infection). If an individual is suspected of having an acute infection, leucocytes can be separated from his or her blood, labelled with the technetium complex and then re-injected. The leucocytes will concentrate in any regions of infection, thus enabling the area to be identified by using a gamma-camera.

In the above example, the specificity of the diagnostic test relies upon the biological properties of the leucocyte; there are many other diagnostic tests that use ^{99m}Tc as the detectable label but where it is linked to other biologically active materials which confer the specificity upon the particular test.

The objective of using radionuclides for diagnostic purposes is to administer enough radionuclides to enable an adequate level of detection, while at the same time keeping radiation doses as low as reasonable. The objective in treatment, however, is to deliver as much radiation as is needed to the target tissue, usually a malignancy, while at the same time keeping the radiation dose to healthy tissue as low as possible. In treatment, radiation is being used to destroy diseased or aberrant cells, or at least severely impair their function. A problem is to avoid at the same time damaging healthy cells.

The techniques that use radiation in treatment can be summarized under three broad headings, as follows:

(1) *External Beam Therapy*. This technique uses a radiation beam, which is generated outside the body and is targeted on the tissue to be destroyed. A variety of means is used to generate the beams, including X-ray tubes and accelerators. Radionuclide gamma-ray sources are sometimes used; these have included ^{226}Ra, ^{137}Cs and ^{60}Co. The latter is the most commonly used radionuclide for this purpose.

(2) *Brachytherapy*. In this technique, a small sealed source of radionuclide is placed in, or in close proximity to, the malignant tissue in order to irradiate it. The sources can be in the form of wires, needles, tubes, or discs. The radionuclides used include ^{192}Ir, ^{137}Cs, ^{125}I and ^{90}Sr. In the past, ^{226}Ra was also used for this purpose.

(3) *Radiotherapy*. This technique uses the elemental properties of a radionuclide or the biological properties of a material to which the radionuclide is linked, in order to target the radionuclide to the desired tissue. One of the most well known examples of this technique is the treatment of hyperthyroidism (excessive activity of the thyroid) with ^{131}I. Iodine is used by the thyroid in the production of the hormone thyroxine and so the element concentrates in this tissue; the radiation dose from administered ^{131}I is intended to produce the required destruction of a proportion of the thyroid tissue. A rapidly advancing research area is the use of tissue-specific antibodies to deliver radionuclides to cancerous tissue in order for the radiation dose to kill cancerous cells with minimal damage to healthy tissue. The technique depends upon production of antibodies that bind specifically to cancer cells. Radionuclides that have been used in this technique include ^{32}P, ^{67}Cu and ^{186}Re. Relatively short-lived radionuclides are used. To be most effective, the half-life of the radionuclide should be of the same order as the half-life of the binding of the antibody to the cancer cells; if it is longer, healthy tissue may be irradiated during any subsequent redistribution in the body, whereas if it is shorter, efficient use is not being made of the time of binding.

7.6 ACCIDENTS

There have been a number of accidents in the non-nuclear power industry. The majority of these accidents have involved over-exposure to radiation sources and are described in IAEA (1996d). A few accidents in the non-nuclear sector have resulted in the dispersion of radionuclides into the environment; the worst of these occurred in the town of Goiania, Brazil, in 1987. Goiania is a city of around one million inhabitants, about 170 km to the south east of the capital, Brasilia.

The IAEA has published a comprehensive account of the accident (IAEA, 1988) to which the interested reader is referred. Briefly, the accident involved a teletherapy unit containing 50.9 TBq of ^{137}Cs in the form of a powder of the soluble chloride salt. The unit had been used by a private radiotherapy institute but when the institute moved premises in 1985, it was left in the abandoned building; the regulatory authority was not notified of this fact. The building was partially demolished. About two years later, in September 1987, the building was entered by two local people looking for saleable scrap; they removed the source from its protective housing and took it home where an attempt was made to dismantle it. The source capsule was ruptured in the process and some of its contents began to be dispersed into the environment. The remains of the source were sold to a scrap metal merchant who noticed that it glowed in the dark (this phenomenon was possibly due to Cherenkov radiation from beta- and gamma-rays passing through a layer of absorbed water); relatives and friends came to see and fragments of the source were distributed among several families.

After around five days, some individuals fell ill with gastrointestinal problems. Although it was not realized at the time, this was the gastrointestinal syndrome caused by high doses of radiation (see Section 4.1). One individual associated the illness with the scrap source and took its remnants to the public health department. One physician and one of the technicians suspected, rightly, that the scrap was radioactive. A medical physicist was called in to investigate and the scale of the accident soon began to unfold.

About 110 000 people were monitored for contamination. Signs were found on 249, of which 104 were contaminated internally. Deterministic effects (see Section 4.1) were observed in a number of individuals; fourteen showed bone marrow depression and eight developed acute radiation syndrome. Twenty-eight people had localized skin burns caused by radiation. Four people died from deterministic effects (Oliveira *et al.*, 1991). Doses must obviously have been high, of the order of grays to cause such effects. Others, not exhibiting deterministic effects, may nevertheless have received relatively high doses; for example, a few people at the public health department probably received doses of a few hundreds of millisieverts. A significant area of the city was contaminated and this had to be dealt with. Initially estimates showed that an area of about 1 km^2 was contaminated significantly and activity was being dispersed by the action of wind, etc. Clearly, it was not possible to remove every Bq of ^{137}Cs; the authorities had to decide which areas to decontaminate. This was decided on the basis of dose to people. Decontamination was carried out if the dose in the first year was higher than 5 mSv or the average lifetime dose was 1 mSv per year. It was found that the first criterion was the most important (Rochedo, 2000). Significant contamination was found in 85 houses, 41 of which were evacuated. Decontamination involved washing buildings, removing contaminated items and internal surfaces, and removing walls, roofs, paved surfaces and topsoil. Several buildings had to be completely demolished. Decontamination took about six months and the quantity of radioactive waste produced totaled around 3500 m^3. The activity in the waste is estimated at 44 TBq, compared with the total activity originally in the source of 50.9 TBq.

Radiation doses to the $\sim$ 750 individuals involved in recovery from the accident were estimated; about 70% had doses of less than 1 mSv, with the highest dose being 16 mSv.

A remarkably similar accident had occurred in Mexico in 1983. A second-hand radiotherapy unit, containing 37 TBq of ^{60}Co, was imported into Mexico in 1977. The purchasers, a medical centre in the city of Jaurez, failed to notify the regulatory authorities and so regulatory control over the source was never established. For various reasons, the source was not used and was stored in a warehouse. In 1983, a technician, who worked at the medical centre, attempted to dismantle the source in order to sell it for scrap. Most of the activity went as scrap metal to a number of foundries where it was incorporated into steel for concrete reinforcing bars and table legs. Some activity contaminated areas in the vicinity of the scrap yard.

By a strange twist of fate, the accident came to light when a lorry carrying some of the reinforcing bars passed close to Los Alamos nuclear laboratory in the USA and triggered the site radiation alarms. The origin of the activity was traced back to contamination of scrap metal and a major survey was initiated to find where the contaminated metal had gone. In Mexico, 814 houses were demolished because they contained contaminated reinforcing bars; in the USA, a search was made for contaminated table legs and some 2500 contaminated items were returned to Mexico for disposal. The total active waste was 16 000 m^3 of soil and 4500 t of metal. There were no fatalities from deterministic effects, although a handful of individuals did receive doses of up to 7 Sv.

There have been a number of other incidents of radioactive sources being accidentally smelted with scrap steel. A survey in 1998 reports 49 confirmed accidental smeltings of radioactive sources world-wide (Lubenau and Yusko, 1998) although the number of instances where discrete radioactive sources have been found with scrap metal is much higher.

Occasionally, the accidental smelting has gone unnoticed for some time. For example, the smelting of a ^{137}Cs source with scrap metal at a steel mill in Quebec in 1985 came to light when furnace dusts were sent for processing in the USA to recover nickel. When the dusts were received, they were found to be radioactive and the origin of the activity was traced back to the Quebec steel mill.

As well as the Mexican incident described above, there have been other instances of the metal product being contaminated by radionuclides. The most common contaminating radionuclide is ^{60}Co, while the other common radionuclide in sealed sources is ^{137}Cs which, unlike cobalt, does not tend to follow the steel during smelting but instead volatilizes and contaminates furnace dusts. One of the incidents giving rise to the most widespread contamination occurred in Taiwan. The actual circumstances surrounding the contamination of steel produced in Taiwan have not been established but it is clear that around 1982–1983 one or more ^{60}Co sources were accidentally smelted in a scrap steel plant in Taiwan. Some of the contaminated steel was used for steel reinforcing bars in buildings. By 1996, around 100 buildings with elevated levels of gamma-radiation due to ^{60}Co contamination had been identified in Taiwan although this probably represents only about one third of the contaminated steel with the remainder still to be found (Chang *et al.*, 1997). Around 4000 people are estimated to have received annual doses of over 1 mSv from the known contaminated steel.

The possibility of contamination has led the scrap metal industry in many western countries to install radiation monitors at the entrances to scrap metal yards.

8

Instrumentation for Radiation Detection and Measurement

The interaction of radiation with matter has already been discussed in Chapter 3. The processes involved, however, form the basis of all methods of detecting radiation and therefore will briefly be summarized here.

Energetic charged particles (electrons or positive ions such as an alpha-particle) give up their energy as they pass through the material by interacting electromagnetically with atomic electrons, causing both ionization and excitation. Gamma-rays also interact with the atomic electrons of matter and in doing so give rise to fast, secondary electrons which can then, in turn, give rise to excitation and ionization in the molecules of the matter in which they are produced. These interaction processes (usually the ionization process) then make possible the detection of the radiation present. It should be pointed out that the term 'ionization' used here also includes the formation of electron-hole pairs in semiconductor detectors (see Section 8.1.3).

In general, two types of detector are normally used for radiation measurements: the first of these simply produces an output which is a function of the number of particles or photons entering the detector and so acts just as a counter, while the second not only responds to the number of particles or photons passing through it but is also able to measure the energy of the particle or photon. The latter is more useful but normally requires much more expensive and sophisticated associated electronics.

Detectors of radiation may be based on ionization in either gases, liquids or solids. When a neutral molecule is ionized, the process results in the production of a positively charged ion and an electron, known as an ion-pair and the collection of these ion pairs produces an electrical pulse or current. Measurement of the pulse or current then provides the basis for the determination of the amount of radiation present and possibly the energy distribution of that radiation. Detectors which essentially rely on excitation processes rather than ionization (although ionization may proceed the excitation in some cases) are the scintillation detectors in which the kinetic energy of charged particles is converted into light pulses or scintillations. Finally, there are detectors which rely on the chemical and physical effects brought about by the passage of ionizing radiation through certain materials such as the production of images in photographic emulsions or the damage tracks produced in certain types of plastic or minerals. Since neutrons are not normally met with in environmental radiation, no mention of the special techniques utilized in their detection will be made here. For further information on detectors and instrumentation, the interested reader is referred to the text by Knoll (1979, 1989).

8.1 TYPES OF DETECTOR

8.1.1 Gas-Filled Detectors

This class of detector represents one of the oldest design concepts in radiation measurements and has been in use for over 50 years. Although there are several different types of gas-filled detector, their fundamental properties are similar. All of them operate on the principle that the energy of the incoming radiation is converted in part to the production of ion-pairs in the gas and the charges so produced (both negative and positive) are collected by applying an electric field across the gas. This field causes the electrons to move towards the positive electrode, i.e. the anode, and the positive ions to move more slowly to the negative electrode, i.e. the cathode. Their main operational difference comes from the magnitude of the voltage applied between the two collecting electrodes. The different regions of operation as a function of applied voltage are illustrated in Figure 8.1.

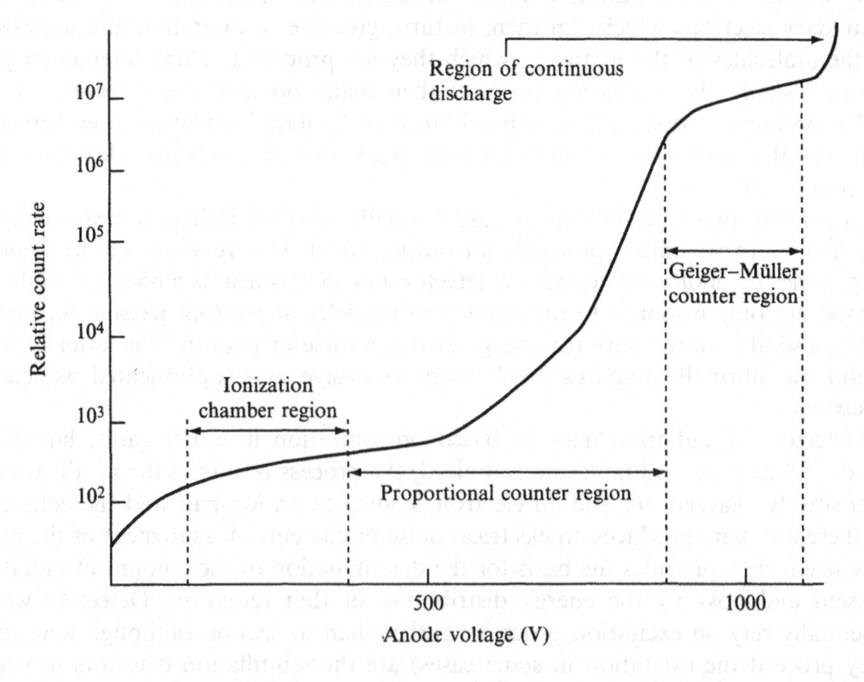

Figure 8.1 Relative count rate of a gas-filled detector as a function of the anode voltage

Ionization chambers

These devices are operated in the voltage region which is sufficient to produce a field that allows for the collection of all of the ions produced. The collection of these charges constitutes an electric current. Since all of the ions are collected for a given energy and level of radiation, the output current from the device should be constant. Measurement of this ionization current is the normal mode of operation of these devices. The voltage required to provide proper operation of the detector depends on the detailed detector design and the type of radiation being measured but is of the order of 10–100 V.

Ionization chambers are fairly simple in both construction and operation (see Figure 8.2). In their simplest form, they consist of a cylinder of gas with an anode wire running along the axis and the outer casing operated at ground potential. The gas filling is normally air, although other gas fillings may be used depending on the particular application. The radioactive sample is placed outside the window of the chamber or in the case of a radioactive gas, e.g. radon, the gas can be incorporated into the gas filling.

Ionization chambers are normally operated in current mode and are referred to as current–ion chambers. As a result, they 'integrate' events and provide information about the average flux of radiation impinging on the detector.

The very small ion current produced ($\sim 10^{-12}$ A) requires very careful measurement. These small currents require very sensitive devices for their measurement. The current can be measured indirectly, using either an electrometer to provide an amplified voltage or alternatively to connect the chamber to a charge-measuring device, where the charge collected in a given interval gives a measure of the radiation exposure in that interval. Used in this manner, these devices find a place in the measurement of environmental radioactivity, such as for measuring gamma-ray exposure, in radiation survey instruments and also as personal dosimeters.

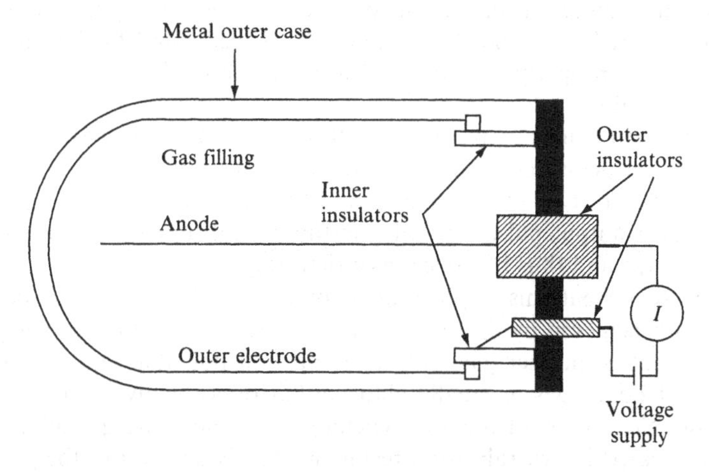

Figure 8.2 Cross-sectional view of a cylindrical ionization chamber showing the basic circuit required to measure the current, I

Geiger–Müller counters

Reference to Figure 8.1 shows that as the voltage is increased between the two electrodes in a gas-filled detector a point is reached at which not only are all the ions formed collected but that the electrons of the ion-pairs are given sufficient acceleration that on colliding with unionized gas molecules they cause ionization of these molecules. The electrons released in this secondary ionization will also be accelerated and will themselves bring about further ionization of the gas molecules. This gas multiplication process is known as an *avalanche*. In the zone designated as the proportional region of Figure 8.1, this avalanche is terminated when all of the ions and electrons are collected at the electrodes. If, however, the voltage is raised even further, into the Geiger–Müller region, then additional effects take place which cause separate avalanches to occur in other regions of the gas or from the walls of the anode, when the electrons strike this. The result is that the initial avalanche spreads throughout the whole volume of the gas within a few microseconds. Consequently, a single ionization event will result in a Geiger discharge in this voltage region and the size of the electronic pulse generated is essentially constant, irrespective of the energy of the radiation.

While the electrons are travelling very rapidly towards the anode, the positive ions formed, being much heavier, travel far more slowly towards the cathode, where they are effectively neutralized.

The current pulse from the avalanches of electrons produces a fast-rising voltage pulse in a series resistor chain. This voltage pulse is normally large enough that it requires no further external amplification and can operate a counting device, e.g. a scaler, directly. For this reason Geiger–Müller (G–M) detectors are relatively cheap and simple to operate.

During part of the period between the initial ionization event and the drift of the positive ions towards the cathode, the detector is unable to respond to further ionizing radiations due to the reduction in the field gradient caused by the sheath of positive ions around the anode. The detector is 'dead' during this period and this is therefore referred to as the *dead time* of the detector and further incident radiation will not be properly recorded. Clearly, the dead time limits the maximum count rate of a tube.

When the positive ions arrive at the cathode, they are neutralized but energy is released in this process and it may be large enough to release a free electron from the cathode surface. This free electron could then trigger another Geiger avalanche and produce an output pulse. Indeed, the tube could now go into a state of continuous discharge and thus be useless as a detector.

In order to prevent this happening in practice, the tube has to be *quenched*. Usually, this is achieved by the addition of a *quenching agent* to the gas filling. The quenching agent is another gas with an ionization potential less than that of the main gas. A halogen gas, normally chlorine but occasionally bromine, is used for this purpose. Now, most of the ions reaching the cathode are quench gas ions and the energy released in neutralization results in the dissociation of the gas molecules rather than release of an electron. Later, the dissociated atoms recombine, thus maintaining the quench gas. Halogen-quenched tubes have a minimum count life in excess of 5×10^{10} counts.

The normal filling gas for a G–M tube is either helium or argon but heavier noble gases are used for certain applications. It is essential to remove all traces of oxygen-containing gases.

All of the major forms of radiation – alpha- and beta-particles, X- and gamma-rays – encountered in the environment can be detected using a G–M detector. As we mentioned earlier, the detector cannot be used for energy discrimination of radiation, although a crude discrimination between particle types can be made. For both alpha- and beta-particles, ionization occurs by direct interaction of these particles with the (main) gas atoms. In the case of X-rays and low-energy gamma-rays, the electrons are produced by the mechanisms of the photoelectric effect or Compton scattering (see Chapter 3) in the gas. For high-energy gamma rays (energy > about 100 keV), the secondary electrons are produced mainly in the cathode material and if the processes occur close enough to the inner surface that the electrons produced can enter the gas, they can then initiate the Geiger avalanche.

Alpha-particles and low-energy beta-particles are very easily absorbed in matter. The range of a 5 MeV alpha-particle in air is about 35 mm but only about 0.02 mm in aluminium. Consequently, if the alpha-particle is to enter the sensitive volume of the detector a very thin window of a suitable material must be incorporated. This usually takes the form of a very thin mica window positioned at one end of a cylinder. Such a configuration is known as an *end-window detector* and is the most typical form for a G–M tube. The lowest alpha-energy normally encountered is about 4 MeV and the range of such an alpha-particle in mica is some 0.0125 mm. Consequently, for such a particle to pass through a window and still have sufficient energy to ionize the gas, the windows must be even thinner than this. A window thickness for an alpha-detector is of the order of 0.005 to 0.007 mm, or in the more usual units used by nuclear scientists, 1.5 to 2.0 mg cm^{-2}.

Ranges for beta-particles are based on the end-point energy, E_{max} (see Chapter 3), and are greater than for alphas of equivalent energy. Thus, a 5 MeV beta would have a range of about 9.5 mm in aluminium. For isotopes with low end-point energies, e.g. ^{14}C ($E_{max} = 155$ keV), ^{35}S ($E_{max} = 167$ keV) and particularly tritium ($E_{max} = 18$ keV), very thin windows must still be used but the constraints on window thickness are much reduced when counting beta-particles with energies greater than a few hundred keV. Indeed, beta-counters may be constructed without the use of a window at all. Figure 8.3 illustrates the relationship between beta end-point energy and absorption in windows of varying thickness.

For X- and gamma-rays, being far more penetrating than even energetic beta-particles, absorption in windows is not a problem. The problem here is one of efficiency of production of secondary electrons to initiate the ionization event. In the case of X-rays, with energies usually of less than 40 keV, secondary electrons are produced mainly in the gas. Since the probability of a photon reaction producing a secondary electron increases rapidly with atomic number (Z), heavy gases, such as argon or krypton, are used and the tube is presented end-on to the radiation and is as long as possible.

G–M tubes constructed essentially to count gamma-rays are normally made with metal walls of thickness in the range 250–1600 mg cm^{-2}. Again, the efficiency of conversion of photon energy into secondary electrons increases with Z, so that a

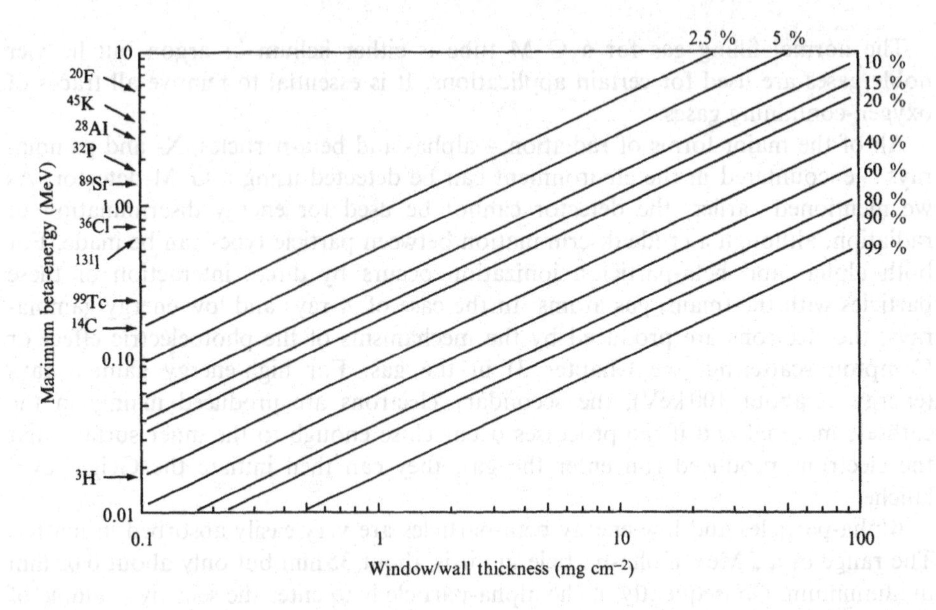

Figure 8.3 The percentage absorption of beta-particles as a function of the window or wall thickness of a Geiger–Müller counter for various end-point energies. The latter energies of some common beta-emitters are shown on the beta-energy axis. Reproduced by permission of Philips Electronics UK Limited from Technical Information Literature No. 40, published by Mullard Limited

metal such as bismuth ($Z = 83$) has been widely used in the past as the cathode material for G–M tubes for gamma-ray counting.

Construction of Geiger–Müller tubes

The basic design of a Geiger–Müller tube is similar to that of an ionization chamber, namely two electrodes in a gas-filled envelope. The most usual form is again a cylinder (acting as the cathode) with a window sealed into one end. The anode is then normally a thick wire mounted along the axis of the cylinder. This type of tube is known as an end-window Geiger tube (Figure 8.4) and is the type most commonly used. Tubes for X- and gamma-ray counting are usually metal-walled and filled with a high-atomic-number gas. Gamma-ray counting efficiencies are normally very low.

Proportional counters

Referring to Figure 8.1 again, we have so far covered detectors which operate at the two extremes of the voltage range, i.e. ionization chambers at the low-voltage end and G–M tubes at the high-voltage one. In between these two extremes is the long voltage region known as the *proportional region*. This is the only voltage region in which any realistic energy discrimination may be made. The reason for this is that in this voltage region gas multiplication occurs but is linear with voltage.

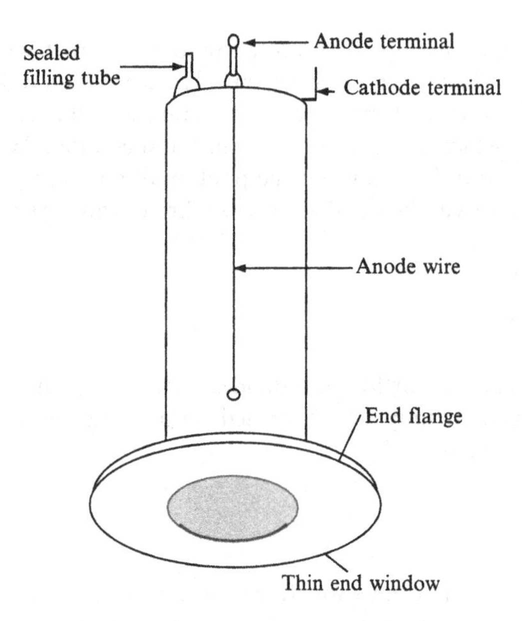

Figure 8.4 Schematic diagram of an end-window Geiger–Müller tube

Consequently, the charge collected will depend on the initial number of ion-pairs produced in the gas and for a given gas this depends only on the energy of the radiation and hence the pulse size will be proportional to the energy of the radiation.

Conventional proportional counters are widely used in nuclear physics but have relatively little application in environmental radiation measurements, particularly with regard to photon counting. Although they have the potential for spectrometry of X-rays or soft gamma-rays, they are rarely used for this type of work due to difficulties in interpreting the spectra and their very low counting efficiency. Consequently, if gamma-ray spectrometry is required then a semiconductor detector (or occasionally, a scintillator detector) would be used, as discussed in the next sections. For simple gamma-ray detection, a G–M tube would be used in preference to a proportional counter because of the greater simplicity of operation.

Proportional counters do have a role in the measurement of alpha and low-energy beta activity. This is mainly in the calibration of sources where the absolute activity of the source is required but such measurements are normally only performed in specialized laboratories.

8.1.2 Scintillation Detectors

Scintillation detectors depend on the fact that when certain solid or liquid materials, called phosphors, are exposed to ionizing radiation, excitation of some kind occurs in the material and de-excitation of the material results in the emission of visible radiation. The prompt emission of visible radiation is termed *fluorescence*. The flashes of light produced are of very short duration and low intensity and are not measured directly. Instead, they are allowed to interact with a photocathode which results in the release of photoelectrons. These are then multiplied in a device known as a photomultiplier tube. The light quanta are thus converted into pulses of electrical charge, which can then be amplified and counted. We shall look first at the scintillator itself and then consider the photomultiplier tube.

Scintillator detectors can be divided into two broad categories, as follows:

- solid scintillators
- liquid scintillators

Solid scintillators can be further sub-divided into (i) organic, and (ii) inorganic scintillators. We shall mainly be concerned with inorganic scintillators but will briefly consider organic ones.

Solid scintillators

Organic scintillators Light emission from these materials results from molecular transitions and the maximum wavelength of emission of most common organic scintillators is around 430 nm. Today, organic scintillators are mainly represented by plastics. These materials are cheap, can easily be fabricated into almost any shape and allow the production of extremely large detectors.

In environmental studies they have limited application in alpha- and beta-counting. By using what is known as *pulse-shape discrimination*, it is possible to discriminate between the pulses from alpha- and beta-particles and it is also possible to count very low levels of both radiations using specific plastic detectors. In fact, survey instruments fitted with a plastic detector have approximately twice the efficiency of a G–M tube for beta-particles. Plastics are not normally used for photon counting and almost all gamma-ray counting involving scintillators is carried out by using inorganic materials.

Inorganic Scintillators This is by far the most important group of solid scintillators and includes materials for counting alpha- and beta-particles, as well as X- and gamma-rays. The detection of gamma-rays, however, is the most important application of this group of materials. A wide range of substances are now commercially available as inorganic scintillators. These usually contain at least one and some of them two elements of high atomic number. It is the presence of these elements which provides the scintillator with its relatively high efficiency, even for high-energy gamma-rays. Secondly, many of the materials are 'doped', i.e. small amounts of a foreign ion have been deliberately introduced into the main crystal lattice. For example, the most commonly used, NaI(Tl), is a NaI crystal doped with small amounts of thallium ions. This introduction of an impurity ion has important effects on the quality of the light output from the scintillator. The impurity ion is called the *activator*.

The scintillation mechanism in these activated inorganic materials is quite different from that in organic ones. When ionizing radiation passes through the scintillator, electrons are promoted from a lower to a higher, excited state. Most of the electrons in the crystal are initially in the lower state. The return of the electron to the lower state results in a fraction of them emitting a light photon. The activator promotes these transitions and also shifts the wavelength of the emitted radiation into the visible range which better matches the requirements of the photomultiplier tubes used in conjunction with the scintillator. This shift in wavelength also means that the crystal is essentially transparent to the light emitted. It is important that the light output, i.e. the number of light photons emitted, should be a linear function of the energy of the radiation. Most scintillators exhibit such behaviour and thus provide a linear response with the energy of the radiation. The energy resolution of a scintillator is set by the fluctuations in the number of light quanta generated by radiation of a specific energy, which varies with crystal type. We shall discuss energy resolution more fully later in this chapter.

The major disadvantage of NaI(Tl) is that the crystal is hygroscopic and so has to be kept perfectly dry and cannot be exposed to the atmosphere. Consequently, it has to be kept permanently inside a suitable can and so cannot be used in 'windowless' mode. The scintillator is normally used in the form of a cylinder but unusual shapes can be fabricated by pressing small crystals together. An important configuration for NaI(Tl) detectors is the *well-type* detector. Here, a hole is drilled centrally through the crystal and the sample is placed inside this well. This provides optimum efficiency of counting but with some loss of resolution. Obviously, only samples small enough to fit in the well can be counted in this way.

Other types of inorganic scintillators are occasionally used in environmental studies such as CsI(Tl), CaF_2(Eu) and bismuth germanate (BGO), $Bi_4Ge_3O_{12}$. The latter has achieved some prominence in recent years as a scintillator, mainly due to its high efficiency for detecting gamma-rays, particularly high-energy ones, due to the presence of the very-high-Z element, bismuth.

Liquid scintillators

In this technique, instead of placing the radioactive source near or on the scintillator, as would be the case for solid scintillators, the source is actually mixed with the scintillator. This has a number of advantages for counting radioactive materials but also poses some problems. The advantages are that it provides (i) essentially a 4π-counting geometry and so maximum geometrical efficiency of counting, (ii) a 'windowless' counting arrangement and so no window absorption for low-energy beta- and alpha-particles, and (iii) essentially zero sample self-absorption. The problems stem from having to incorporate the source in the liquid scintillator. The most obvious problem is that this usually means that the source itself must be a liquid, although techniques have been developed for counting solids (such as filter papers) containing the radioactivity. In general, however, liquid sources are preferred whenever possible. The other problem results from the requirement that the source be mixed with the scintillator, implying that the two liquids are miscible. Since the scintillator is an organic liquid (or, rather a mixture of organic materials), this implies miscibility only with organic sources. In many cases and certainly in many areas of environmental counting, the sources will probably be inorganic materials and this obviously poses problems for directly incorporating the source in the scintillator. We shall return to this problem in the next chapter.

The scintillation process in liquid scintillators Liquid scintillators, being organic compounds, produce their fluorescence in basically the same manner as solid organic phosphors, i.e. by transitions in the electronic levels of the organic molecule. The scintillator solution, however, contains at least two components, i.e. (i) a solvent, and (ii) the scintillator solute or phosphor. Many scintillator solutions also contain a number of other components such as additional scintillator solutes and/or solvents to increase the miscibility between a radioactive sample and the primary solvent. This mixture of substances is usually referred to as a *scintillator cocktail.*

When ionizing radiation passes through the basic solvent–solute system, most of the energy of this radiation is transferred to the solvent molecules which subsequently transfer a fraction of this energy to the solute (scintillator) molecules. The final process is the release of photons which are characteristic of the solute.

The components of a scintillator cocktail

The solvent plays a crucial role in the scintillation process and is a basic component of all scintillator cocktails. Solvents are classified as primary or secondary, depending on the relative amounts present and also their role in the scintillation process. Primary solvents are the primary absorbers of radiation energy, while secondary solvents are used to increase the transfer of energy from the primary solvent to the scintillator solute. The best solvents are aromatic substances, of which toluene is probably the most widely used.

Scintillator solutes again are classified as either primary or secondary solutes and the use of secondary solutes is quite widespread. The role of the primary solute is to produce light photons by de-excitation following interaction with an excited solvent molecule. A secondary solute may sometimes be used when a sample to be counted contains some component that strongly absorbs the light emitted by the primary solute. Then, the secondary solute is used so that the light emitted is at a wavelength not absorbed by this component.

A problem in liquid scintillation counting arises from samples that cause quenching of the light output and so reduce the pulse size ultimately produced by the detector. This is probably the most serious problem associated with this type of counting and will be discussed further in the next chapter.

Photomultiplier tubes

In all forms of scintillation counting, whether it involves the use of solid or liquid scintillators, the light scintillations have to be converted into electronic pulses to allow the radiation to be counted. This conversion of light pulses into equivalent electronic ones is achieved by the use of a *photomultiplier tube*. The combination of scintillator plus photomultiplier tube is known as a *scintillation counter*.

There are three major components of a photomultiplier tube, i.e. (i) the photocathode, (ii) the dynode chain, and (iii) the anode. Their arrangement in a typical photomultiplier tube is shown in Figure 8.5. All of the components are enclosed in a glass vacuum envelope with leads usually to the anode and last dynode. In some systems, a pre-amplifier is also built in.

The photocathode converts the photons into an equivalent number of electrons and is therefore the source of electrons in the photomultiplier tube. The face of the photomultiplier tube is made of glass on which is coated a material (usually a combination of alkali metals and antimony), which releases electrons when irradiated with light photons. These electrons, known as *photoelectrons*, are then accelerated and focused onto the first dynode of the tube.

The chain of dynodes then provides for electron multiplication. When the photoelectrons strike the first dynode, a number of secondary electrons are produced and emitted from the surface of the dynode material. These emitted secondary electrons are then accelerated towards the second dynode, where further secondary electrons are produced. This process is then repeated down the chain of dynodes. The final bunch of electrons emitted from the final dynode is then collected by the anode from which an electron current is generated, which, in turn, can be converted into an equivalent voltage pulse.

Scintillation counter assemblies

The scintillator and the photomultiplier tube have to be brought together to form the scintillation counter. For solid scintillators, such as NaI(Tl), this usually involves mounting the crystal on the face of the photomultiplier tube using an optical grease to both act as a sealant and also provide optical continuity between the scintillator and the photomultiplier tube, with the crystal in a hermetically sealed container of aluminium or stainless steel. For special purposes, the connection between the crystal and the photomultiplier tube may be made with a fibre optic flexible light guide, so that the crystal and tube may be separated by some distance.

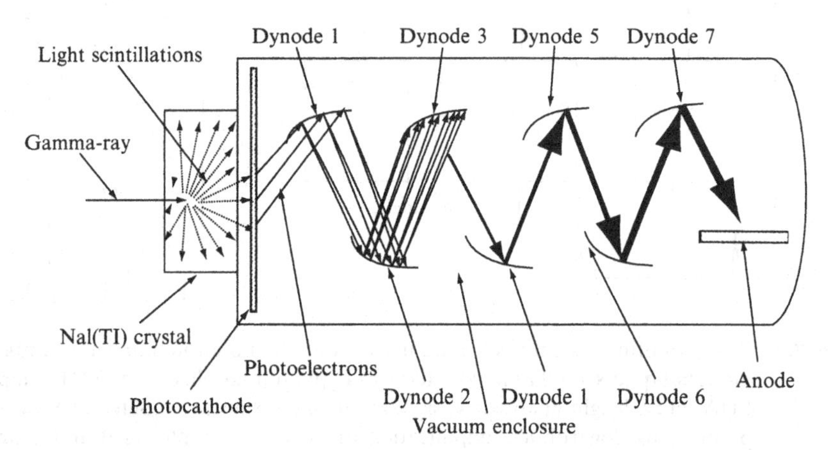

Figure 8.5 NaI(Tl) crystal and photomultiplier assembly, illustrating the multiplication of electrons at each dynode of the assembly

In the case of liquid scintillation counting, a somewhat different arrangement of scintillator and photomultiplier tube is used and a much more complicated electronic system is required. The liquid scintillator and sample are contained in a suitable counting vial, either glass or plastic, which is then positioned between two photomultiplier tubes at 180°, as illustrated in Figure 8.6. Using a pair of photomultiplier tubes in conjunction with a coincidence circuit, as shown in this figure, brings about a considerable reduction in the background count, particularly for soft beta-emitters such as tritium and ^{14}C, where noise from the photomultiplier tubes interferes with the inherent small pulse size.

The two most important quality factors related to liquid scintillation counting are (i) the counting efficiency, E, and (ii) the background count rate, B. For the detection of extremely low levels of activity, maximization of the factor E^2/B, called the 'Figure of Merit', is essential.

8.1.3 Semiconductor Detectors

The semiconductor materials used for detectors are made from crystalline silicon or germanium. In such materials, the atoms are bonded together via valence electrons and each atom is bound to four others. In energy terms, these valence electrons are in the 'valence band'. Above the valence band, lies the 'conduction band'. In metals, the energy gap ('band gap') between these two bands is small and so electrons are easily promoted from the valence to the conduction band, where they can travel through the crystal lattice quite freely. This freedom of movement of electrons in the conduction band is why metals are good electrical conductors. Conversely, in insulators the band gap is very wide and virtually no electrons are found in the conduction

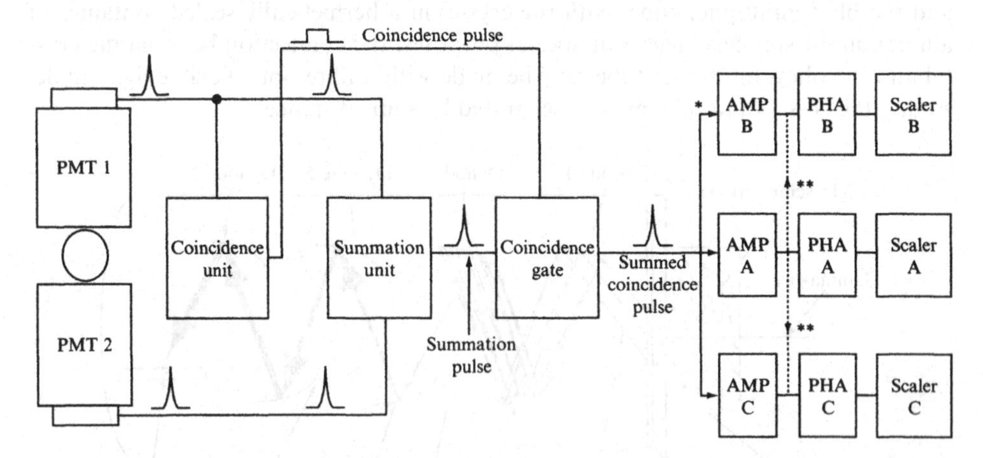

Figure 8.6 Basic counting system electronics used with a dual photomultiplier assembly in a modern liquid scintillation counter: PMT, photomultiplier tube; AMP, amplifier; PHA, pulse height analyser: *, signal pathway for linear amplification; **, signal pathway for logarithmic amplification (in this case, amplifiers B and C are not needed). Reproduced by permission of Packard Instrument Company from Kessler, M. J. (Ed.), *Liquid Scintillation Analysis*, Packard Instrument Company, 1989

band – hence their inability to conduct electricity. Semiconductors, as the name implies, lie between these two extremes. In highly pure forms of both silicon and germanium, the band gap is quite wide and promotion of the outer valence electrons from the valence to the conduction band is only brought about at high temperatures so that these materials have high resistivity at ambient temperatures.

If, however, these pure materials are 'doped' with elements that are either one place above or below them in the Periodic Table, then this alters the electronic structure and effectively reduces the band gap.

N-type semiconductors

Choosing silicon, which belongs to Group IVb of the Periodic Table, as our typical semiconductor material (germanium behaves in exactly the same way), we can dope this with an element from Group Vb, such as phosphorus. The phosphorus atoms replace silicon atoms in the crystal lattice. There is a charge imbalance now, however, with the phosphorus having an extra electron available which is not used in covalent bonding to its neighbouring silicon atoms. This extra electron is easily removed from the phosphorus atom and is free to migrate through the lattice and leaves a corresponding positive charge behind. Since the electron is donated to the lattice, elements like phosphorus are known as 'donor impurities'. Such electrons can occupy positions in the band gap and require little energy in order to be promoted to the conduction band. The overall effect of such doping is to produce a material in which the electrical conductivity is primarily due to the flow of electrons. Since the charge carrier is negative, the doped material is referred to as *N-type*. Under an applied field, a current of electrons will flow and the conductivity of a doped semiconductor can easily be changed by the level of doping up to a certain level. Doping is normally at the level of a few ppm.

P-type semiconductors

If, however, silicon is doped with an element from Group III of the Periodic Table, e.g. boron, then this element will have one too few electrons to satisfy the bonding requirements of silicon. Because of the narrowness of the band gap that now exists, this allows an electron from silicon with sufficient thermal energy to move to the boron atom to fulfill the bonding requirements. Consequently, one of the silicon atoms adjacent to the boron will not have its full-bonding capacity satisfied. This is equivalent to creating a hole in the lattice structure. If an electric field is now applied, an electron will jump from a neighbouring silicon atom into the hole. This leaves behind a 'hole' (positive charge) in the silicon atom from which the electron jumped and so the move is equivalent to the migration of the hole to another site. In this way, the positive holes migrate through the lattice in a completely analogous manner to the migration of electrons and the electrical conductivity is due to this hole migration, with electron-deficient impurities being called 'acceptor impurities'. Hence, the charge carriers in P-type silicon are positive holes and since such material has an excess positive charge it is therefore known as a *P-type*. Obviously, under a given applied field, electrons and holes will move in opposite directions.

P–N junctions

When a piece of P-type silicon is brought into contact with a piece of N-type silicon, then at the junction of the two pieces there is an excess of electrons on one side and positive holes on the other which will move in opposite directions to cancel out the net charges. This occurs in a small region either side of the junction which is now effectively free of charge carriers and thus forms a *semiconducting diode*. This depleted region can be extended by applying a reverse bias to the diode and it will also now act as a detector of ionizing radiation since any ions formed there will be swiftly swept towards the N-type material and the electrons will move towards the P-type material under the field that exists there, thus giving rise to an electrical pulse. The magnitude of the pulse will be directly proportional to the energy of the ionizing radiation and so these detectors can function as *energy spectrometers*. Semiconductor detectors are always operated at reverse bias. Such a detector is analogous to a gas-ion chamber as discussed in Section 8.1.1.

There are basically two types of semiconductor detector used depending on the radiation to be detected. The reverse-bias diode just described is used for charge-particle detection (both alpha- and beta-particles) and detectors for photon counting and spectroscopy, which operate on a slightly different principle and will be discussed later.

Charged-particle detectors

These detectors are reverse-bias diodes normally made from single-crystal silicon although germanium is sometimes used for the detection of very high-energy charged particles. The basic type of charged-particle detector is the surface-barrier detector. Typically, a surface-barrier detector is formed from N-type silicon which has been etched to produce an effective but very thin P-type layer and hence form a junction. A very thin layer of gold evaporated onto the P-type layer forms an entrance 'window'. The application of a reverse bias, as described above, has the effect of extending the compensated region, increasing as the bias is increased. In other words, the active volume of the detector is controlled by the bias applied. Such detectors are known as *partially depleted* detectors. There is a maximum voltage at which these detectors may be operated beyond which catastrophic damage can result. The thin window is transparent to light which would produce pulses and so this must be excluded from them when they are operated. In addition, the thin window makes these detectors delicate to handle.

Surface-barrier detectors are excellent for the measurement of alpha-particles. They have an efficiency of 100% for all alphas that enter the depleted region and this region is thick enough to stop all of the usual environmental alpha-particles. In general, it is usual to use a detector whose depletion layer is not significantly greater than that required to stop the particle since this results in decreased noise, which in turn produces a better resolution. The energy resolution of a good detector can be about 12 keV for an alpha-particle and about 6 keV for a beta-particle. It is interesting to note that a 1 mm thick depleted surface-barrier detector is equivalent to a 100 cm gas counter in its efficiency.

There are surface-barrier detectors produced on lithium-compensated silicon (Si(Li)) but their mode of operation is basically the same as the surface-barrier detectors discussed above. These are frequently used for beta-particle detection and beta-end-point determinations. A more recent development in charged-particle detectors is the Passivated, Implanted, Planar, Silicon (PIPS) detector. These are fabricated by photolithographic techniques with a junction formed by ion-implantation which defines the junction with high accuracy and hence gives potentially better resolution than surface-barrier detectors. Such detectors have a number of advantages over conventional surface-barrier detectors. In particular, the entrance window is extremely thin (usually about 50 nm), stable and rugged since it is formed as a passivated, implanted surface and this also allows it to be cleaned if necessary. The thin window also gives better resolution and detection efficiency for alpha- and beta-particles.

Photon detectors

In principle, if the donor and acceptor impurity concentrations are equal in a semiconductor material then the resulting material should have the properties of the depleted region in a charged-particle detector, with relatively few charge carriers of approximately equal numbers. Consequently, if ionizing radiation passed through such material then the resulting electrons and ions could be swept out by the application of a voltage to produce a suitable field. A material containing such balanced impurities is said to be 'compensated'. Equally, if one could produce extremely pure semiconductor material, containing insignificant amounts of impurities, then again one would have a potential detector. Such a high-purity semiconductor material is said to be an *intrinsic* semiconductor.

Since photons require considerably more material to absorb them than do charged particles, it is obvious that photon detectors must have considerably greater depth than charged-particle detectors and the material for their construction should preferably have a higher atomic number than silicon. In fact, gamma-ray detectors are usually made from germanium, as are some X-ray detectors, although detectors specifically built for X-ray detection are frequently based on silicon.

Based on the discussions of the production of depleted regions in charged-particle detectors, it might be thought that one could produce depletion depths of any size simply by increasing the reverse bias and using a suitably large silicon or germanium crystal. In practice, however, it is not possible to produce depletion depths much greater than 2–3 mm. This is too small even for the efficient detection of quite soft X-rays and hopelessly small for the detection of gamma-rays. Thus, some other method must be used to produce a semiconductor diode of the dimensions suitable for photon counting. The two methods which might be used are either to prepare compensated material or alternatively to produce an intrinsic semiconductor.

Historically, the first useful, semiconductor photon detectors were prepared by the compensation technique. The method that was developed in the early 1960s was to produce P-type silicon or germanium (since P-type material could be produced in the highest purity) and then compensate for the excess of acceptor impurities in this material by diffusing an alkali metal, i.e. lithium, into the P-type crystals. The result is that a P–N junction is formed and the compensated region can extend up to a centimetre or more into the original semiconductor. Hence, detectors produced in this way are known as *lithium-drifted silicon* or *lithium-drifted germanium* detectors, usually abbreviated to Si(Li) or Ge(Li) and colloquially called 'silly' or 'jelly' detectors, respectively. A high voltage is then applied to provide the collector field in the detector.

The mobility of lithium in germanium is greater than in silicon and remains sufficiently high even at room temperatures so that if left at ambient temperature for any significant length of time there is a redistribution of lithium ions and the detector is ruined. Consequently, Ge(Li) detectors have to be maintained at liquid nitrogen temperatures (77 K) at all times and can never be allowed to warm up to room temperature. The lower mobility of lithium in silicon means that these detectors can be allowed to occasionally warm up to room temperature without doing permanent damage to the detector but in general they are normally maintained at liquid nitrogen temperatures.

The standard way of cooling a detector is to mount it in a cryostat with a copper rod immersed in liquid nitrogen in a Dewar to conduct heat away from the detector. Frequently, the first stage of the pre-amplifier is also cooled to reduce noise in the system. The entire cold assembly is maintained under a good vacuum for both thermal insulation and for protection of the delicate assembly. Figure 8.7 shows a cryostat in a Dewar with the detector mounted in the cryostat. Since the detector has its long-axis vertical, this is called a 'vertical dipstick cryostat'. Detectors mounted on horizontal dipsticks are also common. Ge(Li) and Si(Li) detectors are always operated at liquid nitrogen temperatures in order to reduce the leakage current to the lowest possible values and so obtain the highest possible resolution.

The major disadvantage of Ge(Li) detectors is the requirement that the detector is always maintained at liquid nitrogen temperatures. Liquid nitrogen is not an inexpensive item and a large detector will require 5–10 l of liquid nitrogen per week under normal operating conditions. Consequently, there was always an impetus to try to provide intrinsic germanium detectors. The real breakthrough came in about 1976 when the first large-volume intrinsic germanium detectors were fabricated. Since that time, the technology has advanced to the point where detectors much larger than any Ge(Li) crystal have been produced – however, at a price! The ability to successfully produce intrinsic germanium detectors of a wide range of sizes led to the demise of the Ge(Li) detector and in fact, no Ge(Li) detectors for gamma-ray detection have been produced by Western manufacturers for several years. Thus, all gamma-ray spectroscopy is now based on the use of intrinsic or hyperpure Ge detectors.

Hyperpure detectors are produced either as P-type or N-type detectors, depending on the majority carrier concentration of the final detector and these detectors can be stored at room temperature when not in use but must usually be operated at liquid nitrogen temperatures to reduce the leakage current. These detectors can also be cooled by a mechanical refrigeration unit which makes their operation in areas remote from liquid nitrogen sources possible.

Types of photon detector Germanium detectors can be one of two types, depending on their shape. The two types are (i) *Planar*, and (ii) *Co-axial*. Si(Li) detectors, because they are used almost exclusively for X-ray spectroscopy are

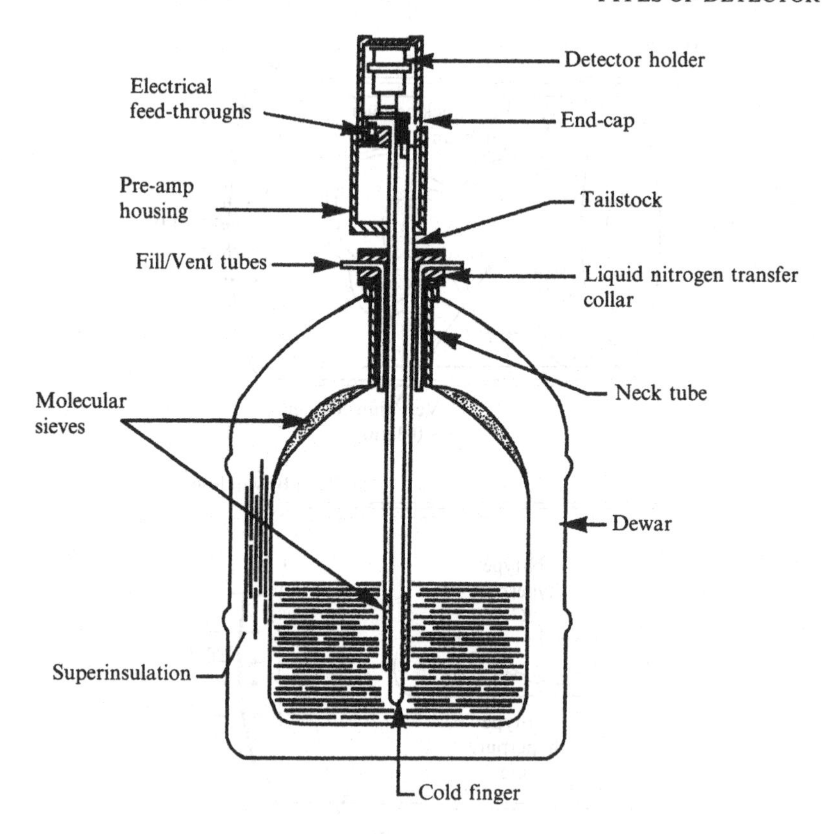

Figure 8.7 Vertically mounted hyperpure germanium detector and liquid nitrogen Dewar. Reproduced by permission of Canberra Industries

always planar. Because of the advances in technology of the fabrication of hyperpure Ge detectors, planar hyperpure detectors are less common than when Ge(Li) detectors predominated.

Planar Ge(Li) or Si(Li) detectors are fabricated from thin wafers of the semiconductor material, with the lithium drifted from one flat surface of the wafer toward the opposite one. Planar detectors are used mainly for the determination of low-energy gamma-rays and X-rays, where their superior resolution is of paramount importance. Si(Li) detectors are preferred over all other types of semiconductor detector for the determination of X-rays in the energy region from 1 to about 30 keV. These low-energy detectors normally have very thin entrance windows made of a low-atomic-number element to reduce absorption of the photons – beryllium is normally used. Planar Ge(Li) or hyperpure detectors have a high intrinsic efficiency for photons up to about 120 keV and have useable efficiencies up to about 300 keV.

The inability to fabricate planar detectors with large active volumes means that a different configuration is required for large-volume detectors with good efficiency for gamma-rays up to several MeV in energy. The configuration used here is the co-axial one, where a highly purified germanium cylinder is produced as P-type material for normal gamma-ray detectors, as shown in Figure 8.8. Detectors with a

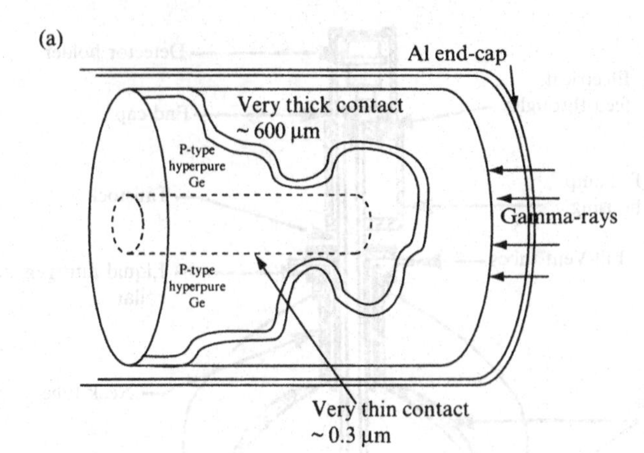

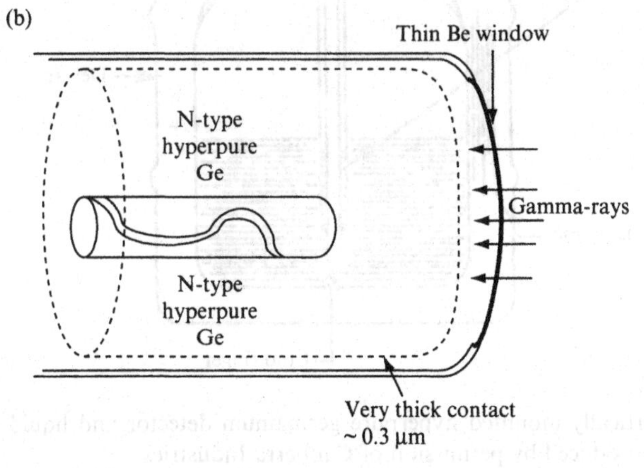

Figure 8.8 Co-axial hyperpure germanium detectors, suitable for counting both gamma- and X-rays: (a) conventional co-axial; (b) reverse electrode

detection efficiency much greater than a 7.5 cm × 7.5 cm NaI(Tl) crystal (see next chapter) are currently obtainable, with the prospect of even larger detectors in the future. Finally, it should be noted that hyperpure germanium detectors can also be fabricated with a 'well' configuration.

More recently, *Cadmium Zinc Telluride* detectors have been used in environmental studies. These have resolutions only slightly inferior to that of hyperpure Ge detectors but unlike the latter do not require liquid nitrogen for cooling. This makes them much more convenient for field measurements of complex gamma-ray spectra although their efficiencies are not high.

8.1.4 Passive Detectors

By this term, we mean those detectors which record and store the events resulting from the passage of ionizing radiation and which are subsequently 'interrogated', normally by the application of some physical or chemical treatment to determine the integrated response. A feature of such detectors is that they require no direct electronics in order for them to operate although they may require quite sophisticated electronics in the production and measurement of the final output, which is usually a visual one.

Photographic emulsions

These represent possibly the oldest radiation detectors, having featured in the very origins of the study of radioactivity with the discovery by Becquerel and Madame Curie (as she was later to become) of the blackening of photographic plates stored in a cupboard next to some uranium salts. This we now know was due to the fact that photographic emulsions contain particles of silver halides suspended in a gelatine emulsion. The (mainly) alpha-particles from the uranium sensitized the silver halides, in the same manner that ordinary light sensitizes them, so that on development the sensitized grains were converted to black, metallic silver, while the unsensitized grains remained unaffected.

The principle use of photographic emulsions in modern environmental studies is that of autoradiography, which is essentially an application of the phenomenon described above. A radioactive sample containing distributed alpha- and beta-emitting activities is placed in contact with a photographic plate or film for a given length of time. The film is then developed to give a negative darkened to an extent corresponding to the strength of the local radioactive 'source'. The method can quite accurately locate local distributions of pure alpha- or low-energy beta-emitters due to the short range of these particles. A good example of this high resolution is shown in Figure 8.9. Higher-energy beta-particles travel much further and reduce the spatial resolution of the method. This is useful, for instance, in looking at the distribution of uranium- or thorium-bearing minerals in a rock sample. Some quantitative information of the concentration of the radioactive element can be gained from the density of the blackening. Probably the greatest present use of emulsions is in the form of the familiar 'film badge' personal beta- and gamma-dosimeter.

Solid-state nuclear track detectors

Any heavily ionizing particles (such as protons or alpha-particles) on passing through insulating materials leave trails of damage. In crystals, this is due to atomic displacements with resulting lattice strain. In plastics, the polymers are disrupted, with the formation of molecular fragments and free radicals.

There is a threshold density of ionization which must be exceeded in any given material before track formation occurs. Certain plastics will register quite low energy protons, whereas most minerals require quite massive particles, such as fission fragments, before any tracks are formed. In this way, it is possible to change the selectivity of the response of a solid-state nuclear-track detector (SSNTD) to a particular particle by careful selection of the detector. A useful feature of these devices is their insensitivity to electrons and gamma-rays which allows them to be used in high radiation fields which would swamp other types of detector. This and

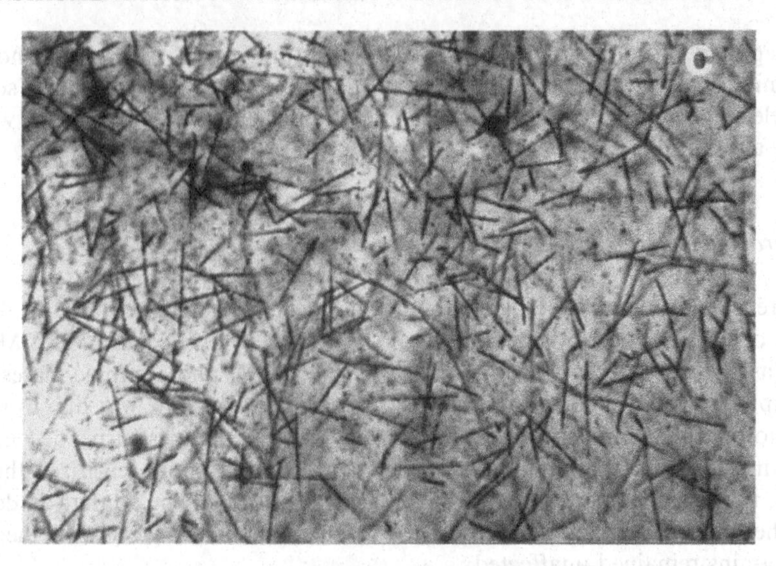

Figure 8.9 Image (magnified × 350) of an autoradiograph showing the alpha tracks from astatine-211

their simplicity of construction (often simply a piece of a plastic sheet or film) makes them very useful as detectors.

In crystals or plastics, the tracks are revealed by the fact that certain chemicals etch the damaged regions at a much higher rate than the surrounding, undamaged material. This produces a hole in the material and if the etching continues long enough the hole will be visible through an optical microscope. In general, until the material is etched the track will remain invisible, i.e. 'latent' in the material and in some cases for periods on a geological timescale.

Most of the environmental applications of SSNTDs are to determine the amount of radiation rather than to identify the particle type and its energy. For such applications, it is the number of tracks which is required and the most common method of track counting employs an optical microscope with a grid graticule in the eyepiece. Manual track counting is, however, very tedious, particularly if track densities are low and statistical accuracy requires the counting of a large number of tracks or where a large number of samples have to be processed. In such circumstances, automatic scanning systems are almost essential and would be useful in most track-counting procedures in order to remove the human error associated with manual counting. Unfortunately, the application of automated techniques has been fraught with technical problems and automation for many applications is still subject to a number of errors. The physical parameters of a number of plastic SSNTDs are presented in Table 8.1.

In recent years, probably the most important application of SSNTDs has been in the measurement of radon in the environment. The two naturally occurring isotopes of radon, derived from the decay of uranium and thorium, present in a wide range of rocks and soils are both alpha-emitters and can therefore be detected by the use of plastic SSNTDs. We shall say more about the measurement of radon in the environment in the next chapter.

Table 8.1 Physical parameters of some plastic SSNTDs.

Material	Trade name	Monomer composition	Density (g dm^{-3})	Particle and minimum energy to produce tracks[a]	Reference
Cellulose nitrate	Daicell, CN-85	$C_6H_8O_9N_2$	1.4	Proton; 0.5 MeV	Fleischer *et al.* (1975)
Cellulose acetate butyrate	—	$C_{20}H_{32}O_5$	1.23	—	—
Bisphenol-A polycarbonate	Lexan, Makrofol	$C_{16}H_{14}O_3$	1.29	^{4}He; 0.3 MeV	Fleischer *et al.* (1975)
Poly(allyl diglycol) carbonate	CR-39	$C_{12}H_{18}O_7$	1.32	Proton; 10 keV	Cross *et al.* (1986)
Polypropylene	—	CH_2	0.9	^{4}He; 1 MeV	Fleischer *et al.* (1975)
Poly(ethylene terephthalate)	Mylar, Melinex	$C_5H_4O_2$	0.93	Ne; 1 MeV[b]	Fleischer *et al.* (1975)

[a] This is the lowest-atomic-number (Z) particle and the minimum energy that this particle must possess in order to register etchable tracks in the given material.
[b] Estimated particle and energy based on theoretical considerations.

Thermoluminescence detectors

Thermoluminescence is the term applied to semiconductors or insulating materials which after exposure to ionizing radiation (or other forms of energy absorption) emit light when heated. Thus thermoluminescence is the thermally stimulated emission of light and like some SSNTDs thermoluminescent materials (phosphors) are capable of storing the absorbed energy over long periods of time until they are heated. For this reason, natural thermoluminescent materials like natural SSNTDs can be used for dating samples over periods of millions of years. In the case of thermoluminescent dating, the 'date' obtained will refer back to the last time an object was heated and so re-set the 'clock'. This re-setting of the clock by heating also allows the phenomenon to be used for archaeological dating for artifacts that have undergone a period of heating, either intentional or otherwise.

The technique of thermoluminescence involves the measurement of the amount of light (visible or ultraviolet) emitted from the phosphor as it is steadily heated. A plot of thermoluminescent intensity versus temperature is known as the *glow curve* and is a characteristic of a particular phosphor. Thus, a thermoluminescent detector combines a suitable phosphor with a device for controlled heating of the phosphor and finally an instrument for measuring the light emitted. The light detector is usually a photomultiplier tube but photodiodes may be used for reading phosphors that have received a high dose.

A large number of inorganic compounds exhibit thermoluminescence but lithium fluoride (LiF) is by far the best-studied and is still widely used although it has a number of drawbacks for many thermoluminescent applications. This material, like most commercial phosphors, contains one or more activators to enhance the thermoluminescent activity and the actual 'recipes' for a particular phosphor are often 'commercial secrets'. The important characteristics of a number of 'environmentally useful' TLD phosphors are shown in Table 8.2.

Thermoluminescent devices find their main application in dosimetry, particularly personal dosimetry. This arises from the fact that the intensity of the light output, as measured by the area under a particular peak in the glow curve is a function of the dose of electrons or gamma-radiation received. The dose response characteristics of LiF, the most widely used phosphor for personal dosimetry, are rather complex and governed by a number of factors, all of which have been extensively studied. For personal dosimetry, the phosphor is either in the form of a loose powder packaged in a suitable container or mixed with polymers and then fabricated as chips or discs for direct use. Discs of LiF in PTFE can be used as finger dosimeters.

8.2 ASSOCIATED DETECTOR ELECTRONICS

So far we have only discussed nuclear detectors. These are the 'front end' of a counting system and in order to convert the analogue output from the detector into signals that can be counted or energy-analysed we require several additional pieces of electronic equipment. A detector plus this electronic equipment constitutes a *counting system*. Before discussing the other components of the counting system we need to briefly outline how modern components are connected together to form a system.

8.2.1 The Modular Counting System

The electronics used in conjunction with nuclear detectors should be designed in accordance with either the 'Nuclear Instrument Standard' (NIM) or the 'Computer Automated Measurement and Control' (CAMAC) standards although the latter has only really found widespread acceptance in large, nuclear research facilities and consequently we shall only discuss NIM modules here.

Table 8.2 Important characteristics of some environmentally useful TLD phosphors (from Mahesh *et al.*, 1989).

Phosphor	Dosimetric peak temperature (°C)	Useful dose range	Thermoluminescence fading time
LiF:Mg, Ti	195	50 μGy–10^3 Gy	10% per month
$Li_2B_4O_7$:Cu, Ag	185	50 μGy–10^4 Gy	10% per month
$CaSO_4$:Tm	210	1 μGy–10^3 Gy	10% per month
$CaSO_4$:Dy	210	1 μGy–10^3 Gy	3% per month
CaF_2:Dy	200	1 μGy–10^3 Gy	12% per month
CaF_2:Mn	260	1 μGy–10^3 Gy	10% per month
CaF_2 (natural)	260	10 μGy–10^3 Gy	—
Mg_2SiO_4:Tb	195	1 μGy–10^3 Gy	3% per month
Al_2O_3	250	10 μGy–10^3 Gy	5% per two weeks

The reason for an NIM standard is that all manufacturers build equipment to this standard and hence it is possible to put a counting system together using modules from more than one manufacturer, rather like one can put together a hi-fi system with a CD player from one manufacturer, an amplifier from a different one, and so on. In addition, one can change a counting system simply by adding or deleting the appropriate NIM modules. Such modules are designated 'single-width', 'double-width', etc. according to the number of slots they occupy in a NIM 'bin' (see below). The modules, except for the high-voltage units, are connected together by the use of standard BNC cables, which are co-axial cables fitted with BNC connectors. The industry standard for connections for high voltage (> 1 kV) is now the super high voltage (SHV) connector.

The NIM modules are powered from a NIM 'bin'. This is a crate fitted with slots to accommodate up to twelve single-width NIM modules, or six double-width bins, etc. and distributes all DC and AC power levels from the bin power supply to the module connectors through a wiring harness at the rear of the bin. Thus it is only necessary to push a NIM module into a bin to make connection at the rear for that module to automatically receive the correct DC voltages to power it.

8.2.2 The Components of a Basic Modular Counting System

A basic set-up for simple radiation counting, that might be used with an ionization chamber, a proportional counter or a scintillation detector, is shown in Figure 8.10. The electronics required for a Geiger–Müller tube are even simpler since the pulse size from the tube is sufficiently large to require no further amplification.

From this figure we see that the four basic components required to convert the signals from the detector into a readable count, related to the radiation intensity (in terms of disintegrations per unit time) are (a) a high-voltage supply, (b) a pre-amplifier, (c) an amplifier, and (d) a scaler–timer unit. If a Geiger–Müller tube is used as the detector then normally no pre-amplifier or main amplifier are required but an electronic quenching unit must be included. In addition, a single-channel analyser (pulse height discriminator) and rate meter may frequently be used.

We shall now briefly look at the operation of each of these devices.

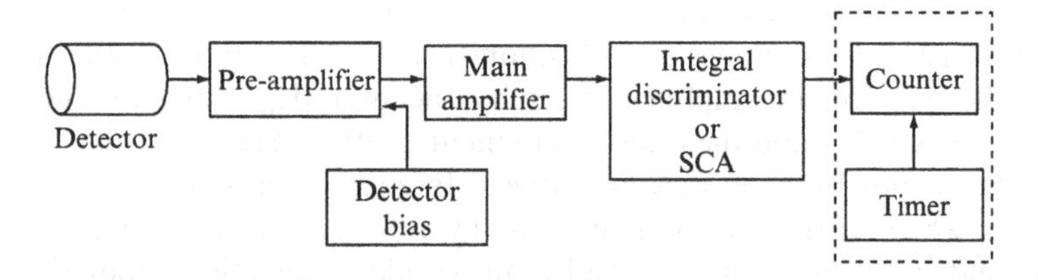

Figure 8.10 Block diagram of the basic electronic components needed for counting with an ionization chamber, Geiger–Müller tube, proportional counter or scintillation detector. The discriminator only passes pulses above a selected threshold value. A single-channel analyser (SCA) would normally be used with a scintillation detector; the timer and counter are normally part of a single unit

The high-voltage unit

All non-passive detectors require a bias voltage in order for them to operate correctly. This can range from a few volts for a surface-barrier detector to several kilovolts for a semiconductor photon detector. The bias voltage is normally applied through a filter network in the pre-amplifier, except for scintillation counters, where the bias is applied to the various components of the photomultiplier tube via a resistive network in the base of the tube. Detectors incorporating photomultiplier tubes require high-voltage units capable of delivering a few milliamps of current, whereas the current requirements of most other detectors are much smaller.

Pre-amplifier

The amount of charge produced by nuclear radiation is very small and depends on the average energy required to produce an ion-pair, i.e. the W value. This is $\sim$ 3–6 eV in solids and $\sim$ 30 eV in most common detector gases. For the production of a photoelectron, the W value is $\sim$ 300 eV. The smaller the W value, then the greater the number of electrons formed (charge) for a given energy of ionizing radiation. Based on the above W values, the largest voltage that could be produced from a one MeV radiation event is about 0.4 mV, which is very small. The two main purposes of a pre-amplifier are (i) to provide an initial amplification of the usually very small signal from the detector, and (ii) to provide a low impedance output to drive the succeeding components. In order to maximize the inherent signal-to-noise ratio of a detector, it is important to have the pre-amplifier as close to the detector as possible, which also reduces the input capacitance. The output pulses from the pre-amplifier consist of a rapidly rising pulse followed by a long, exponential tail.

Amplifier

This is an important item in any counting system but it is often the key component in an energy spectroscopy system. We shall only refer here to linear, pulse-shaping amplifiers.

One of the main functions of an amplifier is to convert an input signal from the pre-amplifier in the range of one to a few tens of millivolts into an output signal in the range 0–10 V. This implies amplification factors of the order of 100 to 1000. In general, the gain of an amplifier is set by the user depending on the range of energies to be studied and the other components in the system.

The other major function is *pulse shaping*. In energy analysis, it is the amplitude of a pulse which carries the inherent energy information about an event in the detector and we assume that the pulse amplitude is directly proportional to the energy deposited in the detector. The output pulse from the pre-amplifier has the shape discussed in the previous section. The aim is to preserve the correct amplitude but considerably shorten the length of the pulse. In practice, a Gaussian-shaped pulse gives the best results and considerable efforts have been made by manufacturers to achieve a pulse shape as close to Gaussian as possible.

There are various strategies available for achieving a viable pulse shape from an amplifier, with most of these involving the use of resistors and capacitors (*RC networks*). The output pulse can either be single-lobed (positive or negative with respect to a zero baseline), referred to as a *unipolar* output, or can be double-lobed with an equal positive and negative lobe, referred to as a *bipolar* output. Which type of output is preferred, as far as energy resolution is concerned depends very much on the other components in the system and so is really system-dependent.

It is found that unipolar pulses do not immediately return to the baseline but rather 'undershoot' it to some extent, as shown in Figure 8.11. This problem of a varying baseline has a component which is rate-dependent but suitable circuitry can minimize this and that of undershoot. Bipolar pulses do not suffer from such baseline shifts.

A further difficulty with any amplifier arises when the count rate is very high (of the order of several thousand hertz). At high counting rates, pulses may be being processed so rapidly by the amplifier that one pulse arrives before the preceding one has reached the baseline. This results in the pulses 'piling up', one on top of another and the resulting output pulse from the amplifier does not correctly represent the original detector pulse and hence a distortion of the spectrum results. Modern circuitry has extended the count rate limit before this occurs to many tens of thousands of hertz. The interested reader is referred to one of the books on nuclear electronics listed at the end of this present text.

Scaler–timers

In a simple counting system, it is only necessary to count the number of pulses leaving the amplifier. This is achieved with a counting unit (*scaler*) where basically, each pulse registered causes a visible display unit to increment by one. In some systems, it is also possible to 'dump' the scaler contents into a computer memory for subsequent storage and manipulation. Noise pulses are discriminated from true signal pulses on the basis of their smaller size by using suitable circuitry. Remember that the output pulses from the amplifier will be in the range of about 0–10 V.

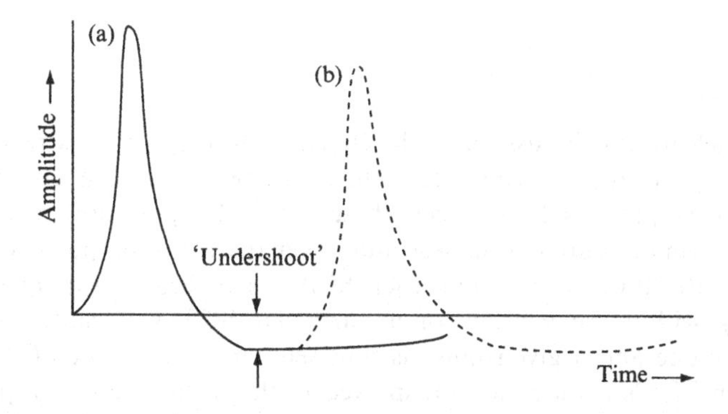

Figure 8.11 Illustrations of the phenomena of (a) 'undershoot', and (b) an amplitude defect

Normally, we need to know the number of counts accumulated in a given length of time. Consequently, it is necessary to have a timing unit with the scaler and this may be either a separate unit or more often an integral part of the scaler. This is then referred to as a *scaler–timer* unit.

Ancillary units in a simple counting system

Two units may sometimes be used with the above basic counting system. These are a pulse-height discriminator (or single-channel analyser (SCA)) and a ratemeter.

The SCA is used when only pulses within a narrow range of amplitudes are to be counted, e.g. when using a scintillation detector for gamma-ray counting and where just one photo-peak, out of several, is to be counted, or when one wants to count all the pulses exceeding a certain amplitude. The provision of user-operated, variable lower- and upper-level discriminators, calibrated in volts allow both aims to be achieved.

The ratemeter is a device which effectively monitors the count rate. The total number of counts which are accumulated over a certain length of time, which is variable and set by the user, is output to a meter, the deflection of which corresponds to the count rate. Usually, the range of the count rate can be continuously varied manually over several decades of counting to increase the sensitivity. The ratemeter is always used in conjunction with some form of discriminator, which may be an integral part of an amplifier. Sometimes, the output of the ratemeter is used to drive a chart recorder for continuous monitoring of the count rate.

8.2.3 The Components of a Modular Energy Spectroscopy System

A block diagram of the components required for a basic energy spectroscopy system is given in Figure 8.12. It is seen that two additional units are required over those used in simple counting, i.e. the *analogue-to-digital convertor* (ADC) and some form of memory/display unit. The two together, plus certain other devices, are known as a *multichannel analyser* (MCA). Before discussing these units we need to digress slightly to understand certain concepts fundamental to energy spectroscopy.

Characteristics of multichannel counting

Referring back to our discussion of the function of an SCA, it is possible to set a lower- and upper-level discriminator (the 'window') and then count only those pulses whose amplitudes lie between these limits. Imagine now a series of such SCAs whose windows are contiguous and cover the entire output voltage range of the amplifier (0–10 V). In addition, each SCA is connected to its own counter. A pulse falling within the window of a particular SCA will cause its counter to increment by one and a given pulse is uniquely counted by one SCA. If the contents of each SCA's counter were displayed in the form of an X–Y plot, with the SCA number (or the equivalent window number) along the X-axis and its equivalent

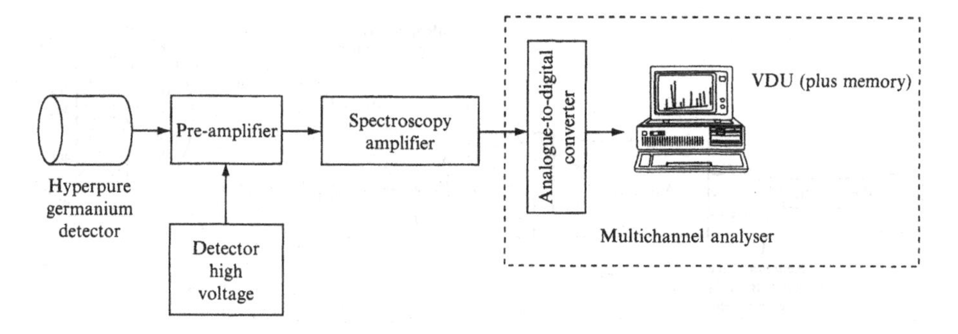

Figure 8.12 Block diagram of the basic system necessary for carrying out energy spectroscopy using a hyperpure (HP) germanium detector. The components in the larger box (dashed-lines) constitute the multichannel analyser (MCA). Nowadays, the memory and display units are normally part of a personal computer system

counter value on the *Y*-axis, one would in fact have a crude energy spectrum of the sort shown in Figure 8.13. Each window is called a *channel*, with the count in each window or channel being the *channel content*. This set of contiguous SCAs is a crude form of a multichannel analyser. If we increase the number of SCAs and correspondingly reduce the window width, we will increase the resolution of the system until ultimately we have an infinite number of channels and see a continuous distribution of pulse amplitudes which exactly mirrors the energy spectrum of the source being analysed. Obviously, we cannot have an infinite number of SCAs, but, in principle, we could have a very large number to approximate as closely as possible the energy spectrum. It is, in fact, completely impracticable to consider building a multichannel analyser based on a large number of SCAs and in practice a different method is used to achieve this end, based on the ADC and memory/display unit.

The analogue-to-digital convertor

The analogue-to-digital convertor (ADC) achieves the requirement of splitting the incoming signals from the amplifier into a corresponding number of channels. It does this basically by converting the output, analogue voltage signal from the amplifier into an equivalent digital signal. Such signals can then be processed by a standard computer to allow for the display and analysis of the spectrum. The most usual type of ADC, known as the *Wilkinson* ADC operates on the principle of converting the analogue signal into an equivalent number of clock pulses, which constitute the digital output.

A disadvantage of the Wilkinson ADC is that the processing time depends on the amplitude of the pulse; the higher the amplitude (higher gamma-ray energy), then the longer the clock pulse gate is open. Processing time is also inversely related to the clock frequency so the aim has been for the highest clock rates possible. Clock frequencies of 450 MHz are now readily available.

(a)

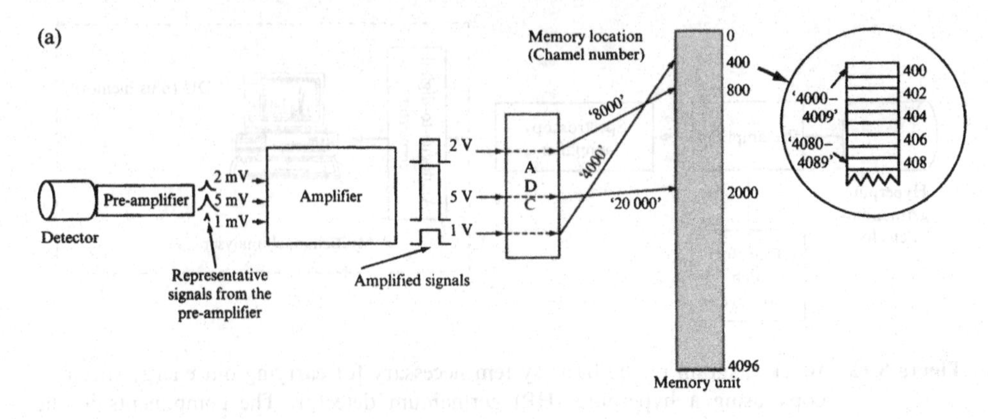

(b)

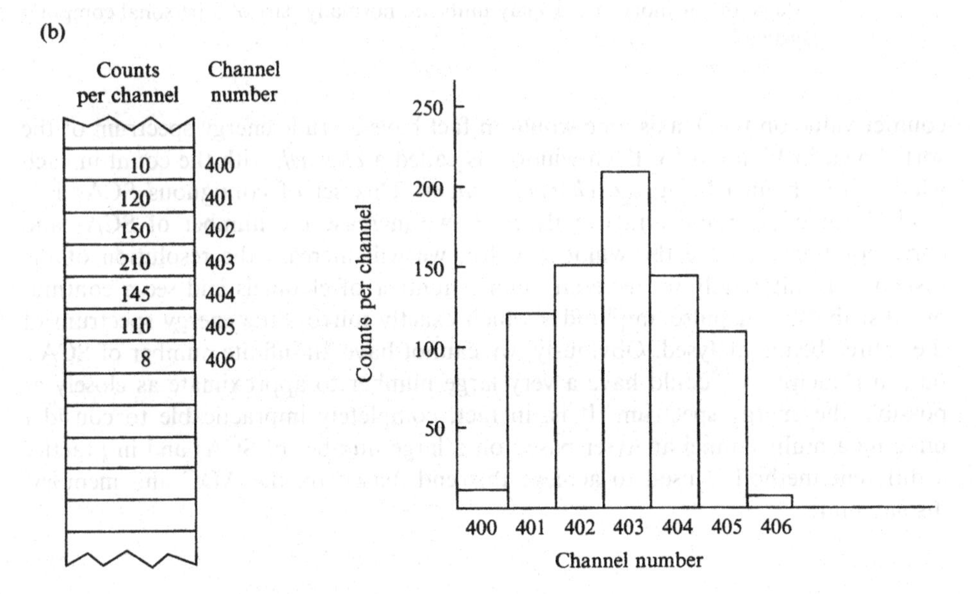

Figure 8.13 (a) Schematic diagram illustrating the amplification, digitization and storage of signals resulting from three gamma-rays initially incident on a germanium detector. The numbers by the arrows after the analogue-to-digital convertor (ADC) represent the digital values assigned by the latter to the analogue voltage signals entering this unit. These digitized values are then stored in an appropriate memory location (or channel number) in a memory unit (computer). In this simplified example, for instance, a value of 4000 is in channel number 400, a value of 8000 is stored in channel number 800, and so on. The circle in the top right of the diagram represents an expanded portion of the memory: values within a given range (here assumed to be 10 units) will be stored in a single channel. This represents the resolution of the system. (b) The left-hand side of this diagram shows an imaginary section of memory where the memory locations are the channel numbers. The numbers inside the memory unit represent the number of counts recorded in a set time in each channel. These can then be represented in the form of a histogram, as illustrated on the right

The number of channels that the ADC operates over, i.e. the ADC resolution, depends on the resolution of the particular detector. The normal measure of detector resolution is that of the width of a peak in the spectrum at a level which is exactly half the maximum peak height (the *centroid* of the peak). The width will normally be in keV and the height in counts per channel. This is known as the *full-width at half-maximum* (FWHM). It is usual to have at least four or five channels covering the FWHM of a peak. Thus, if a gamma-ray detector has a resolution of 1.8 keV, then with five channels covering this range, it would require a resolution of about 0.35 keV per channel. Therefore, the number of channels required to cover the whole gamma-ray spectrum will obviously depend on the maximum gamma-energy to be analysed. For gamma-ray spectroscopy with modern hyperpure Ge detectors, either 4096 or 8192 channels are commonly used, but with the very large Ge detectors, 16384 channels may be required to cover the energy range. This number is known as the *conversion gain* of the ADC. A typical hyperpure Ge spectrum collected over 4096 channels is shown in Figure 8.14. We shall say more about detector resolution and its measurement in Section 8.4.

While the ADC is processing a signal, an input gate is closed and remains closed until the processing is completed. This disables the ADC as far as accepting any further signals from the amplifier for processing. This therefore represents a *dead time* for signals reaching the ADC and such signals are not processed. The period that is available for processing signals by the ADC is the *live time*. It is obvious that the higher the count rate of a source, then the more signals arrive during the dead period and so cannot be processed. If nothing were done about this it would be necessary to correct the data for the dead time. In fact, the ADC has a method of compensating for this dead time by monitoring the live time of the ADC, i.e. the

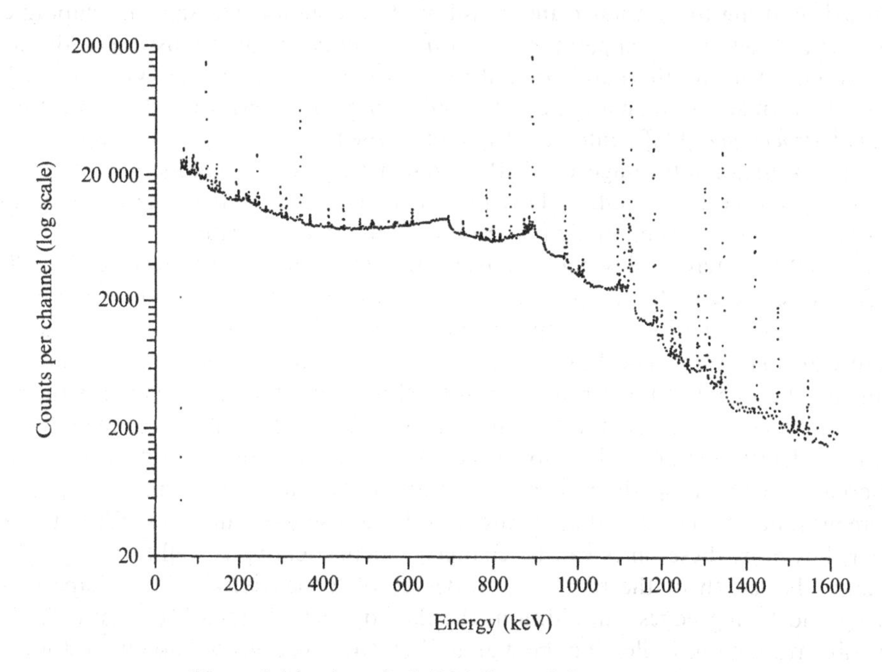

Figure 8.14 A typical 4096-channel Ge spectrum

time the input signal gate is open to receive pulses. As soon as the ADC starts to process a signal, the live time clock is disabled. It only starts again when the gate is opened and a further signal can be processed.

Counts carried out for the same live times are then directly comparable. Thus, it is the norm to count for a pre-selected live time and to determine count rates as counts per unit of live time. As the count rate increases, so the dead time also increases and the longer the 'real time' (time measured on an independent clock); this is also normally recorded. The ratio of *clock time* to live time gives the dead time and the latter is usually indicated on an ADC by a meter or digital output (as a percentage of the live time).

The memory/display

Once the ADC has completed processing a pulse and has assigned it an address (channel number) this enables an equivalent memory location to be addressed. Every time a pulse with that address is received, the count in that memory location is incremented by one. Conventional computer memories are used for storing the digital pulses. Displays are also normally conventional graphic computer displays but dedicated systems may use much smaller display units.

Digital spectroscopy systems

In the conventional analogue system described above, pulse shaping is accomplished by using the amplifier and possibly other components and the shaping can vary as a function of temperature and other operating conditions. Since the mid-1990s, however, an alternative digital method has been available (Vo *et al.*, 1998; Vo, 1999). In such a digital system, the pulse shaping is performed by using *Digital Signal Processing* (DSP) integrated circuits, which results in a much more stable system. A major advantage of DSP is that a very wide combination of shaping parameters become available. This means, in effect, that the detector is always operated in the optimum mode for resolution at any count rate.

In a DSP system, following a conventional (possibly gated) amplifier is a *flash* ADC, which repeatedly samples each analogue signal and converts it to the requisite digital value. This digital output is then processed by a digital filter according to a special algorithm. The function of the filter, as in an analogue system, is to enhance the signal-to-noise ratio. It can be shown that the ideal shape for the filter is a 'cusp', i.e. a pulse with a concave rise and fall time. In practice, the effects of ballistic defects require a flat top to the cusp. For an analogue system, the closest approach to this cusp shape is a quasi-trapezoidal pulse shape and changing the shaping times in such a system changes both the rise and fall times. With a digital filter, however, these times can be changed independently, as well as being able to change the length of the flat top and degree of 'concavivity' of the shape of the rising and falling edges. In addition, the flat top can be 'tilted' to ensure the best possible resolution. Following the digital filter, the functions of baseline restoration, fine gain and spectrum stability are all achieved with digital stability and accuracy.

At low counting rates, a digital system provides an overall resolution comparable to the best analogue systems but gives much better long-term stability in resolution, peak shape and peak position stability. This is of particular importance in environmental counting when counting times are frequently very long (often extending over several days) and stability is of prime importance.

8.4 THE ENERGY RESOLUTION OF A DETECTOR

The energy resolution of any detector is ultimately dependent on the number of 'charge' carriers produced from the deposition of a given amount of energy in the detector, with the resolution improving as this number increases. In the case of semiconductor detectors, the charge carriers are electron–hole pairs, for scintillator detectors it is the number of photoelectrons emitted by the photocathode, and for gas-proportional detectors it is the number of ion-pairs produced. The number of such charge carriers is dependent on the W value of the detector material (see

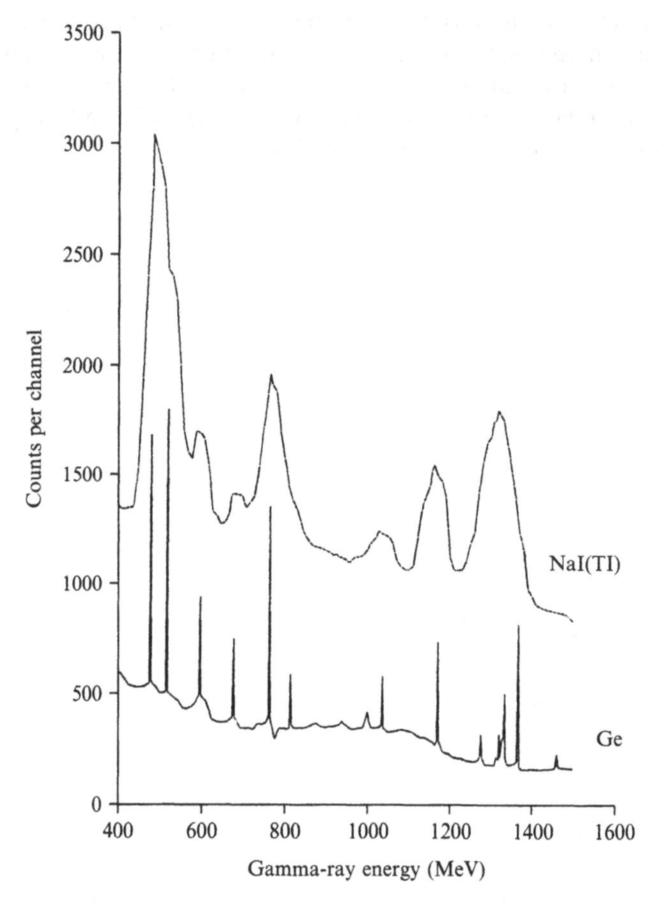

Figure 8.15 Comparison of Ge and NaI(Tl) spectra, showing the greater resolution of the former

Section 8.2.2) and from this it is obvious that semiconductor detectors will have much better resolution than any other type of detector.

Following on from the definition of W, we would anticipate that the number of charge carriers, N, formed in a given substance by the deposition of an amount of energy, E, would be given by the following expression:

$$N = \frac{E}{W} \tag{8.1}$$

In fact, we would expect this number, N, to fluctuate about a mean value and this fluctuation should be governed by Poisson statistics (see Section 11.3 below) with the fluctuation in N given simply by $\sqrt{N}$. This fluctuation in the number of charge carriers formed will set the ultimate energy resolution of any detector, with the resolution, R, being proportional to $\sqrt{N}/N$, i.e. $1/\sqrt{N}$, although other factors also have to be taken into account. The resolutions that we actually measure are not detector resolutions as such but *system* resolutions, containing various noise contributions. The measured resolution of a good Ge detector might be about 1.7 keV at 1.33 MeV. Detector resolutions at other energies are roughly proportional to $\sqrt{E}$, where E is the energy of the radiation. The difference in the resolution of a Ge detector compared to that of a 7.5 cm × 7.5 cm NaI(Tl) crystal is illustrated in Figure 8.15 and indicates the enormous problems that arise in trying to analyse a complex gamma-ray spectrum with a NaI(Tl) detector.

9

Measurement Techniques and Procedures

In Chapter 8, we discussed the detection and measurement of the radiation emitted by the sample being counted. Now we need to look at how we prepare a suitable sample for counting and the requirements of particular types of detector. Before any counting of samples begins, however, it is necessary to decide on whether the measurement is to be a 'relative' one or an 'absolute' one.

In a relative measurement, one is simply comparing a set of measurements under a given set of conditions and the results are reported as counts or counts per unit time. For absolute measurements, the absolute disintegration rate of the sample is required and results are reported in becquerels (for an integrated count) or becquerels per unit time. The constraints on the former are much less rigid than when an absolute measurement is required. In general, for relative measurements all that is required is that the samples have approximately the same counting geometry and radiation absorption behaviour. For spectrometric measurements, the detector need only be energy calibrated (see Section 9.1 below). Absolute measurements on the other hand, require not only a detailed knowledge of the counting geometry but the precise maintenance of that geometry over a series of measurements. It is also necessary to determine or make allowance for absorption of radiation in the sample. In addition, the detector must be calibrated so that its efficiency is known and an energy calibration must also be performed for spectrometric measurements. In this chapter, we will discuss both types of measurement.

9.1 ENERGY CALIBRATION OF DETECTORS

The general approach for calibrating any type of detector is to use a set of sources emitting particles or photons of known energy and fairly evenly spaced throughout the total energy range to be covered. This latter requirement is not always easy to meet and frequently a compromise has to be made between availability of suitable sources and the spacing of the energy intervals. For gamma-ray detectors (NaI(T) and Ge detectors), there are several gamma-ray sources emitting gamma-rays of very precisely known energy. They should cover the energy range from about 60 keV up to approximately 2 MeV. This should be extended to about 3 MeV if gamma-rays in the ^{232}Th decay series are to be determined. Many suitable sources have been suggested for this purpose and some of the most commonly used ones are listed in Table 9.1. These data are based on the compilation of evaluated gamma-ray sources from the International Atomic Energy Agency (IAEA) in Vienna, Austria.

Some of these sources emit just a single gamma-ray, giving rise to just one photo peak in the spectrum, e.g. ^{241}Am and ^{137}Cs. Most of the sources, however, emit several gamma-rays. In any case, it is usually necessary to count several sources to cover the appropriate energy range. A useful source for energies up to 1.5 MeV is ^{152}Eu, which emits a large number of gamma-rays from 122 keV up to 1.4 MeV. It should be noted, however, that modern amplifiers and ADCs are very linear and in theory require only two energies for calibration, i.e. one at the low-energy end of the spectrum and the other at the upper end. Some commercial systems, in fact, only allow for two energies to perform an energy calibration.

For the calibration of alpha-detectors, again a number of sources may be used and some of the most useful are listed in Table 9.2. The restricted energy range of most alpha-emitting nuclides means that the number of alpha-energies required for calibration is usually quite small.

In practice, one or more sources is placed in front of the detector at appropriate distances to give a reasonable dead time for the particular detector concerned (e.g. about 20% for a semiconductor detector but no more than 10% for NaI(Tl) if the FWHM of the peaks is to be determined at the same time, which is often the case). The gain on the amplifier is then adjusted such that the peak corresponding to the highest-energy alpha- or gamma-rays from the calibration sources is in a channel of the MCA that will allow the highest-energy alpha- or gamma-rays from the samples to be determined to fall within the range of the MCA but near to the upper end of that range. It should also be noted that the energy resolution measured will depend on the amplifier time constants and these should be the same as the ones recommended by the detector supplier.

9.2 EFFICIENCY CALIBRATION OF DETECTORS

The ultimate aim of most environmental radiation measurements is to determine the absolute activity of a sample in terms of becquerels per unit time and usually to determine the dose rate attributable to that sample. Absolute activity measurements require a knowledge of the absolute *counting efficiency* of a given detector. The definition of absolute counting efficiency, ε, is given as follows:

$$\varepsilon = \frac{\text{number of counts recorded per unit time}}{\text{number of particles or photons emitted per unit time}}$$

The absolute counting efficiency depends on two factors, namely (a) the absolute *detection efficiency*, i.e. the number of the given radiation entering the sensitive volume of the detector and resulting in a measurable pulse, and (b) the *counting geometry*.

There are two basic approaches to the problem of determining the counting efficiency. The first is essentially a theoretical one in which all of the relevant

Table 9.1 Nuclides useful for the calibration of gamma-ray detectors (from IAEA, 1991).

Nuclide	Half-life (d)	Energy (keV)	Emission probability
^{22}Na	950.8 ± 0.9	1274.542 ± 0.007	0.999 35
^{46}Sc	83.79 ± 0.04	889.277 ± 0.003	0.999 844
		1120.545 ± 0.004	0.999 874
^{54}Mn	312.3 ± 0.4	834.843 ± 0.006	0.999 758
^{56}Co	77.31 ± 0.19	846.764 ± 0.006	0.999 33[a]
		1037.844 ± 0.004	0.1413
		1238.287 ± 0.003	0.6607
		1771.350 ± 0.015	0.1549
		2034.759 ± 0.011	0.077 71
		2598.460 ± 0.010	0.1696
		3253.417 ± 0.014	0.0762
^{57}Co	271.79 ± 0.09	14.4127 ± 0.0004	0.0916
		122.0614 ± 0.0003	0.8560
		136.4743 ± 0.0005	0.1068
^{60}Co	1925.5 ± 0.5	1173.238 ± 0.004	0.998 57
		1332.502 ± 0.005	0.999 83
^{65}Zn	244.26 ± 0.26	1115.546 ± 0.004	0.5060
^{85}Sr	64.849 ± 0.004	514.0076 ± 0.0022	0.984
^{88}Y	106.630 ± 0.025	898.042 ± 0.004	0.940
		1836.063 ± 0.013	0.9936
^{109}Cd	462.6 ± 0.7	88.0341 ± 0.0011	0.0363
^{113}Sn	115.09 ± 0.04	391.702 ± 0.004	0.6489
^{134}Cs	754.28 ± 0.22	569.328 ± 0.003	0.1539[a]
		604.720 ± 0.003	0.9763
		795.859 ± 0.005	0.854
^{137}Cs	11 020 ± 60	661.660 ± 0.003	0.851
^{133}Ba	3862 ± 15	80.998 ± 0.005	0.3411
		276.398 ± 0.001	0.071 47
		302.853 ± 0.001	0.183
		356.017 ± 0.002	0.6194
		383.851 ± 0.003	0.089 05
^{139}Ce	137.640 ± 0.023	165.857 ± 0.006	0.7987
^{152}Eu	4933 ± 11	121.7824 ± 0.0004	0.2837
		244.6989 ± 0.0010	0.0753
		344.2811 ± 0.0019	0.2657
		411.126 ± 0.003	0.022 38
		443.965 ± 0.004	0.031 25
		778.903 ± 0.006	0.1297
		867.390 ± 0.006	0.042 14
		964.055 ± 0.004	0.1463
		1085.842 ± 0.0.004	0.1013
		1089.767 ± 0.014	0.017 31
		1112.087 ± 0.006	0.1354
		1212.970 ± 0.013	0.014 12
		1299.152 ± 0.009	0.016 26
		1408.022 ± 0.004	0.2085
^{203}Hg	46.595 ± 0.013	279.1967 ± 0.0012	0.8148
^{241}Am	157 850 ± 240	26.345 ± 0.001	0.024
		59.537 ± 0.001	0.360

[a]Only the most prominent lines are listed.

Table 9.2 Nuclides useful for the calibration of alpha-particle detectors (from OECD/NEA Data Bank, 1993).

Nuclide	Alpha-particle energy (MeV)	Emission probability (%)
^{226}Ra	4.602	6
	4.785	94
^{228}Th	5.340	28
	5.423	72
^{232}Th	3.954	23
	4.013	77
^{238}U	4.150	23
	4.199	77
^{237}Np	4.640	6[a]
	4.766	8
	4.771	25
	4.788	48
^{240}Pu	5.124	27
	5.168	73
^{241}Am	5.443	13

[a]Only the most prominent lines are listed.

experimental parameters are calculated and then combined with the relevant basic physical data to calculate the efficiency. This is usually very difficult, and often impossible. Consequently, the approach normally adopted is to use an entirely experimental method and usually one which automatically results in the combined determination of both the detector efficiency and counting geometry from a single set of measurements to yield the required counting efficiency.

From the above discussion, it is perhaps not surprising that determining the efficiency calibration for a detector is a rather more complicated operation than that for energy calibration. It also differs markedly according to the detector type and counting system. Two features, however, are common to all efficiency calibration measurements. The first is that the counting geometry of the calibration source must match that of the sample as accurately as possible. The term 'counting geometry' includes the physical shape and size of the actual source, together with its position relative to the detector. For alpha- and liquid-scintillation counting, the geometries used for environmental samples may not differ from those normally used for other types of nuclear measurements but for gamma-ray detectors the counting geometry for an environmental sample may well be quite different. Secondly, sources of known absolute activity must be used and this may present problems for some counting systems.

We will now briefly outline some suitable methods for the calibration of different types of detector and counting systems. Only Geiger–Müller, semiconductor, gas-flow, liquid scintillation, gamma-ray and solid-state nuclear-track detectors will be considered as these are normally the only detectors used for absolute activity measurements.

9.2.1 Geiger–Müller Detectors

Although Geiger–Müller (G–M) detectors are widely used in environmental radiation detection and measurement they are relatively little used for absolute measurements. This is because for most purposes other detectors are more suited to this role. They are, however, sometimes used for the determination of the gross beta-activity of certain materials. The detector will normally be of the end-window type (see Chapter 8).

Gross beta measurements

For gross beta measurements involving a complex source then calibration is often a ^{40}K standard, consisting of a suitable potassium salt or solution (potassium chloride is often used). Two types of source may be used for such measurements. The first is known as an *infinitely thick* source, which is so thick that the beta-activity beyond a certain depth cannot escape to be counted, irrespective of the beta end-point energy. A depth of about 2 cm is sufficient and the material can be packed into a suitable cylindrical container. A source prepared in this way will give a constant count rate, provided that the surface area remains constant. The ^{40}K standard will be similarly infinitely thick and prepared in an identical container. No self-absorption correction is required with this method.

Alternatively, a thin source providing minimum self-absorption can be prepared and methods of preparing thin, uniform sources are discussed in Section 9.4.2. Their obvious advantage is in the counting of beta-emitters with a low end-point energy, since infinitely thick sources severely discriminate against such activity. On the other hand, these thick sources give much higher count rates than thin ones. The other point to note is that these detectors are sensitive to alpha-particles but if these are present they will either be absorbed in the window of the G–M tube or easily stopped with a thin, external absorber.

Quantitative measurements involving a single beta nuclide

Counting of a specific, beta-emitting nuclide almost always follows the radiochemical isolation of the nuclide (see Section 9.3). The absolute calibration of the detector involves preparing *correction curves* in which a series of radioactive sources are prepared, each containing exactly the same amount of the activity of the particular nuclide but progressively diluted with increasing amounts of the inactive matrix material. A plot of the activity of the source as a function of the source thickness (measured in mg cm^{-2}) is then prepared, as shown in Figure 9.1. By knowing the true value of the activity added, one can then correct for the self-absorption at each thickness value from the experimental activity and hence determine the counting efficiency. It is also possible, in principle, to determine the absolute activity of the nuclide by extrapolating to zero thickness. In practice, this is not a straightforward operation due to increasing back-scattering from the source as the thickness decreases (see Figure 9.1).

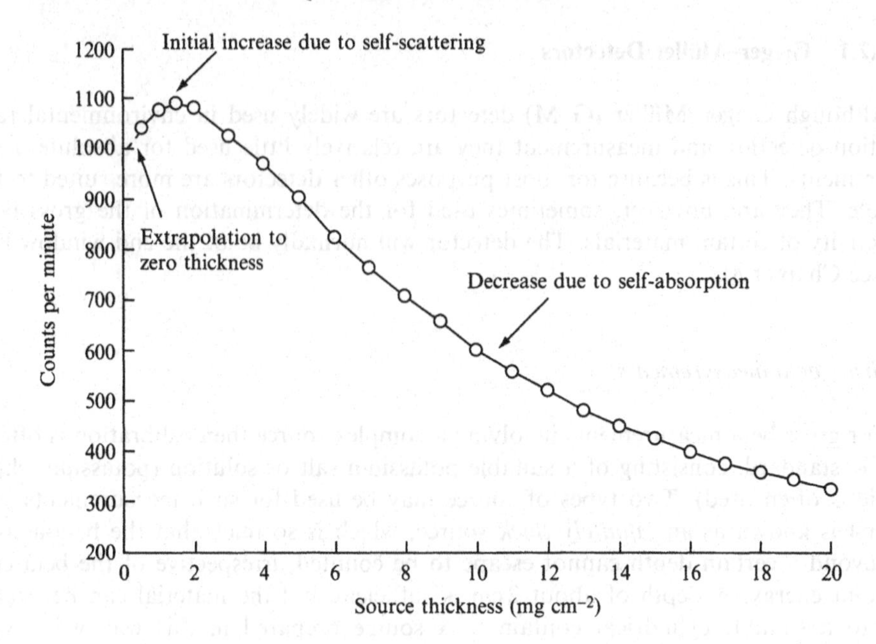

Figure 9.1 A plot of the activity of a source as a function of the source thickness (mgcm^{-2}) when the total activity of the source remains constant

A method for preparing sources and backing supports of negligible mass for the absolute calibration of solid beta-sources involves the preparation of a backing support, on which the source is mounted from an organic polymer known as 'VYNS' (poly(vinyl chloride) – $C_{22}H_{33}O_2Cl_g$). It is possible to prepare self-supporting films as thin as 5–10 μg cm^{-2} on an aluminium annulus (about 2.5 cm internal diameter). This can now be used directly as the source mount. The source is constrained to a given area by pipetting a few drops of 'wetting agent' (e.g. insulin solution) onto the centre of the film and allowing it to dry. If a solution of the nuclide is now carefully pipetted onto this wetted area and the film continuously rotated while the source is drying, then a very uniform and thin source can be prepared. Due to the low atomic number and thickness of the backing, back-scattering is negligible, thus allowing absolute activities (and efficiencies) to be accurately determined.

Even sources prepared by this method may produce significant self-absorption for low-energy beta emitters. Here, a modification of the above method developed by Randle (1967) can be used to produce sources with essentially negligible self-absorption. It is relatively easy to form a double film of polystyrene on top of the VYNS. The combined film thickness is still only about 10–15 μg cm^{-2}. A circular area of the polystyrene is sulfonated with sulfuric acid but only to a depth of a few nanometres (see Figure 9.2). This sulfonated polystyrene now has ion-exchange properties (see Section 9.3.3) and will chemically bind many metal cations. Thus, by placing a small volume of radioactive solution of a metallic nuclide on this sulfonated layer, an exact fraction of the nuclide is bound to the sulfonate groups to produce sources only a few μg cm^{-2} thick.

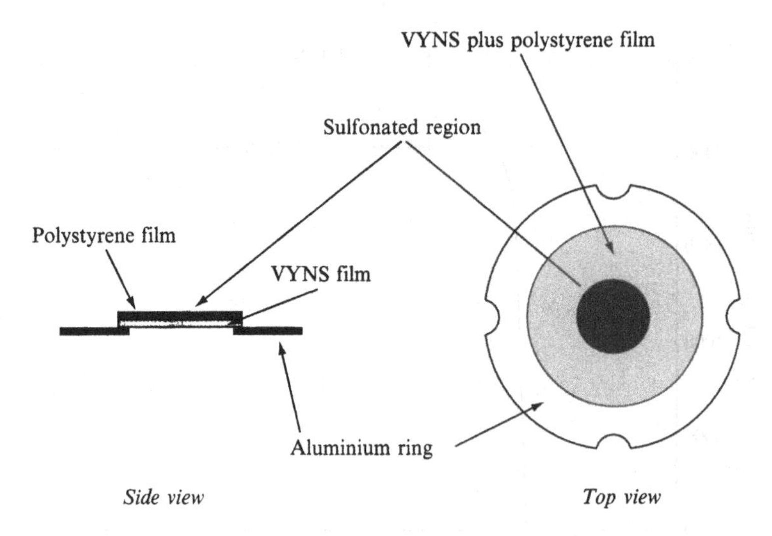

Figure 9.2 VYNS and polystyrene films supported on an aluminium ring; the sulfonated region of the polystyrene is also shown

For either type of measurement it is necessary to find the optimum high voltage to apply to the detector. Referring to the earlier discussion in Section 8.1.1, it was shown that the magnitude of the output pulses from a G–M tube vary with the voltage applied to the tube. There is a threshold voltage (typically a few hundred volts), below which the pulses produced by the detector are too small to trigger the counting system. Due to slight fluctuations in pulse size, the count rate then rises rapidly above this threshold voltage to the point where all the pulses are counted and the count rate versus applied voltage becomes relatively constant but still shows a constant rise (see Figure 9.3). This is known as the *plateau region*. At a sufficiently high voltage, this plateau ends due to the onset of the continuous discharge mechanism and the count rate rapidly increases again. The G–M tube is normally operated at a voltage which lies in the middle of the plateau.

9.2.2 Semiconductor Detectors for Alpha-Emitting Nuclides

Sources for either alpha- or beta-spectrometry using silicon surface-barrier detectors requires the preparation of very thin, uniform sources. The techniques used for producing such sources are discussed later in this chapter (Section 9.4.2). The response of surface-barrier detectors for alpha-emitters is essentially constant over the range of alpha-energies encountered in environmental studies (about 2–10 MeV). The calibration sources for transuranic nuclides are normally prepared by electrodeposition of a mixed radionuclide standard solution obtained from one of the recognized national standards laboratories, e.g. the National Physical Laboratory (NPL) in the UK, the National Institute of Standards and Technology (NIST) in the USA, or from a source traceable to such an organization. It is also possible to obtain single nuclide alpha-sources to check counting efficiency.

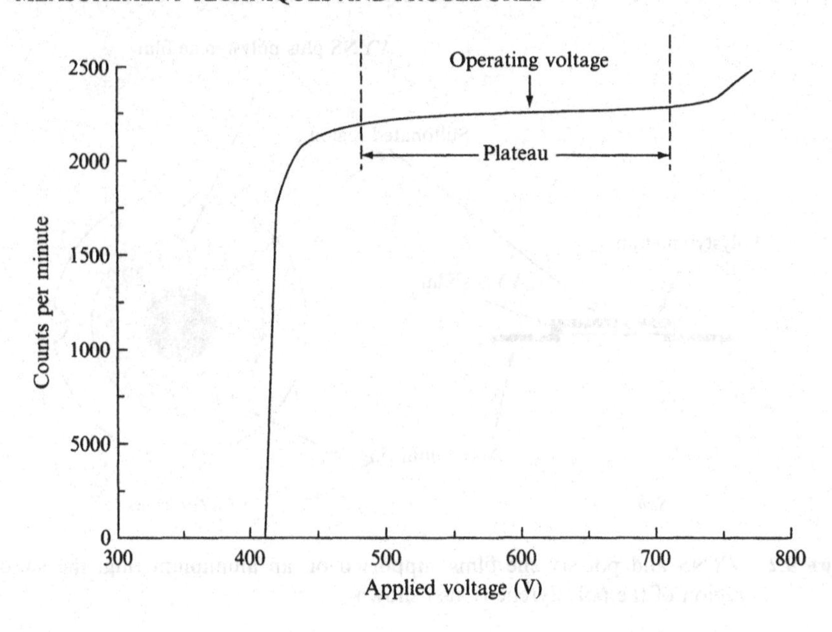

Figure 9.3 The count rate as a function of applied voltage for a Geiger–Müller detector; the plateau and operating voltage are also indicated

9.2.3 Gas-Flow Detectors

Gas-flow counters, operating in the proportional or Geiger region, offer a number of advantages in the counting of environmental beta-activity, such as low backgrounds, high stability and the potential for very thin windows to allow the counting of very soft beta-radiation. These advantages, combined with their high efficiency of counting, offer the best method of determining environmental beta-activity. Proportional counters are also used for the absolute determination of alpha-sources and radioactive noble gases.

Because alpha-sources for counting with a gas-flow counter are usually prepared by electrodeposition of the source onto a metal backing, calibrated, alpha point sources on metal backings, available from several national standards laboratories, may be used to calibrate the detector. Similarly, large-area sources may also be obtained for calibration purposes. Calibration of these detectors for counting beta-sources is exactly the same as that described for G–M detectors.

The problem of determining activities in natural matrices was addressed several years ago and a number of 'natural matrix standards' have been prepared, ranging from sediments to organic tissue. These contain beta- and gamma-emitting nuclides, as well as alpha-emitters. Although they may be used as calibration standards, they were primarily developed to test radiochemical procedures.

9.2.4 Liquid Scintillation Detectors

Liquid scintillation counting of environmental samples, used almost exclusively nowadays for counting alpha- and beta-activity, usually requires the sample to be dissolved in the scintillator cocktail (see Section 8.1.2), although activity on filter papers suspended in the scintillator, suspensions of fine powders in the scintillator, use of emulsions, etc. are alternative counting methods that may be used with liquid scintillators. Here, we shall confine the discussion to samples soluble in the scintillator cocktail.

Beta-activity can include both negatrons or positrons, although the latter are rarely met in routine samples. Since beta-particles are emitted with a spectrum of energies ranging from zero to some maximum energy (see Chapter 3) the pulses produced by the detector have a corresponding range of sizes. The existence of this low-energy 'tail' for all beta-emitters means that theoretically they can never be counted with 100% efficiency. Beta-emitters with high maximum energies (e.g. ^{32}P) can be counted with close to 100% efficiency since the fraction of particles with energies below the counting threshold is very small and effectively lost in the counting statistics. In fact, the counting efficiency for alpha- and beta-particles with energies greater than 5 keV is theoretically 100%.

In general, alpha-emitters are more easily counted than beta-particles, in spite of a lower scintillation yield. This is due to the high energy of alpha-particles emitted and the fact that the alpha-particle is emitted with a line spectrum and hence has no low-energy tail like a beta-particle. One hundred percent counting efficiency is realized for many scintillation systems where no quenching occurs and since the counting geometry is also 100%, the count rate gives the activity directly.

Quenching is the absorption of light photons prior to their detection by the photomultiplier and results in an apparent decrease in the energy deposited by the radiation in the scintillator. This effectively shifts the energy spectrum to lower values. The effect is most marked for beta-emitters such as tritium which have low end-point energy maxima and results in counting efficiencies considerably less than 100%. Quenching can be caused by coloured solutions or the presence of components which absorb the photons, such as many organochlorine compounds. If quenching is occurring in the scintillator, as will frequently be the case for environmental samples, then the efficiency must be determined experimentally.

Several methods are available for determining the counting efficiency in liquid scintillation systems and are fully described in the texts listed in the Bibliography at the end of this present book, e.g. Dyer (1974) and Horrocks (1974). One reliable but very time-consuming method for single-isotope counting is that of *internal standardization* in which the sample is first counted and then counted a second time after the addition of a known activity of the isotope. The counting efficiency is simply the ratio of the count rate to the known disintegration rate of the added activity. There are, however, several disadvantages to this method and consequently instrumental techniques have been developed to determine the counting efficiency and are normally used in modern liquid scintillation counters. The first of these is known as the *sample channels ratio* method and uses two counting channels, with one set to cover the bulk of the spectrum while the other covers just the low-energy portion. Count rates in each channel of a progressively quenched standard allow the counting efficiency to be determined. Certain disadvantages of this method are overcome by the use of the *external standard ratio* method. Again, two counting channels are used but the counts in these are produced by Compton electrons produced by bringing a gamma-ray source close to the counting vial.

A count is taken of the sample of known activity in the normal counting channel. Then, the gamma-ray source is positioned by the counting vial and the counts recorded in the two external channels. The ratio of the counts in the external channels is then plotted as a function of the efficiency derived from the counting channel data.

Modern liquid scintillation counters incorporate microcomputers which allow these quench correction curves to be determined automatically and to allow for automatic quench compensation, although these methods are not suitable for tritium counting. Equally, modern instruments allow for dual-isotope counting, e.g. ^{14}C and tritium.

Very recently, small, portable liquid scintillation counters have been developed. These have the advantage of enabling counting to be performed on the spot at the sampling site. Thus, radon and radium (^{226}Ra) in water samples have been measured in this way. Similar instruments have been used for uranium detection also – in water.

9.2.5 Gamma-Ray Detectors

NaI(Tl) detectors

In general, NaI(Tl) detectors are not much used for gamma-ray spectrometry and find limited application in environmental determinations of activity. Their main use nowadays is in general surveys of large land areas for the estimation of the local uranium, thorium and potassium concentrations, or for a rapid estimation of these concentrations in environmental samples when large numbers of samples have to be measured in the laboratory. With the advent of very large Ge detectors, however, their use in even this role is becoming more limited.

Since the count rates for typical environmental samples are very low, sample size tends to be large (0.5–1.0 kg is usual). Samples are normally packed as powders in cylindrical containers with a diameter approximately that of the crystal (typically 7.5 cm) or in special *Marinelli beakers* which allow the detector to be surrounded by an annular sample. We shall discuss the use of Marinelli beakers more fully in the next section.

For laboratory measurements, the gamma-ray spectrum is divided into three broad energy regions, as shown in Figure 9.4, representing the 'thorium range' (2.38–2.62 MeV), the 'uranium range' (2.2–1.48 MeV) and the 'potassium range' (1.46–0.97 MeV). The efficiency calibration and special spectral factors are obtained by counting separate samples containing known concentrations of one each of the three elements, using the same container as for the sample count. Calibration of NaI(Tl) detectors for gamma-ray surveying is discussed further in Chapter 11.

Ge detectors

As with NaI(Tl) detectors, the low levels of activity usually encountered with environmental samples usually require the use of similarly large sample sizes (0.5–1 kg). The most popular counting configuration for such samples involves the use of a Marinelli beaker (see Figure 9.5). A standard Marinelli beaker has a volume of 0.5

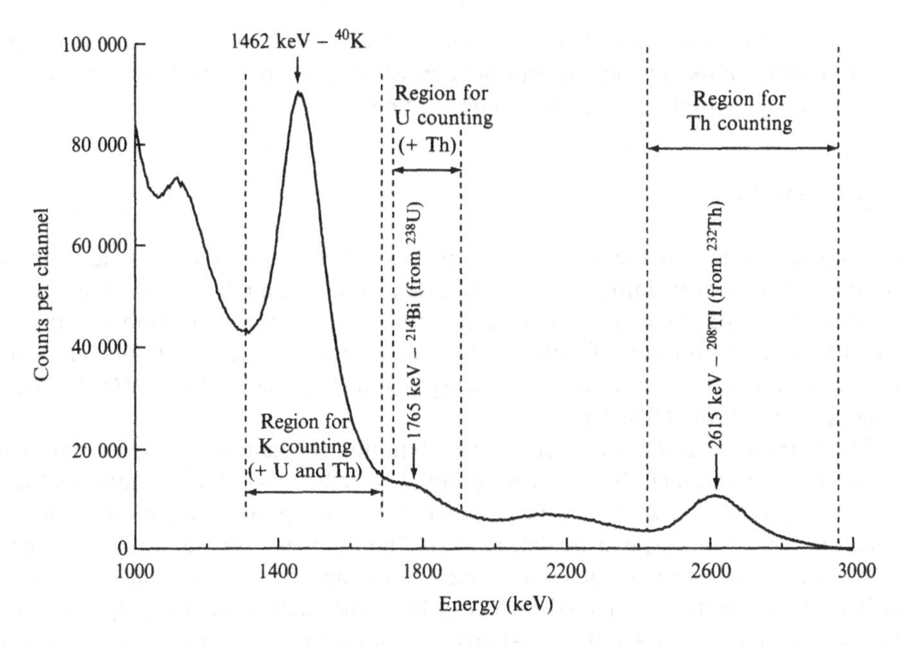

Figure 9.4 Part of a gamma-ray spectrum of a sediment sample obtained with a NaI(Tl) detector showing the peaks from ^{40}K, ^{232}Th and ^{238}U used for the analysis of these radionuclides. The total count in the thorium region ($\sim$ 2.4 MeV– $\sim$ 2.9 MeV) contains only counts from ^{232}Th, whereas the total count in the uranium region contains a contribution from ^{232}Th and that in the potassium region contains contributions from both ^{232}Th and ^{238}U

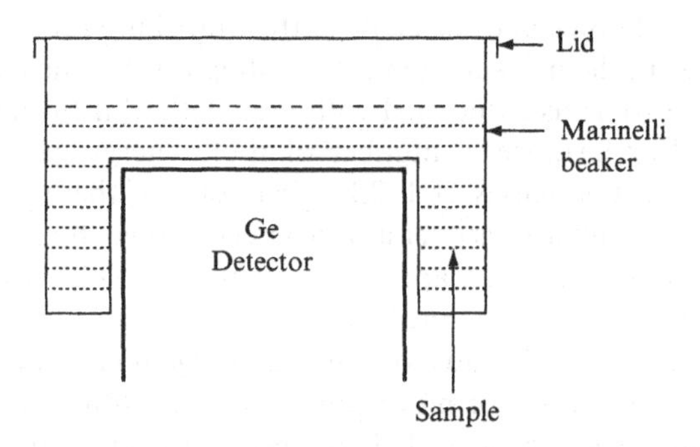

Figure 9.5 Schematic of a filled Marinelli beaker in position on a Ge detector

or 1 l and is always filled to the same height, to maintain a constant counting geometry. The problem lies in producing efficiency standards of the required size and configuration, with the activities uniformly distributed throughout the matrix. The latter for the calibration standard must mimic that of the sample to produce the same overall gamma-ray absorption. This usually requires several standards to be prepared to cover the range of matrices normally encountered in environmental studies. Common matrices that are used include water, suitable for most liquid samples, dried milk powder to represent most dry, organic materials, and one or more sediment samples to represent soils, sediments, etc.

Liquid standards

The liquid standard is the easiest to prepare since it only requires the appropriate dilution of a known amount of a calibrated activity standard, available from a number of companies and from some national laboratories. For example, the National Physical Laboratory (NPL) in the UK supplies a mixed radionuclide standard, with gamma-rays covering the energy region from 60 keV to 1.9 MeV, with a total activity of only 6000 Bq.

The efficiency calibration involves first obtaining the gamma-ray spectrum of the solution in the Marinelli beaker, with good (1% or better) counting statistics for all the relevant photo peaks. From the net area under each photo peak and the counting time, the count rate, R, can be determined. The number of gamma-rays emitted at each photo peak energy is then determined. This can be obtained from the reported activity of the nuclide, corrected for decay from the calibration date, A_P, multiplied by the gamma-ray emission probability, or branching ratio for that gamma-ray energy, B. The photo peak efficiency, PE, is then given by the following expression:

$$PE = \frac{R}{A_P B} \tag{9.1}$$

This procedure is then repeated for each gamma-ray energy.

Apart from counting statistics, the overall error in the photo-peak efficiency determined in this way will depend on the accuracy with which the absolute activity of the calibration source is known and the error in the branching ratio. The latter may well be the largest error in the measurement, depending on the nuclide concerned. There may also be other problems associated with this method if the activity source is not carefully chosen. For instance, a number of multi-energy gamma-ray sources are available with certified activities of 1–2%. Examples of these are ^{152}Eu, ^{226}Ra and ^{56}Co. These all emit gamma-rays over a wide energy range, their branching ratios are well known and at first sight all appear eminently suitable as activity sources for efficiency calibrations. The major problem with such sources is that they emit many gamma-rays in cascade, which leads to *coincidence–summing losses*.

Such losses occur when two or more gamma-rays of different energy arrive at the detector within the resolving time of the detection system. This will result in these gamma-rays being counted as a single gamma-ray and a signal produced corresponding to an energy equivalent to the combined energies of the gamma-rays concerned. This may result in two errors. First, counts corresponding to the individual gamma-ray energies are lost. Secondly, if the combined energy corresponds exactly to the energy of a single, higher-energy gamma-ray, then a spurious count is recorded for this gamma-ray. It is not always realized that such coincidence–summing losses can be quite appreciable even when very low activities are being counted, particularly if a large Ge detector is being used. Coincidence–summing losses may often be significant even for nuclides such as ^{134}Cs, which has a relatively simple gamma-ray spectrum. Even values for ^{60}Co can be low on this account, yet this nuclide emits only two gamma-rays.

It is possible to calculate the coincidence–summing losses and a number of computer programs have now been written to perform these calculations for point sources. Several of the more important ones are discussed in Debertin and Helmer (1988), found in the Bibliography at the end of this present text, but these still require quite an extensive range of experimental input to the programs, including peak and total efficiencies. Summing correction factors for several environmental radionuclides in a counting geometry consisting of a 1 l beaker placed on top of a detector have been published which indicate that correction factors of about 1.05 are typical. Correction factors for a Marinelli geometry would be expected to be similar to this value.

Solid standards

Preparation of solid efficiency standards is more problematical. This results from the requirement to have the activity uniformly distributed throughout the sample. Thus, the IAEA distribute standards for several marine sediments, a lake sediment and a soil containing a limited number of certified nuclides. The lake sediment with a certified content of ^{137}Cs is available in quantities of up to 1.2 kg but the others are of much smaller sizes.

The alternative approach to these certified matrix reference materials and the one usually adopted is to adsorb a known amount of the liquid standard into a suitable matrix. This technique, however, has a number of drawbacks and it is extremely difficult to obtain a homogeneous source in this manner. One of the best methods which has been described is that due to Croudace (1991). A certifiable and traceable multi-isotope liquid, of the type described above, is used as the source. An accurate aliquot of this solution (containing about 1800 Bq of activity) is added to a mixture consisting of a few grams of anion- and cation-exchange resins, together with chromatographic-grade cellulose powder, and the mixture is then thoroughly stirred until the solution has evaporated to near dryness. This ensures quantitative uptake of all of the activity by the solid components, either by ion-exchange or adsorption. This active material is then dried and pulverized and the resulting powder is then thoroughly mixed with a given matrix powder and transferred to a Marinelli beaker. Tests have shown that the homogeneity of standards prepared in this way is within 1–3 %.

It is not always possible to obtain sufficient sample to fill a standard Marinelli beaker. In such cases, smaller standards of comparable volume to the sample will have to be prepared. In order to reduce the effort required in producing large numbers of standards to comply with every possible sample size, it is best to decide on a small number of sample volumes, ranging from those equivalent to a few tens of grams up to Marinelli volumes. The equivalent standards can then be prepared in the manner just described.

In some cases, a completely different approach to the preparation of standards has to be used because of the particular requirements of the method of sample collection. Thus, in air pollution studies samples are collected on filter papers and counted as such. Standards here are produced by adsorbing a known amount of liquid standard into the filter paper to produce a source of the same aerial dimension. Small sources of this nature will result in more severe coincidence–summing corrections.

9.2.6 Solid-State Nuclear-Track Detectors used for Radon Counting

With these devices, it is possible to calculate the counting efficiency, or rather the relationship between the track density and the radon (in equilibrium with its daughter nuclides) activity.

Assuming that a 'can-type' geometry is used, i.e. an open-ended can with the SSNTD located at the other end, radon as a gas is distributed evenly throughout the can volume while the sensitive volume is restricted by the can wall. Remembering that the tracks recorded come from alpha-particles, a nucleus at a distance greater than the alpha-particle range cannot contribute to the track density. If any material other than CR-39 is used as the SSNTD, then there will also be a minimum distance from the detector in which no contributions are possible because the energy of the alpha-particle is greater than the maximum to produce an etchable track. There is also a minimum angle of incidence in which no track registration is possible. These parameters define the sensitive volume in the absence of any other restrictions. From the can geometry and the above parameters, it is possible to calculate the required relationship, if the assumption that the radon daughters are distributed uniformly on the can walls is valid.

9.3 RADIOCHEMICAL SEPARATION TECHNIQUES

Studies of environmental radioactivity usually involve the analysis of large numbers of samples. Consequently, techniques which involve the minimum of preparation are obviously favoured. In particular, the avoidance of techniques and manipulations requiring skilled personnel and expensive consumables are always avoided if at all possible.

The criteria of simplicity and minimum preparation are most easily met with gamma-ray counting. The equipment may be expensive but, particularly for high-resolution gamma-ray spectroscopy, sample preparation is minimal, although the interpretation of the data requires expertise. Thus, apart from simple gross beta measurements utilizing just a Geiger–Müller counter, some form of gamma-counting will usually be the preferred method wherever applicable.

If the information required can only be provided by either alpha- or beta-counting of specific radionuclides then this will inevitably involve some form of radiochemical separation. Alpha-spectroscopy is relatively straightforward but source preparation requires a certain amount of radiochemical separation, as does the avoidance of spectral interferences, in some cases involving a number of different alpha-emitters.

Finally, radiochemical separations may be required even in the case of gamma-emitting nuclides if the sought-for gamma-ray line is close to the limit of detection due to the matrix-induced background or is seriously interfered with by another activity.

Implicit in the dissolution of a sample is the requirement to separate out specific elements or small groups of elements which will allow their determination free from interferences. A number of standard, analytical separation methods have been adopted for the separation of radioactive species and are discussed in most textbooks on radiochemistry (see, for example, Geary (1986) and Tölgyessy and Bujdosó (1991), found in the Bibliography at the end of this present text).

Frequently, an additional aim of a separation is to *pre-concentrate* the nuclide(s). That is, to increase the final concentration of the radionuclide relative to its initial concentration in the sample, in order to increase the sensitivity of the determination. A dissolution procedure such as dry ashing (and to a lesser extent, wet ashing) will tend to provide a significant degree of pre-concentration, which in many cases may be sufficient for the particular analysis.

9.4 RADIATION MEASUREMENTS FOR ENVIRONMENTAL SAMPLES

In this section, we will discuss some of the more routine methods that are used in the measurement of environmental radiation. Such methods are grouped in relation to the type of detector used, and each type of detector is considered in turn. It must always be remembered that since we are measuring environmental radiation, this normally means that we are measuring very low levels of activity. Indeed, we may frequently wish to determine count rates that are little higher than that of the ambient background. In these circumstances, it is essential to reduce the background to the lowest possible value in order to obtain accurate results.

Consequently, before considering particular detectors in detail it is useful to consider some general techniques that are used with a number of counting systems to produce the lowest background possible. These are techniques particularly used in beta- and gamma-ray counting. In the first place, the detector must be essentially free of radioactive contaminants. Secondly, the detector must have a shield to absorb as much of the external radiation as possible. These are mainly cosmic rays and an anti-coincidence shield is also frequently placed around the detector to further reduce the counts due to cosmic rays. Finally, the whole counting assembly may be located inside a chamber constructed from materials with very low natural radioactivity levels. Thus, detectors have been placed in chalk caverns, and in rooms constructed of carefully selected cement, while several groups have favoured ultrabasic rocks such as dunite. We will consider particular assemblies in a little more detail in the appropriate sections.

9.4.1 Ionization Chambers

Ionization chambers are sometimes used in the measurement of gamma-ray dose rates. This is essentially the determination of the relationships between physical measurements of the environment and radiation doses. In Chapter 4, we saw that dose refers to the mean amount of energy adsorbed by a unit mass of matter. The usual unit of dose in environmental radiation is microgray per hour (μGy h^{-1}). An ionization chamber measures the exposure in units known as roentgens (R). The relationship between R and the adsorbed dose, D, in air is given by the following:

$$D = 0.008\,77\text{R Gy} \tag{9.2}$$

When the medium involved is not air, the relationship can be quite complex. Often, it can be shown that Equation (9.2) must be multiplied by the ratio of the mass-absorption coefficient of the medium to that of air.

Ionization chambers designed to measure the dose arising from gamma-radiation consist essentially of a plastic cylinder in which is contained a cavity ('the sensitive volume') filled with CO_2 gas and surrounded by graphite. The sensitive volume may be of the order of 50–100 cm^3. By assuming that the chamber may be approximated by a continuous carbon medium, and from the definition of the roentgen as 2.58×10^{-4} C kg^{-1}, it is possible to relate the ionization current in amperes to the dose rate. For a chamber of 100 cm^3, containing gas at atmospheric pressure, the ionization current corresponding to a dose rate of 1 R h^{-1} is of the order of 10^{-8} mA. This is easily measurable and in fact such an ionization chamber could be used for measurements of dose rates of the order of 0.001 Gy h^{-1}. With a vibrating-reed electrometer, the sensitivity may be improved by one or two orders of magnitude.

9.4.2 Geiger Counters

These are by far the most extensively used of all environmental radiation detectors and are employed for a number of purposes. One is as a general survey instrument for detecting the presence of radioactivity in a gross measurement of both beta- and gamma-activity. Another is for surveys to measure external gamma-dose rates from a particular source or, more commonly, from an areal site. Thirdly, they may be used for the determination of the gross activity of pure beta-emitting sources, either in the form of single isotopes or as more complex sources using an end-window G–M tube, as discussed in Section 9.2.1.

Survey instruments are usually used to establish normal background levels and to monitor any changes in this background as an initial alert of possible radiation hazards. Normally, a site will be chosen for such measurements which is either flat or only gently sloping, free from man-made structures and, as far as possible, unchanging.

Determination of gamma-dose rates with a G–M tube requires certain modifications to the normal detector, since the latter gives an energy-dependent response. An essentially flat response can be obtained above about 200 keV by suitable shielding of the end and sides of the tube with thin sheets of tin and lead. The detector must be calibrated in order to convert a count rate into a gamma-dose rate. This should be done with a source traceable to a national standards laboratory, e.g. the NPL in the UK. A suitable calibration source is ^{226}Ra of accurately known activity, placed at a known geometry to the G–M tube. Since the exposure in air per becquerel is known for this nuclide, it is thus possible to correlate a given count rate to the dose rate in the usual units of μGy h^{-1}. Once absolutely calibrated in this way it is normal to regularly check the constancy of the detector before a set of measurements by using a small gamma-emitting source.

The accurate determination of very low levels of activity using a G–M counter requires a very sophisticated counting arrangement, careful choice of detector and long-term stability of the electronic components. The latter arises because if, for instance, the background has been reduced to about 1 count per minute (cpm), to obtain an uncertainty of 1 % in the counting statistics then requires a total count of 10 000 (see Chapter 11), and hence a counting time of 10 000 minutes, which is almost a week. To achieve counting statistics better than 1 % will require far longer counting times. Since counting statistics are only one source of error, 1 % may represent the maximum allowable error.

As we mentioned earlier, to achieve backgrounds as low as 1 cpm with a G–M tube requires a number of features. It should be remembered that a typical background count for an unshielded Geiger tube is of the order of 100 cpm. In the first place, the detector must be made of low-background materials, avoiding any steel which might contain ^{60}Co (frequently added to modern steels to aid in the production process) and keeping the potassium content (and hence ^{40}K) of any glass to a minimum.

The main factor in the reduction of the background is to surround the detector with a heavy metal shield to absorb external gamma- and other radiation. Lead and steel are normally used as shielding materials, again using the lowest-activity material available. Sometimes, other components are added to the heavy materials to further reduce the background.

Further reduction in the background is then obtained by surrounding the main Geiger counter with an array of Geiger tubes to act as an *anti-coincidence shield*. The idea here is that an external radiation which penetrates the heavy-metal shield will also probably interact with one of the 'shield Geigers' prior to entering the main Geiger counter. In that event, a signal will arise from both the shield and main Geiger. By arranging the electronics to reject the main Geiger pulse, if a coincident pulse from the shield is also detected this considerably reduces the external background count. It may also result in a reduced counting efficiency for the actual sample in the situation that two beta-particles from the sample are emitted within the resolving time of the anti-coincidence circuit, one of which enters the detector while the other triggers a shield Geiger. In this case, the detector count would be rejected even though it was a legitimate count. Methods to increase the efficiency of such beta-counting have been developed. For example, a 'multicounting system' (multichannel system) has been described by Theodorsson (1993) in which a number of detector elements, consisting of gas-flow G–M counters are combined in a single unit. The electronics for such a system are rather complex but significant increases in overall counting efficiency are achieved.

The preparation of solid sources for gross beta measurements

The preparation of sources for gross beta measurements is essentially the same as that discussed for the calibration of end-window Geiger–Müller detectors (Section 9.2.1).

The preparation of solid sources for quantitative beta measurements

A quantitative beta measurement will necessarily follow a radiochemical separation procedure to isolate a specific beta-emitting nuclide. The final step is then the preparation of a suitable source for counting. For nuclides emitting beta-particles of end-point energy greater than about 250 keV, then a solid source is usually prepared. For end-point energies below this value, then liquid scintillation counting is probably the preferred method. A number of methods for preparing solid sources may be used but an overriding factor in all of them is the need for thin sources because of matrix absorption problems. Another feature of all of these solid sources is the need for uniformity of thickness and reproducible geometry. The absolute area of the source is usually not important, as long as it is comparable to the detector window diameter. It is only important to maintain a consistent area throughout a counting sequence. For low-level counting, a special large-area detector may be used to improve the counting efficiency.

There are three steps to the preparation of a source, i.e. (i) preparation of the source mount or backing, (ii) deposition of the source on a defined area, and (iii) covering the source. The material for the source mount should be of low atomic number if possible and thin in order to reduce back-scattering effects; aluminium or filter papers are most often used but stainless steel or Teflon (PTFE, see Section 9.4.3) are used when chemical inertness is required. The deposition of the source may be by one of several different methods. The simplest method, which can be used if the source material is in the form of a dry powder, uses an aluminium disc with a shallow, central depression, known as a *planchet*. This depression is filled with the powder, which defines both the area and thickness of the source. Simple evaporation of a liquid on a planchet may be employed, although we have already seen that a number of precautions must be taken to ensure a uniform, thin source. The area is defined by using a confining barrier of a hydrophobic material such as a hydrocarbon or silicon wax. Alternatively, the liquid is adsorbed on a filter disc, thus allowing a more uniformly thick source to be prepared, although obtaining a reproducible area is more difficult. Filtration is widely used for preparing sources. The source area can be defined by the use of a special 'filter stick' and the filter paper containing the precipitate can then be transferred to a flat planchet prior to counting. Electrodeposition may also be used in some circumstances to deposit a thin layer of the nuclide in the form of a metal, or the hydroxide of a metal, on a planchet. Thin sources of ^{55}Fe, ^{63}Ni and ^{203}Hg can be prepared in this way. The method is discussed more fully in Section 9.4.3. Vacuum evaporation produces thin, uniform sources but is wasteful of material.

The final part of the source preparation is to cover the source. This is very necessary when the source reacts with the atmosphere or is hygroscopic. In any case, it prevents the possibility of contaminating the detector. A covering can be used to absorb alpha-radiation if the source contains a mixture of alpha- and beta-emitters. A suitable covering is 'Sellotape' or a similar material, but any thin film of plastic, or occasionally aluminium foil, may be used.

9.4.3 Semiconductor Detectors

These detectors are used mainly for alpha- and beta-spectrometry. The latter is primarily used to confirm the radiochemical purity of a pure beta-emitting nuclide following an activity measurement with a more conventional detector, as described in the previous section. For alpha-determinations, on the other hand, these detectors are also normally used for the activity determination. We will therefore confine our discussion here to measurements and determinations of alpha-activity.

It is generally agreed by nuclear analysts that the accurate determination of alpha-activity, whether by semiconductor, gas-proportional or liquid scintillation techniques, is the most difficult of all activity measurements. The ability to routinely produce precise and accurate alpha-activity data requires a combination of skills rarely met with and being blessed with innate luck is certainly an advantage!

The most widely used detectors for alpha-measurements are the surface-barrier types but ion-implanted detectors are now also routinely used. The vast majority of environmental alpha-measurements are used to determine the actinides present and particularly the transuranic elements neptunium, plutonium, americium and curium. Even though spectrometry is used, the spectra of most nuclides is sufficiently complex that precise determinations can usually only be made following the radiochemical separation of the specific nuclide being sought.

Source preparation

The major consideration in the preparation of sources for alpha-counting is that they must be extremely thin and uniform because of the very limited range of alpha-particles, of whatever energy. Even so, frequently the overall spectral quality may be as much affected by the presence of contaminating alpha-emitters as the thickness of the source.

It is possible to produce suitable sources for counting by extracting the alpha-emitting nuclide into an organic phase and then evaporating this on a stainless-steel (or other suitable metal) disc and then heating the latter in a flame to destroy the organic matter. This method has a number of problems, however, and is little used.

The most widely used method of source preparation for alpha-emitters is that of *electrodeposition*. This also results in sources of the highest spectral quality. Deposition is normally carried out under slightly acid or alkaline conditions using current densities of a few hundred mA cm^{-1}, at voltages usually between 10 and 20 V.

Cells for alpha-electrodeposition are normally fairly simple. An example is shown in Figure 9.6, in which a cell consisting of a detachable, Perspex base containing the cathode with the upper part of the cell being a cylinder of polytetrafluoroethylene (PTFE) with an inner lining of polyethylene which screws on to the base. The cathode, frequently a stainless-steel (or platinum) disc or counting planchette, is clamped between the base and the polyethylene liner, using a PTFE washer to form a water-tight seal between the two parts of the cell. An essential

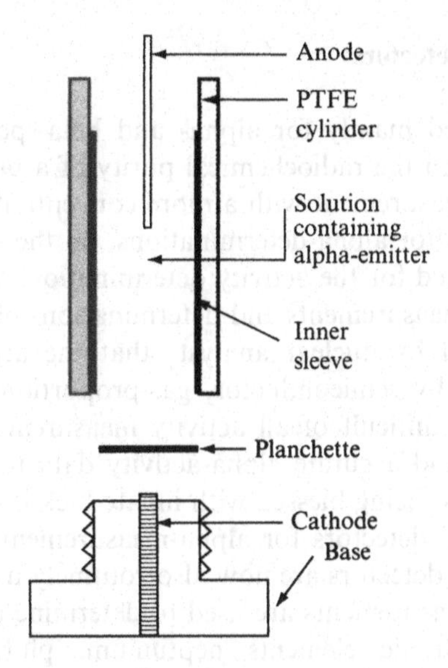

Figure 9.6 Schematic diagram of an electrodeposition cell for preparing alpha-sources

requirement for the production of thin, uniform deposits on the cathode is that it is thoroughly cleaned and polished prior to plating. The anode is normally a length of thick platinum wire, sufficiently long to dip into the electrolyte in the cell. Variations of this general cell type have been developed, including one in which electroplating is carried out on a vertically held cathode, which is claimed to produce a thinner electroplated deposit.

The sample containing the alpha-activity is mixed with a suitable electrolyte, such as an ammonium salt, and the pH, current, voltage and temperature adjusted to optimum values and then electrodeposition is carried out. The cell parameters will usually vary according to the radionuclide being deposited.

9.4.4 Gas-Proportional Counters

For alpha-particle counting of environmental samples, the source preparation is similar to that discussed in the previous section. The requirement for very thin sources is, however, not nearly as stringent and thick sources may also be used. Co-precipitation can be employed to prepare thick sources of specific radionuclides. Thus, barium sulfate will 'carry' radium alone if the precipitating conditions are strictly controlled. There are a few other examples of this type of co-precipitation, although it is not widely applicable. In addition, detectors of this type are useful for counting very low levels of noble gases and $^{14}CO_2$ for dating purposes.

9.4.5 Liquid Scintillation Counting

As always, the major interest in the liquid scintillation counting of environmental samples is in low-level measurements. Hence, the aim is the now familiar one of reducing the background to a minimum. The major contributors to the liquid scintillation background are as follows:

(a) natural radioactivity in the chemicals used for the scintillator solution, and light emission from these chemicals due to the phenomena of chemiluminescence and phosphorescence;

(b) natural radioactivity and chemiluminescence and phosphorescence in the sample;

(c) natural radioactivity in the materials of the counting vial, cosmic-ray-induced background from the vial, chemiluminescence and phosphorescence from the walls of the vial, and build-up of static charge during movement of the sample;

(d) natural radioactivity in the materials of the photomultiplier tubes, cosmic-ray-induced background, and electronic noise.

The natural radioactivity present in chemicals and materials is difficult to eliminate. For the sample vials, the major activities are ^{40}K, ^{232}Th and ^{7}Be present in the materials used to make the glass. Special, 'low-potassium' vials can be obtained or, even better, is to make use of plastic vials. For the other components, the aim is to use the lowest-activity materials available. The presence of ^{3}H and ^{14}C in the scintillator chemicals cannot be avoided and it is interesting to note that $10\,cm^3$ of toluene, for example, contains about 200 disintegrations per minute (dpm) of ^{14}C.

The cosmic-ray-induced background is reduced, as in most counting systems, by the use of massive shielding, usually lead. A thickness of about 5 cm is normal. For the lowest-background systems, this will be specially selected 'low-activity lead' with a graded shield of cadmium and copper. As discussed earlier for G–M and proportional counters, an additional shield, in the form of an anti-coincidence system of either NaI(Tl) crystals or plastic scintillators, may also be used but is not yet common in commercial instruments. In a typical system, the background is of the order of a few cpm for ^{3}H but about an order of magnitude higher for ^{14}C. For specifically designed low-level systems, the background for both isotopes can be reduced to about 1 cpm or less, depending on the volume of scintillant being used.

Sample preparation techniques

The major problem in the liquid scintillation counting of both beta- and alpha-emitters is in sample preparation. It is most acute when counting low-energy beta-emitters (especially tritium) but the accuracy and precision of measurements depend critically on correct sample preparation. In a few cases, virtually no sample preparation is required, with the sample simply dissolved in the liquid scintillator cocktail (including certain gases such as noble gas nuclides, CO_2, etc.), but in general, sample preparation is much more complex and really outside the scope of this current text. We will, however, outline some of the most commonly used techniques but the interested reader is referred to the book on liquid scintillation counting by Horrocks (1974), given in the Bibliography at the end of this present text.

For samples that are not directly soluble in the scintillator cocktail, some method has to be found to render them soluble. For many biological materials, including tissue, blood, proteins, etc., there are a number of commercial *solubilizers* available that achieve this end, although they may introduce a degree of quenching with some scintillator materials. Reacting the compound with a suitable reagent to produce a soluble complex is also possible, as is treatment with certain acids.

A different approach is to use a *secondary solvent* in toluene-based scintillators, wherein this solvent takes up the sample and is also soluble in the given liquid scintillant. This technique is commonly used for aqueous samples which are not directly soluble in organic liquids. Again, a number of commercial cocktails are available which allow the incorporation of varying amounts of sample, depending on whether the sample is simply water or is an acidic or basic material. A well-known, dioxane-based cocktail is 'Bray's Solution'. An interesting variation on this approach has the aqueous sample dispersed in a *micellar structure* containing both the sample and the scintillator molecules. These micellar solutions have an ordered structure in which scintillator and sample molecules are aligned, hence allowing for good energy transfer of the radioactive decay energy to the scintillator. The primary solvent in these systems is xylene and such systems can incorporate up to 40% aqueous sample but again with poorer counting efficiencies for ^{3}H than for normal toluene-based scintillators.

In contrast to the above approaches to the problem of assaying aqueous (or generally polar) samples, is that of forming *suspensions*. At its simplest, this involves dispersing finely divided solids in the scintillator, with efficiency corrections to allow for settling. An improvement on this involves the use of gelling agents to prevent settling but in general these methods, although adequate for energetic beta-particles, are of limited use for counting tritium. More recently, surfactants have been used to prepare much finer emulsions with counting efficiencies comparable to, or better than, those obtained with Bray's Solution.

Pulse Shape Analysis is a relatively recent technique which allows conventional liquid scintillation counters to be used as spectrometers. That is, it has enabled the analyst to distinguish between different types of particle. This has allowed for the counting of, say, alpha-activity, in the presence of beta-particles. The technique is discussed more fully in the next section but it is now possible electronically to distinguish between the two particles and obtain count rates for each of them individually when both are present in a sample.

Specialised techniques for alpha-counting

It has already been noted that the counting efficiency for alpha-emitters is usually 100% and carefully prepared, pure alpha emitting-nuclides can be counted with good accuracy by using conventional liquid scintillation counting equipment. Unfortunately, environmental samples rarely conform to this simple picture and both beta- and gamma-emitters are also usually present. These degrade both the resolution of the alpha and produce a variable background. Methods have now been developed for improved sample preparation and, in particular, in the development of *pulse-shape discrimination* to eliminate the interference from beta- and gamma-contributions to the background in alpha-counting (see the report by McDowell (1986), found in the Bibliography at the end of this present text).

Pulse-shape discrimination is based on the fact that the energy deposited in the scintillator by a beta- or gamma-pulse is dissipated in 1–2 ns, while that from an alpha-particle requires about 300 ns to completely dissipate the energy. Thus, by incorporating a timing system in the electronic circuit, the very short pulses from beta- or gamma-interactions are easily discriminated against, hence allowing just the alpha-signal to be processed. This slow decay of the alpha-pulse allows a large fraction of the deposited energy to be dissipated by non-radiative processes, which accounts for the relatively small light output of an alpha-particle compared to that from a beta-particle of equivalent energy. All alpha-decays, however, occurring in the scintillator are detected, so resulting in essentially 100% counting efficiency, as discussed earlier.

Alpha-spectra can be obtained from liquid scintillators although the resolution, of the order of 200–300 keV, is poorer than that possible with semiconductor or gridded ionization chambers. It is even worse if an aqueous solubilizing agent, of the type discussed in the previous section, is part of the scintillator cocktail. Consequently, the systems with the highest resolution must avoid the use of such agents.

This, however, presents a problem in that most environmental, alpha-emitting nuclides will originally be associated with some form of aqueous solution. The best answer is to incorporate the alpha-emitter in an entirely organic scintillator cocktail, thus eliminating the need for material to take up the aqueous phase. This can be carried out by using certain organic reagents which *complex* with the nuclide, i.e. form a nuclide–organic entity which basically behaves as an organic substance and so will be soluble in many organic solvents. Examples of such organic reagents are bis (2-ethylhexyl) phosphoric acid (HDEHP) or thenoyltrifluoroacetone (TTA). The nuclide can be *extracted* from the aqueous phase into an organic solution of an appropriate organic complexing agent by simply shaking the two, immiscible phases together for a suitable length of time. By incorporating the organic extractant in the scintillator cocktail, this allows the alpha-emitting nuclide to be directly incorporated into the scintillator. Such an arrangement can produce significant increases in the concentration of the nuclide since the volume of organic extractant is usually much smaller than the original aqueous phase. It has also been found that the addition of naphthalene to the cocktail significantly increases the light output of the system and hence an improvement in resolution.

Systems specifically designed for the liquid scintillation counting of alpha-emitters are somewhat different in construction to standard beta-counting systems. A major change is to use just one photomultiplier tube (PMT) since this allows for better light collection between the sample and the PMT than in dual-tube assemblies, which hence improves the energy resolution of the system. In addition, a timing unit is incorporated in the electronic processing system and the output of the amplifier is fed to a multichannel analyser to provide a spectrum of the alpha-emitters present.

A large number of extracting systems have now been investigated for alpha-counting involving liquid scintillation techniques. The actual choice of extractant depends on both the nuclide to be determined and also the presence and nature of possible interferences. Where complex spectra are being analysed it is vital to maintain the exact conditions throughout, in order to prevent any shift in peak positions, so avoiding possible ambiguities in nuclide identification.

Low-level counting of radon and radium

The normal method of determining radon in the gaseous state is by solid-state nuclear-track detection and this technique is discussed in more detail in Section 9.5. Radon is, however, an important radionuclide in drinking water and the level of this nuclide must be carefully monitored. The US Environmental Protection Agency (EPA) now recommends either the Lucas cell method (based on the detection of light scintillations using a ZnS phosphor) or liquid scintillation counting for determining the levels of radon in drinking water (Kinner *et al.*, 1991).

For liquid scintillation counting, a water sample is incorporated into the scintillator cocktail. Toluene was formerly used as the solvent and gave low detection limits for counting times of a few tens of minutes. Problems with the use of toluene have led to it being superseded as a solvent by mineral oil, plus some special commercial cocktails are also used. It has also been shown recently that toluene-based scintillators actually give lower count rates than either those using mineral oil as the solvent or the special commercial cocktails.

The recommended method involves shaking the cocktail and water for 1 min to facilitate the extraction of radon from the water into the cocktail. In addition, the vials are not counted for several hours to allow the short-lived daughters to reach secular equilibrium and so provide a constant disintegration rate. In fact, it has been shown that shaking is not necessary, as the much higher solubility of radon in the organic solution compared to the water means that extraction is rapid in any case and is continuously occurring. Some care has to be exercised in the collection of low-activity water samples, i.e. not to unduly agitate them prior to incorporation into the scintillator, as this can lead to significant losses of radon from the water. Levels of detection of about 0.5 Bq dm^{-3} of water are possible for counting times of 1 h.

A method for the determination of low-levels of ^{226}Ra in water, based on α–γ coincidence counting, has been described. The coincidence system consisted of a 3.5 cm × 3.5 cm NaI(Tl) well-type gamma-detector and a window-less, gas-flow-proportional alpha-counter that fitted within the well. Radium was quantitatively isolated from the water samples by co-precipitation with barium sulfate and the addition of some lead. The barium (radium) sulfate precipitate was mounted on a filter paper between the two detectors for counting the ^{226}Ra. This system gave a lower detection limit of 0.01 Bq for a 12 h counting period.

9.4.6 Gamma-Ray Counting

Sample preparation for gamma-counting, when using either NaI(Tl) or Ge detectors, depends on whether results are to be presented on a '*dry-weight*' or '*as-received*' basis. The former is the norm and is preferred as it allows direct comparison between widely different materials and results from different laboratories.

A number of drying methods are available, depending partly on the matrix and the nuclide(s) being sought. The simplest is oven drying. The sample for drying (usually a solid) is generally spread out on a tray to increase the surface area exposed and is then placed in an oven for drying. Oven temperatures range from about 30 to 100 °C, depending on the volatility of the nuclides present. Lower temperatures are usually preferred, however, for all but the most refractory elements. Although the method is simple, the dried product tends to be very hard and requires some form of mechanical device (e.g. mortar and pestle, ball mill, etc.) to reduce the material to a fine, particulate form suitable for homogeneously packing into a standard counting container. This process is rather tedious and can destroy the original particle size of the material, which can be important in some applications. Oven-drying is mainly confined to sediments, soils and the like.

The most widespread method for drying samples is *freeze drying*. This involves placing the sample in a container (usually a glass flask) and then evacuating the latter with a rotary pump. When the pressure has been reduced to a few mm of mercury, the flask is rotated to expose as large a surface area as possible. The low pressure allows for relatively rapid evaporation of water from the sample with the minimum damage to the matrix. Even samples with high moisture contents can be dried by this technique in periods of 24–48 h. The low temperature of evaporation means effectively no loss of even volatile elements such as mercury or iodine. The method can be applied to most types of samples, including liquids such as milk. Again, the solid formed must be reduced to fairly fine particles but now quite gentle treatment is all that is usually necessary, owing to the friable nature of the solid produced by this drying technique.

The next question to be addressed is the energy range of the gamma-rays to be counted. For gamma-rays with energies in excess of about 100 keV the standard container for counting the samples is a Marinelli beaker (see Section 9.2.5) with a nominal capacity of either 0.5 or 1 l. Obviously, the volume chosen will correspond to the volume used in calibration of the detector, although many laboratories engaged in routine environmental monitoring will have calibrations available for both sizes (and possibly others as well). The solid, in particulate form, is merely poured into the beaker and gently tamped down to produce a uniform packing until the beaker is filled. It is then weighed to determine the mass of counting material present.

In the event of counting for only low-energy gamma-rays or X-rays, a different counting arrangement is used. Due to the significant self-absorption of such radiation, a much lower volume arrangement is desirable. This usually takes the form of a disc of material, with a diameter comparable to that of the detector and some 5 mm–1 cm thick, depending on the lowest energies to be counted. Such a disc can be prepared by compressing suitable material into a pellet with a press or simply packing the material into a suitable box (preferably cardboard or thin plastic) with a sealable lid. Again, the geometry of the counting disc should match that of the efficiency standard.

Systems for low-level counting

For all of the low-level counting systems considered so far, there have been two universal procedures for the reduction in the background that the detector receives.

The first and most important is to place the detector inside a suitable shield, usually made of some heavy metal such as lead. The second is to ensure that all the construction materials for the detector, ancillary equipment and surrounding structures should contain the minimum of radioisotopes. Other factors contributing to background reduction are somewhat detector-dependent. We shall now consider the particular requirements for the construction of a low-level, hyperpure-Ge-based gamma-spectrometry system.

Shielding The aim in designing a suitable shield is to reduce the ambient activity at 1 MeV by a factor of at least one thousand. This could be achieved with almost any metal shield of the requisite thickness, depending on the density of the material. In fact lead (or occasionally, a combination of lead and mercury, or possibly bismuth) is the best material because it has the highest atomic number of any of the easily available metals. In order to understand the importance of atomic number, we need to remind ourselves of how gamma-rays interact with matter. Gamma-rays (of energy less than 1.02 MeV) may interact with the shield either by the photoelectric effect or by Compton scattering. For the former interaction mode, the gamma-ray is totally absorbed and so cannot contribute to the detector background. For Compton scattering, there is a possibility that the scattered gamma-ray may still reach the detector. The cross-section for photoelectric interactions increases as the fourth or fifth power of the atomic number (Z^{4-5}), whereas that for Compton scattering is directly proportional to the atomic number.

In practice, a *graded* shield is used. Here, the lead shield is frequently lined with a thin layer of cadmium followed by a layer of copper. The cadmium effectively filters the lead X-rays produced by the photoelectric interactions of gamma-rays in the lead and copper then attenuates the cadmium X-rays induced by those from lead. Cadmium, in fact, is not the best choice for the lining material since the cross-section for the reaction $^{113}Cd(n, \gamma) \rightarrow {}^{114}Cd$ is high and results in the production of 558 keV gamma-rays. The source of neutrons for this reaction are cosmic rays. Hence, a better choice for this inner liner is a 1 mm thick layer of tin, followed by about 2 mm of copper. A final, plastic liner may also be used to aid in decontamination of the shield, although a minimum thickness should be used.

Construction materials As has been stressed earlier when discussing shielding for detectors, it is most important to use special low-activity lead in the construction. The major culprit is ^{210}Pb, with a half-life of 22 years, which may occur at levels of 500 Bq kg^{-1} of modern lead and is the result of contamination in the ore or in the smelting process. A source of low-activity lead is from ancient church roofs but this is rarely legally available to spectroscopists!

Similarly, modern steels contain ^{60}Co, added to aid control of the production process. Pre-World War Two steel is free from this contaminant and old battleship steel has been used in some shields.

The use of materials for the detector housing and any electronic or other components that are inside the lead shield should have the lowest intrinsic radioactivity possible. The most significant potential radioactive contaminants are the ^{238}U and ^{232}Th decay series nuclides, ^{40}K and ^{137}Cs. Careful choice of materials eliminates most problems and where this is not possible the system is designed so that the active components are sited outside the shield and not in the line of sight of the detector.

Anti-coincidence shield The use of an anti-coincidence shield around the detector to reduce the background is not unique to gamma-ray counting, as we have already seen. For gamma-ray detectors, the shield is usually either a ring of gas-proportional counters or a NaI(Tl) or CsI(Tl) scintillation detector. If external radiation (essentially cosmic rays) interacts with both the anti-coincidence shield and the Ge detector within a pre-set time interval, then the associated electronic circuit is triggered and generates an output that effectively prevents the Ge detector signal from being counted. Thus, a potential background count is not registered. There is also a slight loss of counting efficiency due to random signals from the source being registered in the detector within the coincidence time interval that a signal is registered in the shield detector. Such losses are very small, however, when low-level activity sources are counted. A typical arrangement of a Ge detector and an anti-coincidence shield is shown in Figure 9.7. Although CsI(Tl) has some physical advantages over NaI(Tl) and a smaller diameter is required to achieve the same effective shielding, it is more expensive and is only used in special circumstances.

Considerable effort has been made over recent years in designing gamma-ray spectrometers for obtaining the lowest *minimum detectable activity* (MDA). The latter is inversely proportional to the detector efficiency but proportional to the square root of the background counts under a given peak. We have just dealt with

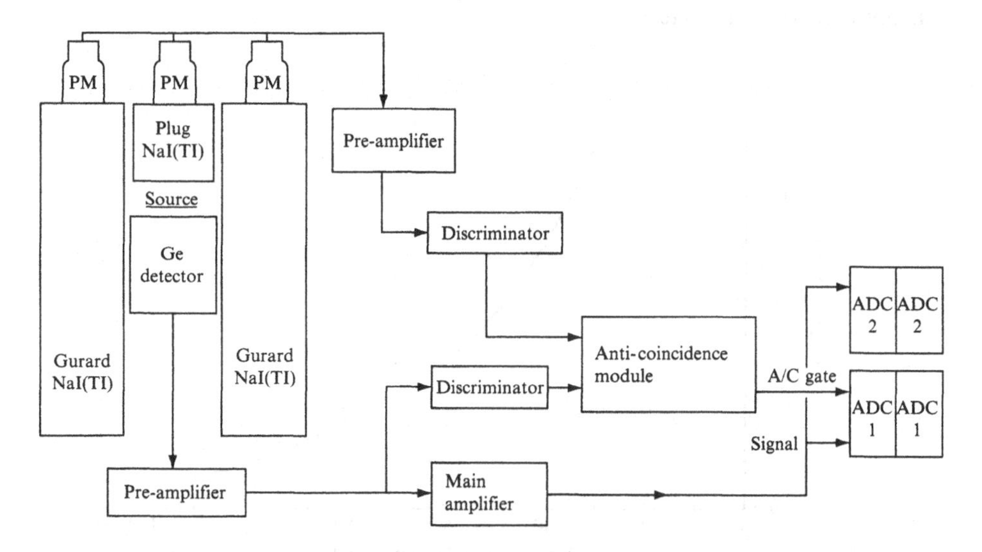

Figure 9.7 Schematic diagram of a NaI(Tl) anti-coincidence shield and associated electronics used for the reduction of the background in a Ge detector: PM, photomultiplier; ADC, analogue-to-digital convertor; MCA, multi-channel analyser. Reproduced by permission of Bicron, a division of Saint-Gobain Industrial Ceramics, Inc. from Helms, H.A., *Anti-Compton Spectrometers for Extreme Low-Level Gamma Counting*, Harshaw Chemie B.V., De Meern, The Netherlands

the ways of reducing the background and having reduced this to a minimum it would appear to pay to use the largest detector possible and have the source as close as possible to the detector.

The argument for using very large detectors for environmental counting has been addressed by several workers and there is some experimental evidence to support this case. It was shown that the 'non-peak' background decreased per percentage of detector efficiency, for a detector inside a lead shield, with or without an external source present. This reduction in background was ascribed to an increase of the peak-to-Compton ratio with increasing relative efficiency of the detector. The relative MDA as a function of detector efficiency is shown in Figure 9.8.

Positioning the source as close as possible to the detector can lead to increased problems from coincidence–summing losses, as discussed above in Section 9.2.5. It was shown there, however, that such losses are a minimum when using a Marinelli geometry.

Background spectrum for a low-level detector

Assuming that we have a high-efficiency Ge detector, constructed from low-activity materials, surrounded by an anti-coincidence shield and inside a graded lead shield, what will the background spectrum look like? Such a spectrum is shown in Figure 9.9. This very low background may allow certain gamma-ray lines to appear that would normally be hidden. These may be lines from ^{222}Rn gas present in the shield enclosure and certain cosmic-ray-induced neutron activation lines from the various construction materials.

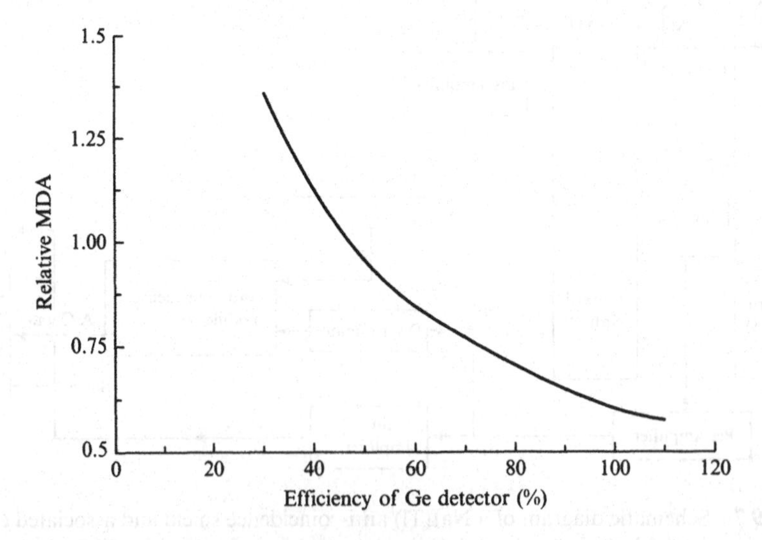

Figure 9.8 The relative minimum detectable activity (MDA) as a function of the efficiency of a Ge detector for gamma-rays with energy of ~ 800 keV. Reproduced by permission of Caretaker Technology, Inc. from Keyser, R.M., Twomey, T.R. and Wagner, S.E., *Radioact. Radiochem.*, (1990) **1**, 47–57

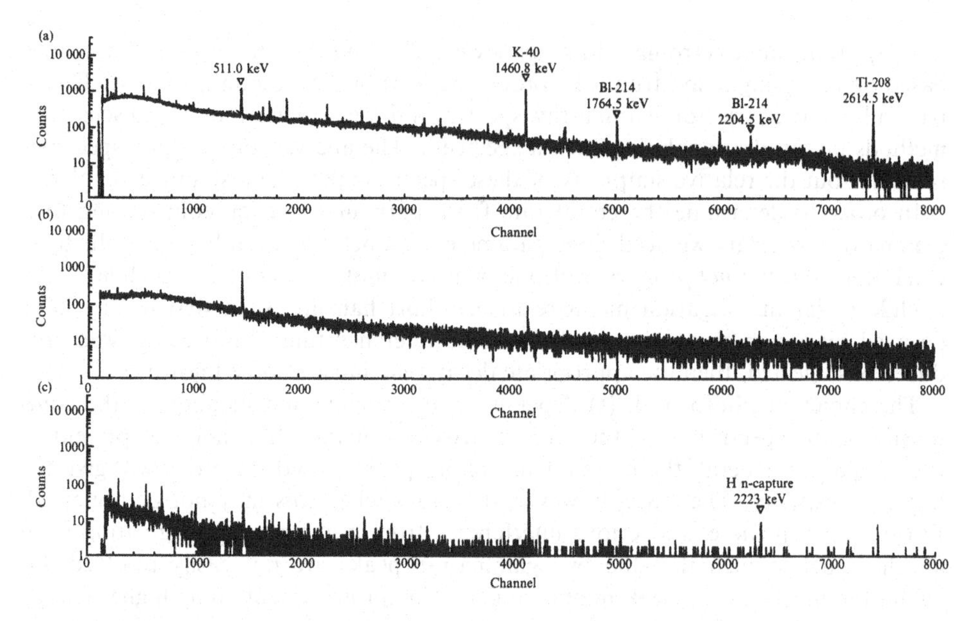

Figure 9.9 Reduction of a background spectrum in a Ge detector: (a) shielded with 5 cm lead; (b) achieved with a graded shield of 10 cm lead and 6 cm copper; (c) using the same graded shield but in conjunction with an anti-coincidence system. Reproduced by permission of H. Petri, Jülich, Germany

The radon gas activity, which is always present in the environment, is reduced by maintaining a slight positive pressure inside the shield, often with a gas such as nitrogen. This may be achieved by using the 'boil-off' from the liquid nitrogen in the Dewar system.

The influence of cosmic rays on the background has been experimentally investigated by Verplancke (1992). This author concluded that certain elements such as Ge, Cu and Pb gave rise to lines resulting from inelastic scattering reactions with fast neutrons arising from cosmic-ray interactions with materials in the vicinity. The presence of large amounts of plastic in the shield lining makes matters even worse, since this material significantly thermalizes the neutrons which can then produce gamma-rays from capture reactions with elements such as B, F, Cd, Li, etc.

A question that is frequently asked is: 'Is it better to use one very large detector rather than two or three smaller detectors of equivalent volume?'. It has been shown experimentally that for a given sample throughput and minimum detectable activity (MDA) the single, very large detector is preferable, assuming that sufficient capital is available for the purchase of such a system!

9.4.7 Analysing Gamma-Ray Spectra

The aim of all of the methods discussed so far is to determine the activity in a given sample. For most of these methods, this essentially involves determining a count rate by using an electronic counting device. For the determination of activities based on analysing a spectrum, the procedure is rather less straightforward. This is particularly the case for gamma-ray spectra and we will now briefly discuss the methods involved in analysing such a spectrum. The analysis of an alpha-spectrum is similar but the relative simplicity of these spectra makes the task much simpler.

In order to determine the count rate for a particular nuclide contributing to a gamma-ray spectrum we need three parameters, namely the total live time, the total clock time and the *net peak area* of one or more photo peaks of the nuclide. If the nuclide under investigation has a relatively short half-life, such that it may have decayed by a significant amount from some reference time, e.g. since it was produced or collected, then the appropriate decay time must also be known.

The choice of photo peak (s) depends on the nuclide and its activity, the complexity of the spectrum and the relative activities of the other nuclides present in the sample. In general, the most intense photo peak is used since this will give the highest sensitivity. There are, however, occasions when this may not be preferred. For example, if there is a serious interference from a gamma-ray originating from another nuclide such that the two gamma-ray peaks are not easily resolved. Or again, the most intense peak may lie on the Compton edge of some higher-energy gamma-ray and so be present on a steeply sloping background, e.g. the 889 keV photopeak of ^{46}Sc which sits on the Compton edge of the 1121 keV gamma-ray from the same nuclide. It is also possible that the most intense peak is part of a multiplet of peaks not totally resolved from one another, when both the peak area and the background under the peak are difficult to determine accurately. This is a situation unlikely to occur in spectra of environmental samples unless fission-product release from a reactor is concerned.

The intensity of a peak is related to the net peak area, i.e. the total peak area minus the background under the peak. The total peak area is obtained by summing all of the counts contributing to the peak. The main problem is to define the beginning and the end of the peak. Figure 9.10. shows a typical photo peak in a gamma-ray spectrum. The beginning (*low-energy side*) of the peak is normally easier to define than the end (*high-energy side*) of the peak due to *tailing* of the latter. Various mathematical techniques can be applied to aid the process, such as determining the first derivative of the spectrum. Once the beginning and end of the peak have been defined, then the background under the peak must be determined and subtracted. Frequently, the background can be assumed to be linear and provided that the peak is not part of a multiplet, the channels either side of the peak are used to determine the background. Up to five channels on each side of the peak are summed (see Figure 9.10) and the mean value each side is determined. The 'mean' of these two means is then taken and multiplied by the number of channels in the peak in order to give the background count to be subtracted. If multiplet peaks are involved, the procedure is a little more complex but essentially similar. The count rate is then determined by dividing the net peak area by the analyser live time. For short-lived isotopes and long counting times, corrections may have to be made for the loss in activity during the counting period, which will also require the clock time.

Nowadays, there are several software packages available for such analyses. Most of these also include algorithms for determining the absolute activity of the sample, provided that the relevant efficiency and decay time data are available.

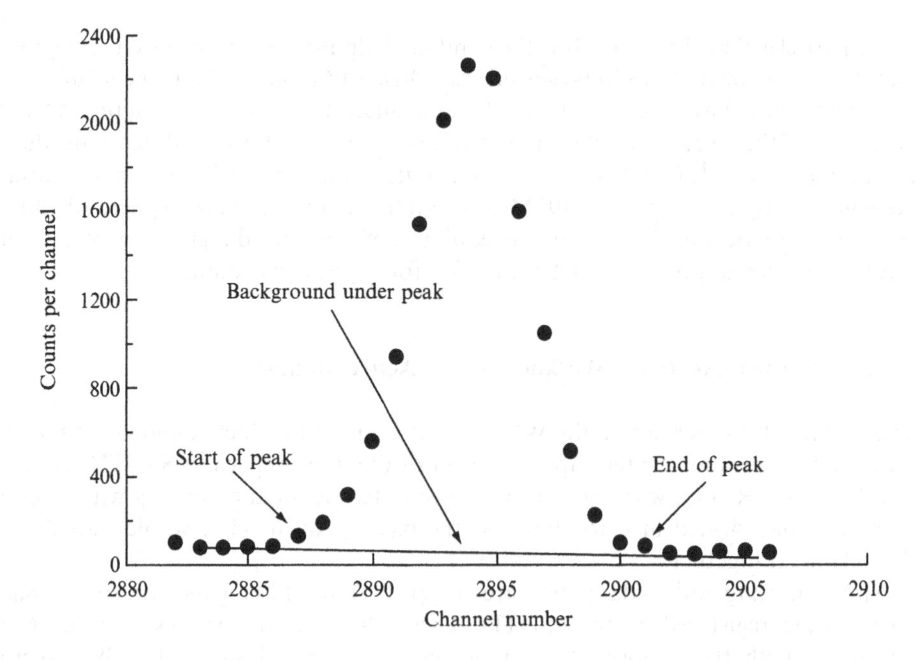

Figure 9.10 The 1333 keV peak of ^{60}Co taken with a Ge detector, indicating the starting and ending channels used for the peak-area determination. The fitted background under the peak is also shown

9.5 DETERMINATION OF RADON

You will recall that there are two isotopes of radon – ^{222}Rn is the predominant isotope and is produced as a daughter product in the decay of ^{238}U, while, similarly, ^{220}Rn results from the decay of ^{232}Th. Although ^{232}Th is more abundant in the earth's crust than ^{238}U, the longer half-life of ^{222}Rn (3.82 d) compared to ^{220}Rn (54 s) produces a greater concentration of the former in the atmosphere. Usually, when one speaks of 'radon' it is the ^{222}Rn isotope that is meant.

We have already mentioned that radon can be determined by liquid scintillation counting (Section 9.4.4). The most widely employed method for the determination of this gas in the environment is by the use of a solid-state nuclear-track detector (SSNTD). These are now used in a wide range of applications as both active and passive detectors but are mainly employed as passive radon monitoring devices.

Radon is considered harmful mainly because of its short-lived daughters rather than because of the gas itself. These daughters are isotopes of polonium, bismuth and lead and are therefore solids which are readily absorbed on fine particles of dust, etc. in the atmosphere. When we breathe in air containing radon and dust the radon gas is fairly rapidly expelled without doing any real damage to the tissues of the lung. The radon daughters, both resulting from inhaled radon gas and absorbed on the dust tend to be much more readily trapped in the lung tissues. The short-lived daughters can then inflict considerable local damage to the tissues by virtue of the alpha-particles that they emit. It is the combined alpha-energies of the daughters per unit volume of air that has to be considered when estimating the damage incurred by the lung tissues. The special unit used for this quantity is the *working level* (WL). A definition of the working level is 'any combination of the short-lived decay products of radon (^{218}Po, ^{214}Pb, ^{214}Bi and ^{214}Po) in 1 dm^3 of air that will result in the ultimate emission by them of 1.3×10^5 MeV of alpha energy'. An atmosphere of radon containing 3.6 Bq dm^{-3} of radon in equilibrium with its daughters would be at 1 WL. It is often important to determine WL for a given environment.

9.5.1 Determination of the Working Level – Active Method

The monitor for determining the WL is an air sampler in which a known volume of air is sucked through a filter paper. In front of the filter paper is a SSNTD, either a CR-39 or an LR–115, with the two being separated by an annular ring with a series of radial holes drilled in it to allow for the passage of air. The whole assembly is then clamped together.

Alpha-emitting radon daughters are trapped by the filter paper and the alpha-particles are registered by the SSNTD. These alphas have energies of either 6 or 7.68 MeV, with the majority having the higher energy. Using a CR-39 detector results in registration of all the alphas intercepted by the detector since this material is sensitive to alpha-particles of these energies. The LR–115 detector, on the other hand, has an upper sensitivity limit of about 4 MeV. Hence, to register any of the daughter alphas requires the use of a degrader foil between the filter and the detector. Using an aluminium foil of 12 μm degrades the 6 MeV alphas to below 4 MeV, and hence they are registered, while those of 7.68 MeV are still above the maximum registration energy. The optimum thickness for the degrader is 24 μm, which, in fact, completely absorbs the 6 MeV alphas but degrades those of 7.68 MeV to below 4 MeV.

In practice, the WL is determined from a piece of LR–115 plastic, in which one half is covered with 12 μm aluminium and registers only 6 MeV alphas, while the other half is covered with 24 μm aluminium to register the 7.68 MeV alphas. The plastic is then etched and the track densities in the two halves are determined after exposure of the detector to a known volume of air. The system is calibrated by determining the sensitivity of the LR-115 detector to known fluxes of 6 and 7.68 MeV alpha particles. From these measurements, one is able to determine the total potential alpha-energy, and so the WL.

9.5.2 Time-Integrating Method for Radon Determination – Passive Method

Active methods of the type just described are not easy to perform on a routine basis and also suffer from lack of sensitivity. They are useful methods for determining levels in, say, a uranium mine, but could not normally be used to determine radon levels in an environment such as a building.

Probably the commonest and simplest integrator detector is the 'can-detector'. This consists simply of an ordinary, domestic can with the closed end replaced by a

SSNTD disc. The open end of the can is then placed against the surface from which the radon is emitted and left in place for a sufficient length of time to record a suitable number of tracks. These can then be etched and counted to determine the radon activity.

In order to convert track density to activity, one must know the efficiency of track registration for a given SSNTD. Remember that the tracks are due to the alpha-particles from radon and its daughters. The energy of the alphas must be within the range that can be registered by the particular SSNTD. Again, the CR-39 and LR-115 plastic detectors are the two most widely used, with the former able to register all of the radon alpha-energies above about 0.1 MeV, while the LR-115 system only records those which lie between about 0.1 and 4 MeV. There is also a *critical angle* of penetration, θ_c, below which a track will not be registered. This is about 10 ° for Cr-39 and about 25 ° for LR-115. For both types of plastic, there is a maximum distance, R_{max}, from the detector beyond which the alpha-particle will be completely absorbed by the air in the can and so cannot register. For the LR-115 detector, there is also a minimum distance, R_{min}, from the detector such that the energy of the alpha-particle at the detector will be greater than the maximum registration energy of 4 MeV. These factors then provide for a *sensitive volume* inside the can in which track registration is possible (Figure 9.11). Because of its superior track registration properties and its great chemical stability, the CR-39 detector is now almost the universal choice as the SSNTD in radon measurements.

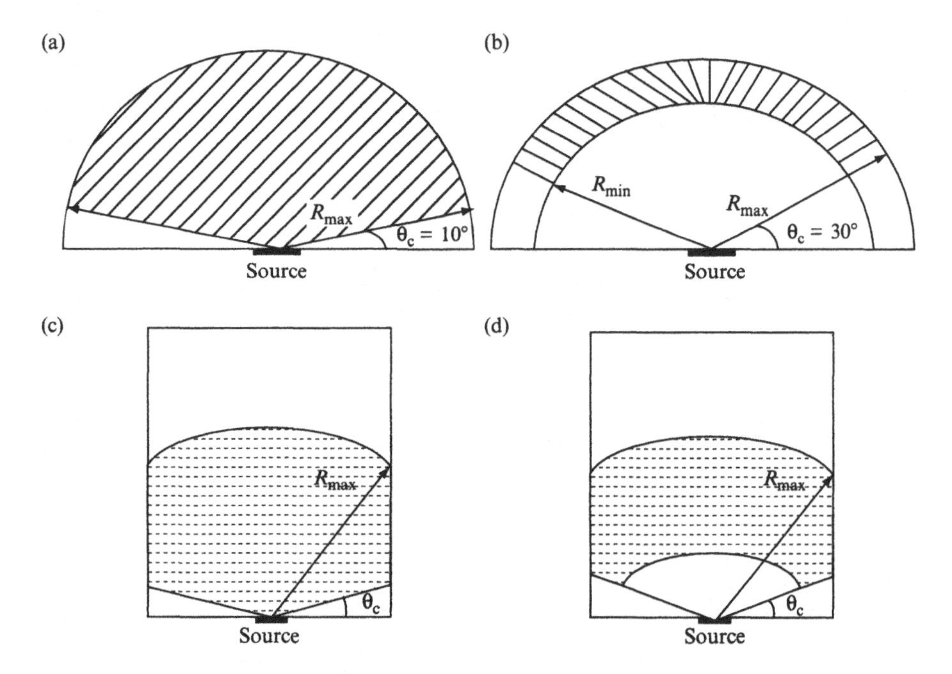

Figure 9.11 Radon sensitivity volumes for CR-39, and LR-115 SSNTDs when the detectors are located in the open air, (a) and (b), respectively and inside a can, (c) and (d), respectively. See the text for details of the meanings of R_{max}, R_{min} and θ_c

A further complication arises when relatively small cans (radii less than about 5 cm) are used as radon detectors. The radon daughters are all solids and their half-lives are sufficiently long that it is probable that they will diffuse to the walls of the can before they decay. In fact, it is usually assumed that the walls of the can, including the top and bottom (represented by the SSNTD) will be uniformly coated with the radon daughters. Thus, the tracks recorded by the detector will come from either the radon gas occupying the sensitive volume or daughter nuclides on the walls of the can. Based on the above, it is possible to calculate the relationship between track density and total radon activity. Hence, from a measurement of the tracks registered in a given time in a CR-39 detector by using a standard etching technique one can determine the radon activity in becquerels.

10

Sampling and Sample Preparation

'Could you analyse this for me, please?'. A request that analysts of whatever sort are often asked. If this is a request for the analysis of an environmental sample then this may frequently consist of a fairly nondescript-looking material that has been taken without much thought from a site e.g. a rubbish tip, the spoil heap from a mine, a beach, etc. about which the information is really required.

In the case of an analysis for radioactivity, the initial request will usually contain the rider 'we think that it might be radioactive'. Thus, we might start by checking with a Geiger counter to see if any activity present is at a level that will be detected by this fairly sensitive instrument. The next step might be to run a gamma-spectrum of the sample, particularly if activity was registered by the Geiger. If gamma-lines are seen apart from those of the normal background, then these lines will have to be identified and assigned to specific nuclides as far as possible and then the absolute activity of these nuclides determined. Depending on the sample size provided, this may simply entail using a standard efficiency source as described in the last chapter, or modifying such a standard to match the sample size, or in extreme cases, preparing a new calibration standard.

If pure beta-emitting nuclides are suspected, such as ^{90}Sr, or alpha-emitters, then the task becomes far more time-consuming and difficult. In the end, however, the analyst will aim to provide as accurate a measure of the activity present as possible. The person who made the request then receives this analytical data. 'Ha!' he cries, 'so the site is radioactive', to which the only honest reply the analyst can make is that the place from which that particular sample was taken is radioactive. As to the site in general, he would have to admit that he did not really know at this stage.

This brings us to a fundamental truth about any form of analysis. This is that *an analysis is only as good as the sample provided.* The case just considered is an extreme, but illustrates the point that if one wishes to gain realistic information about a sample that is too large to be determined in its entirety then careful selection of a sub-sample or samples will be necessary. This is particularly true of environmental analysis when the 'sample' may consist of several cubic metres or even kilometres and be very heterogeneous. The ability to produce analytical data of great precision becomes meaningless unless the sample analysed truly reflects the composition of the site under consideration. A fundamental aspect of any attempt to understand an environment through analysis lies in the sampling and this is one of the most neglected areas of analysis. Thus, in the first part of this chapter we will look at general methods and considerations to enable truly representative samples to be acquired and then go on to look at more specific techniques for various types of environmental sample. To begin with, however, we should understand the basic terminology of sampling.

10.1 SAMPLING TERMINOLOGY

Until recently, there was no recommended terminology for sampling. In 1990, however, the International Union of Pure and Applied Chemistry (IUPAC) set out recommendations for a nomenclature of sampling in analytical chemistry (Horwitz, 1990). These recommendations would also fit sampling for analysis of environmental radioactivity.

We use the term *site* here to represent any specific, geographical area. There are in fact two aspects to sampling: the *design* of a sampling strategy and then the *implementation* of that design strategy. The design strategy aims to specify, from statistical considerations, the number of samples, type of sample, frequency of sampling and its location, etc., in order to obtain portions for analysis. If the design strategy is soundly based then this will result in a minimization of the differences in properties (e.g. nuclide activity) between the portions analysed and the actual parent population or site. In other words, it should result in our having confidence that the results of our analyses truly reflect the situation for that particular site.

The implementation of this strategy obviously requires that it be put into action, i.e. samples are removed by some means or other. This seemingly simple operation is often fraught with difficulties. One not always considered carefully enough is actual access to a site, both legally and physically. For example, sampling an estuarine sediment may encounter both problems but particularly actual physical access due to the very nature of the sediments themselves and the related problems of tides. Having obtained samples, there is still the problem of transporting bulky and heavy samples, often over considerable distances on foot to a vehicle.

Choice of suitable tools also often presents problems. One tool may be fine for one type of sample but almost useless for another one. Then there is the associated problem of the materials for tool construction, the avoidance of contamination, sufficient strength, weight considerations and so on. Some of these considerations, will be discussed again when we consider more specific types of sampling.

The word *sample* itself is also rigorously defined in the 1990 IUPAC document to mean 'a portion of material selected in some manner to represent a larger body of material'. In environmental radioactivity determinations, we are dealing with *dynamic* situations, in the sense that the parent material changes with time. On the other hand, the parent body or *bulk material* may be *static*, as for foodstuffs.

In some situations, e.g. air sampling, there is no permanent record of a particular site as a function of time. If the air is not sampled at a particular period then data for that period are permanently lost. For many solid parent bodies, however, such as sediments or soils, then a record of the past environment is preserved within the material itself. By taking a *core* through such material one is in fact sampling past environments. Such a record, though, is often only an imperfect one as many factors may cause changes of the material over periods of time. When dealing with these types of material, then only the upper portion of that material represents 'present' material.

A large site, e.g. beach, estuary, etc., will be divided by some relevant statistical means into *segments*. The *portion* taken from the parent body for analysis then represents the *laboratory sample*. In environmental radiation measurements, each single portion taken from a given segment of a site is called an *increment*. These increments may be combined in some way to obtain a physically averaged sample, known as a *combined sample*. Normally, however, the increments will be separately analysed in order to estimate the heterogeneity of the sector or the site.

The sample taken to the laboratory for analysis is the laboratory sample and may be used directly for analysis. Frequently, however, the laboratory sample is altered in some way, e.g. drying or homogenizing. In this case, the resulting material is called the *test sample* from which a *test portion* is removed for analysis. If the test portion is dissolved to form a solution, this is called the *test solution*.

There is also the question of how the sample is actually taken in the field. If sampling is done on a completely random basis, so that there is equal probability of selecting any portion from the population, this provides a *random sample*. On the other hand, where the sample is selected on the basis of a sampling plan to yield a sample which is hopefully representative of the population, this is a *representative sample*. Occasionally, one may deliberately discriminate against certain types of material that do not meet certain criteria. In these cases, this yields a *selective sample*. A sample chosen on grounds of accessibility, expediency, cost or any other reason not associated with any sampling parameters, is a *convenience sample*. Such samples are usually of least relevance to the composition of the site. Samples removed from the site successively, in a predetermined manner, are called *sequential samples*.

10.2 GENERAL SAMPLING METHODS

We have already alluded to the problem of our sample being representative of the site we are investigating. Acquiring such samples relies on statistical considerations. In Chapter 11, we will consider various types of statistical distributions. The one that applies to sampling is the *normal distribution*, which is a *Gaussian* distribution. It is shown in Chapter 11 that, based on the normal distribution, there is a 95% probability that a measured value will lie within two standard deviations of the real mean value. The parameter defined in Chapter 11 as the *variance*, V is the square of the standard deviation, σ. In fact, the overall variance of the analysis of a sample will be the sum of the sampling and analytical variances, namely:

$$\sigma_o^2 = \sigma_s^2 + \sigma_a^2$$

where σ_s and σ_a are the sampling and analytical standard deviations, respectively. In general, the sampling standard deviation will be far larger than the analytical one, so that to a first approximation the standard deviation of an environmental radioactivity measurement is mainly due to sampling errors. In fact, once σ_a is one third or less of σ_s further reduction in the analytical standard deviation has little effect on σ_o.

10.2.1 The Sampling Strategy

Let us consider, as an example of environmental radioactivity monitoring, the estimate of the distribution of activity of, say, ^{137}Cs over a section of moorland covering a few square kilometres. Two general strategies may be employed and both have advantages and disadvantages. Systematic sampling may seem the obvious method but in fact is often more biased than random sampling. A typical systematic approach to sampling would be to divide the given area into equal squares, with the number depending on the potential variation of concentration of the activity and the time and manpower available. Then, the increment sample could be taken from the centre of each square.

For random sampling, one would again divide the area up in a uniform way as before, using squares, of about 100 m × 100 m (see Figure 10.1). Using a random number generator, a fraction of the large squares would be selected for sampling. Each of the selected squares would then be sub-divided into smaller squares of perhaps 25 m × 25 m and the random number generator would select the smaller squares where an increment sample would be taken. These increments might then be combined to give a composite sample, from which the test sample would be derived.

Another sampling procedure that is often adopted is that of an *expanding hexagonal sampling* grid. A sample is taken from a specific site and then further samples are taken at the apexes of hexagons centred on the first site. The hexagons increase logarithmically in size, usually either doubling or quadrupling at each stage. Thus, hexagons might have apexes at 2 m, 8 m, 32 m, 128 m, 512 m, etc. from the initial site. In this case, one would collect 31 samples in total.

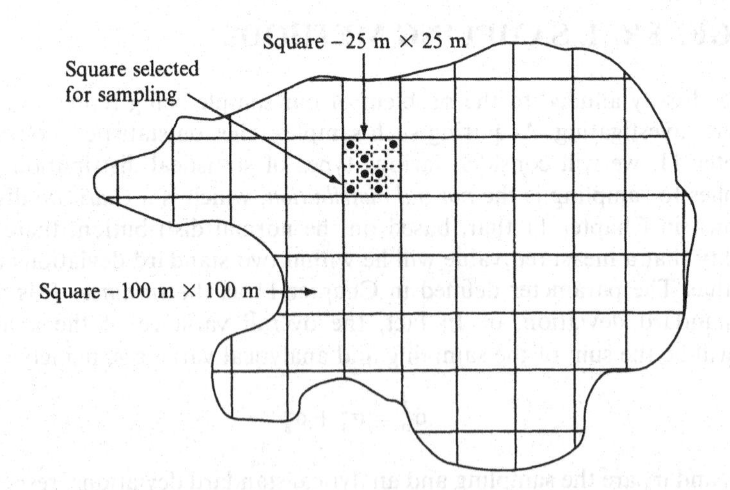

Figure 10.1 Schematic diagram showing how grid sampling could be applied to a large area, such as a moor. The latter is sub-divided into 100 × 100 m squares; squares are then randomly selected for further sub-division into 25 × 25 m squares and finally samples are randomly collected from within these squares

10.2.2 Estimation of Minimum Number of Samples Required

Consideration must also be given as to the number of samples that should be collected in order to remove the effect of large-scale heterogeneity on the overall determination of activity for a square of 100 m × 100 m. An estimation of this number can be made by the application of statistical theory and is done most straightforwardly if it can be assumed that the activity follows a normal distribution (see Section 11.3). The minimum number, n, required is given by the following:

$$n = \frac{t^2 s_s^2}{R^2 \bar{x}^2} \times 10^4 \qquad (10.1)$$

where t is the value for the level of confidence required (Student's t-table, which may be found in most textbooks on statistics – see Chapter 11), and s_s^2 and $\bar{x}$ are the variance of the sampling and the mean value for the activity determination, respectively, both of which are determined from preliminary measurements. The parameter R is the percentage relative standard deviation acceptable as the uncertainty in the determination of the activity. Thus, for a value of $R = 95\%$, $t = 1.95$ (for $n = \infty$) and so an initial value of n can be found. This value then allows a more accurate value of t to be determined from the t-table and by this iterative process a final value for n is determined. For the above example, a value of about six or seven would be expected.

This presupposes some knowledge about the uniformity of the distribution of the activity based on particle size, type, etc. To reduce the variability introduced by such parameters, it is best to take fairly large increments. The increment size would also depend on the optimum size for the laboratory or test sample. Thus, if a coring instrument were used, a core diameter of about 40 to 50 mm might be desirable. A similar sampling strategy could be adopted for any type of site. For counting packaged items of food, say, the sampling simply requires the removal of a randomly selected number of items from the lot, in perhaps, a warehouse or supermarket.

10.2.3. Sampling Hot Spots and Hot Particles

The previous discussion has assumed that an average value for the activity being sought is sufficient to describe a given area. Occasionally, however, the aim of the sampling strategy is to locate any *hot spots* that may be present. A hot spot in this connection describes a locality where the radioactivity is exceptionally high. For instance, hot spots may be sought in a landfill site if there is evidence that radioactively contaminated material was present in the landfill material.

The ability to locate such hot spots with a given level of confidence will depend on the grid spacing used and the size of the hot spot. These questions are discussed more fully by Gilbert (1987) and only an outline of the approach is given here. A number of assumptions have to be made about the hot spot and the sampling grid to be used; it is assumed that the spot is either elliptical or circular in shape and that there is no error in deciding when a hot spot has been located. There is some evidence that a triangular grid gives more information than the square grid.

The grid spacing to locate a hot spot depends on the size of the spot, the expected shape (the spacing becomes smaller as the shape departs from circular to more of a thin lens), the grid shape and the acceptable probability for *not* locating a hot spot that is present, β. This latter parameter will also depend on sample size but will include considerations of time and the financial cost of carrying out the sampling. With all of these parameters defined, it is then possible to determine the appropriate grid spacing based on a set of nomographs constructed for the purpose (see Gilbert, 1987). These nomographs also allow one to find the maximum size of a hot spot that can be located for a given grid size and shape and β value, or the probability of not hitting a hot spot of a given size and shape when using a specified grid size.

A related phenomenon is that of the presence of *hot particles* in a sample. These are defined as particles which are very small (usually less than 150 μm in diameter) but have an extremely high specific activity. They result from nuclear fall-out or a nuclear accident due to either disintegration of the nuclear fuel or condensation of a radioactive vapour. Hot particles from the Chernobyl accident were scattered over a wide area, with particles decreasing in size with distance from the crippled reactor. Obviously, sub-samples taken from a sample containing such hot particles will show variable overall activity depending on whether hot particles are present in the sub-sample or not, which could produce very misleading results. If one took a given number of sub-samples and found no overall difference in activity between them, then this does not necessarily mean that no hot particles are present since the number of such particles and the distribution in their activities could result in each sub-sample having the same activity, each with different numbers of hot particles but with these having an activity distribution resulting in essentially equal activities. Assuming, however, a particular type of distribution of activities for the hot particles (known as a 'log-normal' distribution) and that a difference of more than 30 % in the activities between the sub-sample with the lowest activity and that with the highest can be detected, then Bunzl (1997) has shown that a sample need only be split into three sub-samples to detect the presence of up to five hot particles with a confidence limit greater than 95 %, while four sub-samples would detect up to 20 hot particles with the same degree of confidence. If 100 hot particles were present in the original sample, then splitting into four sub-samples would reduce the confidence limit for their detection to only 82 %.

So far, we have considered sampling from a solid body. Sampling water or air, however, follows these same general principles as regards the sampling strategy. Only the collection of the required increment would differ, depending on the nature of the substance. The next section deals in more detail with sample collection.

10.3 SAMPLING SOLIDS

Under this heading can be included soils and sediments, rocks, vegetation (for consumption by humans or animals), meats of all kinds and foodstuffs in general. The different varieties will require slightly different methods of sampling. Whatever type of laboratory sample is produced, it must be contained in a separate, clean plastic container, when it can be directly counted or further treated prior to counting.

10.3.1 Soils and Sediments

A major problem with sampling soils and sediments is that they contravene a fundamental assumption in sampling, namely that the entire parent body or bulk material is accessible for sampling. Thus, sampling such materials is at best a compromise, but with a careful strategy should yield valid data.

Soils

Due to the high variability in soils it is necessary to obtain a sufficient number of increments to take this into account – probably at least 10. It is possible to obtain a mean value for a segment by 'bulking' the increments – combining several increments from different points within the segment. The number required really depends on variability within the segment. Soils are mainly useful as long-term accumulators of long-lived, air-borne radioactivity, mainly derived from atmospheric weapons testing. It is not recommended, however, that soils be used for periodic, short-term estimations of radioactive fall-out (Environmental Measurements Laboratory, 1992).

Often, if only a surface sample is required then this is easily obtained by cutting out a piece of material down to a depth of about 1 cm. For material from a greater depth, but where a core is not required, then a tool such as an *auger* (for boring) can be used. When cores are required to investigate the radioactive history of a site then the use of tubes or *hole cutters* is necessary. The depth to which soil can be usefully sampled is slightly problematical. Yamamoto *et al.* (1980) have suggested that the bulk of the activity is confined to the top 10 cm. There is evidence, however, that for ^{90}Sr at least, activity may be found in some soil types at depths in excess of 30 cm (Environmental Measurements Laboratory, 1992). Hole cutters are only used for short cores.

Sampling sediments

The term 'sediments' is usually applied to the solid material that accumulates at the bottom of reservoirs, lakes, and the seas and oceans. These materials have to be sampled through an overlying water column, which presents a number of difficulties. Sediment can also refer to the material in the inter-tidal zones of estuaries and which is only covered by water at certain times of the day. These *estuarine sediments*, which have both marine and riverine characteristics and are valuable sites for trapping radionuclides transported through coastal seas, can be sampled in a similar manner to soils. Sampling sediments, however, tends to be a more specialized procedure than for most other materials due to its relative inaccessibility and the fact that all sampling has to be carried out remotely.

Most sediment sampling requires the production of cores, although sometimes removing a surface layer of the order of 3 cm thick may be useful in some contexts. Cores provide an almost unique chronological record of radionuclide deposition. Unfortunately, unless the deposition rate is very rapid (several centimetres per year), it is necessary to produce slices from the core only a few millimetres in length. Unless the core diameter is large (20 cm or so), then the amount of material from such a slice will not be adequate for many analytical purposes. This problem is most acute if direct gamma-ray counting is to be used, which usually requires several hundred grams of sample. Obtaining cores of this diameter and up to one metre in length can only be achieved with specialized equipment. It is important to retain the top layers of sediment intact for studies of *vertical profiles* of the concentration of radionuclides.

A further problem arises if the transuranic elements are sought. There is evidence that these elements are preferentially associated with the finest sediments and organic detritus (such materials rapidly remove plutonium from the water column). Thus, sample collection must not discriminate against such material or otherwise the transuranics are not properly represented in the samples ultimately analysed. Certainly for fall-out material, however, careful sampling has been shown to provide a representative sample.

10.3.2 Vegetation

An important consideration in the collection of vegetation for any type of analysis is that the sample variation within a collection area probably greatly exceeds all other analytical variations. This necessitates very careful consideration being given to the collection strategy and will often require proper statistical design for sample collection if meaningful results are to be obtained. Such considerations are beyond the scope of this present book but are treated by, for example, Barnett (1974) and Webster (1977). Vegetation in this context could include grasses, vegetables of all sorts, leaves of trees or needles from conifers and fresh water and marine plants, particularly seaweed. It is particularly important to record all site data carefully in the collection of terrestrial vegetation, such as the density of vegetation, types growing in the area, drainage conditions and local topography.

10.3.3 Foodstuffs

In general, the aim in sampling foodstuffs is to determine the potential dose to either individuals or to particular groups of individuals through their intake of food (and drink). This may be carried out by sampling at the point of manufacture but the most common, and probably most realistic, method is to sample at the point of sale, involving purchases of foodstuffs from a range of outlets from supermarkets to small corner stores. Milk may also be taken from bulk tanks at farms in the event of any accident or if the farm is located close to a nuclear installation.

The sampling should take into account the normal dietary habits of the population at large, together with that of special groups where their dietary habits are likely to differ significantly from that of the general population. Thus in the UK, a number of studies have been made of populations in South Wales who consume lava bread on a regular basis. This is because lava bread is based on seaweed, known to heavily concentrate iodine and hence may prove to be a potential radiological hazard if radioiodine is present in the environment.

When sampling foodstuffs, one should bear in mind the major categories of these products and obtain representative samples within these categories, e.g. staple goods (bread, milk, sugar, flour, butter, margarine, etc.), breakfast foods, meats, fish, vegetables, sweets (candy), etc. It may also be important to include samples of foodstuffs that are both home-produced and imported on a large scale, particularly if this is from countries in the former Soviet Bloc, where the possibilities for radioactive contamination may be higher than similar foods produced in the Western World.

Care should be taken not to unduly bias the sampling towards those foodstuffs known to be potential sources of radionuclides, e.g. shellfish, but which form a relatively minor part of the general diet. On the other hand, it would be important to include, say, bottled mineral water, which frequently originates from volcanic regions and may therefore have a relatively high content of uranium and thorium and their daughter nuclides and which now figure quite prominently in the diets of many groups within more affluent Western populations.

10.3.4 Excreta

Urine and faecal samples may be required from humans, particularly if there is suspicion that abnormal amounts of activity may have been ingested. Thus, urine is frequently tested if a population has been exposed to tritium. This nuclide has a biological half-life of about 12 d (assuming that it is in aqueous form) and the activity in urine is readily checked by using liquid scintillation counting to determine the amounts excreted. Samples are normally taken in the morning, preferably just after waking, to give a sample which has 'collected' for a relatively long period, although this will tend to give a 'pessimistic' result. Faecal samples from animals should be collected in such a manner that they do not come into contact with the ground where they may be contaminated by activity from the soil or vegetation.

10.4 SAMPLING LIQUIDS

Superficially, this appears to present fewer problems than the sampling of solids. In practice, however, this is probably only true if the bulk sample is fairly small and constitutes a fairly homogenous whole. For many systems, great care is required in the sampling. Environmental sampling of liquids will usually involve liquids flowing in open systems (rivers, canals, effluent streams, etc.), liquids in large, open bodies (seas, lakes, ponds, etc.) and thirdly, liquids in containers (bottles, drums, etc.). We shall look at each of these in turn.

10.4.1 Liquids Flowing in Open Systems

Two major difficulties of sampling this type of system are that it may vary in two dimensions, i.e. both with depth and along its length. There are also frequently significant variations of activity with these systems. For example, if there is a known sewage outlet into a river, then sampling may be necessary close to the input and at various places downstream, together with some samples collected from upstream locations which will provide a 'background' level of activity. The sampling will probably have to take into account changes in flow rate – both seasonal and on a shorter timescale.

Samples can be collected in wide-mouth bottles or small buckets, probably of plastic construction. If samples are required at various depths, then the container is fitted with a removable stopper and weighted to descend to the appropriate depth. The stopper may then be removed, the container fills and can then be brought back to the surface. If the samples are not to be taken back to the laboratory for some time, then they are best stored in a refrigerated container. It may also be useful to record the pH of the water from the point where it was taken.

10.4.2 Liquids in Large, Open Bodies

Again, it will probably be necessary to obtain samples from various parts of the body and at various depths. The strategy for sampling such a body will be along similar lines to that discussed for large land masses, i.e. based on some form of grid and random sampling within the grid areas. The samples can be taken by using similar devices to those discussed in the previous section. Special devices have also been constructed for sampling relatively shallow lakes and ponds on a regular basis to quickly obtain samples from several fixed, depths.

Sample collection of transuranics in seawater may give rise to problems, due to the fact that most are alpha-emitters and hence difficult to detect. In the case of spills or accidents, there may be a temptation to monitor closely allied species which are gamma-emitters and so more easily detected rather than the alpha-emitting nuclide itself and then to assume that they will accurately reflect the distribution of the alpha-emitting nuclide. This could be a dangerous policy since even some isotopes of the same element may show different behaviour. Thus, ^{238}Pu is much more mobile that ^{239}Pu, due to the greater tendency of the latter to form polymers.

10.4.3 Liquids in Closed Containers

If the container is small enough to be shaken or the contents can be thoroughly stirred so that the contents are homogenous, then sampling the liquid is straightforward. If, on the other hand, the container is very large, such that the liquid may be stratified with depth, then a rather more complex sampling strategy will be required. Samples will have to be taken at various depths in the container to determine any variation of activity with depth by using suitable sampling devices.

10.5 SAMPLING AIR

Due to the influence that air quality has on the quality of human life, probably more effort has gone into air sampling than all of the other types of sampling put together. It has also been realized that a contaminant (radioactive nuclide) in air can vary by orders of magnitude over a fairly small radius from the point of emission. This variation is due to a number of parameters, such as air flow (whether turbulence is present or not), temperature, humidity, wind speed and concentration of the nuclide (see Chapter 14 for more detailed information.

There are really two different types of air sampling, depending on whether one is trying to determine a purely *gaseous* component of the air or if the radioactivity is present in *aerosol* form, where the latter is defined as a dispersion of liquid or solid particles in a gaseous medium. Gaseous sampling usually consists of adsorbing the gas into a suitable material, while sampling an aerosol may involve precipitation, gravity settling or centrifugal collection. The majority of environmental, radioactivity air sampling involves aerosol collection, with the most widely used methods of collection being by *filtration* and *impingement*.

10.5.1 Collection of Aerosols

Filtration is simple and provides sufficient material for subsequent radionuclide analysis. The type of filtration unit and the filter medium used depends on the nature of the radioactivity to be analysed. A basic filtration unit is shown in Figure 10.2. This consists of essentially a filter holder, filter, an air mover (pump) and a flow-rate meter. The air mover is always placed downstream of the filter to prevent contamination of the latter by the former. A uniform flow of air across the filter is desirable since measurements of radioactivity are nearly always made directly on the filter. Accurate measurement of flow rate is also important in order to determine the total volume of air from which the radioactivity was collected.

A wide variety of filter media are available and a list of the most commonly used ones is given in Table 10.1. Of these, the two types most routinely used for sampling environmental radioactivity are *fibrous filters* (mainly cellulose paper or glass fibre) and *membrane filters* (gel-type or 'Nuclepore'). Cellulose papers are relatively

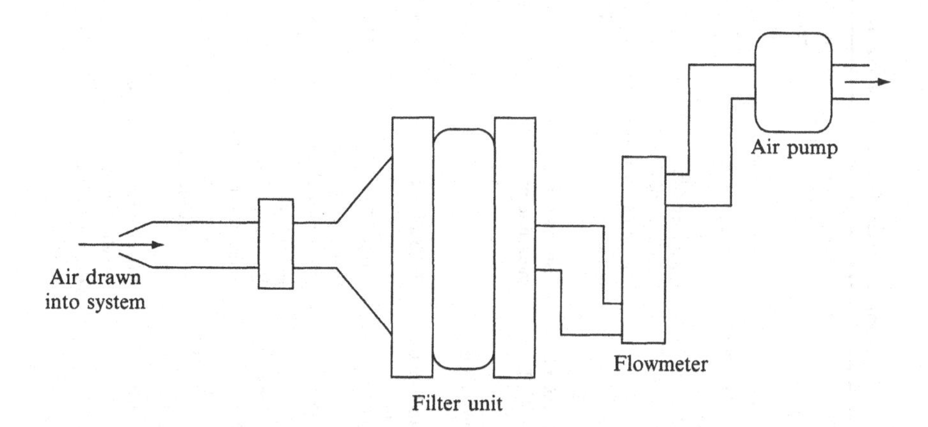

Figure 10.2 Schematic diagram of a basic filtration unit used for collecting aerosol samples from air

Table 10.1 Relevant characteristics of filters used for air sampling.

Filter manufacturer and type	Composition	Particle retention size (μm)[a]	Thickness (μm)	Weight/unit area ($g\ m^{-2}$)	Ash content (%)	Air flow rate ($dm^3\ min^{-1}\ cm^{-2}$)[b]	Comments
Cellulose fibre filters							
Whatman[c]							
1; 5	Cellulose fibres	> 2	180	87	0.06	—	—
40; 42	Ashless cellulose fibres	1–2	200–210	95–100	0.01	—	—
50	Hardened cellulose	1	120	97	0.025	—	—
540; 541	Hardened ashless	> 4	160	78	0.008	—	—
Millipore[d]							
ACE1; ACE2; ACE3; ACE4	Ashless cellulose with melanine binder	—	—	—	—	—	—
Glass fibre filters							
Sigma[e]							
F-4000; F-5000	Borosilicate glass	1.5–2.3	270–650	137–530	~ 100	Not applicable	—
Whatman[c]							
Grade 72	Glass fibre with activated charcoal	Not applicable	—	—	—	—	Used for high adsorption of radioiodine
934–AH	Borosilicate glass – no binder	1.5	330	64	—	—	—
EPM 2000	Borosilicate glass	—	430	80	—	—	Low levels of iron and zinc
QM-A	Pure quartz (SiO_2)	—	450	85	—	—	Extremely low in metals
GMF 150	Graded-density borosilicate glass	1–2	730–750	150	—	—	—
Millipore[d]							
AP15; AP25	Borosilicate glass with acrylic binder	—	380	80	95	—	—
AP40	Borosilicate glass – no binder	—	410	69	100	—	—

APFA; APFC; APFD	Borosilicade glass – no binder	—	—	—	—	—	—
Membrane filters							
Whatman[c]							
Cyclopore	Polycarbonate	0.4–10	7–20	7–20	0.03–0.09	—	Accurate pore size for particle size analysis. Very thin, useful for alpha-counting
Cyclopore	Polyester	0.4–5	9–23	9–23	0.01–0.26	—	Accurate pore size for particle size analysis. Very thin, useful for alpha-counting
WCN	Cellulose nitrate	0.45–5	—	—	—	—	—
WME	Mixed cellulose esters	0.45	—	—	—	—	—
Nylon	Nylon	0.45–0.8	—	—	—	—	—
PTFE	Polytetrafluoroethylene on a polypropylene support	0.2–1.0	—	—	—	—	—
Millipore[d]							
GVHP; HVHP	Poly(vinylidene fluoride)	0.22–0.45	125	—	—	3–6	—
VC; PH; HA; AA; RA; SS; SM; SC	Mixed cellulose nitrate and acetate esters	0.1–8.0	105–135	28–56	< 0.045	0.4–65	—
FG	PTFE bonded to polyethylene	0.2	175	22	—	3	—
FH	PTFE bonded to polyethylene	0.5	175	22	—	8	—
FA	PTFE bonded to polyethylene	1.0	145	22	—	16	—
FS	PTFE bonded to polypropylene	3.0	200	—	—	20	—
LS; LC	PTFE without backing	5.0–10.0	125	80	—	9–14	—

(*continued*)

Table 10.1 *(continued)*

Filter manufacturer and type	Composition	Particle retention size (μm)[a]	Thickness (μm)	Weight/unit area ($g\ m^{-2}$)	Ash content (%)	Air flow rate ($dm^3\ min^{-1}\ cm^{-2}$)[b]	Comments
GT; HT; DT; AT; RT; TT; TS; TM	Polycarbonate	0.2–5.0	9–11	—	—	1–50	Accurate pore size for particle size analysis. Very thin, useful for alpha-counting
PVC0; PVC2; PVC5	Poly(vinyl chloride)	0.8–5.0	60–98	—	—	16–22	Developed for air monitoring
Nalgene[f]							
200	Cellulose nitrate	0.2	—	—	—	—	—
252	'TurboCellulose' nitrate	0.2	—	—	—	—	—
215	Nylon	0.2–0.45	—	—	—	—	—

[a] Minimum particle size retained by filter.
[b] Flow rates are at 20 °C, with a differential pressure of 7×10^4 Pa and an exit pressure of 1.01×10^5 Pa.
[c] Whatman International Ltd, Maidstone, Kent, UK.
[d] Millipore Corporation, Bedford, MA, USA.
[e] Sigma–Aldrich, Poole, Dorset, UK.
[f] Nalge Company, Rochester, New York, USA.

cheap and robust but suffer from variability in collection efficiency. Nevertheless, for the rapid collection of sufficient material for an analysis they are quite adequate. Glass-fibre filters have more uniform collection efficiencies and tend to collect more material at the surface of the filter, which may be advantageous for analysis of alpha- and soft-beta-emitting nuclides.

Membrane filters are quite different in construction to fibrous filters. In the latter, the pores of the filter are formed simply from the matting together of the fibres and hence are non-uniform. The gel-type membranes are made from one of several cellulose esters, poly(vinyl chloride) (PVC), nylon, etc. These essentially consist of a membrane of interlocking pores. A significant difference to the fibrous type is that the pore size is both controllable and variable and they can be used for the collection of particles down to 0.1 μm. Their collection efficiency is better than for fibrous filters and for a pore size of $\leq 0.6\ \mu$m they collect essentially all particles. Nuclepore membranes also have a very uniform pore size but are produced by etching the holes (pores) produced following nuclear bombardment (see Section 8.1.4).

The most significant difference between membrane and fibrous filters is that in the former the particles collected concentrate at or close to the surface. Hence, they are ideal for use with alpha-emitting nuclides since absorption losses are minimal, in contrast to fibrous filters where absorption within the filter is a serious drawback. It is also possible to obtain autoradiographs with membrane filters for the same reason. In some respects, the high uniformity of Nuclepore membranes makes them the choice for alpha-counting.

On the other hand, the concentration of sample at the surface of membrane filters severely limits the mass of sample that can be collected and once a surface layer of material has been collected they exhibit high air flow resistance. Consequently, for the collection of relatively large amounts of material (several milligrams), which is usually the overriding factor if alpha-counting is not required, makes the fibrous type the usual choice. The expense of the membrane type and their fragility also limits their applicability.

Impingement for the collection of aerosols relies on a stream of air being forced through a nozzle at the back of which is some form of obstruction. Particles carried in the air stream then collide forcibly with the obstruction and tend to collect on it. The collection efficiency is usually increased by attaching a ‘sticky’ film to the surface of the obstruction, such as coating it with a film of ‘Vaseline’.

The main use of impingers is the ability to collect *fractions* of different particle size so that one can determine the range of particle sizes associated with the majority of the airborne radioactivity. The major disadvantage of impingement devices, however, is the relatively small mass of material that can be collected. The *cascading impactor* is perhaps the most useful design of impingement equipment. It consists of a series of stages, each of which collects successively smaller particle fractions.

10.5.2 Collection of Gases

Although most airborne radioactivity is in the form of aerosols there are a few nuclides, including radiologically important ones, that exist as gases or in vapour form. These include fission products that occur in gaseous form such as the noble gases (krypton and xenon), most forms of iodine, together with tritium, $^{14}CO_2$ and radon.

The usual method for determining the activity of a gas involves absorbing it in a material such as *activated charcoal* or *silica gel* and then counting this absorber directly. Occasionally, the gas may be released from the trap for analysis and this is the procedure normally adopted for the determination of tritium.

Collection of radioactive gases from the air involves pumping the air through a cartridge of activated charcoal or silica gel (Figure 10.3). The collection time depends on the concentration of the radionuclide in the air and the accuracy required but is usually in the range of a few hours to a few days. The total volume of air pumped through the cartridge must be determined as usual.

Cartridges for the collection of gases other than tritium or radon are normally counted with a hyperpure Ge detector in order to identify all of the nuclides present. Sodium iodide detectors may be used in certain circumstances when there are no significant interferences expected but Ge detectors are usually preferred.

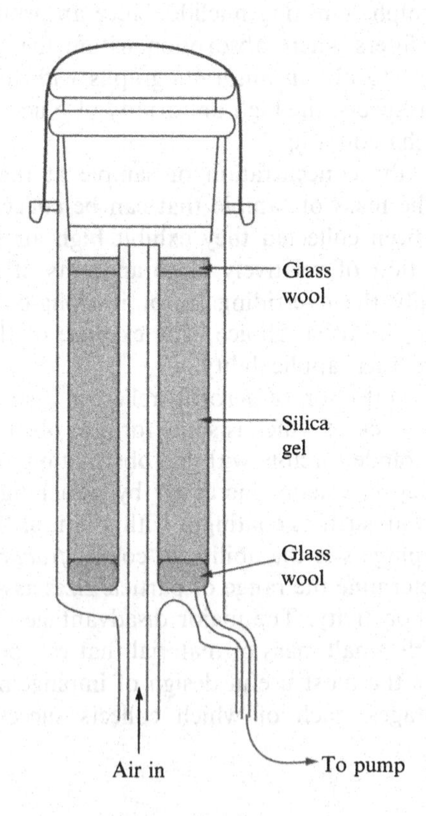

Figure 10.3 A typical adsorption cell used for tritium or other radioactive gases. Reprinted with permission from Lodge, Jr, J. P., *Methods of Air Sampling and Analysis*, 3rd Edn, 1989. Copyright Lewis Publishers, an imprint of CRC Press, Boca Raton, Florida

Sampling and measurement of iodine

Iodine presents a special case in that although most forms of this element likely to be released from a power reactor are gaseous or vaporous, caesium iodide (CsI) is a solid and must therefore be collected as an aerosol. In addition, some gaseous iodine species may also be absorbed on particulate matter in the atmosphere. Thus, the determination of the total radioiodine present in the atmosphere requires a combined aerosol (usually a filter system) and gaseous collection system. The usual method is to use a pre-filter, followed by an activated charcoal cartridge. If the charcoal is treated with potassium iodide or triethylenediamine then this ensures a collection efficiency close to 100%.

Sampling and measurement of radon

The determination of radon has already been briefly discussed in Section 9.5, based on the use of solid-state-nuclear-track detectors (SSNTDs). This method is primarily used for the routine determination of radon in buildings, since it is simple and the apparatus unobtrusive. Because of the perceived radiological importance of radon, a number of other methods for its determination have now been developed. Some of them determine the radon (and thoron) activities directly (as in the SSNTD method), while others are based on the determination of the short-lived daughters of ^{222}Rn and ^{220}Rn.

Another method for determining radon is based on its absorption in activated charcoal, followed by the measurement of the resultant gamma-ray activity of the ^{214}Bi and ^{214}Pb daughters. The methods used are basically similar to those described for iodine and tritium, except that both passive and active (using an air mover) sampling may be used. Passive sampling is more common but active sampling may be used for the rapid collection of radon. The only other difference is the possibility of radon diffusing out of the cartridge if long collection times are used; consequently, the collection time is normally restricted to less than 48 h. This activated charcoal method is well suited to field measurements of radon but is not particularly accurate.

If the daughter nuclides are to be determined as an indirect measure of radon activity, then a filter-based method is normally used. Membrane filters with a pore size of about one micron are preferred and these are used in conjunction with a pump, with a short collection time being employed. The determination is based on the ^{222}Rn decay series shown in Figure 10.4. This indicates that ^{218}Po, ^{214}Pb and ^{214}Bi are the principle daughter activities present after a short collection time and it is assumed that their concentrations do not change during the collection period. The measurement involves determining the alpha-activity during three different time intervals in the range 2–30 min. A final count at least 4 h after the end of collection is used to correct for long-lived alpha-activity. The efficiency of the detector is determined in the usual way. From these data, together with the volume of air sampled, it is possible to determine the activity of each daughter nuclide and hence the activity of the parent ^{222}Rn. The sensitivity of the method is about 0.04 Bq l^{-1} of air.

The advantage of this technique over most other methods of determining radon is that the short collection time allows an 'instant' determination of radon activity

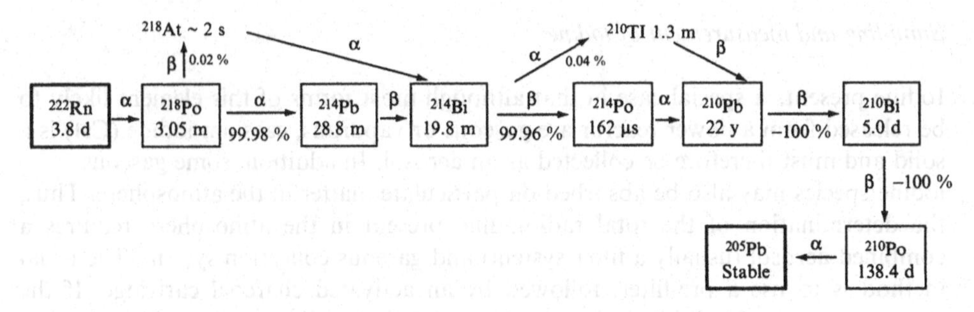

Figure 10.4 The ^{222}Rn decay series – only the most prominent decays are shown here: s, seconds; d, days; m, months; y, years

at a specific location, with the possibility of following changes of radon concentration with time. In general, however, it is the long-term radiological impact of radon on a given population which is of importance so that integrating methods, which are the basis of SSNTD and charcoal absorption determinations, are generally of greatest use.

10.6 LABELLING AND DOCUMENTATION

An important aspect of sampling is proper recording and *labelling* of the samples, usually at the time of collection. Thus, samples should be given a unique but simple code that can be used at all stages of storage, preparation and analysis. In the event that a sample must be re-analysed (rather than a normal replicate to check on the reproducibility of the procedure), there must be no doubt which sample is to be used. This is by no means a trivial consideration in laboratories where hundreds of similar-looking samples are stored.

In field sampling, records should be taken at the time of sampling, such as recording all relevant information about the site, depth of sampling, and maybe the weather conditions. Some of this information may not be particularly relevant to the specific analysis envisaged but it should always be borne in mind that some later and perhaps quite different analysis on the same material may be necessary and require more complete information regarding the collection site than would otherwise have been provided. Sufficient information should also be available that would allow a similar sample to be collected at a later date if this should prove necessary.

10.7 ON-SITE TREATMENT AND STORAGE OF SAMPLES

Samples taken in the field will normally be packaged immediately on-site, prior to their return to the laboratory for analysis. For many samples, particularly water, soils and sediments, polythene containers with screw-top lids are widely used. These have the advantages of being light, non-breakable and easily cleaned. Containers must be thoroughly clean, particularly if they have been used previously and due to dangers of *cross-contamination*, containers should be used for one type of sample only, as far as possible. For some samples and some types of analyses, polythene is not a suitable container material and glass is preferred. This is the case if tritium is to be determined since this is known to permeate through polythene.

In the case of soils and sediments, the *in situ* pH of the material is often important and should be recorded before the sample is taken. Modern solid-state pH probes are useful for this purpose. Checks on the pH of the sample later will indicate any change that has occurred and possible problems.

Acid (usually nitric) is normally added to solutions to prevent *adsorption* of radionuclides on the walls of containers. Acidification may not completely eliminate adsorption problems with lead and slight adsorption may occur even at pH 2. Carriers (the inactive form of the element) may also be added to solutions to prevent adsorption if particular nuclides are sought. The mass of carriers used is usually of the order of 10 to 20 mg of each carrier element used. Occasionally, *preservatives* may be added to a sample to stabilize it. An example is the addition of formalin to milk to prevent coagulation. Preservatives, however, should only be used with caution as they may have unexpected side effects if the sample is to be chemically processed later.

Soil and sediment samples, particularly if stored in plastic bags, may undergo biological changes due to attack by micro-organisms. This may lead to unacceptable chemical changes and/or release of some radionuclides in volatile form. Storage below freezing temperatures may be required for soils that must be kept for more than a day or so although some workers have suggested storage in moist and aerobic conditions.

Vegetation and other types of biological material may be cooled or frozen in the field by using liquid nitrogen or solid carbon dioxide, especially if radioiodine is present since it is possible to lose some of this during transport. In general, however, rapid transport back to the laboratory is probably the best method of preserving the material intact.

10.7.1 Problems Associated with Colloids

Colloids (particles with sizes between 1 nm and 1 μm) are ubiquitous in all natural waters and soil and sediment systems. They are in the form of particles so fine that they normally pass through any filter and so contribute to the water fraction. Many metal oxides form colloids. Because of their high specific surface areas, they are efficient adsorbents for radionuclides. In addition, many actinides form colloids in solution via hydrolysis and these tend to adsorb on naturally occurring colloids and so may separate with the solid fraction. The addition of acid helps prevent this process.

If colloidal material is present in water, filtration (to separate solid, particulate matter from the actual water fraction) must be carried out as quickly as possible after sampling to prevent any coagulation of particles occurring, which would then be retained as part of the precipitate (Stumm, 1992). Colloids may be removed from the water fraction during the filtration process if they attach to the filter material. In addition, colloidal particles smaller than the filter pore size may still become attached to the particles of the precipitate and hence retained as part of the solid fraction. This process is enhanced due to the non-uniformity in the pore sizes of most filters. Even soluble radionuclides may become adsorbed on both the filter medium and the precipitate on the filter. This problem is reduced by the prior addition of acid. On the other hand, acid can cause leaching from solid particles, thus providing another reason for early filtration. Any measurements of pH, however, should be carried out prior to the separation.

10.8 STORAGE OF SAMPLES IN THE LABORATORY

Once samples are returned to the laboratory, they will almost certainly have to be stored prior to analysis. In general, this storage time should be as short as possible. This is particularly important in the case of radioiodine analysis, where the most important isotope is ^{131}I which only has a half-life of 8 days and must be analysed soon after sampling. Since many types of samples may still require further treatment prior to analysis, storage times will vary according to the extent and complexity of other treatment.

Water is prone to alteration in chemical composition due to the presence of micro-organisms, which can occur within a few hours or less if the water is heavily polluted. It is best stored by rapid cooling and storage at about 4 °C. At this temperature, microbiological activity will still continue but at a considerably reduced rate. If iron or manganese are allowed to precipitate out of the water (more likely if the water is frozen), the resultant precipitate may well act as a 'scavenger' for any radionuclides present, thus resulting in a transfer of activity from the liquid to the solid phase. Acidification of the samples helps reduce losses due to iron or manganese.

Soils and sediments should be stored at temperatures below 0 °C, if possible, in order to prevent microbiological activity, although there have been claims that this may affect the release of some constituents in subsequent chemical processing. Vegetation should also be stored cold wherever possible. Prolonged storage in polythene bags at room temperature should be avoided due to considerable breakdown of the organic matter and possible release of some nuclides. If storage for more than a few days is required, then freezing at about 0 °C is recommended, followed by freeze-drying. An alternative approach, which is preferred for some types of analysis, is rapid air-drying at about 35–40 °C, followed by grinding of the material. The ground material may then be stored at room temperature.

Other types of biological material, such as meat, milk and faecal samples, should normally be freeze–dried soon after collection in order to prevent decay processes affecting the material and the release of nuclides such as tritium, ^{14}C and ^{35}S. Preservatives such as formalin are occasionally used but are best avoided, particularly if radiochemical analysis is required.

10.9 PRETREATMENT PRIOR TO ANALYSIS

10.9.1 Sub-Sampling

Almost inevitably, if field sampling has been used and carried out correctly the sample returned to the laboratory will be much larger than is necessary for the actual activity determination, except possibly in the case of direct gamma-ray counting, where up to 1 kg of material may be required per determination. Certainly, for any analysis involving chemical processing of the material prior to the activity measurement, the amount of material collected will be greater than that required for the determination. In all such cases, some method of *sub-sampling* is required.

Sub-sampling of liquids and gases is fairly straightforward, merely requiring the removal of an aliquot (integer portion) of the liquid or gas. Sub-sampling of solids, however, is usually less simple. For instance, the particulate matter separated from water may contain a rather inhomogeneous distribution of nuclides due to the presence of colloids and adsorbed material. Dividing such material in the solid state is unlikely to provide reproducible samples. A better method would be to dissolve the entire precipitate, make the solution up to a constant volume and then remove aliquots from this for the actual analysis. This method could also be used for other types of non-homogenous solids.

Normally, however, sub-sampling of solids is carried out on the dried, solid material and the first question to be answered is what size the sub-sample should be in order to provide a representative sample of the original material collected and brought back to the laboratory. The answer to this question depends on the particle size of the material. Thus, Kleeman (1967) showed that, starting with a sample of about 1 kg, the particle size could be about 0.4 mm in diameter to provide a sample of 100 g (based on a sample density of 2.7 g cm^{-3}), but this would have to be reduced to about 0.21 mm in order to split the sample down to 20 g. For a 5 g sample, the particle diameter would have to be only about 0.14 mm. If there is any evidence that any particular grains resist crushing but contain some fraction of the radioactive nuclide, then they will tend to be concentrated in the coarser fraction and the above figures would have to be increased accordingly. Few analysts seem to be aware of the above restrictions on sub-sampling, or certainly do not incorporate them in their sub-sampling procedures.

10.9.2 Weight of Sub-Sample Required

In the context of sub-sampling, a frequently quoted formula is that due to Ingamells and Switzer (1973). This again was based on geochemical analysis and relates a sampling constant, K_s, to the required relative standard deviation (the standard deviation as a percentage of the amount of activity present) in the determination of the activity, R (%), and the weight of sample, w, to give that value of R as follows:

$$w = \frac{K_s}{R^2} \text{ (g)} \tag{10.2}$$

The parameter K_s is the sub-sample weight necessary to ensure a relative standard deviation of ± 1 % at the 68 % confidence level in a single determination of activity. This has to be determined experimentally by carrying out a series of activity measurements on sets of sub-samples of different weight. The R value for each set is then calculated and K_s found from Equation (10.2). Plotting K_s as a function of w allows an estimate of the correct sub-sample weight to be determined. The use of

Equation (10.2) then allows the appropriate sub-sampling error to be determined for the sub-sample actually being employed in a determination, assuming a Gaussian distribution of activity values.

Soils and Sediments

Soils are usually crushed or milled prior to obtaining the sub-sample. This is the process of *comminution*. In practice, the soil is crushed in a suitable device, e.g. an impact mortar or jaw crushers, down to about 2 mm diameter. The material is then thoroughly blended by using a rotary cylinder and the sub-sample taken from this material. The preparation of a sub-sample can be by a standard technique. One such method is 'cone and quartering', in which the material is formed into a cone, quartered and opposite quarters taken. These are then combined, mixed and the process repeated until a suitable sample weight is obtained. Care has to be exercised in order to provide a completely unbiased sample – with respect to particle size – due to the tendency for larger particles to collect near the base of the cone. Mechanical methods can be based on *riffling* (using a series of baffles), or even better, with the sample poured onto a solid cone and the material then thrown from the cone and collected in a series of pots moving round the cone. This sub-sample may then be ground to a much finer size ($<\sim 1$ mm) by using a *ball mill* or rotary pestle and mortar and then rolled thoroughly to ensure a homogeneous mixture just prior to analysis.

Sediment cores are normally required to be cut into sections, usually after the core has been extruded from the coring device but while still wet. The sections can then be treated in the same way as for soils.

Vegetation

Sub-sampling of vegetation is also normally performed on dried material, although fresh material may be used. The latter is more usual for vegetables and for these cases industrial blenders or homogenizers are frequently used for reducing the material to a fine mass. The resultant material is often liquid enough that aliquots can be taken directly for analysis.

In the case of dried material, this will be reduced in size by some form of mill or grinder and particle sizes of < 0.5 mm are recommended. Due to the fibrous nature of much vegetation, the fibres may have to be subsequently broken down to < 0.5 mm lengths in a ball mill. This material may then be sub-sampled by using the cone and quartering technique described above for soils. Dried faecal samples may be treated in the same way as dried vegetation.

Meat and other biological material

Meat and other animal tissues can be dealt with in pathological tissue homogenizers which produce a homogeneous material, or cooled to liquid nitrogen temperatures and then ground to a fine powder, so allowing sub-sampling by some form of cone and quartering. In large animals, individual organs may be removed prior to the comminution of the bulk and the former treated as a sub-sample.

10.10 SOLUTION PREPARATION

In the case of direct gamma-ray counting and gross beta- and alpha-counting, such measurements may be carried out on a suitable sub-sample of dried material. The particle size has to be very fine for alpha-measurements due to the very short range of alpha-particles, but apart from this, no further sample preparation is required.

For alpha-spectrometry, the determination of pure beta-emitting nuclides or when beta-counting is preferred, and even for some gamma-ray measurements, it is necessary to dissolve the material in order to carry out a radiochemical separation. Methods of preparing solutions from any type of environmental sample are the subjects of this section.

10.10.1 Leaching

An initial consideration concerns whether complete dissolution of the sample is required or whether only part of the sample, containing the nuclides of interest, needs to be brought into solution – the process of *leaching*. The primary consideration in leaching from a sample is that the extraction of the desired nuclides is quantitative. If there is any doubt as to the efficacy of the leachant, then total dissolution should be used.

Leaching is usually brought about with mineral acids (hydrochloric acid (HCl), nitric acid (HNO_3), sulfuric acid (H_2SO_4) and hydrofluoric acid (HF), either singly or in suitable combinations. If the silicate portion of a material is to be leached, then hydrofluoric acid will have to be used. Thus, ^{90}Sr in soils can be determined by leaching with 1:1 hydrochloric acid following the addition of Sr carrier and ^{85}Sr as yield monitor. A good deal of controversy has arisen over the dissolution of transuranic nuclides in sediment samples but careful acid leaching (using 8 M HNO_3 appears to be sufficient and better than a mixture of 8 M HNO_3 and 12 M HCl for some sediments.

10.10.2 Total Dissolution of Samples

If complete dissolution of the sample is required (as is usually the case), then there are four basic techniques that can be used; i.e. (a) *acid digestion*, (b) *fusion*, (c) *wet ashing*, and (d) *dry ashing*. One or other forms of ashing, however, are the most widely used methods. The aim in ashing a material (both wet and dry) is to decompose the organic matter and remove it as CO_2, H_2O, etc. For the determination of actinides, it is important to destroy all of the organic matter in order to prevent any possibility of the formation of stable complexes which are very insoluble.

One of the major objections to dry ashing is the potential loss of radionuclides of volatile elements, such as caesium, ruthenium, iodine and lead. Surprisingly, it is possible to avoid the loss of ^{131}I from vegetation before analysis if it has previously been contained in the presence of a solution of sodium iodide and sodium hydroxide. After drying, it can then be dry ashed, finally leaching the iodine species from the ash with water (Foti, 1977).

In general, vegetation, biological samples and various types of air-filtered products are easier to get into solution than are soils, sediments and rocks, due to the presence of silicates and insoluble compounds of metals such as chromium, titanium and zirconium in the latter. All of these methods for total dissolution of the sample are discussed in most standard analytical chemistry textbooks.

11

Statistical Treatment of Radioactivity Measurements

The ultimate aim in any measurement of radioactivity is to obtain the activity of the given sample, either as an absolute activity value, e.g. becquerel (Bq) per unit weight or volume, or as a relative value, e.g. counts per unit time per unit weight, or volume. Simply reporting a value for the activity, however, is of very limited usefulness. This is because such a datum tells us nothing about how *accurate* or *precise* this value is. In addition, under certain circumstances of measurement the registration of counts may not by itself be sufficient to say with complete confidence whether a particular source is actually present or not. The degree of confidence that we have for believing a source is present depends on the quality of the data. One of the aims of this chapter will be to look at the statistical devices that have been developed to make such judgements and how to take into account systematic effects from the uncertainties in the parameters defining the quantity that we are trying to determine.

11.1 ACCURACY AND PRECISION

We use the terms 'accuracy' and 'precision' in everyday speech, often in a way which suggests that they are actually the same. In statistics, however, they have very explicit meanings. These are best understood by reference to Figure 11.1. *Accuracy* refers to how close the value that we obtain for a parameter is to the 'true' or 'correct' value. In practice, we would normally be using the *accepted* true value. We shall say more about this topic later. *Precision* indicates how repeatable is the value that we obtain and indicates the spread in the values for the parameter about its mean value. Figure 11.1(a) illustrates the situation where the darts are accurately thrown into the bull and are also tightly bunched, and hence precise. In Figure 11.1(b), we have the position where the darts are precise but not accurate (not in the bull), while Figure 11.1(c) illustrates the case of poor precision but good accuracy, since the centre of the distribution of the darts is still in the bull.

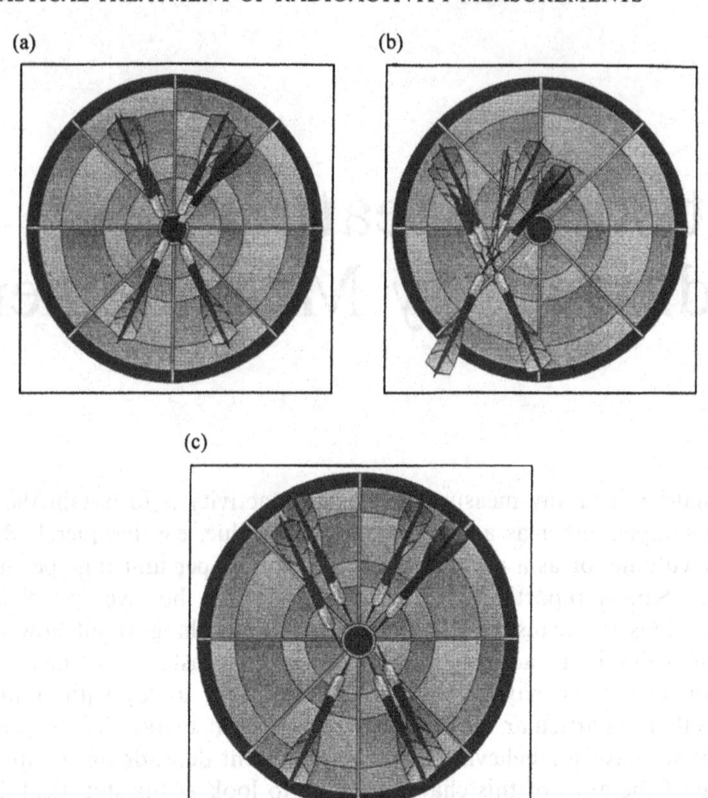

Figure 11.1 Illustration of the terms *accurate* and *precise*: (a) the darts are both accurate and precise (closely bunched and all in the bull); (b) the darts are precise but not accurate (closely bunched but far from the bull); (c) the darts are accurate but not precise (close to the bull but not tightly bunched)

11.2 MEAN AND STANDARD DEVIATION

Radioactive decay is a statistical process. By this, we mean that if we have a large number of radioactive atoms we know that after a period defined as the *half-life* we will have half the original number of atoms remaining. We cannot say, however, at any given instant in time which atoms will actually decay and which will remain intact. There is only a *probability* that any given atom will decay and the laws of probability can only be applied to a large number or *population* of radioactive atoms to predict some outcome, e.g. the half-life of the population.

Bearing this in mind, if we determine the activity of this population (assuming that it is sufficiently long-lived for the activity to remain constant over practical counting periods) and we repeat the measurements a number of times we will come up with slightly different values each time. If we assume that we carry out N measurements of the activity, resulting in values $x_1, x_2, \ldots, x_N$, then we can define a *mean* value for the activity, $\bar{x}$, where:

$$\bar{x} = \frac{x_1 + x_2 + \ldots x_N}{N} = \frac{\sum_1^N x_i}{N} \tag{11.1}$$

The difference between this mean value and the true ('accepted') value is the *absolute error*, expressed in the same units as the mean value. Another common way of expressing this difference is by the *relative error*, which is the absolute error divided by the measured value (the mean value in this case) and multiplied by 100 to give the relative error in percentage terms.

From our measured values, x_1, x_2, etc., we also want to know what is the precision of these measurements. There are various ways of expressing this but the usual one is to determine the *standard deviation*, σ of these values about the mean. The standard deviation for the above data is given by the following:

$$\sigma = \sqrt{\frac{\sum (x_i - \bar{x})^2}{(N-1)}} \tag{11.2}$$

The standard deviation, as discussed so far, is an estimate of the error for a single measurement. If we consider the mean value, $\bar{x}$, however, taken from N separate measurements, then this will be more precise than any single measurement. Hence, we would normally quote the mean value and its standard deviation, σ_{m}, where σ_{m} (the standard deviation of the mean) is related to σ by the following:

$$\sigma_{\mathrm{m}} = \frac{\sigma}{\sqrt{N}} \tag{11.3}$$

If we examine Equation (11.3), we find that it has an interesting property – namely, that the larger the number of experiments we perform (larger N), the value of σ remains essentially constant, whereas σ_{m} decreases, producing essentially a more precise result. Taken to extremes, this leads to the illogical conclusion that by performing a large enough number of experiments one can reach any desired degree of precision. Thus, if one were using a pair of kitchen scales to weigh a parcel by performing the weighing operation a sufficiently large number of times, one could weigh to a precision of a microgram! This is clearly absurd and this paradox is resolved by realizing that the extreme precision assumes that we can record infinitesimally small differences between values, whereas in practice any piece of measuring equipment has a finite resolution and in the case of kitchen scales this resolution is grams rather than micrograms.

The mean and its standard deviation, as expressed by Equations (11.1) and (11.2), assume that all the data points are equally reliable. This is by no means always the case and in such situations we have to modify these two equations. In the case of the mean, this is carried out by assigning each data point a statistical *weight*, W and then determining the *weighted mean*, $\bar{x}_{\mathrm{W}}$. The weighted mean can be expressed in the following form:

$$\bar{x}_{\mathrm{W}} = \frac{\sum_1^N W_i x_i}{\sum_1^N W_i} \tag{11.4}$$

The uncertainty in the mean $\bar{x}_W$, i.e. $\sigma_{\bar{x}W}$, is given by the following:

$$\sigma_{\bar{x}W} = \sqrt{\frac{\sum[W_i(x_i - \bar{x}_W)^2]}{(N-1)\sum W_i}} \tag{11.5}$$

The next question that arises is how to assign an appropriate weight to each value. The most common method is based on the standard deviation, σ, with the weight being inversely proportional to σ^2. This latter function is known as the *variance*, v and so W is inversely proportional to the variance, or:

$$W_i = \frac{1}{\sigma_i^2} \tag{11.6}$$

$$\sigma_{\bar{x},w}^2 = \frac{1}{\sum_1^N \frac{1}{\sigma_i^2}} = \frac{\sigma^2}{N} \tag{11.7}$$

$$\therefore \sigma^2 = \sigma_{\bar{x},w}^2 N \tag{11.8}$$

Note that Equation (11.8) is equivalent to Equation (11.3). In addition:

$$\sigma^2 = \frac{N\sum[W_i(x_i - \bar{x})^2]}{(N-1)\sum W_i} \tag{11.9}$$

11.2.1 Propagation of Errors (Uncertainties)

We have confined our discussion to considerations of the uncertainties that arise in the measurement of a single quantity or parameter. Frequently, however, it is necessary to combine two or more quantities, each with its own uncertainty. What is the uncertainty of the combined quantity? This is dependent on how the quantities are to be combined and the arithmetic operation being carried out. Consider two quantities, x and y, with respective uncertainties $x \pm \sigma_x$ and $y \pm \sigma_y$. Combining these quantities and their uncertainties by the operations of addition, subtraction, multiplication and division, gives the following results:

$$\text{Addition} \quad (x \pm \sigma_x) + (y \pm \sigma_y) = (x + y) \pm \sqrt{\sigma_x^2 + \sigma_y^2} \tag{11.10}$$

$$\text{Subtraction} \quad (x \pm \sigma_x) - (y \pm \sigma_y) = (x - y) \pm \sqrt{\sigma_x^2 + \sigma_y^2} \tag{11.11}$$

$$\text{Multiplication} \quad (x \pm \sigma_x)(y \pm \sigma_y) = xy \pm xy\sqrt{\left(\frac{\sigma_x}{x}\right)^2 + \left(\frac{\sigma_y}{y}\right)^2} \tag{11.12}$$

$$\text{Division} \quad \frac{(x \pm \sigma_x)}{(y \pm \sigma_y)} = \frac{x}{y} \pm \frac{x}{y}\sqrt{\left(\frac{\sigma_x}{x}\right)^2 + \left(\frac{\sigma_y}{y}\right)^2} \tag{11.13}$$

We can also write:

$$\text{Exponential } e^{a(x\pm\sigma_x)} \simeq e^{ax} \pm e^{ax}\sqrt{(a\sigma_x)^2} \qquad (11.14)$$

Consider a typical situation where we are trying to determine the count rate of an unknown source. We can easily do this by carrying out two, separate counts. Let us say that we believe the source to have a count rate of about 100 counts per minute (cpm). If we count the source for 10 min, this will give us a gross count for the source and the background (C_T). Let us say that in 10 min, C_T is 1080 counts. We now count for a further 10 min period with no source in place to give us the background count, C_B. This we find is 80 counts. The net count for the source, C_S, is then simply given by the following:

$$C_S = C_T - C_B = 1080 - 80 = 1000$$

It will be shown later that $\sigma_T = \sqrt{C_T}$ and $\sigma_B = \sqrt{C_B}$.

From Equation (11.10), the uncertainty for C_S is as follows:

$$\pm\sqrt{\sigma_T^2 + \sigma_B^2} = \pm\sqrt{1080 + 80} = \pm 34.1$$

$$= 1000 \pm 34.1 \text{ counts}$$

$$= \frac{1000 \pm 34.1}{10} \text{ cpm} = 100 \pm 3.4 \text{ cpm}$$

A slightly more complex example relates to the following situation. Here, we wish to determine the concentration of potassium in a sample from the activity of the ^{40}K present, based on the 1460 keV gamma-ray peak. This is frequently carried out by comparing the activity of the sample with that of a standard containing an exactly known concentration of potassium. Since the abundance of ^{40}K is assumed constant in all terrestrial samples of potassium, the ratio of the activities in the sample and standard will be equivalent to the ratio of potassium for the two samples. By using a single channel analyser with an energy window set at around this energy, we measure the count rate due to ^{40}K. If the mass of the sample and the standard are assumed to be equal, then this count rate is equivalent to the activity of ^{40}K. The ratio of the count rates for the sample, divided by that of the standard multiplied by the concentration of potassium in the standard, gives the concentration of potassium in the sample. Thus, if the count rate in the sample is A_{sam} and that in the standard is A_{std}, with the standard having a potassium concentration of C_{std} %, then the potassium concentration of the sample, C_{sam} is given by the following:

$$C_{sam} = C_{std}\left(\frac{A_{sam}}{A_{std}}\right) \qquad (11.15)$$

Now suppose that the sample, standard and background are counted for 1000 min each. The total count for the standard, containing 3.05 ± 0.05 % K, is $7800 \pm \sqrt{7800}$, while that of the sample is $5200 \pm \sqrt{5200}$. The background count was $250 \pm \sqrt{250}$. We then have:

$$(\text{Gross count rate})_{\text{std}} = \frac{7800 \pm 88}{1000}\ \text{cpm} = 7.8 \pm 0.088\ \text{cpm}$$

$$(\text{Gross count rate})_{\text{sam}} = \frac{5200 \pm 72}{1000}\ \text{cpm} = 5.2 \pm 0.072\ \text{cpm}$$

$$\text{Background count rate} = \frac{250 \pm 16}{1000}\ \text{cpm} = 0.25 \pm 0.016\ \text{cpm}$$

Thus, by using Equation (11.11):

$$A_{\text{std}} = (7.8 \pm 0.088) - (0.25 \pm 0.016) = 7.55 \pm \sqrt{0.088^2 + 0.016^2} = 7.55 \pm 0.089$$

$$A_{\text{sam}} = (5.2 \pm 0.072) - (0.25 \pm 0.016) = 4.95 \pm \sqrt{0.072^2 + 0.016^2} = 4.95 \pm 0.074$$

With $R = A_{\text{sam}}/A_{\text{std}}$, and from Equation (11.13), we then have:

$$R = \frac{4.95 \pm 0.074}{7.55 \pm 0.089} = 0.656 \pm 0.656\sqrt{\left(\frac{0.074}{4.95}\right)^2 + \left(\frac{0.089}{7.55}\right)^2} = 0.656 \pm 0.0125$$

$$C_{\text{sam}} = C_{\text{std}}\left(\frac{A_{\text{sam}}}{A_{\text{std}}}\right) = (0.656 \pm 0.0125)(3.05 \pm 0.05)\,\%$$

$$= 2.00 \pm 2.00\sqrt{\left(\frac{0.0125}{0.656}\right)^2 + \left(\frac{0.05}{3.05}\right)^2} = 2.00 \pm 0.050\,\%$$

11.3 PROBABILITY AND DISTRIBUTIONS

The distribution of data points about the mean value is not random but will follow a particular pattern or *distribution*. For the case of radioactive nuclei, if we have N nuclei then the *probability* $P(n)$ that n disintegrations will occur in time t is given by the *binomial distribution*, which can be expressed as follows:

$$P(n) = \frac{N!}{(N-n)!n!}(1 - e^{-\lambda t})^n e^{-\lambda t(N-n)} \tag{11.16}$$

Fortunately, this rather awkward equation can be simplified if the counting time is short compared to the half-life of the nuclide (which is usually the case for environmental monitoring conditions), n is very small compared to N, and N is large. Under these circumstances, the distribution approximates to the *Poisson distribution*, given by the following:

$$P(n) = \frac{m^n e^{-m}}{n!} \tag{11.17}$$

where $m(=\bar{n})$ is the mean value. The standard deviation, σ, of the Poisson distribution is $\sqrt{m}$, so that in general one assumes that the standard deviation of a single count is the square root of the count. The Poisson distribution where n is 10 and $\bar{n}$ is 4 is shown in Figure 11.2. If n is sufficiently large, then the Poisson distribution can be shown to tend towards the *Gaussian* or *normal distribution*, as follows:

$$P(n) = \frac{1}{\sqrt{2\pi\bar{n}}} e^{-(n-\bar{n})^2/2\bar{n}} \tag{11.18}$$

Even for values of $\bar{n}$ as small as 10, the Poisson and normal distributions are very similar, as shown in Figure 11.3. The normal distribution is usually represented as a bell-shaped curve, as illustrated in Figure 11.4. One should note, however, that if the counting time is, in fact, long compared to the half-life of the nuclide, i.e. $\lambda t \geq 1$, then Poisson statistics do not apply (Matthews *et al.*, 1980). Here, it is better to use the so-called Ruark–Devol statistics. Fortunately, this situation rarely arises in environmental counting.

11.4 CONFIDENCE LEVELS

If we know *a priori* what probability distribution applies to our data, then we can calculate the probability that a particular measurement lies within a given range of the true or mean value. The probability that a value lies in a specified interval from c_1 to c_2 is simply the area under the curve within that interval.

Referring to Figure 11.4, the abscissa is divided into units of σ from the mean, both positive and negative. The area under the part of the curve represented by positive σ values is given in published tables (see, for example, Bevington, (1969) and Anderson *et al.* (1981) in the Bibliography at the end of this present text). The area in the shaded part of the curve in Figure 11.4, from 0 (the mean value, $\bar{x}$) to

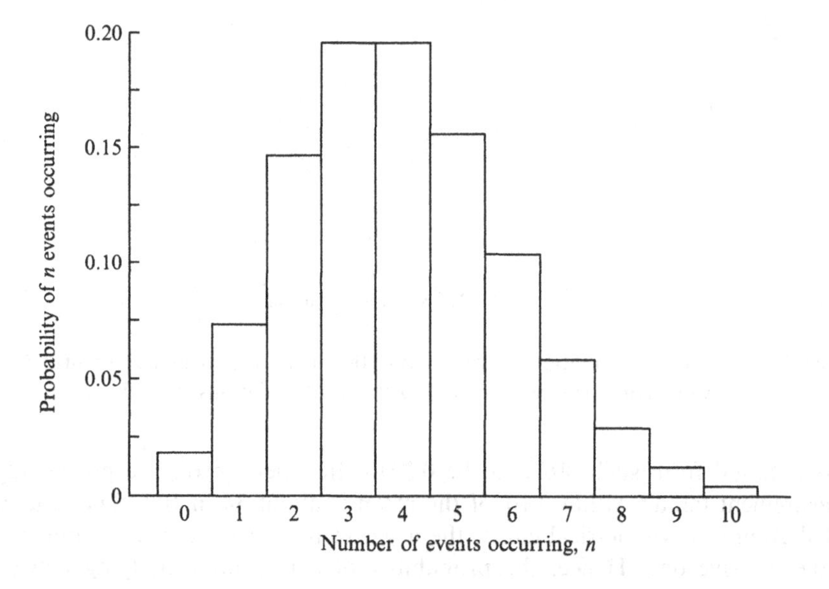

Figure 11.2 The Poisson distribution where $n = 10$ and $\bar{n} = 4$

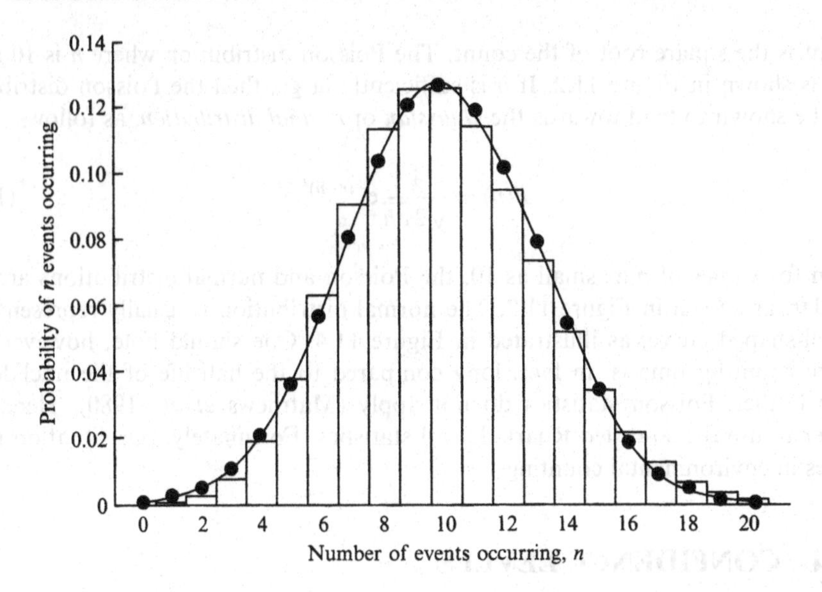

Figure 11.3 Poisson and normal distributions for $n = 20$ and $\bar{n} = 10$; (–●–), normal distribution

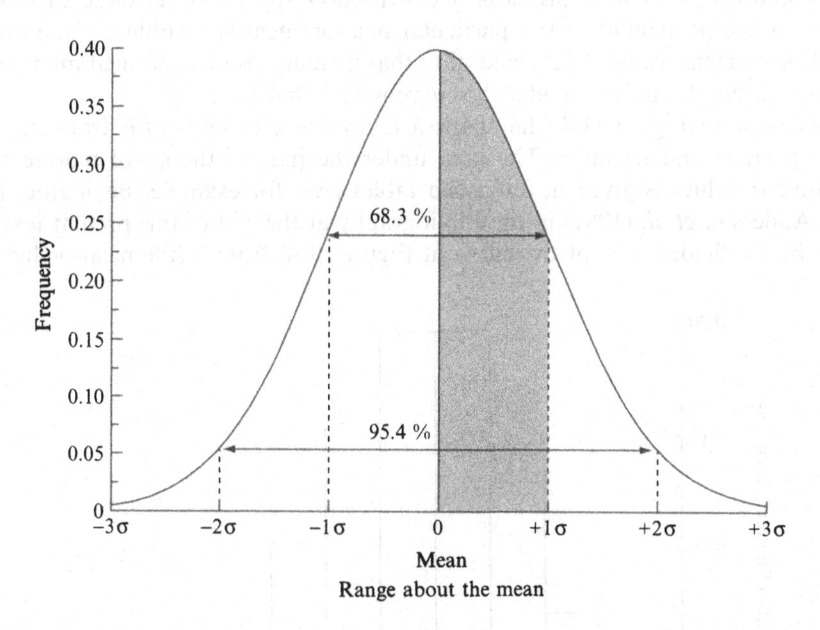

Figure 11.4 Normal distribution curve showing the area under the curve or probability for a value lying between one or two standard deviations from the mean value

$+1\sigma$ is found from such tables to be 0.3413. In other words, the probability of a measurement falling within $+1\sigma$ of the mean is about 34 in 100. Since the normal distribution is a symmetrical curve, the negative part of the curve is a mirror image of the positive one. Hence, the probability of a measurement lying between the mean and -1σ of the mean is also 0.3413. Consequently, the probability of a measurement lying between $+1\sigma$ and -1σ of the mean is about 68 in 100. The area values lying between $\pm 2\sigma$ and $\pm 3\sigma$ are 0.9544 and 0.9972, respectively. Since the total area under the curve is 1, the figure of 0.9972 for $\pm 3\sigma$ indicates that a measurement is almost certain to be within these limits, provided that the data follow a normal probability distribution.

Knowing the form of the continuous normal distribution function allows us to determine the probability that a measurement will fall between any given limits. From the above, we know that there is a 68 chance in 100 that a given measurement will lie between + and -1σ, which are known as the 68% *confidence levels*. Similarly, there is a 95 chance in 100 that it will lie between + and -2σ, with these being called the 95% *confidence levels*. Another way of expressing the probability of values lying between given multiples of σ is to quote the x% interval. The intervals normally quoted are 10, 5 and 1%. These correspond to measurements lying between $\pm 1.64\sigma$, $\pm 1.96\sigma$ and $\pm 2.58\sigma$, respectively. Müller (1979) has pointed out that confidence limits should be treated with caution. Such limits assume that the uncertainties follow a particular probability distribution but it is usually impossible to test this assumption due to the lack of available data.

Usually, one quotes the value for some quantity as a mean with an 'error' (more correctly, the 'uncertainty'). The error most often quoted is that at the 1σ level. Thus, a peak area of 100 ± 5 counts infers that the expected value is 100 counts and that the standard deviation is 5 counts. In other words, we believe that there is a 68% chance that the true peak area lies between 95 and 105 counts.

Sometimes, the error quoted is the 2σ error. For the above peak area, this would give us 100 ± 10 counts, with a 96% confidence that the value lies between 90 and 110 counts. If the error were to be quoted at the 95% confidence level, then 100 ± 5 counts at the 95% confidence level would be interpreted as an expected value of 100 counts with a standard deviation of 5/1.96 counts, i.e. 100 ± 2.6 counts. A report on 'Environmental Health Monitoring' by a working party from the UK Institution of Environmental Health Officers (1988) recommends reporting results from counting experiments at the 95% confidence level.

We have already said that the standard deviation for counting data is $\sqrt{n}$. It must be emphasized that n is the total number of counts during a counting period, and *not* the counting rate. Thus, if we are measuring a radioactive source with an activity equivalent to 100 counts per minute (cpm) and we count for 100 min, then we expect a total count of 10 000. The standard deviation is based on the square root of the total counts (10 000) and not on the activity (100 cpm). In this case, we have the following:

$$\text{Count rate} = \frac{10\,000 \pm \sqrt{10\,000}}{100} = \frac{10\,000 \pm 100}{100} = 100 \pm 1\ \text{cpm}$$

In addition, note that if we have a source giving a count rate of R counts per unit time and we count the source for T units of time then we will have a total count of RT counts. The standard deviation of this total count is then just $\sqrt{RT}$. The standard deviation of the count rate, σ_R is then:

$$\sigma_R = \frac{\sqrt{RT}}{T} = \sqrt{\frac{R}{T}} \tag{11.19}$$

If we are recording the 95% confidence level ($\pm 1.96\sigma$), then the error on the count rate is given by the following:

$$\frac{10\,000 \pm (1.96 \times 100)}{100} = 100 \pm 1.96\,\text{cpm}$$

One should take care in quoting errors. Indicating an error with, say, three significant figures is practically never justified and in most cases one is enough.

11.4.1 Variations in Counting Period

A question that may occur to the reader is the following: Is it better to count a source for one, long period and accumulate as large a number of counts as possible, or rather split the total time available into a number of smaller periods, obtaining fewer counts in these shorter periods but carrying out repeated counts?

Let us consider counting our source of 100 cpm again. Suppose that we want to know which is the more favourable – counting the source once for 100 min or 10 times for 10 min each.

Consider first the case of 10 successive counts, each of duration of 10 min. In each counting period, we will detect about 1000 counts. Thus, the standard deviation for each interval is $\sqrt{1000} = 31.6$. So, we predict a count rate of $(1000 \pm 31.6)/10$ cpm:

$$= 100 \pm 3.2\,\text{cpm}$$

However, we have 10 measurements. From Equation (11.3), the standard deviation of the mean is:

$$\frac{3.2}{\sqrt{10}} = 1.0\,\text{cpm}$$

Thus, the final result is 100 ± 1.0 cpm.

We have already seen that counting for 100 min gives a final result of 100 ± 1.0 cpm. In other words, the two methods give exactly the same result. This should come as no surprise, though, since it is consistent with the concept that one cannot increase the precision of a measurement simply by carrying out larger and larger numbers of measurements. Other factors may play a part in deciding on which method to use in a given situation. If, for instance, there was a significant interval between each of the short counts while the total time available was only 100 min, then perhaps only 7 or 8 counting periods could be used, which would slightly degrade the final precision of the count rate. The single-count method is obviously the simplest to carry out. On the other hand, when using just a single count to estimate the count rate one could not detect possible errors in the system such as instabilities in the amplifier or high-voltage supply. These would show up in the other method by producing inconsistent count rates. Thus, the decision on which method to use has to be evaluated in the light of the specific situation being evaluated.

11.4.2 The Division of Total Counting Time between the Source and Background Counting Times

If only a limited measuring time is available for counting samples, perhaps because a large number of samples have to be processed, then one will be faced with the decision of how to divide this limited time between counting the source i.e. source plus background (C_A, T_A, count rate R_A), and counting the background alone (C_B, T_B, count rate R_B), in order to determine the correct source count (C_S) and source count rate (R_S). Since $C_S = C_A - C_B$, the standard deviation of C_S, i.e. σ_S, from Equations (11.10) and (11.19) is given as follows:

$$\sigma_S = \sqrt{\sigma_A^2 + \sigma_{B^2}} = \sqrt{\frac{R_A}{T_A}} + \sqrt{\frac{R_B}{T_B}} \tag{11.20}$$

If the total counting time i.e. $T_A + T_B$, is T, then it can be shown that T_A/T for a given T has to be chosen such that σ_S/R_S is a minimum (de Soete *et al.*, 1972). This leads to an optimum for the ratio of T_A to T_B, given by the following:

$$\frac{T_A}{T_B} = \sqrt{\frac{R_A}{R_B}} \tag{11.21}$$

Thus, the apportioning of time between counting the source and counting the background depends on the relative count rates of the source and background.

11.4.3 Statistics of Rate Meters and Ionization Chambers

The statistics for these devices have to be treated slightly differently from the previous discussion. This is because the net output current at any instant involves an integration of all previous counts, weighted exponentially according to the time elapsed since they occurred, and this condition requires special statistical theory for evaluating the probable error in terms of the fluctuations observed (Kip *et al.*, 1946).

For a device which develops a voltage across a circuit containing a resistor of R ohms and a capacitor of C farads, this will have a time constant of RC units of time. The error, σ, in any single, instantaneous reading of the output signal, which has an average value of x counts per unit time, caused by a purely random counting rate, is given by the following:

$$\sigma = 0.477\sqrt{\frac{x}{RC}} \tag{11.22}$$

where σ, x and RC are all in the same time units.

The probable error, σ_T, of the average value, x, observed for a period of T, is given by the following:

$$\sigma_T \le 0.6745\sqrt{\frac{x}{T}} \tag{11.23}$$

11.4.4 Statistics with an Unknown Probability Distribution

Unfortunately, we do not always know the distribution to apply to a given situation. In these cases, we have to rely on the *central limit theorem*, which states that the sampling distribution of the mean can be approximated by a normal distribution whenever the sample size is large. The next question is obviously: what do we mean by large? It turns out that statistical theory indicates a size of 30 or more can be regarded as large. For smaller sample sizes, one has to decide whether the use of the normal distribution is appropriate or not.

11.4.5 Outliers and their Treatment

If one produces a set of observations for a particular quantity based on, perhaps, repeated measurements of that quantity or from a series of separate experiments to derive that quantity, then often one finds one or more values that do not seem to conform with the rest of the set. That is, values that appear to be much greater or smaller than those in the rest of the set. Such values are usually called *outliers*. The question arises as to whether these outliers are to be used in determining the mean and standard deviation of the set. There is a great temptation, particularly on the part of inexperienced scientists, to simply discard these outliers as 'wrong'. This is reinforced if the suspect observations seem to go against a conclusion that the rest of the data support or perhaps against a conclusion that one has anticipated for one reason or another.

One should strongly resist this temptation unless there is a proven error in the suspect data or some valid statistical procedure allows one to discard such values. Examples of errors that would permit the discarding of a particular value are if it is known that an instrument was faulty during the time the data were recorded, or the value was incorrectly recorded or mistyped, or perhaps transcribed incorrectly.

If there is no evidence to suspect an erroneous value has been recorded, then one has to look to statistical methods to investigate the validity or otherwise of discarding the outlier. These may involve significance tests or more sophisticated graphical procedures to decide whether the data are consistent with a given hypothesis – known as the *null hypothesis*. Most work on testing for outliers has assumed a normal distribution for the observations other than the outliers. A comprehensive treatment of the problem of outliers is given by Barnett and Lewis (1985).

In the case of a data set which has a normal distribution with a known standard deviation σ, then one may test for outliers by determining the value of K, which is given by the following:

$$K = \frac{(x_{\max} - \bar{x})}{\sigma} \quad \text{or} \quad K = \frac{(\bar{x} - x_{\min})}{\sigma} \tag{11.24}$$

where $x_{\max}$ and $x_{\min}$ are, respectively, the largest and smallest values in the set, i.e. the suspected outliers in a sample of size n. The parameter K is called the *standardized extreme deviate*. The value of K derived from Equation (11.24) is compared with a critical value obtained from tables (see, for example, the *Biometrika Tables for Statisticians*, Vol. 1, as given in the Bibliography at the end of this present text) to test if the outlier can be rejected or not.

If a normal distribution is valid but there is no further information on the distribution of the observations, then a method proposed by Ferguson (1961) may be used. This author looked at two different cases – one where the outlier (*s*) lies on one side of the rest of the data, i.e. either higher or lower than the other data, and one where the outliers are on both sides of the data set. In the first case, one uses the following criterion:

$$c_1 = \frac{m_3}{3\sqrt{m_2}} \tag{11.25}$$

and in the second:

$$c_2 = \frac{m_4}{m_2^2} \tag{11.26}$$

where:

$$m_r = \sum \frac{(x_i - \bar{x})^r}{n}; \; r \geq 2, \; n \geq 25 \text{ for } c_1, \; n \geq 200 \text{ for } c_2 \tag{11.27}$$

The criteria, c_1 and c_2 are usually referred to as the *skewness* and *kurtosis*, respectively, of a distribution and are calculated by using all of the data, including the suspect values (see, for example, Snedcor and Cochran, 1980). These are again compared with critical values obtained from standard tables (e.g. the *Biometrika* tables) to test if the outliers are to be rejected or not.

Possibly the best solution to the problem of outliers, particularly for small sets of data (10 values or less), is that advocated by Chatfield (1988). He suggests that one repeats the analysis with and without the suspect values. If the conclusions are similar, then the suspect values 'do not matter', whereas if they differ substantially one should be wary of making judgements which depend so crucially on one or two observations.

A convenient, and more modern approach to the problem is through the use of *robust statistics* (AMC Technical Brief, 2001). Again, all of the data are considered but the procedure provides a model of the data set which identifies suspect values for further investigation. The basis of one widely used method involves a process called *winsorisation*, in which we assume initial estimates of the mean and standard deviation, $\widehat{\mu}_0, \widehat{\sigma}_0$, which can be the normal arithmetic mean and standard deviation of the data. Each datum is then compared to either $\widehat{\mu}_0 + 1.5\,\widehat{\sigma}_0$ or $\widehat{\mu}_0 - 1.5\,\widehat{\sigma}_0$ and if it falls outside one of these ranges then the specific datum point is changed to that value, i.e. $\widehat{\mu}_0 + 1.5\,\widehat{\sigma}_0$ or $\widehat{\mu}_0 - 1.5\,\widehat{\sigma}_0$. If it falls within these ranges, then the value is unchanged. Following this treatment, improved estimates of the mean and standard deviation are calculated. This procedure is now iterated by using the new estimates of mean and standard deviation and continued until the process converges to an acceptable degree of accuracy. The resulting values for the mean and standard deviation, i.e. $\widehat{\mu}$ and $\widehat{\sigma}$, are the robust estimates. We can then transform the data points, x_i, by $z = (x_i - \widehat{\mu})/\widehat{\sigma}$. Any value greater than about 2.5 can be regarded as a possible outlier and treated as such, or the data set investigated further.

The procedures just outlined assume that the data set has an approximately normal distribution but with outliers and a long tail. The methods are not reliable if the data set is markedly skewed or a large proportion of the data are identical in value.

11.5 RANDOM AND SYSTEMATIC ERRORS

The discussion of errors so far has been concerned with what are known as *random errors*. By this, we mean errors over which we have essentially no control and which arise from the very act of making a measurement. Fortunately, as we have seen they are amenable to estimation through statistical theory.

There are also other types of error which may occur in any experimental situation and which are either controllable to some extent or other by the experimenter or can be corrected for in principle. These are known as *systematic errors*. The nature of the systematic errors depends very much on the actual experimental system. For instance, in determining the absolute activity of a source from gamma-ray counting one requires to know the efficiency of the detector at the relevant gamma-ray energy. This is usually carried out by experimental calibration over a range of gamma-ray energies and if it is inaccurate will give rise to an inaccurate activity for the source, no matter how good the counting statistics. Obviously, one could remove or at least reduce this systematic error by performing the efficiency calibration more carefully or determining the extent of the error in the calibration. Alternatively, the errors may be variable but of such a nature that they can still be accounted for and corrected. An example would be a ratemeter which gave different count rates at different range settings.

Systematic errors would include both instrumental errors, of the type just discussed, along with operator errors. These are normally reduced by experience and attention to detail when performing an experiment. In addition, systematic errors may arise from poor design of the experimental procedure. These may give rise to the largest errors in the experiment, yet at the same time be the most difficult to detect. Systematic errors may be random in sign but are often unidirectional, i.e. either positive or negative, but not both. Such errors tend to *skew* the distribution, so making it asymmetric.

This implied clear-cut distinction between random and systematic errors may not be so apparent in practice. There are opposing views in the literature regarding how random and systematic uncertainties should be combined. Some authorities feel that since systematic uncertainties are apparently less well known than random ones then they should be combined separately and linearly with the random uncertainties. This is consistent with the use of a normal distribution to describe the random errors but a rectangular one for the systematic. A variation on this suggested by Croarkin (1985) is to combine systematic errors thought to be independent in quadrature and then add linearly those thought to be non-independent, finally adding random and systematic errors linearly. By dependent, we mean that the probability of an event B occurring depends on whether or not event A has occurred.

They are independent if the probability of event A occurring does not affect the probability of event B occurring. On the other hand, others believe that both types of uncertainties should be treated in the same manner.

Müller (1979) has given an interesting account of this problem. He argues against treating the two types of uncertainty differently and believes that a possible unification can be based upon an approach in which all uncertainties are given in the form of estimated standard deviations, thus allowing the general application of the laws of error propagation. This avoids a number of awkward problems that can otherwise arise and the clear-cut distinction between 'measurable' and 'estimated' errors, based on their perceived different degree of reliability is impossible because such a separation does not exist.

11.6 LIMITS OF DETECTION

Frequently in environmental counting we are dealing with very low levels of activity, that is, levels which are comparable to the measured background. Consequently, in order to obtain a net count for the source we have to subtract the background from the gross count for the source and background and since these two quantities are of similar magnitude the potential for error in the source activity is large. This is particularly true when counting for alpha-activity, where the levels of activity which are environmentally significant may be a very small fraction of one becquerel.

When dealing with low levels of any quantity the concept of a *limit of detection* arises, that is, the lowest level of that quantity that can be measured under a given set of circumstances. This concept of a detection limit may appear fairly clear-cut. This is not always the case and as Currie (1968) pointed out in an influential paper, there are numerous definitions of this concept in the literature – some based on statistical theory, while others are not. Since the publication of Currie's paper, however, many workers in the field of radioactive counting have followed his definitions. We shall also do so here to a certain extent but will also use others which we feel are more appropriate for environmental radioactive counting.

Methods exist to help one make balanced judgements when the activity level is close to or even below the background level, using a branch of statistics known as *hypothesis testing*. Basically, this involves proposing a hypothesis, called the 'null hypothesis' and seeing if the measurements confirm or disagree with it. In the case of radioactivity, the null hypothesis would be 'is a radioactive source present?'. The answer to such a question may seem fairly obvious but can lead to certain types of error. The first is if the measurements lead us to reject the hypothesis when in fact it is true. This gives rise to a Type 1 error. On the other hand, the measurements may lead us to accept the hypothesis when in fact it is incorrect – giving a Type 2 error. These errors arise from the fact that both the background and source counts will have a distribution (normal) about a mean value. The probability of a Type 1 error is denoted as α and that of a Type 2 error as β. Then $(1-\alpha)$ or $(1-\beta)$ represent the probabilities of making the correct judgement (see Figure 11.5). The values assigned to α and β are pre-selected by the experimenter on the basis of the maximum allowable probability of making either error. In practice, α and β are usually made the same, with a value of 0.05. In other words, we allow a 1 in 20 probability of making either error. In Figure 11.6, we illustrate the overlapping of the distributions of counts when the source and background counts are similar.

Consider the situation in which we have a sample with a gross count of C_{S+B} and a background for the system of C_B. The net count for the sample is C_S. Thus, we have:

$$\textit{Background}\text{: count} = C_B\text{; standard deviation} = \sigma_B = \sqrt{C_B}$$

$$\textit{Gross count of sample}\text{: count} = C_{S+B}\text{; standard deviation} = \sigma_{S+B} = \sqrt{C_{S+B}}$$

$$\textit{Net count of sample}\text{: count} = C_S = C_{S+B} - C_B\text{; standard deviation} = \sigma_S = \sqrt{\sigma_{S+B}^2 + \sigma_B^2}$$

For qualitative analysis, we define a *critical level*, L_c and if $C_S \geq L_c$, we accept the null hypothesis 'signal (count) detected'. If $C_S < L_c$, then we reject the hypothesis and decide that no source is present. The value of L_c is based on the predetermined value of β, the probability of a Type 2 error.

Thus, when we carry out a measurement on a low-activity source, if the count, C_S, exceeds L_c we conclude that we can count this source under the given conditions and obtain a value for its activity. Normally, we will repeat the measurement several times and thus obtain a mean value for the activity, $\bar{C}_S$. Again, statistical theory can lend a hand in allowing us to predict an error for either C_S or $\bar{C}_S$. This error estimate arises from the fact that both C_S and $\bar{C}_S$ differ from the 'true' value for the count, μ_S and the error involved, known as the *sampling error*, is given by the following:

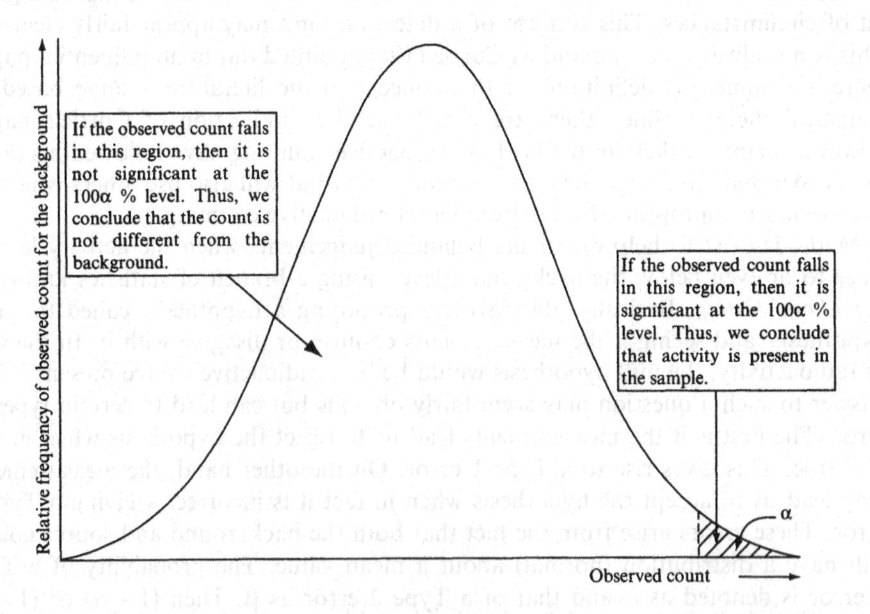
Relative frequency of observed count for the background
If the observed count falls in this region, then it is not significant at the 100α % level. Thus, we conclude that the count is not different from the background.
If the observed count falls in this region, then it is significant at the 100α % level. Thus, we conclude that activity is present in the sample.
α
Observed count

Figure 11.5 The area of a background count distribution used to make a probability statement about whether activity is present or not in a sample being counted, i.e. the procedure for setting the decision level

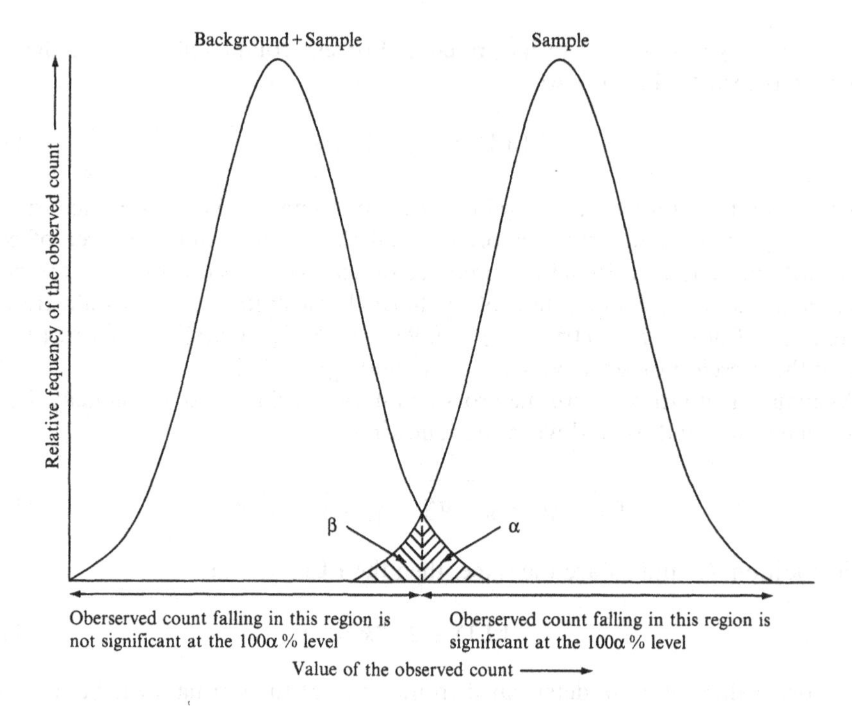

Figure 11.6 The overlapping distributions of a background and low-level sample count and the procedure for setting a detection level

$$C_S(\text{or } \bar{C}_S) - \mu_S \tag{11.28}$$

It can be shown that for a given probability (e.g. 0.95) the sampling error will be $z_{\text{prob}}\sigma_S$ or $z_{\text{prob}}\bar{\sigma}_S$. The parameter z_{prob} is obtained from tables of standard normal probabilities (e.g. see the text by Anderson *et al.* (1981) in the Bibliography at the end of this present book). For example, for a probability of 0.95 ($z_{\text{prob}} = z_{0.95}$), we obtain a value of 1.96 from such tables. In other words, there is a 95% probability of C_S or $\bar{C}_S$ lying within 1.96 standard deviations of μ_S. Thus, when reporting a value of C_S (or $\bar{C}_S$) we quote a value of C_S (or $\bar{C}_S$) $\pm z_{\text{prob}}\sigma_S$ (or $z_{\text{prob}}\bar{\sigma}_S$). The interval ($\pm z_{\text{prob}}\sigma_S$) (or $z_{\text{prob}}\bar{\sigma}_S$) is known as the *confidence interval.*

If, however, C_S (or $\bar{C}_S$) is less than L_c then we can say that no count is detected, i.e. no source is present. In these circumstances, however, we try and report an *upper limit*. In other words, the maximum value the count could be and yet be undetected by our system. In this case, we would report an upper limit of C_S (or $\bar{C}_S$) $+ z_{\text{prob}}\sigma_S$ (or $z_{\text{prob}}\bar{\sigma}_S$).

Currie (1968) provides values for L_c according to two different counting regimes. When we have equal observations of the gross sample and the background counts, L_c is $2.33\sqrt{C_B}$, but if the background is counted repeatedly so that σ_B is negligible, then L_c is $1.64\sqrt{C_B}$.

Of more interest for quantitative analysis is the *detection limit* (L_D in the nomenclature of Currie) or the *lower limit of detection* (LLD), defined by Pasternack and Harley (1971) as the smallest amount of sample activity that will yield a net count sufficiently large as to denote its presence. For environmental analysis, the LLD can be approximated as follows:

$$\text{LLD} \cong (k_\alpha + k_\beta)\sigma_S \tag{11.29}$$

where k_α and k_β represent the values corresponding to the pre-selected risk for concluding falsely that activity is present and the pre-determined degree of confidence for correctly identifying the presence of activity, respectively. The subscripts, α and β, are derived from the tails of the normal distribution curve, as illustrated in Figure 11.5. For $\alpha = \beta = 0.05$, $1 - \beta = 0.95$ (i.e. 95% probability of correctly identifying the presence of activity), $k_\alpha = 1.645$ and $k_\beta = 1.282$.

Assuming a situation where the gross count and background are similar, then we can equate σ_{B+S} and σ_B and write the following:

$$\sigma_S = \sqrt{\sigma_{S+B}^2 + \sigma_B^2} = \sqrt{2\sigma_B^2} = \sqrt{2}\sigma_B \tag{11.30}$$

If, in addition, k_α and k_β are the same, then the LLD becomes:

$$\text{LLD} \cong 2\sqrt{2}k\sigma_B \tag{11.31}$$

As before, values of k are determined from tables of the normal distribution curve, using the appropriate value of α. Some values of k for some common values of α are given in Table 11.1. As noted previously, an α value of 0.05 is commonly used in practice for calculating the LLD.

To give a specific example, let us assume a background count rate of 0.0025 counts per second (cps) and a counting time of 100 000 s for the sample and background. This gives a background count of 250. Thus, we have:

$$\sigma_B = \sqrt{C_B} = \sqrt{250} = 15.81$$

If we have an α value of 0.05, then from Table 11.1, $2\sqrt{2}k$ is 4.66 and the LLD is as follows:

$$\text{LLD} \cong (4.66)\ (15.81) \cong 74 \text{ counts}$$

or $\cong$ 0.000 74 cps.

Table 11.1 Values of the parameter, k, used to determine the lower limit of detection (LLD) of a sample for specific values of α (see text for further details).

α	k	$2\sqrt{2}k$
0.01	2.327	6.59
0.02	2.054	5.81
0.05	1.645	4.66
0.10	1.282	3.63
0.20	0.842	2.38
0.50	0	0

For the special case where no counts from either sample or background are recorded in the counting interval – a situation not uncommon in very low-level alpha-counting – it can be shown that the upper limit for the sample activity in Bq at the 95% confidence level, A_{lim}, is given by the following:

$$A_{\text{lim}} = \frac{3}{TG} \tag{11.32}$$

where T is the counting time in minutes, and G is the detector efficiency expressed as a fraction.

11.7 REGRESSION ANALYSIS

Quite frequently, in any form of analysis we have two variables that are known or thought to be related. This relationship can often be clarified by plotting the two variables. Take the case of a radioactive source of a single nuclide. We know that if we determine the activity (count rate) of the nuclide at regular time intervals and then plot activity as a function of time we will obtain a distribution of points, through which we can draw a curve, since activity varies exponentially with time (Figure 11.7(a)). Regression analysis of these data provides a mathematical equation that relates these two variables. In this case, one that would allow us to estimate the activity of the source at any given time after the start of counting. If instead of activity we plot the logarithm of the activity as a function of time then the distribution of points now clearly approximates a straight line (Figure 11.7(b)). In other words, regression analysis would predict a *regression function* taking the form of a

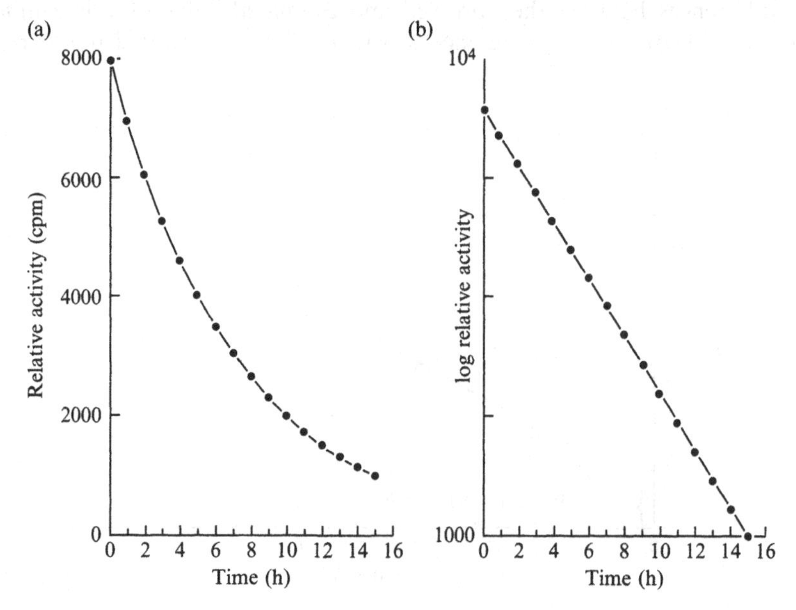

Figure 11.7 Relative activity (a) and log relative activity (b) of a nuclide with a half-life of 5.0 h as a function of time

straight line. Hence, such a regression function is known as a *regression line*. We shall only be concerned here with such linear regression functions.

In any regression analysis, one of the two variables will be *dependent* and the other *independent*. The independent variable predicts a particular outcome for the value of the dependent variable. In our case, the independent variable is time and the dependent variable is activity. We predict the activity at a given time but not vice versa. By convention, the independent variable is plotted along the abscissa (x-axis) and the dependent one along the ordinate (y-axis).

Having obtained a plot such as that shown in Figure 11.7(b), we could try to fit a straight line through the points by eye. This is fine if the points are not too scattered but if the scatter is quite large then this fitting by eye will become very subjective. We obviously require some more rigorous, objective method for fitting the line through the points. How we arrive at this objective fit will be the subject of the next section.

11.7.1 Least-Squares Analysis

The general equation for a straight line is of the following form:

$$y = mx + b \tag{11.33}$$

where y the estimated value of the dependent variable, x is the value of the independent variable, m is the slope of the line, and b is the intercept of the line on the ordinate.

Obviously, a large number of lines could be drawn through the data, each with a different value of m and b. The *least-squares method*, however, shows that the *best fit* to the data, i. e. the best line through the points, is when the sum of the squares of the differences between the observed and estimated value of y is a minimum. This is more clearly seen by reference to Figure 11.8. The plotted points represent

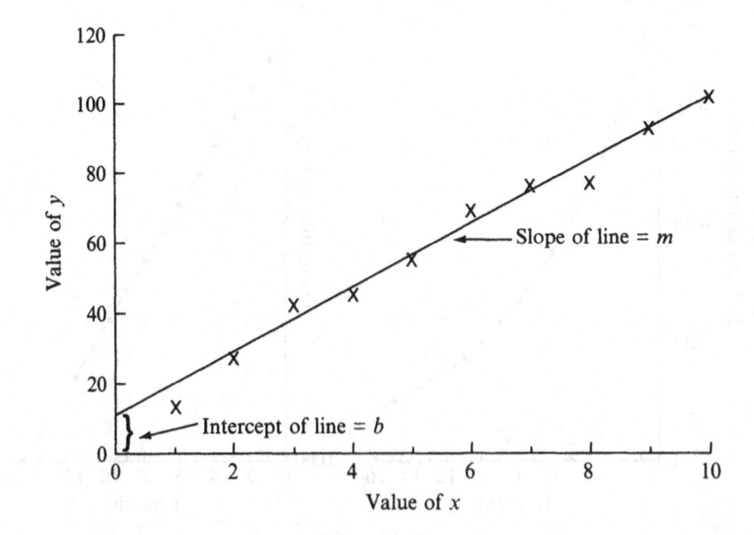

Figure 11.8 Distribution of points (×) as values of y are plotted as a function of values of x. The line through the points is the best fit by least-squares analysis

the observed values of y, i.e. y_i, at the corresponding x values. We have drawn a line through these points, estimated by eye. The estimated y value, y_{ie}, is the value of y read from the line at that x value. The difference between these two y values is $(y_i - y_{ie})$. Thus, the least-squares method predicts that:

$$\sum (y_i - y_{ie})^2 = \sum [y_i - (mx_i + b)]^2 \text{ is a minimum} \tag{11.34}$$

for all of the observations. This equation assumes that there is no error in x, the independent variable.

Equation (11.34) leads to the values for m and b for the best fit to the data as follows:

$$m = \frac{n\sum x_i y_i - \sum x_i \sum y_i}{n\sum x_i^2 - (\sum x_i)^2} \tag{11.35}$$

and:

$$b = \bar{y} - m\bar{x} \tag{11.36}$$

where x_i is the value of the independent variable for the ith observation, y_i is the value of the dependent variable for the ith observation, $\bar{x}$ is the mean of all the values of x_i, $\bar{y}$ is the mean of all the values of y_i, and n is the total number of data points.

Note that the slope, m, will have a positive sign if the y values increase with increasing values of x. Where the y values decrease with increasing values of x, then the slope will be negative. This is the case when plotting the count of a decaying nuclide (y) as a function of time (x).

We expect the y values for any given x value to vary, e.g. in counting a series of identical radioactive sources we would expect to obtain slightly different count rates (y_i) after the same decay time (x_i). The assumption is that these values of y_i will follow a normal distribution and that the variance $(\sigma_y)^2$ of each y is the same for every x. By analogy with Equation (11.2), the standard deviation, σ_y, of each of these y values is of the following form:

$$\sigma_y = \sqrt{\sum \frac{(y_i - y_{ie})^2}{n-2}} \tag{11.37}$$

We use $n - 2$ degrees of freedom, since one each is used for estimating the slope and intercept of the line.

Since m and b are themselves estimates, they will also have their own sampling distributions – these we assume are normal distributions. From the least-squares formulae for m and b, we can determine the standard deviation, σ_m and σ_b for the slope and intercept, respectively. For the slope, we have:

$$\sigma_m = \sigma_y \sqrt{\left(\frac{1}{\sum x_i^2 - n\bar{x}^2}\right)} \tag{11.38}$$

and for the intercept:

$$\sigma_b = \sigma_y \sqrt{\frac{1}{n - \frac{(\Sigma x_i)^2}{\Sigma x_i^2}}} \tag{11.39}$$

11.7.2 Weighted Regression

We stated at the outset of this section that we would regard all of the y data as having the same standard deviation, that is, that we are using *unweighted* methods. Here, every y value has to be treated equally in fitting the regression line to the data. In fact, this situation is quite rare in practice and is usually only found when a particular step in the analytical procedure is dominant. Much more common is the situation where the errors on the y values vary in some way with the x value. Here, we place more reliance on those points which have smaller errors than those where the error is larger and we force the regression line to pass closer to the 'small-error' y values. This gives rise to a *weighted* regression line.

The weighting to give any particular y value depends on the particular case and is often determined from experience. Frequently it is found that the error on the y value increases proportionally as the x value. If we call the weighting factor, w_i at the y value, y_i, then it can be shown that w_i is proportional to $1/x^2$. Such a weighting factor cannot be assumed, however, and each case must be carefully analysed to discover the appropriate weighting factor.

When the correct weighting factor has been determined and the corrected error on each y value calculated, then the slope for the weighted regression line can be calculated from an equation similar to Equation (11.35) and the intercept from the fact that the line passes through the weighted mean value of y. The main advantage of using weighted errors is in the calculation of the standard deviation of an x value interpolated from the regression line, which will vary according to the value of x.

11.8 CORRELATION ANALYSIS

In the discussion so far we have assumed that we know that the two variables are related and we have only been concerned with determining the correct form of the function that relates these variables. Quite often, however, we have the situation in which we wish merely to know if two variables *are related* and if so what is the extent of this relationship. The question that we are asking is, in statistical terms, how well are the two variables *correlated*. Correlation analysis is a technique for measuring the extent of the connection by providing a quantity known as the *sample correlation coefficient*, r. This is a number having a value between zero and one. The closer the value of r is to unity, then the more strongly, linearly correlated are the two variables. In the case of two variables that are related in a negative linear sense, then the value of r will be -1. A value of zero indicates that the two variables are not linearly related.

As an example, consider a situation in which a number of plants are analysed for ^{137}Cs and one species, which we may call 'Plant X', is usually found to have a somewhat higher concentration of the radionuclide than the others. If we now sample a number of tracts of land (soil plus vegetation) in order to determine the mean ^{137}Cs concentration in each tract and we find that those tracts having a high ^{137}Cs content also tend to have high concentrations of Plant X, then we may feel that the activity at a given site is correlated to the amount of Plant X at that site. In order to determine the extent of this association we would have to determine the sample correlation coefficient between, say, the activity per m^3 of soil plus vegetation and the weight of Plant X per m^3 of the same sample for samples taken from the various tracts of land.

If we let the weight of Plant X per m^3 of the sample be x and the activity per m^3 of sample be y, then it can be shown that the correlation coefficient, r, is given by the following:

$$r = \frac{n\sum x_i y_i - \sum x_i \sum y_i}{\sqrt{n\sum x_i^2 - (\sum x_i)^2}\sqrt{n\sum y_i^2 - (\sum y_i)^2}} \tag{11.40}$$

It must be stressed that we are assuming a linear relationship between the two variables.

Quite often the correlation coefficient, r, derived from Equation (11.40) is used to test for linearity. This, however, is potentially misleading. The reason is that a set of data may well be clustered strongly around a straight line and give a value of r close to unity. On the other hand, a value of r could also be close to unity from a set of points that are clustered around a slight curve and the assumption of linearity could result in seriously erroneous results.

It turns out that testing for linearity is not, strictly speaking, possible. The best that can be done is to show that any deviation from linearity is too small to measure. One method of estimating this deviation is to fit a line by linear regression and then examine the residuals. The residuals are the distances of the individual points (y) from the fitted regression line. If the system is actually linear, then a plot of the residuals as a function of the x values should be essentially a normal distribution. Deviations from a normal distribution will indicate that the assumption of linearity is incorrect. In such a case, further statistical tests are possible to determine if the deviation is statistically significant (AMC, 2000).

12

Radioactive Surveying and Remote Sensing

It is becoming increasingly important to determine the radioactivity over large areas of the earth's surface, both in terms of the concentration of radionuclides present and their contribution to the radiation dose to the human population of these areas. In Chapter 10, we discussed common methods of sampling in order to estimate the radioactivity in a specific location. These methods are not, however, generally suited to radioactive surveys covering large areas such as whole countries or even continents, even though they may be required in the calibration of instruments used in large-scale surveys or in surveys of alpha- and beta-emitting nuclides.

Radioactive surveys usually refer to methods involving gamma-ray measurements since the highly penetrating nature of gamma rays allow their detection some distance from their point of emission. This allows detector systems to be carried in cars or even aicraft and measurements made while the vehicle is moving. By such methods, large tracts of land or water can be surveyed relatively rapidly and with the aid of modern electronics and computer technology the data can be rapidly output in a variety of forms, including contour maps of the radioactivity or dose rates.

12.1 SURVEYS BASED ON ALPHA- AND BETA-EMITTING RADIONUCLIDES

Carborne or airborne methods of surveying are not possible if some or all of the nuclides present are pure alpha- or beta-emitting nuclides. The very limited penetration of these particles through soils, rocks or even air mean that large-scale surveys involving these nuclides have to be based on the techniques already discussed in Chapter 10. This will involve the establishment of a suitable grid system covering the relevant area, followed by the collection and analysis of samples taken from within this grid. Such methods are obviously very tedious, time-consuming and expensive and tend to be limited to smaller areas than gamma-ray surveys.

A good example of a survey of alpha- and beta-emitting nuclides was that of determining the environmental radioactivity in Caithness and Sutherland in north-eastern Scotland (Cawse, 1988). This comprised a study of actinides and fission products in the environment due to the presence of the Dounreay Nuclear Establishment on the north-eastern coast of Scotland (see Figure 12.1). A network of sampling sites was established with the aims of determining levels of ^{137}Cs, plutonium and ^{90}Sr accumulated in these soils as a result of weapons fall-out, obtain information on the Dounreay contribution to 'baseline' levels and additionally to provide a network of sampling sites unlikely to be disturbed in the foreseeable future so that re-sampling in the future will permit an assessment of changes during that period.

Sampling was based on a grid of 10 km sides covering a total area of 2200 km^2 of the far north-east of Scotland (see Figure 12.1). Twenty four permanent grassland sites were chosen, one in each of the grid squares. Eight grid areas also included separate sites sampled for peat. The grassland sites were selected on the basis of a number of criteria – undisturbed for at least 20 years and unlikely to be disturbed in the near future, occupying at least 2000 m^2 and were close to the centre of the grid area wherever possible. All of the sampling sites were on level ground and were not subject to flooding; for the peat sampling sites, there also had to be a minimum of 1.5 m depth of deposit.

Sampling of soils to a depth of 0.3 m was required and hence core samples had to be taken. At each soil site, 10–15 cores were taken to a depth of 0.15 m and a similar number at a depth between 0.15 and 0.3 m. Cores, including grass and the organic matter, were randomly taken over an area of about 30 m × 30 m. Samples from the upper 0.15 m were bulked to form one composite sample and a similar procedure was applied for those from 0.15–0.3 m. Thus, each site yielded two composite samples for analysis.

Peat sampling was carried out with a special peat corer to a maximum depth of 2 m. The corer design prevented 'smearing' of the samples. At each peat site, five core samples were collected, each 0.5 m long and spaced about 10–15 m apart.

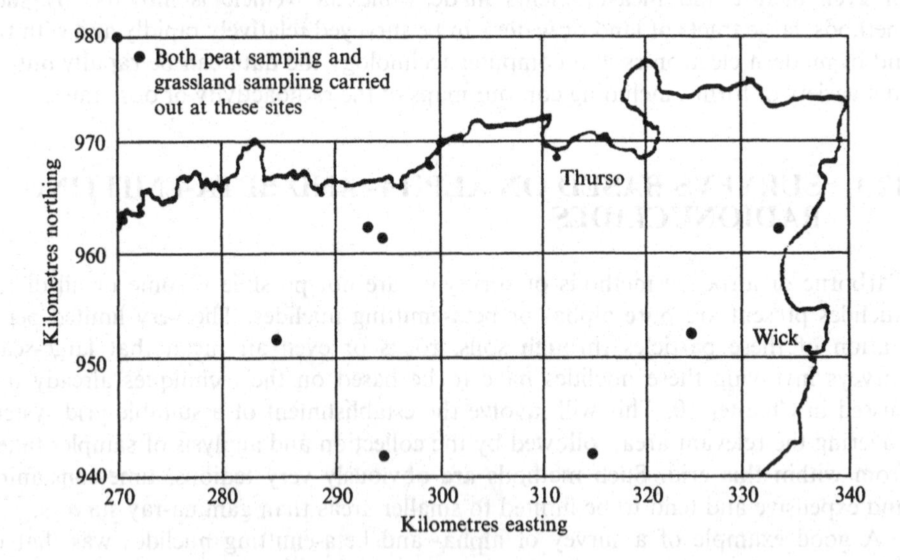

Figure 12.1 Caithness and Sutherland (Scotland) survey of radioactive fall-out accumulation (May–June 1979), illustrating distribution of sampling sites on grassland. Reproduced by permission of the British Nuclear Energy Society from Cawse, P. A., *Nucl. Energy*, **27**, 193–213 (1988)

Cores were cut into 0.1 m sections and the respective sections from the five cores were bulked to form five composite samples.

In addition, samples from the most important crops in the area were also taken, including hay, barley, swedes and potatoes. Samples collected from an area in England remote from any nuclear establishment acted as controls.

All of the samples were then treated as discussed in Chapter 10, prior to analysis for radionuclide concentrations. In the grinding process for the composite samples, all vegetation and stones were included. The stones were not removed as the limestone in the area was known to absorb up to 44% of the plutonium in solution which passed through it. After thoroughly homogenizing the samples, sub-samples of about 100 g were taken for radionuclide analysis.

Soil data were reported in the form of contour maps of the concentration of each radionuclide determined ($Bq kg^{-1}$) for each of the two depths. By comparison with data from other British sites and those from elsewhere it was concluded that all of the ^{137}Cs and plutonium present in these Scottish samples arose from nuclear fallout with no real evidence for any significant contribution from the Dounreay plant. All of the data were corrected for rainfall based on the known correlation of deposition of radionuclides and rainfall from various British sites. The lower levels of ^{90}Sr found in the survey (compared to those expected) were assumed to be due to transport of ^{90}Sr by drainage water. Similar low levels of ^{137}Cs in the peat samples was taken as evidence for the migration of this nuclide to surface waters. None of the vegetation samples taken were found to have harmful levels of radionuclides, although the concentration of ^{137}Cs and plutonium in plants was found to be highest from farms close to the nuclear establishment.

12.2 GAMMA-RAY SURVEYS

The majority of large-scale surveys that have been carried out have been gamma-ray surveys for the reasons already discussed. Most existing gamma-ray surveys have been conducted for geological/mineral exploration purposes, involving just the measurement of the natural radionuclides, ^{40}K, ^{232}Th and ^{2328}U. Even so, there has also been a continuing interest in determining the levels of anthropogenic radionuclides present in the environment. In either case, it is now seen as vital to maximize the information available on the environmental impact that can be derived from these surveys. These data provide a baseline of radioactivity information for both natural and anthropogenic radionuclides and improve the very important role of monitoring global changes in levels of radiation and their potential impact on the health of human populations.

The existence of such baseline maps would also enable the impact of major nuclear accidents, such as occurred at Chernobyl, to be more rapidly and better assessed. Airborne surveys of an entire country could reveal changes of radiation levels and the production of relevant radioactivity maps within a few days of an alert being given, as was demonstrated by the rapid national mapping exercise undertaken in Sweden immediately after the Chernobyl accident (see Section 12.2.8).

High-quality maps of individual radioelements also assist in the earth sciences and the understanding of regional geology. It is also important to realize that maps of radioelement concentrations not only aid in uranium exploration but can also reveal the existence of economic deposits of other important elements, such as molybdenum, tungsten and gold. These elements are often associated with copper and certain types of copper mineralization are often accompanied by alterations in the concentration of potassium. These local changes in the level of potassium are often clearly delineated on potassium distribution maps produced as a result of airborne gamma-ray surveys. Finally, radioelement maps prepared on regional, continental or even a global basis may indicate the presence of structural or tectonic features that are not visible on the ground, in the same way that aerial photographs may reveal the presence of ancient walls, settlements, etc. that cannot be seen by a ground-based archaeologist.

Gamma-ray surveys have also been used to locate lost or stolen medical or industrial application sources (e.g. Mexico (1985), Brazil (1987) and Nigeria (1991)) and the fall of satellites powered by nuclear reactors (e.g. Canada (1978)).

12.2.1 Concentration Units Used in Gamma-Ray Surveying

The units of concentration used for the natural radioelements are g kg^{-1} or mg kg^{-1} ('parts per million, ppm'). Potassium normally occurs in abundances where % K is the more appropriate unit. The average concentration for these radioelements in soils, seawater and fresh water are given in Table 12.1. For normal rocks and soils, 90 % of gamma-rays measured by a spectrometric survey originate from the upper 30 cm. Thus, a gamma-ray survey is a surface-measuring technique and since in most cases the bedrock is covered by soil the actual survey reflects the activity of the soil rather than the bedrock. In addition, natural radioactive elements are not normally present in plants or snow at detectable levels in surveys but they do have a shielding effect. Forested areas, for instance, may cause up to a 15% underestimation of the true exposure rate. Water has a profound effect on the activity observed and as this varies with the time of year so does the observed count rate.

A study has been made of the time variations of the natural gamma-ray intensity at one remote site over a period of years (Burch *et al.*, 1964). The detector used was a composite of four identical ionization chambers filled with argon gas at slightly different pressures. This, together with suitable processing of the signals from the various chambers, eliminated the cosmic-ray contribution to the measured gamma-ray intensity. Thus, the main detector responded only to gamma-radiation

Table 12.1 Average concentrations of potassium, thorium and uranium in soil, sea and river water.

Sample system	Potassium	Thorium	Uranium
Soil	1.3 %	6.2 ppm	1.8 ppm
Seawater[a]	380 ppm	1×10^{-5} ppm	0.003 ppm
River water[b]	2.3 ppm	Negligible	~ 0.001 ppm

[a]Data from Open University Course Team (1989).
[b]Data from Livingstone (1963).

emitted from the ground. The main variations in intensity could be correlated with variations in radon emanation from the soil, which is subject to a number of meteorological and other parameters, such as rainfall, snow cover, presence of dust, etc.

A recurring problem with estimations of the concentration of uranium and thorium in a given area is the assumption of *radioactive equilibrium*. In geological situations, uranium, in particular, is frequently not in equilibrium with its daughter nuclides due to chemical effects such as the differential leaching of radium and uranium associated with percolating ground waters over long periods of time or the separation of components during a partial melting of minerals in the past. The result is that the gamma-ray activity in a non-equilibrium system is not the same as that when true equilibrium is present. Thus, field determinations of uranium and thorium concentrations are always expressed as ppm *e*U or ppm *e*Th (ppm *equivalent* U or Th).

Another unit is also used in gamma-ray surveys. If a detector system is used which gives only a total gamma-ray count and does not distinguish between the various radioelements and nuclides potentially present (mainly ^{40}K, ^{238}U and ^{232}Th, but possibly fall-out activities also), then the comparison of readings taken with different instruments becomes difficult. Consequently, it was decided to use an inter-calibration based on the count derived from a source containing 1 ppm U (this stems from early gamma-ray surveys being mainly used to locate uranium ore bodies). The total count, however, obviously does not refer to uranium alone and so one cannot derive even an eU concentration from such a count. Therefore, it was decided to introduce the term U*r*, whose definition is that it is the count rate produced in a detector identical to that from a source containing 1 ppm of uranium in equilibrium. Thus, a detector which gives a count rate of, say, 10 counts per second (cps) from a source of 1 ppm U and which in the field gave a gross count of 50 cps would be monitoring a source of 5 Ur. The problem is that a completely different detector could give the same count rate for the 1 ppm U source but would give a different gross count when monitoring the same field source and hence a different Ur value for this source due to differences in response to thorium and potassium, and environmental radionuclides in particular. Thus the Ur is of limited usefulness as a unit of concentration.

12.2.2 Types of Gamma-Ray Surveys

There are basically three different types of gamma-ray surveys, namely (i) *hand-held ground surveys*, (ii) *carborne surveys*, and (iii) *airborne surveys*. *Regional* gamma-ray surveys are most frequently airborne surveys but sometimes carborne. In exceptional cases, ground survey methods are used. The design of a particular regional survey depends on many factors, i.e. the area to be covered, the available techniques and equipment, and financial and personnel resources. Corrections for the influence of different effects, such as cosmic radiation, instrument background, altitude effects, etc. must be applied.

Regional ground surveys include both carborne and hand-held gamma-ray measurements. In general, large areas are covered by carborne total count or gamma-ray spectrometric surveys. The major drawback of such surveys are that they are restricted to existing roads and tracks and are influenced by the road materials and geometry.

An important aspect of the data collection is to ensure accurate geographical positioning of the data. Airborne surveys are particularly suitable for regional compilations because they usually cover large areas. Carborne and other regional ground surveys that cover at least 100 km^2 are also suitable. In general, gamma-ray surveys cover only a few percent of the area being surveyed. In airborne or carborne surveys, the data are provided continuously and thus there is a high density of sampling along the sampling path. Regional ground surveys consist of point measurements at densities of one measurement every few km^2.

12.2.3 Instrumentation used In Gamma-Ray Surveys

The instruments used can be either devices which record intensity and/or energy of gamma-rays directly to give readings in counts per unit time and/or counts per unit time per unit energy, or devices which indicate the dose rate directly.

Ground-based detectors for intensity and energy determinations

Instruments which record the numbers and/or energies of gamma-rays are usually portable forms of scintillation or hyperpure germanium detectors. Scintillation counters used for simply measuring the gross gamma-ray field at some point are sometimes called *scintillometers*. They usually have relatively small NaI(Tl) crystals and normally record all of the gamma-rays with energies above a threshold of about 30–50 keV. If the threshold changes in such an instrument, then the response also changes so that scintillometers must be equipped with good stabilization. They are battery operated and can easily be carried by hand or mounted on a vehicle when it is required to cover large distances fairly rapidly.

Scintillation spectrometers

When the detector is used in conjunction with either an energy window system or some form of multi-channel analyser (MCA) system, then this allows spectra to be recorded and analysed and considerably increases the usefulness of the instrument. Scintillation detectors normally incorporate just energy windows, set to record the full energy peaks from the natural radionuclides, ^{40}K, ^{232}Th and ^{238}U (see Section 9.2.5), along with the total count above a threshold. The detectors typically used for survey work tend to be large, having volumes of 350 cm^3 or greater to give sufficient efficiency for high-energy gamma-rays. The suggested window settings by the IAEA (IAEA, 1989) are shown in Table 12.2, where the recommended full-energy peaks are 1.46 MeV for ^{40}K, 1.76 MeV for ^{238}U(^{214}Bi) and 2.62 MeV for ^{232}Th(^{208}Tl).

A typical portable window spectrometer consists of a cylindrical or prismatic NaI(Tl) (or occasionally a CsI(Tl)) detector housed in a battery-operated assembly containing all of the necessary electronics (see Chapter 8). The minimum size for a

Table 12.2 Recommended window settings for a NaI(Tl) gamma-ray spectrometer (from IAEA, 1989).

Energy window	Detected nuclide	Recommended window limits (keV)
Total count	More than one	400–2800
K	^{40}K	1370–1570
U	^{214}Bi	1660–1860
Th	^{208}Tl	2400–2800

cylindrical detector to achieve the required sensitivity for geochemical exploration (of the order of 14 counts min^{-1} per ppm U and 4 counts min^{-1} per ppm Th) is 44×51 mm, but for environmental studies in general a detector of 76×76 mm is normally used, with the crystal at the bottom of the probe. Again, for geochemical work the probe is located on the ground but for environmental determinations it is commonly supported on a wooden tripod 1 m above the surface. The detector is normally calibrated in the laboratory prior to the start of a survey. In order to maintain a check on the energy calibration in the field, a small source (usually ^{137}Cs or ^{133}Ba) is mounted close to the crystal so that the gain of the system can be continually monitored. In some instruments, a feedback system is also incorporated so that any drift from the pre-set value is automatically corrected for.

If the system is to be mounted on a vehicle, then often the spectrometer assembly is mounted on a telescope mast which allows the detector to be elevated to a height of 3 m above the surface. This increases the field of view and helps reduce the gamma-ray contribution from the road itself, which may be quite different to that from the surrounding terrain which is being surveyed.

Hyperpure Ge detectors

In situ gamma spectroscopy using germanium detectors has been in routine use for over three decades. The detectors have to be in a portable, rugged housing, complete with their liquid nitrogen supply. Modern systems can have holding times of up to 72 h. They are normally mounted, unshielded and positioned 1 m above the ground. The MCA must also be portable, rugged and self-contained. Systems have their own internal 12 V battery supplies but usually with provision to also run off-road vehicle and aircraft electrical supplies.

Instruments for dose-rate measurements

Monitoring of environmental radiation levels for the assessment of population doses is typically performed at a specific site by using either an integrating dosimeter or a device which can record dose rates directly and can be easily moved from site to site. Devices have also been developed which respond rapidly enough to be used for carborne surveys.

The integrating dosimeter most commonly used is a thermoluminescent dosimeter (TLD) (see Chapter 8). This is typically attached to a tree or post, etc., about

1 m above the ground in a location where it is unlikely to be disturbed. It is left in place for periods of up to a few months, after which it can be read in the normal way and an integrated dose then determined.

For dose-rate measurements and systematic estimates of doses to populations, the usual instrument is a pressurized-ion chamber (Chapter 8). Typically, this is constructed from stainless steel, with an active volume of about 8 dm^3, and filled with argon to a pressure of about 2×10^6 Pa to obtain a sufficiently large ionization current. The voltage across such a chamber is about 300 V. A more portable ionization chamber has been described, consisting of a 25 dm^3 polythene bottle coated with a special graphite lacquer and filled with air, and which has an operating voltage of only 30 V. This is surrounded by an absorbing layer of polythene and aluminium. Such a device is easily transported by car, along with the measuring equipment. Modern ionization chambers usually contain an analogue-to-digital convertor (ADC) so that a series of measurements can be stored on a floppy disk or similar recording device.

Some modifications to the basic design of ionization chambers have been described for instruments used exclusively for environmental monitoring. An example is the use of thin-walled chambers. These are essential for the estimation of the total environmental radiation, including the very low energy portion of the spectrum, and avoid cumbersome wall corrections which must be used for thick-walled chambers. There is a problem of residual ionization due to contamination in the chamber walls, mainly due to alpha-emitters. To counteract this, chambers of perspex, coated with 'Aquadag' are used which have very low inherent alpha-contents. The chambers are operated with 'Freon-12' (CCl_2F_2) at atmospheric pressure.

The measuring instrument should ideally have a response independent of the gamma-ray energy, down to a cut-off of about 0.05 MeV. For measurements of low dose rates, one can use ionization chambers, G–M tubes or scintillators. Although ionization chambers and G–M tubes have responses independent of energy, particularly ionization chambers, they have slow responses and so cannot be used for measurements taken from a moving vehicle. Large-volume plastic scintillators (17.5 cm × 17.5 cm), on the other hand, have very high sensitivities and short time constants when used in pulse mode. Thus, they respond rapidly and sensitively and are also relatively light and easily portable. Their main disadvantage is that they have no flat response at any energy of use in dosimetry. This problem, however, is solved by the use of a set of discriminators, each set to a different level and the pulses are 'weighted' according to the discriminator setting. The sum of these weighted counts then constitutes the response of the detector. For gamma-ray energies above 100 keV, the response is essentially flat, as required.

It is also possible to use conventional scintillation detectors to determine does rates if these are suitably calibrated. Factors for the calculation of dose rates from natural activities and anthropogenic nuclides are available for the determination of the gamma-ray dose in air from these data. A device to record the integrated 24 h dose a person receives at a given site utilizes a combination of a sodium iodide crystal in contact with X-ray film in a light-tight box. The crystal intensifies the light reaching the film by several orders of magnitude and allows environmental doses to be measured. This system has been used to measure high background dose levels in parts of Brazil containing deposits of monazite sands which contain high levels of uranium and thorium (up to 560 pGy s^{-1}).

Airborne systems

By elevating a gamma-ray spectrometer to a height of between 25 and 100 m above the ground, the field of view of the detector is enormously increased. This is achieved by using either a helicopter or fixed-wing aircraft. The latter is normally flown at a 100 m altitude for surveying measurements but higher altitudes may be used in emergency situations (e.g. a Chernobyl-type accident). At 100 m above the ground, the signals received are spatially averaged over some $1.2 \times 10^5\,m^2$ but at 50 m elevation this is reduced to approximately $3 \times 10^4\,m^2$. A detector about 1 m above the ground is only viewing some 100 m^2. Thus, airborne detectors provide the potential for large-scale surveys, effectively sampling the total radiation environment. In this manner, maps of relative radiation covering whole countries at an effective grid scale of a few hundred metres may be produced in many cases. A major use, however, is in surveys over hazardous or remote terrain, such as polar or jungle areas or where access is difficult by conventional means, as in mountainous terrain.

It is perhaps worth noting that the field of view of an airborne detector is not easily defined (Grasty *et al.*, 1978). At first sight, it might be thought of as a circular area on the ground beneath the detector with the area suitably defined. This would be fine for a uniform, infinite source but for a non-homogeneous source a strong radioactive region outside the defined area could still contribute to the detected radiation. It can be shown, however, that for a given height of aircraft, strips or bands parallel to the flight path contribute a fixed, relative percentage of the total radiation detected in terms of that for an infinite, homogeneous source. In addition, although many of the detectors used in airborne gamma-ray surveys exhibit significant angular sensitivity variations, in practice this angular dependence only introduces small errors to the overall count rate in the detector for large-source bodies. In the case of small sources, then both the energy of the gammas and the angular sensitivity have to be taken into consideration.

Since the gamma-ray flux from ground sources is obviously low at elevations of 100 m or so, and since the aircraft will be moving at a speed of about 200 kmh^{-1}, sampling times are short (normally of the order of 0.5 to 2 s) and it becomes necessary to employ large detector systems to obtain adequate count rates. Arrays of scintillator detectors are normally used but more recently, *hybrid detector systems*, with both scintillator and semiconductor detectors, have been built.

Pure scintillator systems are usually arrays of prismatic NaI(Tl) detectors with typical dimensions of 10 cm × 10 cm × 40 cm. A complete array may have a total volume of 16 dm^3 (four prismatic detectors). Two or more of these arrays may then constitute the complete detection system. The outputs from the photomultiplier tubes connected to each prismatic NaI(Tl) detector are fed to a summing amplifier where they are linearly combined to produce a single output. Linked to a microcomputer, the data can then be fed into an MCA to provide complete spectral information or integrated into the four standard energy windows. A schematic diagram showing the interlinking of a complete airborne scintillator-based gamma-ray survey system and the associated navigational and flight data systems is shown in Figure 12.2.

The advent of fast, PC-based systems has allowed the use of multiple distributed pulse-height analysers using multi-channel buffer (MCB) architecture (Sanderson *et al.*, 1995). Here, the MCB provides the basic MCA in the form of multiple, high-speed inputs, the ADC and memory, and then uses the display and peripheral storage of the associated computer to produce the complete system. This device lends itself to the flexibility in using either a single or multi-detector system. In addition, the necessary altimeter and navigational (latitude and longitude) data from the aircraft can also be stored at the same time as a spectrum is collected. Thus, a file is created for each spectrum recorded which contains information on the time of collection, altitude and the mean position of the aircraft at the time of collection. Real-time analysis of the spectra during the flight are possible so peak area integration, for instance, can be carried out and the data displayed. Collection times of 5–10 s or more are now often used in order to improve statistics to allow the determination of environmental gamma-ray peaks.

When a hybrid detector system is used, combining both scintillation and semi-conductor detectors, the same operating procedures are adopted as previously described. A schematic diagram of such a system is shown in Figure 12.3. This combines two large prismatic arrays of NaI(Tl) detectors with a set of 6 n-type

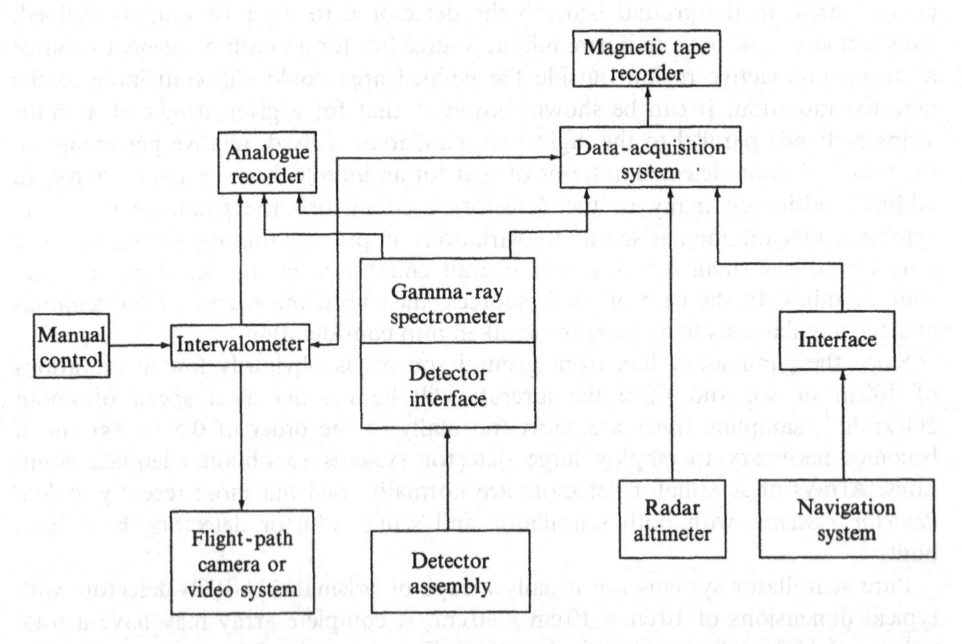

Figure 12.2 Block diagram of an airborne gamma-ray survey system showing the signal flow. This system includes the detector (scintillator), navigation and flight components. Reproduced by permission of IEC, Geneva, Switzerland from CEI IEC61134 1992–05, *Airborne Instrumentation for Measurement of Terrestrial Gamma Radiation*, IEC, 1992

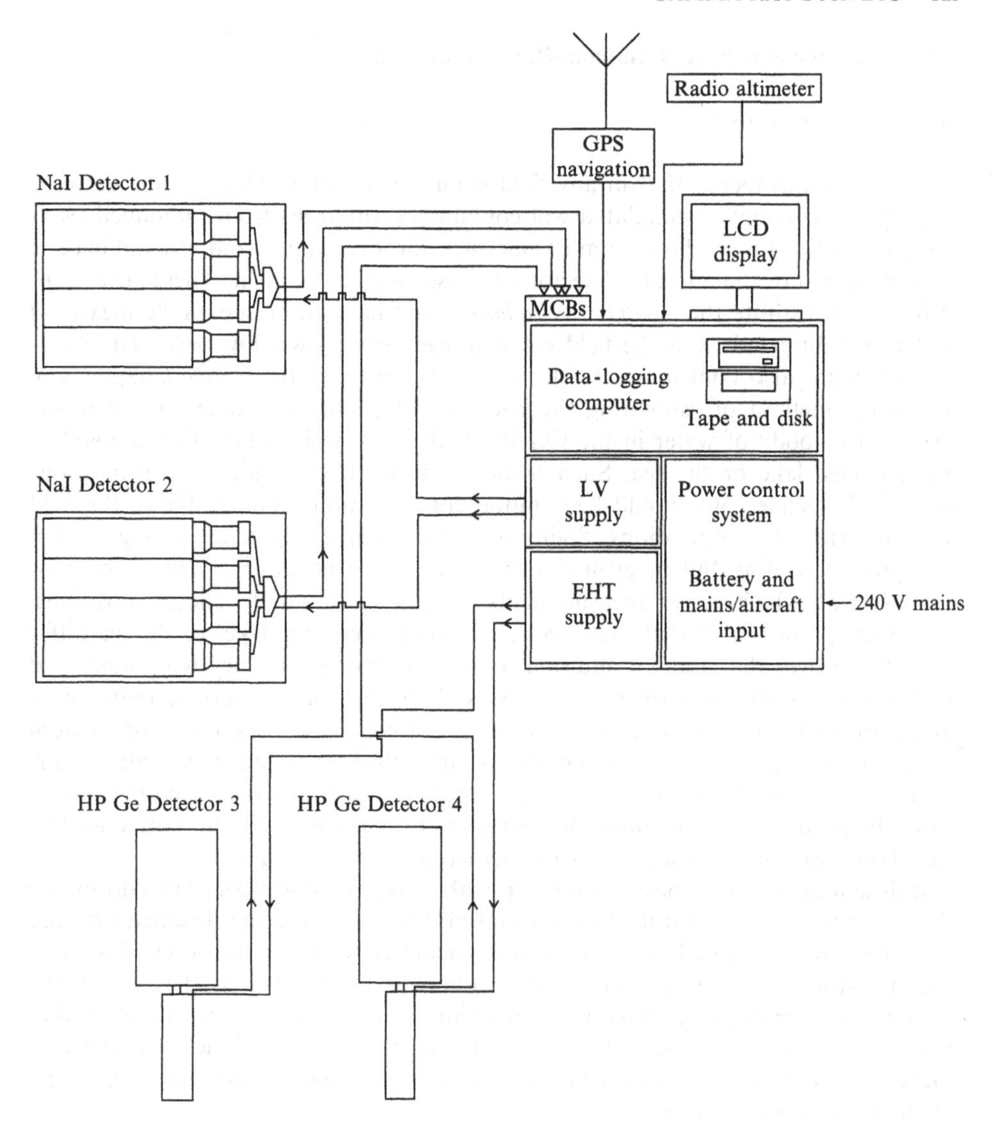

Figure 12.3 Schematic diagram of an airborne gamma-ray spectrometer used at the Scottish Universities Research and Reactor Centre (SURRC): GPS, global positioning system. The system can be configured for one to four detectors, comprising NaI(Tl) and/or germanium detectors. Reproduced by permission of the International Atomic Energy Agency from IAEA-TECHDOC-827, *Applications of Uranium Exploration Data and Techniques in Environmental Studies*, IAEA, 1995

hyperpure Ge detectors (see Chapter 8). Two of these are large (> 50% relative efficiency for the detection of relatively high-energy environmental gamma-rays (^{134}Cs, ^{137}Cs, ^{234m}Pa, etc.)), while the other four are low-energy or X-ray detectors. These can be used for mapping nuclides such as ^{241}Am, ^{234}Th and ^{226}Ra. The measurement of these low-energy gamma-rays using germanium detectors for actual survey work is still in the developmental stage.

12.2.4 Corrections to Raw Gamma-Ray Survey Data

Background corrections

It must be remembered that in any field gamma-ray count, whatever the detector employed, the count accumulated will contain contributions from radionuclides in the ground along with those from cosmic-rays, airborne radon and any radioactive contamination of the aircraft, detector and associated equipment. These latter contributions constitute the *background radiation* and have to be separately measured and then subtracted from the field count in each energy window used. The background count also contributes to the overall counting statistics (see Chapter 11). The usual method of estimating the background count is to take measurements over a large body of water in the vicinity of the survey site where this is possible, e.g. an open lake or the sea. Such bodies of water have negligible contributions from radionuclides but should have radon concentrations comparable to the field site. In airborne investigations, flights over the water can give an average radon background reading but in ground surveys it will be necessary to use a boat or similar craft which limits measurements to a few, randomly chosen positions. A problem, however, is that radon concentrations over large lakes or the sea differ from those over the land. In addition, radon concentrations vary temporally and with climatic conditions over land masses. It is possible to correct radiometric maps for such variable conditions by using mathematical procedures or modern image-processing techniques. Alternatively, the effect of radon in the air on airborne surveys has been estimated by the use of upward-looking detectors shielded from the ground. It is not normally necessary to make background corrections to the photo peaks in spectra taken with a hyperpure Ge detector.

If dose measurements have been taken with a thermoluminescent (TL) dosimeter then an estimate of the transit dose received while bringing the dosimeter to and from the site is required. This is normally achieved with a control TL dosimeter which is stored in a lead container but is otherwise treated exactly the same as the field device. For measurements with ionization chambers, the cosmic-ray contribution to the measurement may be estimated from the latitude of the field device or more rarely can actually be measured at the site by a suitable experimental arrangement of ionization chambers.

Corrections due to spectral interferences in ground-based gamma-ray spectrometers

If a scintillometer is being used and the measurements are based on an energy window system, then additional corrections are required to the background-subtracted gross window counts before these can be converted into concentration or dose-rate data. These additional corrections are due to the fact that the counts in a given window are not just the result of the gamma-rays from the nuclide corresponding to that window but also to the presence of other radionuclides present (see for example, IAEA, 1989, or Dickson *et al.*, 1981). These corrections are referred to as *stripping ratios*.

The stripping ratios arise from the spectral shape of a particular nuclide and the effect of Compton scattering effects in the ground, the air between the ground and detector (negligible for ground-based systems) and in the detector itself. Compton scattering causes high-energy gamma-ray photons to produce a continuum of lower energies below the full energy peak. This results in counts being recorded in the energy windows below those set for that specific nuclide. Thus, the 2.62 MeV gamma-ray from ^{208}Tl (used to determine thorium) contributes to both the uranium and potassium window counts. Similarly, the 1.76 MeV gamma-ray from ^{214}Bi (U) contributes to the potassium window counts. In addition, lower-energy gamma-rays in each of the decay series may also contribute counts to these lower-energy windows. The width of the windows also means that there may be contributions to a given window at a higher energy from a radionuclide normally recorded at a lower-energy setting, e.g. counts in the thorium window due to the presence of uranium.

Stripping ratios are defined as 'the detected counts from an energy window other than that designated for that element to those detected in the normal window for that element when a pure source of the element is counted'. Thus, the stripping ratio of thorium into uranium is the ratio of the counts detected in the uranium window to those detected in the thorium window from a pure thorium source. Similarly, one defines the stripping ratio of thorium into potassium and uranium into potassium. The *reversed stripping ratios* are the ratio of counts in a higher window to those in a lower window normally used for a specific radioelement when only that element is present, e.g. the reversed stripping ratio of uranium into thorium is equal to the ratio of counts detected in the thorium window to those detected in the uranium window from a pure uranium source. In general, reversed potassium stripping ratios are zero, while those for uranium are also usually assumed zero although it may not be so for regions with very high U/Th ratios.

There is also a more general procedure for correcting window counts due to spectral interferences, known as the *matrix procedure*. This procedure would normally be used if there were gamma-ray contributions from anthropogenic nuclides as well as the three natural radioelements. In the matrix procedure, the basis of the approach is that the count rate due to a particular element (or nuclide) is the product of the concentration of that element in the ground and a *sensitivity factor* for that element in the window being considered. Sensitivity factor refers to the count rate in a given element window for a known concentration of that element under the given counting conditions. On the other hand, a sensitivity factor in a window other than the one set for that element is equivalent to the stripping ratios discussed above. These count rates, sensitivity factors and concentrations are then formed into a matrix. Inversion of this matrix then allows corrected window counts to be determined from the net background-corrected counts in that window.

Having determined all of the appropriate stripping ratios for a detector, the corrected counts in a given energy window can then be determined following the appropriate background subtraction in that window, giving the so-called *background-corrected count rate*. The stripped count rate then allows the element or activity concentration to be determined, based on the appropriate sensitivity value for that radioelement.

Corrections due to spectral interferences in airborne gamma-ray spectrometers

A further complication is introduced into these corrections owing to the fact that the stripping ratios and the window count rates are a function of the altitude above the ground. Thus, the stripping ratios depend on scattering effects of the emitted gamma-rays, which are influenced by air scattering and which in turn depend on the length of the air column between the ground and the detector. One approach is to determine stripping ratios at ground level from suitable, large calibration sources and then to calculate the increased values as a function of altitude using both theoretical and experimental studies. The latter will be discussed in more detail in the next section.

12.2.5 Calibration of Gamma-Ray Survey Instruments

To determine the concentration of a radioelement or its contribution to the dose rate in air from a gamma-ray measurement requires the detector to be calibrated. Inherent in such calibration techniques is the ability to determine the energy of the relevant gamma-ray(s) in a spectrum. Energy calibration was discussed in Chapter 9 and exactly the same techniques are used for energy calibration of survey instruments. In addition, a calibration system enables a detector read-out (counts/unit time, ionization current, etc.) to be converted into more practical units such as element concentration, activity concentration, dose rate, etc. This will involves a standard containing a known concentration of the appropriate radioelement and having the same geometry as the source (as we saw in Chapter 9). In the case of field gamma-ray measurements, the source is essentially infinite and may not be homogeneous.

To construct an exact standard to match the source is virtually impossible and so the construction of practical standards always requires some approximations and assumptions. Ideally, however, it should be planar and sufficiently large that it at least approximates an infinite source in the gamma-ray distribution. It should also have a uniform distribution of a single radioelement and for nuclides with a chain of daughter nuclides, e.g. ^{232}Th and ^{238}U, the parent should be in equilibrium with the daughters. This condition of equilibrium may, of course, not be met in the field source for the reasons discussed in Section 12.2.2. As with any gamma-ray standard, the matrix should match as closely as possible that of the source in its absorption properties. Fortunately, although geological sources vary widely in type the majority have overall elemental compositions that are not too different and therefore have absorption properties which are similar.

Calibration pads

These are large slabs of concrete containing a known and uniformly distributed concentration of a particular radioelement. Concrete provides a robust calibration source which will remain physically intact over long periods of time. It also provides an excellent sealant for the added radioactive spikes (uranium, thorium, etc.) so that the gamma-ray fluxes also remain constant over long periods of time.

Generally, the pads are constructed for the calibration of detectors for the natural radioelements, potassium, uranium and thorium. Calibration pads represent a physical approximation to an ideal geological surface source in which the concentration of a specific radioelement and its distribution within the surface are accurately known. These pads are used for the calibration of both ground-based and airborne detectors but for the latter, the physical dimensions are normally greater in order to mimic the greater field of view of these detectors. Even so, pad facilities are of limited use for the calibration of airborne detector systems and part of these calibrations, such as window sensitivities and count rates, will always require flights over a ground strip, as previously described.

The purpose of a pad calibration facility is to provide information on the stripping ratios and window sensitivities for ground-based, portable γ-ray detector systems and the provision of ground level stripping ratios for airborne systems. In addition, information on the relationship between γ-ray counts and the air dose rate can also be obtained from these facilities. Originally, these calibration facilities were designed for the calibration of instruments for use in geological surveys and mapping and hence, involved only the detection and measurement of the three natural radioelements. More recently, attempts have also been made to extend the facilities to the calibration of some environmental radionuclides, particularly ^{137}Cs. We will say more about such facilities later.

Calibration facilities for portable ground-based spectrometers

The normal practice is to use four pads in total, with one each spiked with known concentrations of potassium, uranium and thorium, while the fourth contains no spike and serves as a background to permit subtraction of all counts not contributed by the gammas from the spike radioelements. For a number of reasons pads of mixed radioelements are not recommended and are not required for calibration purposes – they would only serve as a check on the calibration and this is achieved more realistically by other procedures.

The pads are circular and about 3 m in diameter, which for a detector 10 cm above the ground, gives over 90% of the response of an infinite source for 2.6 MeV gamma-rays. Since detectors are normally much closer to the ground, their response is essentially that of an infinite source. Typically, the pads are about 40–50 cm thick, which also provides essentially the same response as that of an infinitely thick one. The ideal pad concentrations are 8% K, 50 ppm eU and 125 ppm eTh (IAEA, 1989). These concentrations give counting rates of the order of 2000 cps from each of the radioelements in a 76 mm × 76 mm NaI(Tl) detector. Each of these pads is sunk into the ground and spaced about 5 m apart, giving negligible interference between neighbouring pads.

Calibration facilities for airborne spectrometers

Again, only four pads are required for the calibration of detectors for the natural radioelements and for practical reasons these are normally square rather than circular. There are good theoretical and practical reasons for making the pads as large as possible. Modern airborne detector arrays are normally very large (16 dm^3 volume for prismatic NaI(Tl) arrays) and the pads must simulate an infinite source as closely as possible. In addition, changes in the position of the detectors over the pads have smaller effects on the count rate if the pad is large. The recommended pad size for airborne equipment is a square with 8 m sides, which gives a response equivalent to 80% of an infinite source for a 76 mm × 76 mm NaI(Tl) detector. Doubling the length of the sides only increases this to 84% and is not justified by the expense and problems constructing such a large pad. Sometimes, however, much smaller pads are used (only 1 m square), particularly if they are to include environmental radionuclides. The ideal concentration ratios for the natural radioelements is 2% K to 5 ppm eU and 10 ppm eTh, which produces nearly equal total counting rates in each of the pads in a 76 mm × 76 mm NaI(Tl) detector (IAEA, 1989).

For an airborne gamma-ray spectrometer system, the detectors are normally mounted in the body of the aircraft. This means that with the aircraft on the ground there is a minimum clearance from the ground of about 1 m. Consequently, detectors calibrated using the concrete pads are typically situated 1 m above the pad.

If environmental radionuclides are also to be determined in surveys, then the calibration problem becomes more complex. The stripping of net spectral data can be extended to many radionuclides, provided that a pure spectrum for each nuclide is available. As with natural nuclides, energy windows for each nuclide must be set and the contribution of each nuclide to the net counts in all of the other windows has to be determined to provide stripping and reverse stripping ratios. These are then combined into an overall stripping matrix and the background-subtracted window counts are multiplied by the inverse of this matrix to provide the relevant net window counts.

The use of concrete pads with uniformly distributed radioactivity does not deal with the problem of stratification of the radioelements frequently found in soils and rocks, nor do these pads mimic the variable moisture content of geological sources. Environmental radioactivity also usually shows a variation with depth and to reduce mathematical problems it is usual to assume an exponential decrease of nuclide concentration with depth, which may not be valid at many sites. The major shortcoming, however, in using the pads for airborne calibration is the absence of the air column between the source and the detector and the subsequent modification of the spectral shape by this air column. For sources containing only the natural radioelements emitting high-energy gamma-rays, the main effect of this air column is to increase the low-energy continua and so the count rates in the lower-energy windows. In the case of environmental radionuclides, such as ^{137}Cs and ^{134}Cs emitting much lower-energy gamma-rays, the effects are more complex. It is found, for instance, that stripping ratios determined without taking the air column into account result in understripping of ^{134}Cs and overstripping of ^{137}Cs.

It is possible to simulate an air column with these calibration pads by introducing sheets of Perspex or plywood between the pad and the detector, above the detector (to simulate back-scattering effects), or both. Perspex and plywood have similar mass attenuation coefficients to both air and soils and are easily cut into sheets to fit the pads, with each sheet equivalent to a few metres of air. This allows the simulation of air columns comparable to those found in actual aerial surveys, or in the case of anthropogenic radionuclides the effect of burial of the activity below the surface. Such experiments, although limited due to the incorrect geometry and other factors, have shown that the stripping ratio for ^{40}K into the lower-energy windows of ^{134}Cs and ^{137}Cs increases linearly with altitude but there is a much lower dependence of these ratios for the 1.76 MeV gamma-ray from uranium and the 2.62 MeV thorium gamma-ray into the lower-energy windows.

Calibration of detector systems, particularly airborne ones, based only on these calibration pads, even those modified with absorber sheets, is always going to be limited due to the impossibility of accurately matching all of the parameters found in a real aerial survey. Furthermore, as techniques develop to monitor ever more environmental radionuclides (especially if short-lived), the expense and time required to construct suitable calibration pads for each nuclide becomes prohibitive. An alternative approach is to use the power of modern microcomputers to both simulate the gamma-ray spectra and calculate the necessary correction parameters, such as stripping ratios, by using gamma-ray transport algorithms. Once a code is developed it has to be validated by comparing the results with those from sources measured in a survey. These codes also allow for the rapid calibration of a detector system for multi-radionuclide sources that might be present following an accidental release of radioactivity from a nuclear plant.

Validation of the calibration procedures

The overall validity of any detector calibration is to check the results of a given survey with the known concentration and distribution of radionuclides over a given survey site. Several validation methods are possible but the most common method is a comparison of airborne- with ground-based surveys using detectors calibrated on the same calibration pads, since this provides traceability of calibration and is also quite rapid. It does not, however, eliminate any systematic errors inherent in the calibration, nor does it fully take into account possible non-uniform distribution of activity over the site.

The most rigorous validation procedure involves the comparison of airborne survey results with those from the analysis of core samples taken from the comparison site. This is a much more time-consuming process but the only one which can provide accurate information on both the spatial and vertical distribution of radionuclides. By comparing spectra with actual radionuclide depth profiles, it is possible to estimate the mean depth of the activity from the spectral characteristics.

12.2.6 Calibration of Dosimeter Survey Instruments

Exposure rates ($\mathrm{pGy\,s^{-1}}$) can be calculated for unit concentrations of a radionuclide at variable heights above a calibration pad based on gamma-ray transport calculations using Monte Carlo techniques. This then allows TL dosimeters and ionization chambers to be calibrated by using these pads. For example, for the calibration of TL dosimeters with the natural radionuclides, four TL dosimeters are irradiated for several weeks on the surface of the pads (three spiked with a given radioactive element and one background pad). The total exposure in pGy is simply the irradiation time multiplied by the exposure rate for a height of zero above the pad. The light output when these dosimeters are read can then be correlated to the known exposure. Ionization chambers are usually handheld or carborne and therefore readings are normally taken about 1 m above the ground. Hence, they are well suited to calibration on airborne calibration pads where the instrument is set 1 m above the pad. In addition, the pads are large enough to produce acceptably large calibration exposure rates at this height.

12.2.7 The Presentation of Gamma-Ray Survey Data

The results of gamma-ray surveys are usually reported initially as radioelement or radionuclide concentrations. Quite often, however, the data are required to determine or estimate the radiation dose to a population, since the elements and nuclides are sources of ionizing radiation in the environment. The effect of gamma-rays from radionuclides in soil or rocks in air is expressed in terms of the exposure rates or *adsorbed dose rate* in air (ADR). The conversion factors from concentrations in the soil to ground level ADRs are estimated from photon transport calculations applied to infinite soil and air media and are dependent on both the concentration of the radionuclide and the energy of the gamma ray(s) emitted. Factors for the conversion of natural radioelement concentrations into radiation exposure rates are practically identical for both soils and rocks. The conversion factors for the three natural radioelements are given in Table 12.3, assuming an infinite source, and have been verified by experiment. The presence of significant anthropogenic radionuclides, of course, adds to the total ADR, with the latter from all sources simply being the sum of the individual components.

A survey involving only total count measurements makes the conversion of these data into a useable form for survey purposes more difficult (IAEA, 1990). The problem arises from the lack of proportionality between the concentration of the individual radioelements and the total ADR. The response of a total-count instrument depends on detector parameters, such as its dimensions, and also the energy discriminator threshold used. Thus, if only the natural radioelements contribute to the total count rate, n_{tot}, then the latter can be expressed in the following form:

$$n_{tot} = s_K c_K + s_U c_U + s_{Th} c_{Th} \tag{12.1}$$

Table 12.3 Conversion factors for the three natural radioelements.

Conversion to	Conversion factors		
	1 % K	1 ppm eU	1 ppm eTh
Air dose rate (pGy s^{-1})	3.633	1.576	0.693
Exposure rate (μR h^{-1})	1.505	0.653	0.287
Ur	2.5	1.0	0.50

where s_K, s_U and s_{Th} are, respectively, the sensitivities of the instrument to K, U and Th, and c_K, c_U and c_{Th} are the concentrations of K, U and Th, respectively.

The sensitivities are expressed in terms of the count rate per unit concentration of the radioactive species in an infinite plane source and it is these that depend on the detector and threshold set. Total-count instruments typically have a discriminator threshold set at 30–150 keV but if a gamma-ray spectrometer is used for surveying then the threshold for the total-count window is usually set at a much higher energy, typically 400–800 keV, and so total count readings even from the same detector will be different.

Equation (12.1) can be rearranged in terms of the uranium sensitivity factor, s_U, as follows:

$$n_{tot} = s_U\left(\frac{s_K}{s_U} c_K + c_U + \frac{s_{Th}}{s_U} c_{Th}\right) \tag{12.2}$$

where s_K/s_U is the uranium equivalent of potassium and s_{Th}/s_U is the uranium equivalent of thorium. The term 'uranium equivalent' refers to the concentration of uranium that would give the identical count rate as 1 % K or 1 ppm Th in a total-count measurement for a particular system. Therefore, n_{tot} will provide values of concentration in Ur units when the values of s_K/s_U *and* s_{Th}/s_U are determined. The ratio s_{Th}/s_U is approximately constant for any total-count instrument, whereas the s_K/s_U ratio varies with changes in detector volume and energy discrimination threshold. It turns out that for a 76 mm × 76 mm NaI(Tl) detector and an energy threshold of about 400 keV, the total-count response of such a system is closely proportional to the ADR, while for instruments with different detector sizes or thresholds this relationship does not hold.

If a total-count detector system is calibrated over a large source containing only uranium or over an infinite rock plane containing the other radioelements but whose concentration is expressed in uranium equivalents, then Equation (12.2) reduces to the following:

$$n_{tot} = s_U c_U \tag{12.3}$$

and the sensitivity factor, s_U, can be expressed in terms of count rate (e.g. cps) per ppm eU. Hence, the equivalent uranium concentrations can be easily obtained directly from a total-count rate. Furthermore, if a 76 mm × 76 mm NaI(Tl) is used with a threshold set at 400 keV, then multiplying the equivalent uranium concentration by the factor 1.576 (see Table 12.3) will give the ADR value directly. Instruments with different parameters to the above do not give ADR values proportional to the concentrations of potassium, uranium and thorium present in the ground. With increasing concentration ratios, K/U and Th/U, the errors in the calculated ADR values also increase. On the other hand, instruments in which s_K/s_U equals 2.30 and s_{Th}/s_U equals 0.44 give correct ADR values irrespective of the concentration ratios of potassium and thorium to uranium. In general, the K/U ratio has a greater effect on the results than the Th/U ratio.

12.2.8 Selected Case Studies of Gamma-Ray Surveys

Although, as has been pointed out earlier, the majority of gamma-ray surveys have been carried out for geochemical purposes, there are a number of other applications of this technique, particularly for airborne surveys. Such applications have increased in recent years, partly as a result of developments in detectors and their associated electronics (with the increasing availability of small but powerful computers being of particular significance) and partly due to increasing concerns about the impact of low levels of radioactivity on the general population. It is with some of these less routine but nevertheless important applications that we shall deal with in this section.

The use of airborne surveys in the search for lost missiles and satellites containing nuclear materials

This first case study is rather an exotic one but one which illustrates the power of airborne surveys utilizing large detector arrays. One of the earliest uses of airborne gamma-ray surveying to locate potentially hazardous nuclear material was in 1970 when the US Air Force had to try and locate an Athena missile which crashed while carrying ^{57}Co as part of its payload (Deal *et al.*, 1972).

The missile, containing some 3.5×10^{10} Bq of ^{57}Co as part of its instrumentation, was routinely launched from Green River in Utah, USA. Its target area was White Sands in New Mexico, but unfortunately, an error in its propulsion caused it to crash land in the desert of northern Mexico. The exact location was not known but radar tracking narrowed it down to a relatively small area and it was decided to use an aerial gamma-ray survey to try and locate it.

The aircraft used was equipped with an array of 14 (10 cm × 10 cm) NaI(Tl) detectors. Calculations based on this system indicated that a source of 3.5×10^{10} Bq emitting 122 keV gamma-rays from ^{57}Co on the surface should be detected by flying lines about 300 m apart at an altitude of 100 m. In practice, slightly narrower flight lines were used.

As the aircraft flew along its trajectory, counts in the 122 keV photo peak were recorded along with the position and altitude and these data plotted continuously on a strip chart. An increase in activity of about a factor of two above the background and of about 6 s duration was found along one of the flight lines. Subsequent flights above this area at various altitudes confirmed the presence of a point source although no crater due to the impact of the missile was visible at the time. Subsequently, bags of white flour were dropped over the source location to guide a ground recovery team. A crater some 15 m from the source was then observed to which the ground recovery team were directed. This team confirmed that the nose cone of the missile carrying the source had produced this crater but that the source was not intact but distributed over a relatively large area.

A similar operation was mounted by the Canadians in 1978 to try and locate the Russian satellite Cosmos 954 (Bristow, 1978; Grasty, 1978). This satellite re-entered the earth's atmosphere and disintegrated, scattering debris over a wide arc of land in the desolate North-Western Territories of Canada. The concern was due to the fact that Cosmos 954 carried a nuclear reactor on board and hence the debris would contain both fission and activation products. The probable impact trajectory was known from monitoring stations and eye-witness accounts. This allowed a search area to be established which was a few hundred kilometres in length and about 100 km wide. Because of the terrain, aerial surveying was the only practical option to try and locate the radioactive debris and estimate the activity on the ground.

The gamma-ray spectrometer used was that developed by the Geological Survey of Canada and consisted of an array of 12 square cross-section NaI(Tl) crystals (102 mm × 102 mm × 406 mm) giving a total volume of over 50 dm^3. Both complete 256 channel spectra could be taken and stored together with counts from up to six window settings. The aircraft carrying the detector and associated electronics also carried conventional navigation equipment but once a potential source area was located it was pinpointed with greater accuracy by using a microwave-based system employing two ground-based beacons.

The form of the radioactive debris from Cosmos 954 gave rise to gamma-rays from about 300 keV to 2 MeV but with the majority lying below 1 MeV. A significant problem in the search area was the high natural background due to the granitic rocks that compose much of this part of Canada. Consequently, total-count data were of limited use since anomalous high count rates were frequently attributable to local anomalies in the geological background.

In order to overcome this problem, two windows were set, i.e. a low-energy one from 300–900 keV and a high-energy one from 900–1500 keV. Two simple diagnostic functions were found, based on the counts in these two windows, that could discriminate between natural and anthropogenic radiation: one a ratio of the counts in the two windows and the other the difference in the counts in the low-energy window and a certain multiple of those in the high-energy one. Experience indicated that the difference technique was slightly more sensitive. Statistical analysis indicated that a point source of about 5.6×10^9 Bq could be located over land or about 4.8×10^9 Bq could be located over ice with 95% confidence (the difference being due to differences in the background over land and ice), with the actual figures being temperature-dependent. These figures were confirmed by flights over a ^{137}Cs source of known activity.

These data and flights over the projected impact trajectory resulted in evidence for most of the debris from the satellite being on the frozen surface of the Great Slave Lake but confirmation from ground surveys was not possible. The results also indicated complete disintegration of the reactor upon entering the earth's atmosphere but that the radioactivity was less than that observed from natural rocks in the surrounding area.

Surveys following the accident at the Chernobyl reactor in the former Soviet Union

Accidents have occurred at a number of reactor facilities in many parts of the world over the past 50 years since the beginning of nuclear operations. On the whole, these have been relatively small-scale and have predominantly affected the country in which they originated e.g. the fire at the Windscale reactor in the UK in 1957 or that at Three Mile Island in the USA in 1979. The accident which occurred at Reactor Number 4 at Chernobyl on April 26, 1986 was on an altogether different scale. The radioactive debris from this accident affected, or was detected over, much of the Northern Hemisphere.

The scale of this accident and the subsequent potential hazards to the affected populations, both animal and human, meant that many countries organized (mainly) airborne surveys to both detect and measure the radioactive plume as it moved over their countries. One of the country's most seriously affected was Sweden, the country that actually alerted the Western World to the accident following an alert at a nuclear power plant north of Stockholm on the morning of 28 April. Following this, both carborne and airborne surveys were quickly put into effect but the emphasis was on the airborne surveys because of their ability to rapidly survey the entire country.

The first preliminary airborne measurements in Sweden were made on 1 May, using an aircraft equipped with four prismatic NaI(Tl) detectors, each 10 cm × 10 cm ×40 cm, giving a total volume of 16.8 dm^3 (Mellander, 1989). Gamma-spectra, each of 256 channels, with the first 255 channels covering energies up to 3187.5 keV and the 256th for monitoring the background from 3187.5 keV to 6 MeV. Measurements were all taken along flight lines at an altitude of 150 m.

Surveying actually commenced on 2 May and concentrated on a region between Stockholm and the town of Gävle, some 170 km to the north of Stockholm. This survey indicated very high levels of activity of the order of 0.5–1.5 mR h^{-1}. In spite of these high values, attention had to be switched to southern Sweden (most of the country south of Stockholm). The reason for this was that this was just the time of the year when farmers in this region were about to send cattle out to graze on the fresh spring grass and it was vital to know if this could be allowed in these circumstances. Thus, for the next three days all of this southern area was surveyed, using east–west flight lines with a separation of 50 km. Fortunately, the results indicated that most of this area had only low levels of contamination and grazing could go ahead.

A problem at this stage of the programme was that the detector system was only calibrated for the natural gamma-ray emitters. No calibration was available for the large number of (mainly) fission-product radionuclides now being detected. Rough dose-rate calibrations were carried out by flying over an area that had been accurately monitored with a calibrated hand-held scintillometer and then comparing these measurements with the total response of the airborne system. Fortunately, the problem of calibration for the detected anthropogenic nuclides was considerably reduced by accepting that only the two caesium nuclides (^{134}Cs and ^{137}Cs) were sufficiently long-lived (half-lives of 2 and 30 years, respectively) and as a consequence these were the only two nuclides whose absolute activities were required. The relevant data for these determinations were obtained at a later stage by using an area on the ground at which the deposition of ^{134}Cs and ^{137}Cs could be measured accurately with a calibrated hyperpure Ge detector. It was assumed that at this stage most of the other nuclides had decayed sufficiently to make only a minor contribution. The aircraft was now flown over this area and the counts corresponding to the channels representing the two nuclides were determined and compared to the known activities of ^{134}Cs and ^{137}Cs from the Ge data. The lower limit of detection for both nuclides was about 500 Bq m^{-2}.

Between 5 and 8 May, the rest of Sweden was covered by east–west lines 100 km apart, together with a few north–south lines to complete the grid. It was found that the most contaminated areas were in a coastal strip extending from Gävle for a further 300 km north and extending to the Norwegian border in the northernmost section. All of the data collected were rapidly available in map form with a grey scale to indicate the level of contamination.

At the end of 8 May, a second cloud of activity struck Sweden and a second complete survey of Sweden was carried out from 9 to 24 May, using a 50 km line spacing, revealing increased activities in many areas due to the new cloud. Following this survey, additional measurements were made in selected, severely contaminated areas using a line spacing of only 2–10 km, lasting until mid-June 1986. In October of that year, more accurate calibrations of the detector system were also carried out.

The end result of all these surveys was a series of contour maps of various scales, plotting either the exposure dose rates or the activities of ^{134}Cs and ^{137}Cs for the whole country. These surveys also allowed the changes of levels of activity over the whole of Sweden to be followed in the future.

Environmental applications of airborne gamma-ray spectrometry

The examples discussed so far have utilized gamma-ray surveying techniques in emergency or accident situations and have had fairly specific aims, such as the location of lost nuclear materials or tracking the movement of a radioactive plume. More frequently, however, these techniques are used in much more routine circumstances when one requires to simply monitor the radioactivity over a wide area. An example is the monitoring of an area around a nuclear facility to ensure that discharges are not exceeding prescribed limits or that there is no long-term buildup of activity, etc.

Sanderson *et al.* (1995) have described such a situation in regard to surveying the Ribble Estuary in the northwest of England. Close to the head of the estuary is the Springfields fuel fabrication plant operated by BNF plc which discharges various low-level nuclides into the estuary. Both the estuary and salt marshes were surveyed, comprising an area of 12 km × 20 km, using an aircraft flying with a 300 m line spacing and carrying an array of prismatic NaI(Tl) detectors, each of 10 cm × 10 cm × 40 cm in dimensions, with a total volume of 16 dm^3. Part of the discharge from this plant is in the form of low-level uranium series waste, giving rise to ^{234m}Pa which binds to the mud particles and so tends to accumulate in the mud flats. Although this nuclide was only expected to be present at very low levels, it was felt that it might be detected by setting a suitable window corresponding to the 1101 keV gamma-ray emitted by ^{234m}Pa. Appropriate stripping ratios were also determined by using a depleted uranium source.

The gamma-ray spectra from the estuary and mud flats did indeed show a background-stripped peak corresponding to ^{234m}Pa. Rather surprisingly, one of the lower salt marshes which lie along parts of the estuary also gave a signal which was initially attributed to ^{234m}Pa. A more detailed examination of the spectra, however, revealed peaks at 911 and 969 keV, rather than at 1101 keV (remember that the resolution of these systems is not very good), and these peaks were ascribed to ^{228}Ac (see Appendix 3 – ^{232}Th decay series, Figure 1 and Table 2). The presence of this nuclide was attributed to a 'dis-equilibrium' situation in which ^{228}Ra was preferentially extracted into solution from material bound to the estuarine sediments during the high water flows at spring tide. The ^{228}Ac (half-life of 6.1 h) from the β^- decay of ^{228}Ra (half-life of 5.75 years) was then deposited on the salt marsh further down the estuary owing to its high affinity for the small particles in these marshes.

The use of aerial radiometrics for epidemiological studies of leukaemia

This last example, also from Sanderson *et al.* (1993) is rather unusual in that it was used not only to ascertain the dose to a large population from a gamma-ray survey but to then infer a connection with the incidence of leukaemia in this population. This arose from the concern expressed at leukaemia clusters, some of which are associated with the presence of a nuclear facility but where an even larger number have no such association. Very recently, work in the UK has shown an increased incidence of leukaemia associated with the presence of industrial and/or highway pollution.

A problem with low-dose studies is that the risks of radiation-induced leukaemias are small and hence it is necessary to have studies covering areas in excess of 10^4 km^2 and populations in excess of 10^6 to enable reliable correlations between low-dose exposure and leukaemia to be established. Such an area and population size could be covered by an aerial survey which would involve effectively determining the radiation dose from the natural emitters (K, U and Th) over such an area but on a scale that would allow small, local variations to be measured.

This was the basis of the study. The area chosen in south-west England contained some of the highest environmental gamma-ray dose rates in the UK, due to the presence of granitic rocks and the associated high levels of all three natural radioelements, particularly uranium and thorium. The purpose then was to see if the measured gamma-ray dose rates could be correlated with the local incidence of leukaemia cases.

In this preliminary study, an area of only 2500 km^2 was covered but 4800 gamma-ray spectra were taken by using the prismatic array of NaI(Tl) detectors discussed in the previous case study. Two areas of roughly 1000 km^2 were surveyed, with a 1 km line spacing, and a further area of 250 km^2 was surveyed, with a 500 m line spacing. The specific activities of potassium, uranium and thorium were then determined from the gamma-ray spectra and these values converted into the equivalent dose rates (as described in Section 12.2.3 above). Finally, these data were used to construct dose-rate maps for the area surveyed. As expected, local dose-rate anomalies could be correlated with the local geology. An interesting aspect of these results was that the Devonport Dockyard, which handles nuclear submarines and was located in the area of study, showed no radiation-dose anomaly, nor did Hinckley Point nuclear power station (also in this region), apart from radiation due to gaseous ^{41}Ar which is emitted from this Magnox-type station.

Two methods were used for linking the epidemiological studies to the radiometric data. The first was to divide the population into groups exhibiting cases of leukaemia and those that did not, with the latter acting as a control group. The aim was to try and match the cases and controls for all factors other than differences in radiation dose. Due to spatial variations in the radiation field, this was very difficult in practice. The second method was to investigate the incidence rate for leukaemia as a function of the radiation variables, a method which is free from problems associated with departures from normality. Weighted regression analysis (see Chapter 11) was used to assess any association between leukaemia cases and the ambient radiation fields.

Unfortunately, as in many previous studies, the results were limited due to the low number of leukaemia cases within the study area and also subject to challenge on statistical grounds. Bearing in mind these restrictions, it was found that there was a positive correlation between leukaemia and the concentration of equivalent uranium, using either method of assessment. On the other hand, there was a negative correlation with equivalent thorium concentration in the incident-rate method although not for the case-control method.

By extending such studies to larger areas and populations it should be possible to ascertain with greater certainty any linkage between leukaemia and the ambient radiation field to a given population. The contribution of other factors, such as radon concentration could also be included in such studies.

13

Modelling the Dispersion of Radionuclides in the Environment: An Introduction to Modelling Concepts

This chapter provides an introduction to the main principles of mathematical modelling in relation to the dispersion of radionuclides in the environment. It discusses the purpose as well as the advantages and limitations of environmental modelling. Critical aspects of model validation and evaluation are also addressed. It is hoped that this introductory chapter will assist those readers who have little or no knowledge of modelling to understand the more detailed treatment contained in the final part of the book. Specific modelling approaches and their applications for estimating concentrations of radionuclides in atmospheric, terrestrial and the aquatic environments are discussed in Chapters 14 and 15. The final chapter describes how doses resulting from exposure to radioactivity can be calculated.

13.1 INTRODUCTION TO MODELLING

When radionuclide pollutants enter the environment they are subject to various pathways and processes. As an example, Figure 13.1 shows the main pathways involved in transferring radionuclides released into the environment to people. As this figure illustrates, there are numerous routes by which radionuclides can move in the environment and reach people. Superimposed on this is the web of physical, chemical and biological processes that control the movement and fate of radionuclides in the environment. As an example, Figure 13.2 shows the processes that affect the transport of pollutants, such as radionuclides, in the atmosphere. In addition to dispersion processes such as advection, diffusion and turbulence (see Chapter 14 for more details), the radionuclides will also be subject to various removal processes. These figures demonstrates the complex nature of the environmental processes which, however, need to be taken into account if one is to reliably simulate the behaviour of pollutants in the environment with any degree of confidence.

As modelling approaches have developed over the past decades, mathematical models have become generally accepted as useful and reliable methods for simulating the transport and behaviour of radioactivity in the environment. In particular, they have played an increasingly important role in the field of environmental assessment. A notable example was the case of the Chernobyl accident where numerous studies involving the use of models to assess the impact of the released radionuclides were reported in the literature. Some of these are discussed in Chapter 14. The applications of computer models for use in nuclear emergencies are

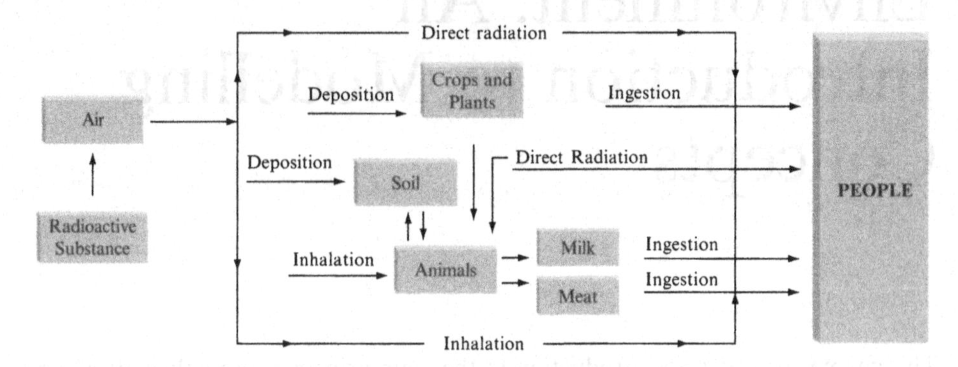

Figure 13.1 The main pathways involved in transferring radionuclides released into the environment to people (adapted from Eisenbud, 1987)

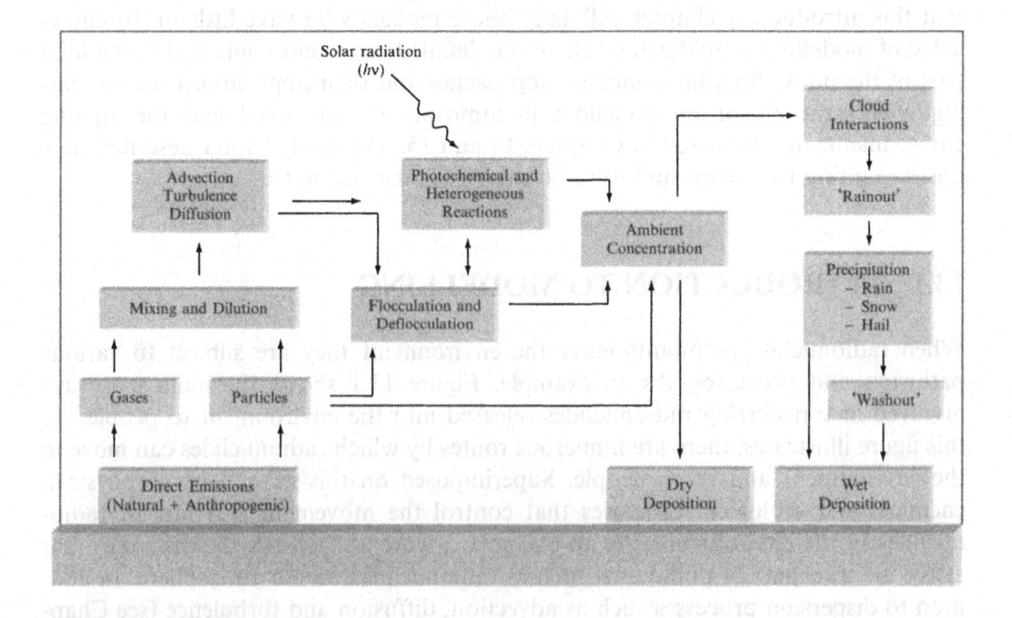

Figure 13.2 The main processes that affect the transport and mixing of pollutants in the atmosphere

discussed by ApSimon (1986) and examples of emergency response models are given in IAEA (1986).

In addition to describing the transport characteristics of radionuclides in the environment, models play a crucial role in estimating doses (see Chapters 4 and 16). Collective doses from discharges of radionuclides to the environment can be delivered to populations remote in time or space from the discharge location. In such cases, they have to be calculated using mathematical models. Because the temporal and spatial scales can be so large, the models used to calculate collective doses are often very different to those used to calculate doses to individuals in critical groups close to the discharge point. This is particularly so for radionuclides that circulate through the environment rapidly and globally; such radionuclides include tritium, carbon-14, krypton-85, and iodine-129. A number of models for calculating collective doses from globally circulating radionuclides have been published. Some examples are tritium (NCRP, 1979; Simmonds *et al.*, 1995), carbon-14 (Titley *et al.*, 1995), krypton-85 (Kelly *et al.*, 1975), and iodine-129 (Kocher, 1979; Titley *et al.*, 1995).

There are various reasons why mathematical modelling has gained popularity in tackling environmental problems. These include:

1. In many circumstances, it is not possible to conduct costly monitoring exercises and mathematical models of radionuclide behaviour have to be used. Hence, models offer an effective alternative where no or limited measurement data are available.

2. Models can be used to assess the impact of possible future practices that may lead to a release of radionuclides into the environment. Similarly, models enable scenarios to be analysed and this is particularly important in areas of policy and decision making.

3. Models are versatile and flexible in that inputs can be changed at will to assess their influence on the final outputs. Different approaches can be applied to the same scenario or a number of scenarios can be analysed with the same model. Furthermore, models enable calculations to be made of a range of parameters, ranging from pollutant concentration to determination of dose and risk. Such flexibility is not easily possible with physical models.

4. The tremendous increase of computer power over the past decade has meant that mathematical models can run on personal computers and hence provide possible solutions quickly and at relatively low cost.

5. Models have now been incorporated into user-friendly graphical interfaces, thus allowing 'non-experts' to use them.

6. They can be used as an aid in the design of monitoring campaigns and networks in terms of determining the optimum location of stations. This can avoid expensive mistakes in implementing monitoring campaigns and programmes. They also play a crucial role in formulating pre- and post-accident control or abatement measures.

7. They are an invaluable research tool and enable fundamental environmental processes to be studied, so leading to improved models.

In addition to the obvious advantages there are also several limitations with mathematical models that need to be recognized. These include the following:

1. Models, especially operational assessment models, tend to reply on assumptions to simplify the problem in question and make it more manageable.
2. Models can instil a false sense of confidence in the user and this can possibly lead to their inappropriate use. Hence, the use of a particular model needs to be proportionate to the complexity of the approach of that model – this leads to the concept of 'fitness of purpose' (RMS, 1995).
3. The output of a model does not only depend on the approach and the underlying science but also on the quality of the input data. This is often unavailable, scarce or unreliable.

The limitations of the models, therefore, are just as important as their advantages. It follows that is imperative that any model is comprehensively evaluated and validated with good quality measured data and with agreed protocols.

13.2 TYPES OF ENVIRONMENTAL MODELS

There are various types of environmental models (Beck *et al.*, 1993) and they can be categorized not only according to their approach but also their use. Most models used in the field of environmental assessment can be divided into the following types.

13.2.1 Physical Models

These consist of mesocosms or microcosms that model ecosystems to isolate a representative segment of the environment. The behaviour of a pollutant is observed in order to predict its environmental fate. Wind tunnels are also an example of physical models.

13.2.2 Mathematical Models

These attempt to define the controlling processes that act on a pollutant in the environment. The processes are a function of the chemical species and of variables that characterize the environmental processes. Various environmental models are discussed further in Chapters 14 and 15.

13.2.3 Statistical Models

These models use statistical techniques to investigate correlations between large datasets and may involve the determination of probability or risk of an incident occurring.

This chapter will focus on the basic concepts of mathematical models used for environmental radioactivity assessment. Such models use systems of equations to describe physical, chemical and biological interactions that affect the pollutant present in the environment. These models are based on concepts that consist of assumptions and approximations designed to simplify a complex problem into a manageable one. Mathematical models themselves can be categorized into the following types:

- *Deterministic models* – these employ fundamental physical and chemical laws that control and govern the fate of pollutants in our environment.
- *Stochastic models* – these incorporate the probabilistic nature of radioactive interactions and may involve the concept of risk, probability or some other measure of uncertainty.

13.3 EXAMPLES OF MODELLING APPROACHES

13.3.1 Source–Receptor Models

Mathematical modelling, for example, for describing the transport of radionuclides through the atmosphere, will require the following:

(a) identification of sources of pollution, whether point or area emission sources;

(b) emission rates for pollutants;

(c) understanding of the atmospheric transport processes which transfer these pollutants from the sources to the downwind locations;

(d) knowledge of the chemical and physical transformation processes that occur during the transport of the pollutants.

Such an approach is adopted in dispersion models that describe the movement of pollutants from the source to a receptor. An example of this type of model is the Gaussian Plume Model, which is described in detail in Chapter 14. In these models, the major dispersion processes need to be represented, as well as processes that remove pollutants from the environmental system. Such models have proved to be very useful to describe the dispersion of releases of radionuclides to the atmosphere.

Mathematical models also vary in their complexity. Whereas Gaussian models are of modest complexity in terms of the coding, algorithms and input data requirements, numerical models, on the other hand, can involve complex solutions to three-dimensional partial differential equations. An example of this approach is the *K theory*, also described in Chapter 14 for atmospheric applications. These models provide a more realistic description of the environment but generally are computationally intensive.

13.3.2 Receptor Models

Receptor-oriented approaches are useful mainly to:

- identify sources
- apportion the pollutants to the sources

These methods, referred to as *receptor models*, focus on the point of impact or receptor site as opposed to the source-oriented dispersion models which focus on transport of the pollutants from the source to the site of impact.

Receptor models rely on the assumption that the composition of the source emissions can be correlated to the composition of the ambient pollutant, such as an aerosol, at the sampling or receptor site. Therefore, a tracer or a marker can be used to identify a source. These types of model are based on the principle of mass conservation. For example, the total airborne concentration of a particular particulate pollutant (P_T in ng m^{-3}) can be considered to be a linear sum (assuming no chemical transformation) of the contribution of the individual sources. This can be expressed as follows:

$$P_T = P_1 + P_2 + P_3 \ldots \tag{13.1}$$

where P_i is given by the gravimetric concentration of pollutant (P) in source i particulate emissions (a_{P_i} in ng mg^{-1}) multiplied by the mass concentration of the source i particles in the atmosphere (mg m^{-3}) (f_i).
Therefore, we have:

$$P_i = a_{P_i} f_i \tag{13.2}$$

A key feature of such an approach is that for these models to work reliably to identify and apportion sources, one needs comprehensive datasets measured at the receptor site at temporal resolutions of no more than a few hours. The two main approaches used as part of receptor modelling are as follows:

(i) Chemical Mass Balance (CMB)
(ii) Multivariate Statistical Techniques

These methods quantify the interrelationships among the concentrations of the chemical species by analysing the data from a series of samples collected at a receptor site. These relationships are then used to identify the polluting sources. Statistical techniques that are often used for receptor modelling include Principal Component Analysis (PCA) or/and Multiple Linear Regression Analysis (MLRA).

13.3.3 Compartmental Models

These models offer a computationally simple method to investigate processes and the transfer of pollutants between main environmental compartments. For example, a river system can be envisaged as consisting of compartments between which radionuclides are transferring. This is shown schematically in Figure 13.3. A major advantage of this approach is that it can be applied to study any environmental problem, including dose estimation.

The general form of a compartment model is shown in Figure 13.4. The transfer of material between the compartments is described by the following general equation:

$$\frac{dY_i}{dt} = \sum_{j=i}^{n} k_{ji} Y_j - \left(\sum_{j=i}^{n} k_{ij} \right) Y_i - \lambda_i Y_i + P_i \qquad (13.3)$$

where $2 \leq i, j \leq n$ and $k_{ii} = 0$. The transfer coefficients between the two compartments are represented by k_{ij}, and the inventories in the compartments are given by, for example, Y_i. The parameter λ_i represents the loss of material from a compartment without transfer to another by radioactive decay, for example; P_i represents a continuous input of material into compartment i.

This approach has been applied for environmental radiological assessment at various geographical scales. Examples of applications of compartmental models are reported in Simmonds *et al.* (1995). An example of an application on a global scale is that of the circulation of tritium, as represented in Figure 13.5. In this model,

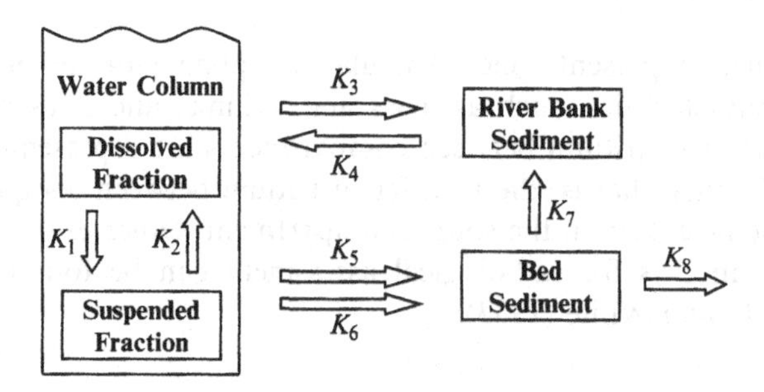

Figure 13.3 An example of a compartmental modelling concept applied to a river system: K_1, equilibrium sorption; K_2, equilibrium sorption; K_3, deposition; K_4, erosion; K_5, net deposition; K_6, mass movement of the water column at velocity V_1 ms^{-1}; K_7, dredging; K_8, downstream movement of bed sediment at velocity V_2 ms^{-1} (adapted from Simmonds *et al.*, 1995)

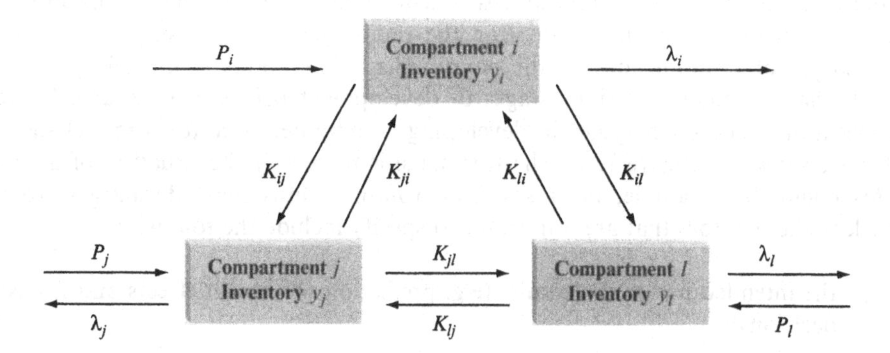

Figure 13.4 A general form of a compartment model showing the interactions between the sub-systems

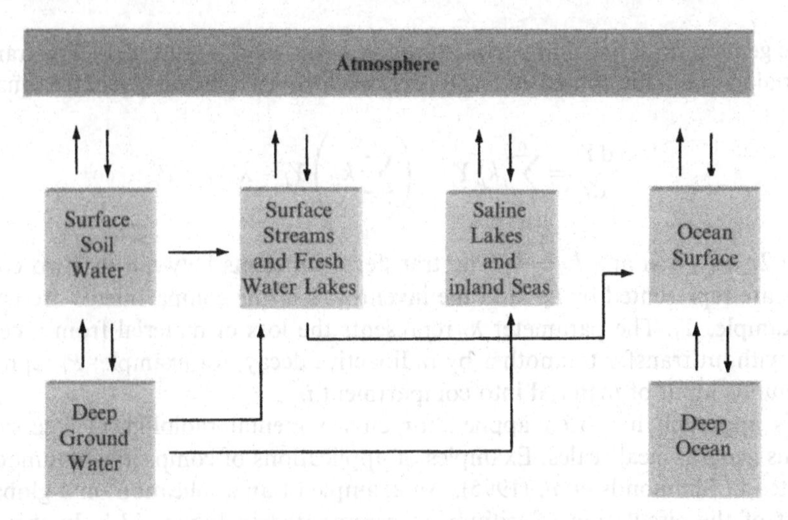

Figure 13.5 A compartmental model for the global circulation of tritium (adapted from Simmonds *et al.*, 1995)

each compartment represents one particular component of the environment, and has its own characteristics, such as mass and volume, and is assumed to be well mixed. Movements of tritium between compartments are represented by first-order exchange coefficients, that is, the transfer of tritium between compartments is proportional to the inventory in the source compartment. Other examples of the use of compartmental models for radiological assessment can be found in Titley *et al.* (1995) and Smith and White (1983).

13.4 STAGES IN THE DEVELOPMENT OF RADIOLOGICAL ASSESSMENT MODELS

The elements of modelling the behaviour of radionuclides in the environment will be described in Chapters 14 and 15. This section briefly considers the main stages of model development for estimating dose and risk from exposure to radiation. A notable example of a radiological assessment model is that of the European Commission's methodology for evaluating the radiological risks arising from routine discharges of radionuclides to the environment (Simmonds *et al.*, 1995). Such a model has to undergo various stages of development before it is suitable for real applications. The main stages in developing a computer code for a model such as this are shown in Figure 13.6. The first stage is to specify the situation of interest. This could, for example, be doses from routine atmospheric discharges from a nuclear site. Factors that are important to specify include the following:

(i) the intended use of the results (e.g. predictions of health effects and licensing decisions);

(ii) the degree of spatial and temporal resolution required;

(iii) the characteristics of the release of radionuclides (e.g. continuous or accidental).

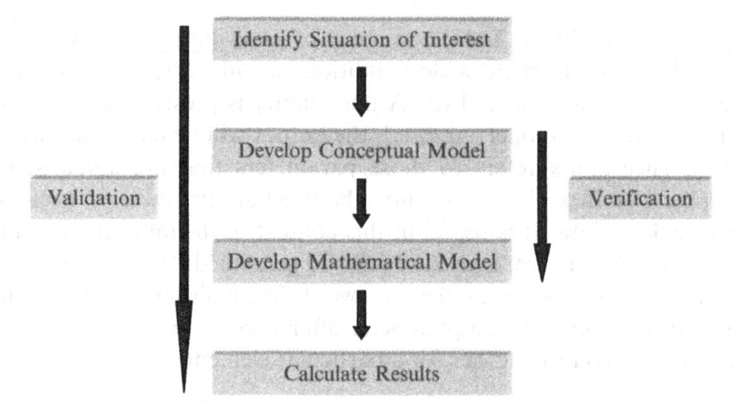

Figure 13.6 The main stages of model validation and verification

The next stage is to develop the conceptual model. At this stage, the exposure pathways of importance are established, the number of environmental compartments under consideration is specified and the processes transferring radionuclides through the environment are identified. The degree of detail in the model will reflect the initial specification. The next stage is to represent the conceptual model in mathematical and computational terms through a computer language code. It is important that such a code is fully verified and that the model outputs are validated. Before considering these aspects, we will first briefly discuss uncertainties in environmental assessment models.

13.5 UNCERTAINTY IN MODEL PREDICTIONS

Predictions from models will have some uncertainty associated with them. In simple terms, there will be error bars on the results. How wide these are will depend upon a number of factors. The sources of uncertainty in the predictions from models can be grouped into four broad categories, as follows:

- *Measurement uncertainty*. This is the uncertainty in the field data on which models are based.
- *Conceptual uncertainty*. This is the uncertainty surrounding the choice of conceptual model derived from the field data. For example, a number of different conceptual models may be consistent with the available data.
- *Modelling uncertainty*. This arises from the representation of the conceptual model in mathematical and computational terms. It includes the use of simplifying assumptions, discretization and numerical methods of solution.
- *Parameter value uncertainty*. This is caused by not knowing the most appropriate values to select for the various parameters of a model. Parameter value uncertainty is often divided into two types, namely **objective** (Type A) uncertainty and **subjective** (Type B) uncertainty. Objective uncertainty refers to parameters which intrinsically have no single correct value but for which a probability distribution of values can be specified. One example is the meteorological conditions at the

time of a future accident. There is no way of knowing precisely what these may be but a distribution of possible conditions could be specified based on past experience. It follows that if Type A uncertainty is present then there is no one single result from the model – instead, the output will be a distribution of values. Subjective uncertainty refers to those parameters that are assumed to have a single value but there is uncertainty about what the value is because of, for example, lack of data. It is useful in this context to distinguish *uncertainty* from *variability*. The latter term is used to refer to genuine differences in the value of a parameter when a model is applied to two different situations. For example, the soil-to-plant transfer factor, a parameter often used in foodchain models, may be different for one soil to another for a particular element.

A fifth category of uncertainty, namely *completeness uncertainty*, is sometimes identified. This is the uncertainty arising from the omission of an important, but hitherto unidentified, process from the model. This can be considered part of conceptual or modelling uncertainty.

It can be difficult to quantify the uncertainty arising from these various sources. For example, there is a considerable amount of work being undertaken into parameter value uncertainty, although conceptual and modelling uncertainties are not easy to quantify.

Two terms are commonly used when discussing the topic of uncertainty in the predictions from models, i.e. *uncertainty* and *sensitivity*. There has, however, been some confusion surrounding the use of these terms.

Sensitivity analysis is the study of the effect of changes in input values on the output from the model. It enables the user to identify the parameter or groups of parameters to which the model is most sensitive and, as such, can be used to direct research programmes. As an example, Desiato *et al.* (1998) have investigated the sensitivity of the APOLLO model to parameters such as horizontal diffusivity and mixing height as part of a study to evaluate long-range atmospheric dispersion Lagrangian particle models.

Uncertainty analysis is the quantification of the uncertainty in the model output. Typically, uncertainty in the values of the model parameters is taken into account. Distributions of parameter values are estimated, usually by expert judgement, and these are used with the model in order to produce a range of model predictions. The important point is that the output is a result of varying all of the values of the parameters across their ranges. Often, the parameters in a model are dependent. For example, when the value of one parameter is large for physical reasons, the value of another parameter has to be small, or vice versa. It is important that such relationships between uncertain parameters are identified and taken into account in the analysis. Uncertainty is often expressed as the ratio of the 95th percentile to the 5th percentile of the model output.

13.6 MODEL EVALUATION AND VALIDATION

Model evaluation and validation are perhaps the most crucial aspects that decide how much confidence one can have in a particular model and, thus, help to decide which model to apply to a specific problem or to a range of similar problems. The use of models to support regulatory decisions and policy making is continually increasing. As decisions based on the predictions of such models can have serious economical and environmental implications for the parties involved, their evaluation and examination is all the more necessary. 'Over-prediction' can lead to unnecessarily strict emission controls or the adoption of expensive pollution control technology, so increasing the economical burden on a given company or causing costly delays in regulatory controls. Further discussion on costs resulting from uncertainties in models used for regulatory purposes in the USA can be found in Dupuis and Lipfert (1986).

Evaluation of model performance can take various stages. These include verification of the code itself, comparison with measured data and model intercomparison studies. The UK Royal Meteorological Society has released a document (RMS, 1995) on the criteria that models should meet. In addition to the aspects considered here, the suggested criteria include documentation on the model and its use, information on the input data, presentation of output data, and quality assurance/quality control (QA/QC) procedures, as well as an audit trail.

Before considering the more detailed aspects of model evaluation and validation, it is important to understand terms such as *validation* and *verification* that are commonly used in the literature.

Verification is the process of deciding whether the computer code is an adequate representation of the conceptual model. Procedures for code verification include peer review by individuals not directly involved in the particular coding process, checking the results from the computer code against known solutions, and intercomparison of model results from a range of different assessment problems.

Validation is the process of deciding whether the conceptual model and its associated computer code is an adequate representation of the situation of interest; in other words, how well does the assessment system represent the 'real world'? In doing this, it is important to remember that the assessment system has only to be suitable for its intended purpose. For example, a model may be adequate for evaluating annual doses from routine discharges of radionuclides to estimate annual average concentrations of radionuclides in the environment but it may not be adequate for representing short-term fluctuations. This, however, may not be a requirement and hence the model is 'fit for the purpose'. Figure 13.6 above shows the main stages of model validation and verification.

13.6.1 Model Validation Studies

Validation is important for establishing the credibility of the model and for establishing the domain of applicability of the model, that is, the range of situations for which the model can make predictions that are sufficiently accurate for their purpose. Validation in its purest sense involves comparing predictions from the model with data sets of measurements that have not been used in developing the model. This is often difficult to achieve because all available data will often have been used in the process of model development. Nevertheless, data, which can be used in model validation exercises, may become available later. The Chernobyl accident is one example. During the course of and following the accident, a large body of data was collected of levels of particular radionuclides, mainly caesium-137, in the environmental materials taken from both close to Chernobyl and much further afield. There were also a large number of studies into the subsequent behaviour of Chernobyl-derived radionuclides in the environment. Data collected following the Chernobyl accident has been used in a number of model validation studies.

There have been three major international model validation studies using Chernobyl data, while a fourth was conducted in 2000. The Biospheric Model Validation Study (BIOMOVS) was organized by the Swedish Radiation Protection Institute (BIOMOVS, 1993). This was extended into BIOMOVS II which was funded by five organizations from Canada, Spain and Sweden. The IAEA set up a Validation of Model Prediction Project (VAMP) (IAEA, 1993). The work started within BIOMOVS and VAMP has been continued in a new model validation project. These projects have provided insights into the problems associated with validating models. Problems include ensuring that data sets of measurements refer to quantities that are, or can be, calculated by the models. Processing of measurement data may be required before this is used in validation studies; conversion of measurement data from a net weight basis to a dry weight basis is one example. There is also the problem of defining what constitutes an acceptable level of agreement between model predictions and experimental data, given that there is uncertainty surrounding the measurement data and also surrounding the model predictions. These are complex issues and a considerable amount of subjective judgement may be involved in their resolution. The issues involved in model validation are the subject of a recent review (Peterson and Kirchner, 1998).

13.6.2 Use of Experimental Data for Model Evaluation

Once a model has been chosen, it is important that the input data are selected carefully. Data can be site-specific, for example, and hence not easily transferable from one problem to another. It is also important to evaluate the performance of the model in conditions that are operating in the real world. In the latter, however, comprehensive data sets are expensive to collate and this can impose a limitation to the validation process. In 1987, the Atmospheric Model Evaluation Study (ATMES) was initiated by the European Commission in collaboration with the International Atomic Energy Agency (IAEA) and the World Meteorological Organization (WMO) (Klug *et al.*, 1992; van Dop *et al.*, 1998). A subsequent, experimental study was sponsored by the same organizations and called the European Tracer Experiment (ETEX) (van Dop and Nodop, 1998). This study involved the controlled releases of tracers, with monitoring over distances of 2000 km. Twenty four institutions participated in this project. Although these experimental data sets were obtained from field situations, this is still a controlled experiment and will only test the ability of part(s) of the models to describe some of the processes. For example, in a real accident there will be a mixture of pollutants, with additional processes such as deposition taking place (van Dop *et al.*, 1998). As an example of model evaluation, Figure 13.7 shows the comparison of ETEX observations and the UK NAME model predictions (Ryall and Maryon, 1998). The authors

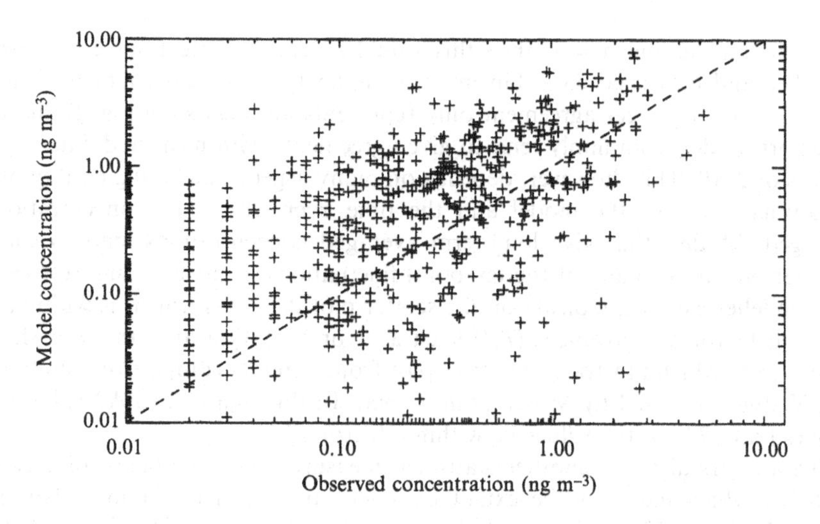

Figure 13.7 An example of the UK NAME model performance evaluation with measured data from the ETEX study. Reprinted from *Atmos. Environ.*, **32**, Ryall, D.B. and Maryon, R.H., 'Validation of the UK Meterological Office's NAME model against the ETEX dataset', 4265–4276, Copyright (1998), with permission from Elsevier Science

concluded that although the model reproduced the overall trends, it 'over-predicted' the observed concentrations.

13.6.3 Statistical Measure of Model Performance

Specific statistical measures can be calculated by comparing model results with measurement data. These measures allow the model performance to be assessed against statistical criteria, as well as to be compared with another model. Some of the most common statistical tests are briefly described below. More details can be found in USEPA (1992).

The fraction bias (*FB*) is usually treated as the minimum standard for testing the performance of a model. This parameter is calculated from the following relationship:

$$FB = 2\left(\frac{C_o - C_p}{C_o + C_p}\right) \tag{13.3}$$

where C_o is the observed value, and C_p represents the predicted value. Fraction bias of the average is normally computed for the highest 25 values. The values range between −2.0 (extreme 'over-prediction') and +2.0 (extreme 'under-prediction'). As it is a dimensionless parameter, it is useful when comparing performance of different models.

Scatter plots are commonly used to compare modelled (y-axis) and measured (x-axis) data. They are a useful visual guide to the model performance. If all of the points lie on the line, $y = x$, then this would represent perfect agreement between modelled and measured data. Given the complexity of the environment, it is obvious that such a perfect agreement only represents an ideal situation. Radionuclide transport models commonly tend to give agreements with measured data to within factors of 2–10. This, however, depends on many aspects, including quality of data, conceptual basis of the model and the time over which the concentrations are averaged. Models that give long-term averaged concentrations tend to show the best agreement, as many of the temporal fluctuations in the data are averaged out. The modeller can also indicate on the scatter plot if the agreement is within a given accepted factor of agreement (*FA*) such as 2 or 5. A typical example is shown in Figure 13.8, which illustrates a scatter plot from a model comparisons study for the ETEX study reported by Mosca *et al.* (1998). In this figure, an FA2 value of 37% implies that 37% of the values lie within a factor of 2.

The authors also use another statistical measure called the factor of exceedance (FOEX), which indicates the extent of 'over'- or 'under-prediction'. This ranges from −50% to +50%, where −50% means that all values are 'under-predicted' by the model (all points lie below the $y = x$ line) and +50% means that all values are 'over-predicted' (all points lie above the $y = x$ line). The best value for the *FOEX* is, therefore, 0%, meaning that the values are equally 'under'- and 'over-predicted'.

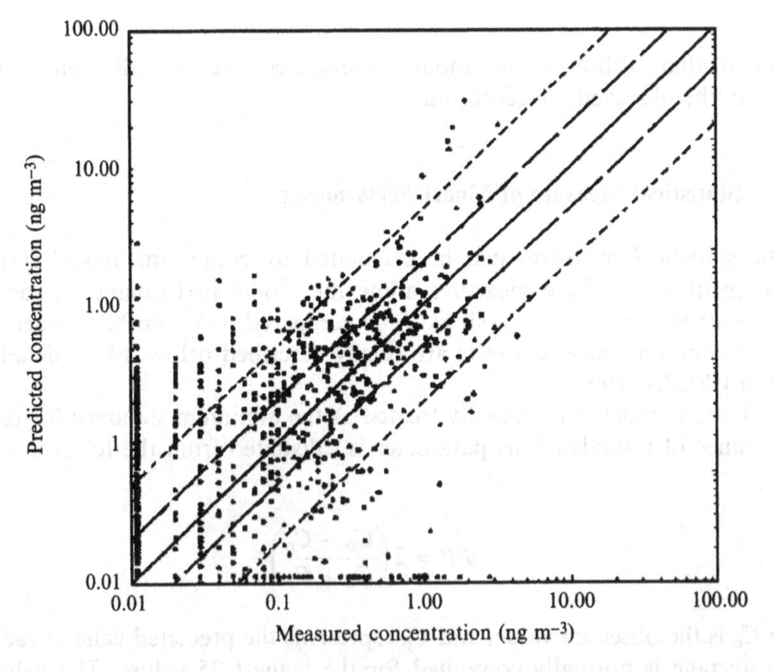

Figure 13.8 An example of a scatter plot for the comparison of model predictions with a dataset from the ETEX study: *FA*, 37 %; *FOEX*, 1 %. Reprinted from *Atmos. Environ.*, **32**, Mosca, S., Graziani, G., Klug, W., Bellasio, R. and Biaconi, R., 'A statistical methodology for the evaluation of long-range dispersion models: an application to the ETEX exercise', 4307–4324, copyright (1998), with permission from Elsevier Science

Other statistical measures commonly used for evaluating model performance include the *mean fraction bias, normalized mean square error* (*NMSE*) and *correlation coefficient*. These are given by the following:

$$\text{Mean fraction bias} = (\bar{C}_o - \bar{C}_p)/[0.5(\bar{C}_o + \bar{C}_p)] \tag{13.4}$$

$$NMSE = \overline{(C_o - C_p)^2}/\bar{C}_p\bar{C}_o \tag{13.5}$$

$$\text{Correlation coefficient} = \overline{(C_p - \bar{C}_p)(C_o - \bar{C}_o)}/\sigma_{cp}\sigma_{co} \tag{13.6}$$

where $C_{o,p}$ represents observed or predicted concentrations and $\sigma_{co,cp}$ are the standard deviations for these respective concentrations. The mean fraction bias is similar to the *FB*, except that the mean values are used to assess the model performance. The *NMSE* gives positive values which indicate the extent of the deviation between the predicted and the measured data. Unlike the *FOEX*, it does not provide information on the 'under'- or 'over-prediction' of the model values. As the numerator is squared (Equation (13.5)), this index is very sensitive to the differences between the modelled and measured values. A low *NMSE* value indicates that the model is performing well. The correlation coefficient (also called the *Pearson's Correlation Coefficient*) ranges from +1 to −1. The sign depend on the slope of the graph. A positive unity value indicates a complete or perfect correlation between the predicted and observed values and hence all the points on the scatter plot will lie on the $y = x$ line. A value of zero indicates that there is no correlation. A review of these and other statistical tests of model performance is given by Mosca *et al.* (1998).

Although such statistical approaches are useful for describing the performance of a model, they should not be treated as the only or the major aspect of model evaluation. It is important to include other evaluation aspects to provide a comprehensive treatment of the performance of the model. Hence, areas such as examination of the structure of the model and its code and sensitivity analysis also play an important role.

14

Modelling Dispersion of Radionuclides in the Atmosphere

Pollutants that enter the atmosphere become subject to various meteorological processes that determine their transport and distribution characteristics. In the event of a radioactive effluent release, say, from a nuclear power station, it would be essential to predict the trajectory of the plume and the downwind concentrations. In addition to meteorological processes, factors such as emission rates and half-lives of the pollutants will also be crucial in determining their concentration in the air and when deposited on the ground. Other factors, including terrain characteristics, also play a crucial role in influencing the transport of radionuclides through the atmosphere.

This chapter begins with an introduction to the most important meteorological processes and then provides an outline of the main theoretical approaches employed for describing the movement of radionuclides in the atmosphere, before discussing selected applications.

14.1 AIR POLLUTION METEOROLOGY

The primary process that is responsible for dispersion is *advection* (flow of wind) and hence wind speed and direction are critical parameters in predicting the transport of radioactive pollutants in the atmosphere. The pollutants also suffer dispersion through processes such as *diffusion* and *turbulence* which dilute the pollutant concentration. Factors such as solar radiation determine the vertical stability of the atmosphere and hence the dispersion characteristics of the pollutants. As will be discussed in Section 14.1.3 temperature *inversions* cause stable atmospheric conditions and create poor dispersion conditions, so leading to high concentrations of pollutants. Other important variables include mixing depth, precipitation and half-lives, which also need to be known in order to estimate the concentration of radionuclides in the air. Processes such as precipitation and radioactive decay are primarily responsible for the removal of the pollutants from the plume. With regard to particles, gravitational settling is also an important mechanism by which pollutants can be removed from the air. This is especially true for larger particle sizes.

14.1.1 Structure of the Atmosphere

The atmosphere can be conveniently divided into layers according to the variations in the temperature profile, as shown in Figure 14.1. The lowest layer, the *troposphere*, extends to a height of about 15 km and is of most importance when considering air pollution and its fate. The region below the *tropopause*, which is the upper boundary of the troposphere, contains nearly 90% of the total mass of the atmosphere and is the region in which most of the atmospheric circulation and weather phenomena occur.

Within the troposphere, lies the *atmospheric boundary layer* (ABL). This layer is directly influenced by the earth's surface and hence suffers frictional drag from the latter. The energy exchanges between the atmosphere and the surface is largely responsible for the heat and moisture profiles (Lyons and Scott, 1990). Pollutants, from power stations and from accidents, are directly emitted into this layer which then become subject to atmospheric processes that transport them away from the source to the receptor.

As a result of the energy exchanges between the atmosphere and the earth's surface, large convective eddies can be created by heating and evaporation in a region called the *convective boundary layer* (CBL). Turbulence within the CBL is generated by the strong insolation and, as a result, properties such as energy and momentum are exchanged vertically. The upper boundary of the CBL is often set by an inversion layer which then limits the vertical extent of the mixing.

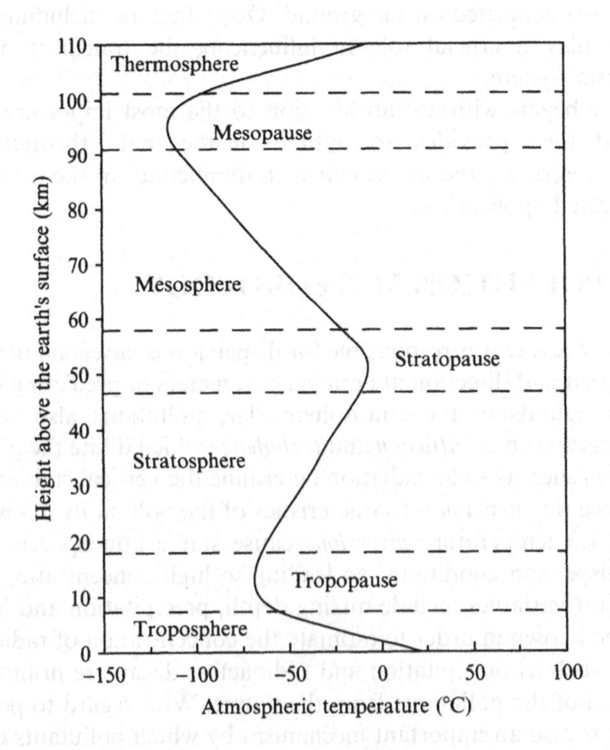

Figure 14.1 Vertical temperature profile of the atmosphere

The lowest 5–10% of the CBL is most affected by the topography and is called the *surface layer*, over which lies the mixed layer which extends until the base of the inversion. Within the surface layer, strong lapse rates are observed (see Section 14.1.3), as well as significant wind shear effects (Seibert *et al.*, 1998). The stability of the atmosphere within these lower regions has been described by the Pasquill–Gifford (P–G) approach and by the Monin–Obukhov similarity theory. Whereas the P–G approach provides discrete stability classes for the atmosphere, the Monin–Obukhov theory makes use of three fundamental flow parameters, i.e. the buoyancy parameter, the friction velocity and the surface temperature flux, to describe the surface layer processes. Although both of these are considered further in Section 14.2, greater details on this subject can be found in Monin and Yaglom (1971), Panofsky and Dutton (1984) and Lyons and Scott (1990), plus other similar texts.

14.1.2 Dispersion Processes

As mentioned earlier, advection is a major process by which air pollution is dispersed horizontally and vertically. Due to the roughness of the surface of the earth, frictional drag causes the transport of the pollution contained in the air parcels to be slowed down. Near the surface, the frictional drag causes the wind speed to decrease to zero, while above the surface layer shearing forces are responsible for slowing the parcels of air. As one might expect, the frictional drag decreases with height and above about 1 km the value reduces to effectively zero. Through these processes, the velocity of the pollutants becomes a function of the wind velocity and of the distance above the surface. In general, the greater the wind velocity, then the greater the dispersion of the contaminants, and the lower their concentration. In approximately the first 10 m of the surface layer, the wind speed increases dramatically, after which its rate of increase slows down.

Mixing of the air pollution is enhanced by *turbulence*. This is a random process that causes eddies to exchange and mix the air in the adjoining parcels of the atmosphere. The extent of turbulent mixing, therefore, depends on factors like the surface topography and buoyancy forces associated with the temperature of the surface. Unlike molecular diffusion, *atmospheric diffusion* involves the movement of large parcels of air from one point to another and is usually called *turbulent diffusion*. In practice, atmospheric diffusion tends to include all of the above processes, and will be influenced to a large extent by the number of sources and the complexity of the surrounding topography, as well as the prevailing meteorology.

14.1.3 Influence of Air Stability on Dispersion of Pollution

The *lapse rate* is a very useful parameter to describe the stability of the atmosphere and can be stated as follows:

$$\Gamma = -\frac{\mathrm{d}T}{\mathrm{d}z} \tag{14.1}$$

where Γ is the lapse rate, T is the air temperature and z is the altitude. The negative sign implies that temperature is decreasing with altitude.

It is useful to consider stability in terms of *adiabatic* processes under which the expansion of an air parcel takes place with no or little exchange of heat with the surrounding air. The air parcel, therefore, undergoes an adiabatic expansion with the temperature of the parcel decreasing with height. Under dry conditions, if the air is expanding adiabatically the temperature of the air parcel decreases at a constant rate called the *dry adiabatic lapse rate* (Γ_d). The value for Γ_d is equal to 9.8 K km^{-1} or approximately 1 °C per 100 m. Under more realistic situations, where air will contain moisture, a lapse rate of about 6.5 K km^{-1} is observed. This occurs because the latent heat released during condensation of the water vapour offsets the cooling experienced by the parcel during expansion. The reader should note that according to the convention followed here (Equation (14.1)), a positive lapse rate implies a decrease in air temperature with altitude. Conversely, a negative lapse rate implies an increase in air temperature with height. Figure 14.2 shows the various lapse rate situations that can arise. The parameter Γ_d is treated as an idealized reference value to compare with the environmental lapse rate (Γ_e). The stability of the atmosphere can be defined in relation to the dry adiabatic lapse rate. Table 14.1 summarizes the relationship between lapse rate and the associated stability of the atmosphere. Under thermal inversion conditions (temperature gradient is positive, but Γ_e is negative – see Equation (14.1)), vertical mixing is negligible and hence the dispersion of pollutants is also minimal. During such conditions, which are particularly prevalent during winter periods, the concentrations of air pollutants can be very high. The mechanisms of inversion formation are not dealt with here and the reader is referred to texts such as Henderson-Sellers (1984) and Colls (1997).

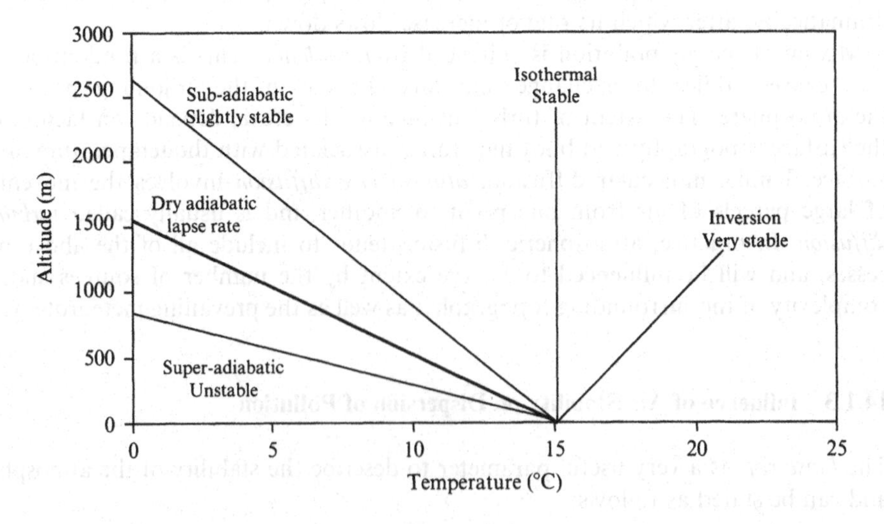

Figure 14.2 Various temperature profiles in the lower atmosphere and the associated stability conditions

Table 14.1 Relationship between lapse rate and atmospheric stability.

Lapse rate	State of stability	Remarks
$\Gamma_e = \Gamma_d$	Neutral or adiabatic	Common condition in the UK, light winds and cloudy skies
$\Gamma_e > \Gamma_d$	Strong lapse rate or super-adiabatic	Unstable conditions, good vertical mixing of pollution
$0 < \Gamma_e < \Gamma_d$	Weak lapse rate or sub-adiabatic	Stable conditions and vertical mixing is limited
$\Gamma_e = 0$	Isothermal	Stable conditions and vertical mixing is very limited
$\Gamma_e < 0$	Thermal inversion	Very stable conditions and generally there is no vertical mixing

14.1.4 Mixing Layer

The mixing layer refers to the region above the ground in which the emitted pollutants achieve complete mixing over a period of time. As mentioned earlier, the temperature profile of the atmosphere can change dramatically with altitude and hence the stability of the air changes with height. During the day, as insolation increases, the temperature of the ground rises, thus causing a warming of the adjacent air. This can lead to a surface inversion, or if one has already formed during the previous night, then it can change into an *inversion aloft*. The latter can behave as a barrier to pollutants below the layer, so limiting the range of vertical mixing.

For the purposes of atmospheric dispersion modelling, the concept of *mixing height* or *depth* can prove to be useful. Although there are a range of definitions of this quantity in the literature, the COST 710 Working Group on Mixing Height Determination (Seibert *et al.*, 1998) have suggested the following:

> *The mixing height is the height of the layer adjacent to the ground over which pollutants or any constituents emitted within this layer or entrained into it become vertically dispersed by convection or mechanical turbulence within a time scale of about an hour.*

Other sources will use a simpler explanation for the mixing height, that is, the distance from the ground to the base of the inversion aloft. During the day, the incoming solar radiation increases the surface temperatures and this leads to the degrading of the inversion layer and an increase in the mixing height. This process continues until midday with the layer eroding and hence providing a large region of mixing (Henderson-Sellers, 1984).

14.1.5 Implications for Plume Dispersion

The stability of the atmosphere, and hence the prevailing meteorological conditions, will affect the shape characteristics of plumes. Plumes from chimney stacks, therefore, will display several different shapes, including looping, coning, fanning, lofting, trapped and fumigations. For example, as shown in Figure 14.3, looping occurs during super-adiabatic (unstable) conditions where large vertical turbulences cause the plume to loop in the vertical direction. Fanning shape, on the other hand, results when the plume is enclosed in an inversion (surface or aloft). As a result of the inversion, there is very little mixing and the plume slowly fans out with distance from the stack. More details on these plume shapes can be found in Beychok (1979).

14.1.6 Wind Profile

The critical parameters when estimating the concentrations of air pollution, including radioactive pollutants, are wind speed and wind direction. Generally, the

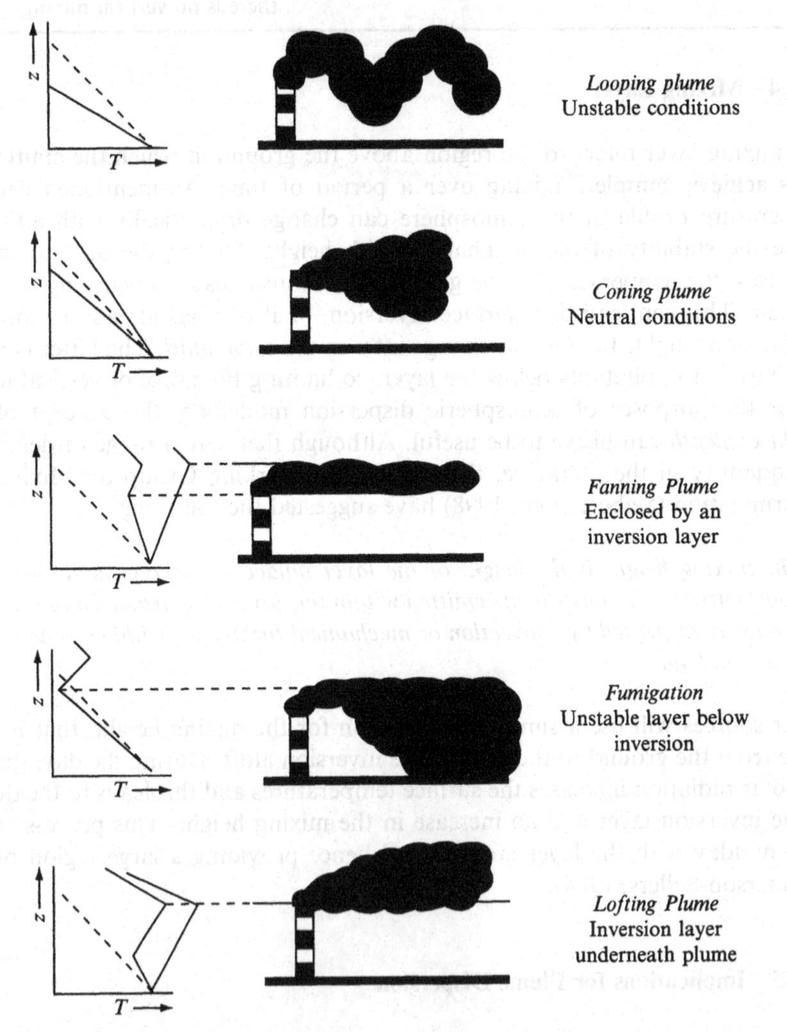

Figure 14.3 Relationship between stability of the lower atmosphere and the resulting plume shapes (adapted from Beychok, 1979). In the inset z vs T diagrams, the dashed lines represent Γ_d and the solid lines Γ_e

concentrations of gaseous pollution will decrease with increasing wind speed. The direction of the wind is obviously important in calculating the spatial distribution of the pollution from the emission source. Due to the effects of frictional drag and wind shear, wind speed will change dramatically with altitude. In general, depending on the roughness of the terrain, wind speed will increase significantly in the first 10 m above the ground and then the rate of increase will slow down with height. Near the ground surface, the wind direction may also be very different to that at higher altitudes where the influence of obstacles is negligible. Above about 1 km, the effects of frictional drag is very small.

The following expression relates wind speed to height above the ground through parameters that reflect the influence of the surface roughness:

$$U(z) = \frac{u^*}{\kappa} \ln\left(\frac{z}{z_0}\right) \tag{14.2}$$

where $U(z)$ is the wind velocity at height z, u^* is the friction velocity and κ is known as the von Karman constant and has a value of about 0.4; z_0 is the *surface roughness length* and is dependent on the topographic characteristics of the ground surface. This length is effectively the height below which the average wind speed is zero. In the above equation, z_0 essentially scales the height, or the height above the ground is expressed in units of z_0. Hence Equation (14.2) is only valid where $z > z_0$. The parameter u^* represents the effective velocity of the wind in the turbulence layer ($\sim$ 20–200 m of the atmosphere) where the shearing stress is approximately constant (Lyons and Scott, 1990). Some values of z_0 are shown in Table 14.2 for a range of land types.

14.1.7 Importance of Wind Direction

When assessing the impact of pollution arising from a nuclear power station, for example, the direction of the wind is of obvious importance. Along with wind speed, the wind direction will have a major influence in determining where the fall-out of a potential incident is likely to be and the magnitude of ground level concentrations. The wind direction frequency at a given location can be represented by an 8-, 16- or

Table 14.2 Surface roughness lengths (m) by land use type and season (Sheih *et al.*, 1979; USEPA, 1996a, 1999).

Land use	Winter	Spring	Summer	Autumn
Water surface	0.0001	0.0001	0.0001	0.0001
Deciduous forest	0.50	1.00	1.30	0.80
Coniferous forest	1.30	1.30	1.30	1.30
Swamp	0.05	0.20	0.20	0.20
Cultivated land	0.01	0.03	0.20	0.05
Grassland	0.001	0.05	0.10	0.01
Urban	1.00	1.00	1.00	1.00
Desert shrub land	0.15	0.30	0.30	0.30

32-sector wind rose diagram. An example of this is shown in Figure 14.4. Lines are used to schematically represent the direction of the wind and the frequency of occurrence in a particular direction. It is generally understood that the direction of the lines corresponds to the direction *from which* the wind is blowing. Sometimes, the wind speeds can be expressed in discrete classes and then represented on the wind rose according to the width or the shading of the lines. Calm situations (non-directional wind with zero speed) and very low wind speeds (for example, below 1 m s^{-1}) are normally distributed among the sectors.

14.2 COMMON ATMOSPHERIC DISPERSION MODELLING APPROACHES

The behaviour of atmospheric pollutants with respect to a *fixed* co-ordinate system is referred to as the *Eulerian* approach. The alternative, i.e., the *Lagrangian* approach, describes their behaviour with respect to *moving* fluids. Measurements made in the Eulerian frame of reference are conducted at a fixed point and thus would represent measurements of many parcels of air that pass the fixed point. In the Lagrangian frame of reference, the motion of a parcel of air is followed and measurements are made relative to that parcel. Most measurements made of pollutant concentrations or of meteorological parameters are Eulerian in nature. The lack of a single coherent theoretical model which describes all of the major aspects of atmospheric diffusion has lead to the development of numerous modelling strategies (Pasquill and Smith, 1983). The following section considers the numerical method of the *K Theory* and then through various assumptions to eventually arrive at a simpler and more operational alternative approach of the *Gaussian Plume Model*.

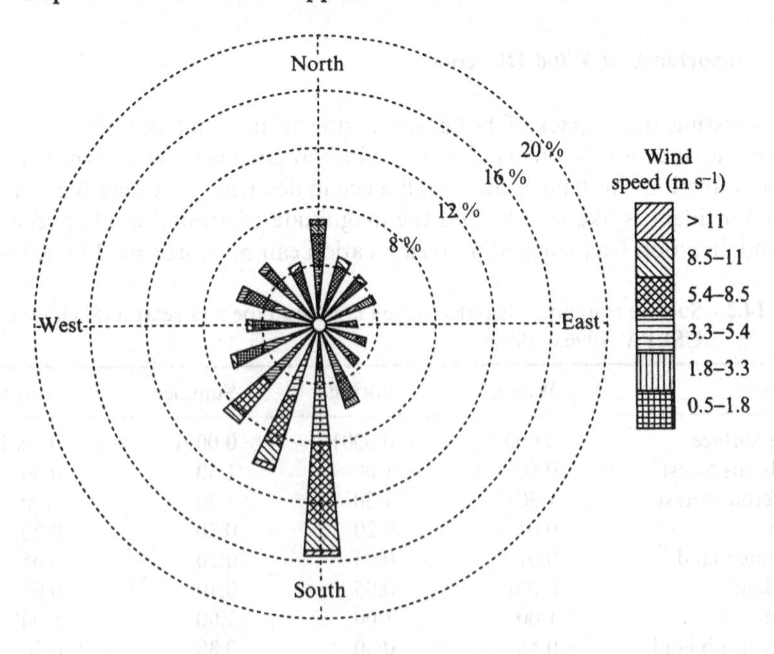

Figure 14.4 An example of a wind rose

14.2.1 K Theory – Three-Dimensional Model

The main equation shown below (Equation (14.3)) is called the *atmospheric diffusion equation* and forms the basis of the K Theory:

$$\frac{D\bar{\chi}}{Dt} = \frac{\partial}{\partial x}\left(K_x \frac{\partial \chi}{\partial x}\right) + \frac{\partial}{\partial y}\left(K_y \frac{\partial \chi}{\partial y}\right) + \frac{\partial}{\partial z}\left(K_z \frac{\partial \chi}{\partial z}\right) \quad (14.3)$$

where t represents time, $\bar{\chi}$ is the mean concentration of the pollutant and $K_{x,y,z}$ represent the eddy diffusivities in the x-, y- and z-directions, respectively. The operator $D\bar{\chi}/Dt$ is the Lagrangian time (or substantial) derivative and represents the following:

$$\frac{D\bar{\chi}}{Dt} = \frac{\partial \chi}{\partial t} + u\frac{\partial \chi}{\partial x} + v\frac{\partial \chi}{\partial y} + w\frac{\partial \chi}{\partial z} \quad (14.4)$$

The first partial time derivative represents the Eulerian derivative taken at a fixed point in space, while the other terms represent the Eulerian spatial derivatives taken at a fixed point in time (advection terms). The total derivative, $D\bar{\chi}/Dt$, denotes the total change observed when travelling with a parcel of air at a velocity characterized by the components u, v and w (Lagrangian). The partial time derivative, $\partial\chi/\partial t$, represents the change observed with time at a fixed spatial point and thus denotes the local rate of change (Eulerian). More details of the derivation of Equation (14.3) can be found in Lyons and Scott (1990) or Jacobson (1999).

In the case of anisotropic diffusion, Equation (14.3) accounts for the different diffusivities in the three spatial directions. If we assume that the eddy diffusivities are constant, then Equation (14.3) simplifies to:

$$\frac{D\bar{\chi}}{Dt} = K_x\frac{\partial^2 \chi}{\partial x^2} + K_y\frac{\partial^2 \chi}{\partial y^2} + K_z\frac{\partial^2 \chi}{\partial z^2} \quad (14.5)$$

One can further assume that the vertical diffusion is much greater than the vertical advection and that horizontal advection is the dominant process compared to horizontal diffusion for dispersing atmospheric pollutants. The eddy diffusivities can be expressed as functions of stability, surface roughness, space and time, thus making the model very versatile. In this case, however, the equations cannot be solved analytically and numerical solutions have to be sought.

The K Theory model gives a comprehensive description of the transport of air pollutants and takes into account realistic variations in the wind and diffusivity fields. A limitation of this model, however, is that the increase of K_z with time as the plume size increases, is not taken into account. This can lead to an overestimation of the dispersion near the source of the emission in turbulent conditions. The effect is considerably less when the atmosphere is stable as then the range of eddies is small.

A simple model based on the K Theory has been described by Lauritzen and Mikkelsen (1999). This model has been used for studying long-range transport of radionuclides resulting from the Chernobyl accident. While such a simple model will not be as accurate as more complex real-time models, especially those coupled to numerical weather prediction models, it does have the benefit in that it can provide reasonable values for ensemble means without recourse to extensive meteorological data. Figure 14.5 illustrates the comparison of the model results with measured data and shows reasonable agreement for ensemble means of surface air ^{137}Cs activity. A more complex modelling example is the operational emergency response model used for assessing the impact of nuclear accidents (Langner *et al.*, 1998). As an example of the latter, Figure 14.6 shows the measured and model predictions for ^{137}Cs surface concentrations. The figures demonstrate that good agreement can be obtained with this modelling approach and that it can be used to study the dispersion of radioactive clouds emitted as a result of accidents. An analogous level of agreement with measured data has also been reported by D'Amours (1998), using a similar modelling approach.

14.2.2 Fickian Diffusion Assumption

Equation (14.5) can be simplified considerably if the eddy diffusivities, $K_{x,y,z}$, are assumed to be constant and independent of spatial directions (*isotropic diffusion*). This then simplifies Equation (14.5) as follows:

$$\frac{D\bar{\chi}}{Dt} = K\nabla^2\chi \tag{14.6}$$

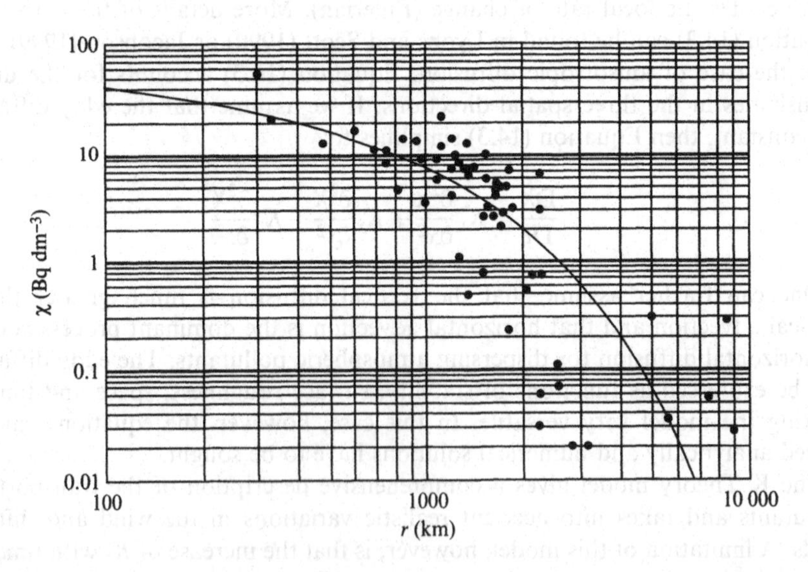

Figure 14.5 A comparison of time-integrated ^{137}Cs activity in surface air with Chernobyl fall-out data. Reprinted from *Atmos. Environ.*, **33**, Lauritzen, B. and Mikkelsen, T., A 'probabilistic dispersion model applied to the long-range transport of radionuclides from the Chernobyl accident', 3271–3279, Copyright (1999), with permission from Elsevier Science

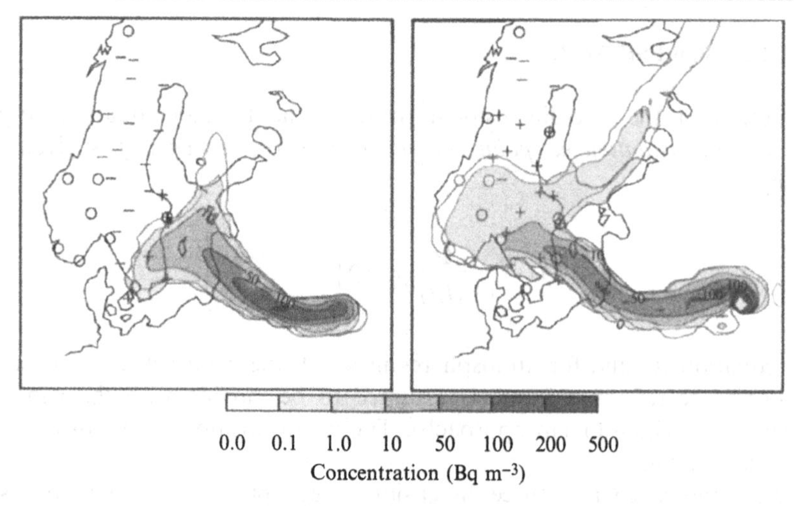

Figure 14.6 Observed and calculated surface ^{137}Cs concentrations (Bq m^{-3}) for two consecutive days in April 1986, where the same shading is used for both concentrations. The open circles indicate that no concentration was detected at that particular location and time. Gamma-radiation was also measured; plus signs indicate an increase in gamma-radiation since the previous day, while minus signs indicate a decrease or no change. Reprinted from *Atmos. Environ.*, **32**, Langner, J., Robertson, L., Perrson, C. and Ullerstig, A., 'Validation of the operational emergency model at the Swedish Meteorological and Hydrological Institute using data from ETEX and the Chernobyl accident', 4325–4333, Copyright (1998), with permission from Elsevier Science

where:

$$\nabla^2 = \frac{\partial^2}{\partial x^2} + \frac{\partial^2}{\partial y^2} + \frac{\partial^2}{\partial z^2}$$

and:

$$K = K_x = K_y = K_z$$

Here, ∇^2 is the Laplacian operator. As Equation (14.6) is analogous to Fick's law of molecular diffusion, the atmospheric diffusion described by this formulation is termed *Fickian diffusion*.

Equation (14.6) can be used to model atmospheric diffusion in one, two or three dimensions. For example, in *one dimension*, the Fickian diffusion equations takes the following form:

$$\frac{\partial \bar{\chi}}{\partial t} + \bar{u}\frac{\partial \bar{\chi}}{\partial x} = K\frac{\partial^2 \bar{\chi}}{\partial x^2} \tag{14.7}$$

where $\bar{v} = \bar{w} = 0$. The first term of this equation represents the rate of change of $\bar{\chi}$ at a selected fixed point in space, while the second term represents the advection of the concentration ($\bar{\chi}$) at velocity $\bar{u}$.

14.2.3 Gaussian Puff Model

The solution of the one-dimensional form of the Fickian equation (Equation (14.7)) for an *instantaneous point source* of emission strength Q is given by the following:

$$\bar{\chi}(x, t) = \frac{Q}{(4\pi K t)^{1/2}} \exp\left[-\frac{(x - \bar{u}t)^2}{4Kt}\right] \tag{14.8}$$

This equation is valid for atmospheres in which the wind velocity $\bar{u}$ is constant. The co-ordinate reference frame is assumed to be moving with the wind at the mean velocity $\bar{u}$ (Lagrangian approach). The mean position of the distribution at time t is given by $\bar{u}t$.

If we are interested in a three-dimensional description of *isotropic* atmospheric diffusion (Equation (14.6)), the solution then is given by the following:

$$\chi(r, t) = \frac{Q}{(4\pi K t)^{3/2}} \exp\left(-\frac{r^2}{4Kt}\right) \tag{14.9}$$

where:

$$r^2 = (x - \bar{u}t)^2 + y^2 + z^2$$

and:

$$K = K_x + K_y + K_z$$

For *non-isotropic* diffusion ($K_x \neq K_y \neq K_z$), the concentration profile of the pollutants is given by the following (solution of Equation (14.5)):

$$\chi(x, y, z, t) = \frac{Q}{(4\pi t)^{3/2}(K_x K_y K_z)^{1/2}} \exp\left[-\left(\frac{(x - \bar{u}t)^2}{4K_x t} + \frac{y^2}{4K_y t} + \frac{z^2}{4K_z t}\right)\right] \tag{14.10}$$

This equation is valid for an instantaneous point source and is employed when predicting the dispersion of puffs of pollutants. The above equation assumes that the distribution of the mean concentration of material being transported as a result of atmospheric diffusion is *Gaussian.* Equation (14.10) forms the basis of the so-called *Gaussian puff model.*

The following boundary conditions apply to equation (14.10):

(i) Concentration (χ) for all points tends to zero as time after release approaches infinity;

$$\chi \to 0 \text{ as } t \to \infty, \quad \text{for all co-ordinates } (-\infty < x, y, z < +\infty)$$

(ii) Concentration tends to zero as the time after release approaches zero for all points except at the source;

$$\mathrm{X} \rightarrow 0 \text{ as } t \rightarrow 0, \qquad \text{for all } x, y, z \text{ except where } x, y, z = 0$$

(iii) Total mass of pollutant present is equal to the amount released:

$$\int_{-\infty}^{\infty} \bar{\mathrm{X}} \mathrm{d}x \mathrm{d}y \mathrm{d}z = Q$$

where Q is the source strength.

An example of the skewed Puff modelling approach for investigating the dispersion of radionuclides in a sub-tropical area of Brazil is described in Oliviera *et al.* (1998). These authors report on a study using a hypothetical release of radionuclides. Another example of a model that is based on this approach is the DIFTRA model (Wendum, 1998). In this model the concentration of radionuclides in the atmosphere is predicted by superimposing three-dimensional Gaussian puffs.

If Equation (14.10) is integrated spatially, then it yields a solution for *instantaneous volume sources*, such as, explosion bursts and if it is integrated with respect to time it then yields a solution for *continuous point sources*. Equation (14.10) can also be rewritten in terms of variances (σ) if we define the following:

$$\begin{aligned} \sigma_x^2 &= 2K_x t \\ \sigma_y^2 &= 2K_y t \\ \sigma_z^2 &= 2K_z t \end{aligned} \qquad (14.11)$$

Hence, Equation (14.10), thus becomes:

$$\mathrm{X}(x, y, z, \; t) = \frac{Q}{(2\pi)^{3/2}\sigma_x\sigma_y\sigma_z} \exp\left[-\left(\frac{(x-\bar{u}t)^2}{2\sigma_x^2} + \frac{y^2}{2\sigma_y^2} + \frac{z^2}{2\sigma_z^2}\right)\right] \qquad (14.12)$$

When solving Equation (14.12), the logarithmic profile expression for the wind velocity (Equation (14.2)) makes the mathematical manipulation very complex. A simpler form of a power law for the wind profile, is therefore, employed, where the eddy diffusivities are proportional to the height z, as shown by the following:

$$\begin{aligned} K_z(z) &= K_i\left(\frac{z}{z_i}\right)^n \\ \bar{u}(z) &= \bar{u}_i\left(\frac{z}{z_i}\right)^m \end{aligned} \qquad (14.13)$$

where $\bar{u}_i$ and K_i are values for $\bar{u}$ and K_z at the reference height of z_i.

14.2.4 Gaussian Plume Model

In the case of continuous sources, the plume can be considered as an infinite number of puffs (instantaneous point sources) superimposed on each other. Mathematically, this is represented by integrating Equation (14.12) with respect to time from $t = 0$ to $t = \infty$. This integration, however, is complicated as σ_i are all functions of time and hence of x, as $t = x/\bar{u}$, and in order to simplify the integration the diffusion along the x-axis is considered to be negligible compared to the transport of the plume by advection (that is, $\sigma_x = 0$). Under the *slender plume approximation*, the spread of each 'puff' is assumed to be small compared to the downwind distance (x) it travels:

$$\frac{\sigma\left(\frac{x}{\bar{u}}\right)}{x} \gg 1 \tag{14.14}$$

In reality, most continuous sources are elevated, with the earth's surface acting as a physical barrier to the plume at ground level. We can, therefore, assume that the earth's surface at the $z = 0$ boundary is totally reflecting, that is, there is no absorption of the plume or deposition of the material. This 'reflection' of the plume can be accounted for by adding the concentration resulting from an imaginary source of an exact image of the real source. If the effective elevation of the source is H, then the imaginary source is assumed to be at a distance $-H$ below the ground surface. After modifying Equation (14.12) and performing the integration, we obtain the following:

$$\chi(x, y, z, t) = \frac{Q}{2\pi\sigma_y\sigma_z\bar{u}} \exp\left(-\frac{y^2}{2\sigma_y^2}\right)\left[\exp\left(-\frac{(z-H)^2}{2\sigma_z^2}\right) + \exp\left(-\frac{(z+H)^2}{2\sigma_z^2}\right)\right] \tag{14.15}$$

Here Q is the continuous source strength in units of mass per unit time, for example, kg s^{-1}; σ_i are the standard deviations in the y- and z-directions and are both functions of x. Note that if σ_x was not ignored then the π factor would have been replaced by $(2\pi)^{3/2}$, i.e. one $(2\pi)^{1/2}$ for each standard deviation. Equation (14.15) is the main expression for the *Gaussian plume model* (GPM) and predicts that the pollutant concentration is inversely proportional to the mean wind speed. Thus, the greater the wind speed, then the lower the concentration or greater the dilution of the plume. Arguably, the GPM has been the most widely used atmospheric dispersion model. The main reasons for this include the following:

(i) The model has a relatively simple formulation and hence is adaptable for computerized codes.

(ii) The model expression is versatile as it can be easily modified for more specific cases such as ground-level or centre-line concentrations.

(iii) Considerable effort has been devoted to validating and evaluating models of this approach (see below for some examples of relevant studies).

(iv) Compared to more complex numerical models, the GPM is less computationally intensive and hence provides estimates of concentrations quickly.

One of the most commonly used Gaussian plume models is the USEPA Industrial Source Complex Long-Term (ISCLT) model. An example of the use of this model for predicting concentrations of radioactive emissions from nuclear fuel plant is given by Al-Khayat *et al.* (1992). The application of the Gaussian plume model to radioactive releases in the atmosphere is also described in Simmonds *et al.* (1995). Various reports published by the National Radiological Protection Board (NRPB), UK, have discussed the use and limitations of this approach. For example, the interested reader is guided towards Clarke (1979) and Jones (1981) for descriptions of these. The earlier version of this model, used in the UK, and commonly known as the R91 model, is reported by Jones (1981) and has undergone numerous changes and developments. The PLUMES model was based on R91 and has been employed for modelling short- and long-term pollution. The DISTARS and DISTRAL models were also based on R91 and allowed the prediction of short- and long-term pollution but also accounted for deposition and plume rise effects. Further detailed treatment of the Gaussian models is given in several textbooks, such as Seinfeld (1986), Lyons and Scott (1990) and Zannetti (1990), as well as IAEA (1982a).

14.3 DESCRIPTION OF ATMOSPHERIC STABILITY

The atmospheric stability conditions ranging from super-adiabatic to inversions have been conveniently categorized into discrete classes by Pasquill (1961). This approach was further developed as reported by Gifford (1976) and hence the classes are normally referred to as Pasquill–Gifford (P–G) stability categories. Each class represents a specific atmospheric condition with class A being 'unstable' and class F being 'stable'. Class G was later added to describe 'strongly stable' conditions (Turner, 1994). Class D corresponds to neutral conditions (near-adiabatic stability) where the day or night is heavily overcast. Within the UK, this is the most common P–G class. Table 14.3 lists the P–G classes of atmospheric stability according to ranges of ambient temperature gradients. A more common way of listing the stability classes is according to wind speed and solar radiation (see Table 14.4). As one can clearly see, stable conditions (classes F and G) occur during night time and when the wind speed is low ($\leq 2\,m\,s^{-1}$). During day times, when the solar heating of the ground is at its maximum, we get unstable conditions as the warms air parcels near the ground level rise and then fall as they expand and cool.

Table 14.3 P–G stability classes according to ambient temperature gradient (°C per 100 m) (adapted from Colls, 1997).

Class	Occurrence in UK (%)	Description	Gradient
A	1	Very unstable	< -1.9
B	5	Unstable	-1.9 to -1.7
C	15	Slightly unstable	-1.7 to -1.5
D	65	Neutral	-1.5 to -0.5
E	6	Slightly stable	-0.5 to $+1.5$
F	6	Stable	$> +1.5$ to $+4.0$

Table 14.4 Meteorological conditions defining the P–G stability categories[a] (adapted from Pasquill, 1961, and Turner, 1994).

Wind speed ($m\ s^{-1}$) measured at 10 m above ground	Day time insolation			Night time conditions	
	Strong	Moderate	Slight	Thin overcast, or ≥ 4/8	Cloudiness, or ≤ 3/8
< 2	A	A–B	B	G	G
2–3	A–B	B	C	E	F
3–4	B	B–C	C	D	E
4–6	C	C–D	D	D	D
>6	C	D	D	D	D

[a]A, extremely unstable; B, moderately unstable; C, slightly unstable; D, neutral; E, slightly stable; F, moderately stable; G, strongly stable.

The dispersion coefficients are key input parameters and are dependent on the atmospheric stability and the distance from the release. The relationships between the dispersion coefficients in Equation (14.15) and the P–G stability classes are shown in Figure 14.7. This figure shows how σ_y and σ_z vary with downwind distance from the source of emission. These dispersion coefficients were derived from the experiment conducted on a flat terrain ($z_0 \approx 0.03$ m), where a non-buoyant tracer gas was released near the surface and measured every 3 min in the downwind direction up to 800 m from the source (Gifford, 1976; Zannetti, 1990). The dispersion coefficients, σ_y and σ_z, can be calculated from the following:

$$\sigma_y = \frac{k_1 x}{[1 + (x/k_2)]^{k_3}} \qquad \sigma_z = \frac{k_4 x}{[1 + (x/k_2)]^{k_5}} \tag{14.16}$$

where x is the distance (m) downwind of the source and the constants k_{1-5} are given in Table 14.5 (Zanetti, 1990; Bualert, 2001).

Briggs (1973) proposed a method to calculate the dispersion coefficients separately for urban and rural areas (for downwind distances from 100–10 000 m). Separate urban and rural equations are given for calculating σ_y and σ_z, depending on the P–G class (see Table 14.6). Modified coefficients are also reported in USEPA (1992, 1995) and these have been used in Versions 2 and 3 of the Industrial Source Complex model for rural areas. For this method, σ_y and σ_z are calculated from the following:

$$\sigma_z = ax^b \text{ and } \sigma_y = 465.116\,28(x)\tan(\theta) \tag{14.17}$$

where:

$$\theta = 0.017\,453\,293[c - d(\ln(x))]$$

where a, b, c and d are constants which depend on downwind distance (see Tables 14.7 and 14.8). One major advantage of such approaches is that these equations can be easily incorporated into mathematical dispersion models to take into

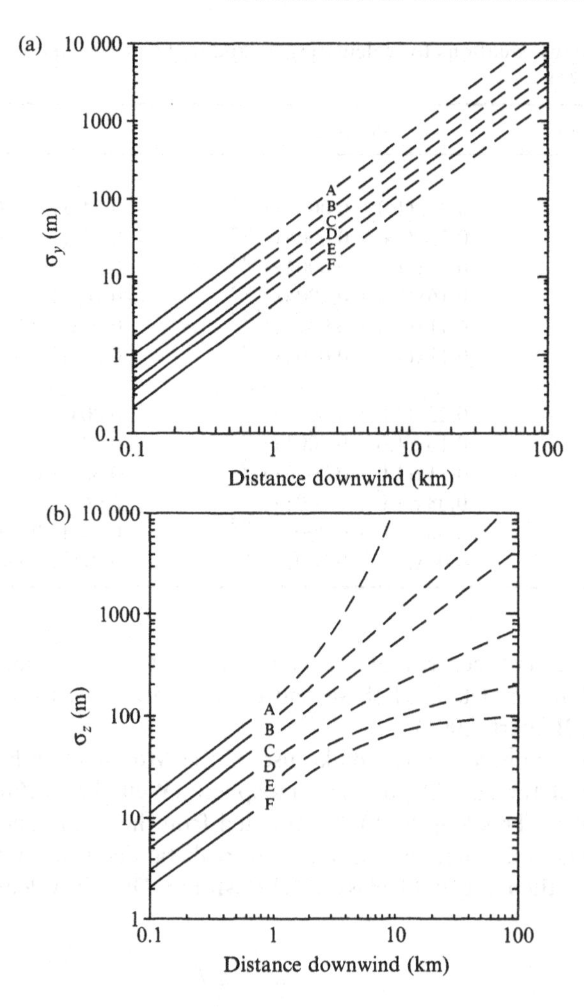

Figure 14.7 Variation of the (a) σ_y, and (b) σ_z dispersion coefficients with downwind distance for different P–G stability classes (adapted from Turner, 1994)

Table 14.5 Constants for calculating the dispersion coefficients σ_y and σ_z as a function of the P–G classes (Zannetti, 1990; Bualert, 2001).

Stability category	k_1	k_2	k_3	k_4	k_5
A	0.250	927	0.189	0.1020	−1.918
B	0.202	370	0.162	0.0962	−0.101
C	0.134	283	0.134	0.0722	0.102
D	0.0787	707	0.135	0.0475	0.465
E	0.0566	1070	0.137	0.0335	0.624
F	0.0370	1170	0.134	0.0220	0.700

Table 14.6 Briggs equations for calculating σ_y and σ_z for urban and rural areas (Briggs, 1973).

Stability category	σ_y (m)	σ_z (m)
Urban		
A	$0.32x(1.0+0.0004x)^{-0.5}$	$0.24x(1.0+0.001x)^{0.5}$
B	$0.32x(1.0+0.0004x)^{-0.5}$	$0.24x(1.0+0.001x)^{0.5}$
C	$0.22x(1.0+0.0004x)^{-0.5}$	$0.20x$
D	$0.16x(1.0+0.0004x)^{-0.5}$	$0.14x(1.0+0.0003x)^{-0.5}$
E	$0.11x(1.0+0.0004x)^{-0.5}$	$0.08x(1.0+0.00015x)^{-0.5}$
F	$0.11x(1.0+0.0004x)^{-0.5}$	$0.08x(1.0+0.00015x)^{-0.5}$
Rural		
A	$0.22x(1.0+0.0001x)^{-0.5}$	$0.20x$
B	$0.16x(1.0+0.0001x)^{-0.5}$	$0.12x$
C	$0.11x(1.0+0.0001x)^{-0.5}$	$0.08x(1.0+0.0002x)^{-0.5}$
D	$0.08x(1.0+0.0001x)^{-0.5}$	$0.06x(1.0+0.0015x)^{-0.5}$
E	$0.06x(1.0+0.0001x)^{-0.5}$	$0.03x(1.0+0.0003x)^{-1}$
F	$0.04x(1.0+0.0001x)^{-0.5}$	$0.016x(1.0+0.0003x)^{-1}$

account the different degrees of dispersion present in rural and more complex urban areas. Further details of these and other approaches for estimating σ_y and σ_z are given in Bualert (2001).

More recent advanced models make use of the Monin–Obukhov length to estimate the state of the stability of the atmosphere (Weil, 1985; Zannetti, 1990; Oliveira *et al.*, 1998). This length (L) is estimated from heat flux and other boundary layer parameters and hence provides a more realistic description of the stability of the atmosphere then the P–G classes. The Monin–Obukhov length is defined as follows:

$$L = -\frac{u^{*3} C_p \rho T}{kgH} \tag{14.18}$$

where L is the Monin–Obukhov length (m), u^* the friction velocity (m s^{-1}), C_p the specific heat capacity of air at constant pressure (1004 J kg^{-1} K^{-1}), ρ the air density (kg m^{-3}), T the ambient temperature (K), k the Von Karman constant (0.4), g the acceleration due to gravity (9.81 m s^{-2}) and H the heat flux (W m^{-2}). The values of L can be used to estimate the stability of the atmosphere, as shown in Table 14.9.

This approach has been recommended in the literature for use in atmospheric dispersion models (e.g. Weil, 1985) and is routinely used in new-generation environmental models such as AERMOD (Cimorelli *et al.* 1998), as well as other systems reported by, for example, Bualert (2001), McHugh *et al.* (1997) and Sokhi *et al.* (2000). The Monin–Obhukov lengths (L) and P–G classes have been related by Golder (1972) and a more complex scheme for using L to determine σ_y and σ_z can be found in Draxler (1976). The puff model reported by Oliveira *et al.* (1998) for simulating radionuclide dispersion, for example, uses this approach to determine atmospheric stability.

Table 14.7 Constants used to calculate σ_z (USEPA, 1992, 1995).

Stability category	x (km)	a	b
A[a]	< 0.10	122.800	0.944 70
	0.10–0.15	158.080	1.054 20
	0.16–0.20	170.220	1.093 20
	0.21–0.25	179.520	1.126 20
	0.26–0.30	217.410	1.264 40
	0.31–0.40	258.890	1.409 40
	0.41–0.50	346.750	1.728 30
	0.51–3.11	453.850	2.116 60
	> 3.11	—[b]	—[b]
B[a]	< 0.20	90.673	0.931 98
	0.21–0.40	98.483	0.983 32
	> 0.40	109.300	1.097 10
C[a]	All	61.141	0.914 65
D	< 0.30	34.459	0.869 74
	0.31–1.00	32.093	0.810 66
	1.01–3.00	32.093	0.644 03
	3.01–10.00	33.504	0.604 86
	10.01–30.00	36.650	0.565 89
	> 30.00	44.053	0.511 79
E	< 0.10	24.260	0.836 60
	0.10–0.30	23.331	0.819 56
	0.31–1.00	21.628	0.756 60
	1.01–2.00	21.628	0.630 77
	2.01–4.00	22.534	0.571 54
	4.01–10.00	24.703	0.505 27
	10.01–20.00	26.970	0.467 13
	20.01–40.00	35.420	0.376 15
	> 40.00	47.618	0.295 92
F	< 0.20	15.209	0.815 58
	0.21–0.70	14.457	0.784 07
	0.71–1.00	13.953	0.684 65
	1.01–2.00	13.953	0.632 27
	2.01–3.00	14.823	0.545 03
	3.01–7.00	16.187	0.464 90
	7.01–15.00	17.836	0.415 07
	15.01–30.00	22.651	0.326 81
	30.01–60.00	27.074	0.274 36
	> 60.00	34.219	0.217 16

[a]If the calculated value of $\sigma_z > 5000$ m, then $\sigma_z = 5000$ m.
[b]σ_z is equal to 5000 m.

Table 14.8 Constants used to calculate σ_y (USEPA, 1992, 1995).

Stability category	c	d
A	24.1670	2.5334
B	18.3330	1.8096
C	12.5000	1.0857
D	8.3330	0.723 82
E	6.2500	0.542 87
F	4.1667	0.361 91

Table 14.9 Atmospheric stability in terms of the Monin–Obukhov length (Seinfeld, 1986).

L (m)	Stability condition
$-100 < L < 0$	Very unstable
$-10^5 \leq L \leq -100$	Unstable
$\|L\| > 10^5$	Neutral
$10 \leq L \leq 10^5$	Stable
$0 < L < 10$	Very stable

14.4 MODIFICATION OF THE GAUSSIAN PLUME MODEL

The basic Gaussian plume model, as presented by Equation (14.15), is only applicable to continuous emissions and to pollutants that are inert. As a consequence, several modifications have to be included for 'real' applications of the model. This section considers some of these modifications briefly, while citing various references that provide a more detailed treatment for the interested reader.

14.4.1 Radioactive Decay

A plume of radioactivity will be depleted by various processes, including decay processes. As the plume disperses downwind, the concentrations will fall exponentially according to the following equation:

$$C = \exp\left(-\lambda \frac{x}{u}\right) \tag{14.19}$$

where C is the correction factor, λ the radioactive decay constant for the radionuclide, x the receptor distance from the release point and u the wind speed at the height of release. The correction is applied by multiplying C to the emission rate (Q) or to χ directly in Equation (14.15). As the parent nuclide decays, the daughter products will increase in activity and thus this effect needs to be taken into account (see Section 2.6).

14.4.2 Plume Rise

Stack plumes are normally released with a high exit velocity or are buoyant by virtue of having a lower density than the surrounding air. Buoyancy can be caused either by the plume being at a higher temperature than its surroundings or if its natural density is lower than air. This can cause the plume to 'rise' well above the stack exit and, consequently, the effective release height can be much higher than the physical height of the chimney. For a given emission rate, receptor co-ordinates and wind speed, the concentration of a radioactive pollutant will decrease with release height (see Equation (14.15)) as the plume will travel further before it reaches the ground and hence has more time for it to be diluted. It is common to use this plume rise phenomenon to ensure that the ground level concentrations are as low as possible, especially if there is a risk of environmental or health hazards. When predicting the concentrations of pollutants in the atmosphere, therefore, it is vital to estimate the effective height of release. For example, a common method for correcting for this effect is the Briggs Plume Rise Model, described in Lyons and Scott (1990) and Simmonds *et al.* (1995).

14.4.3 Complex Terrain and Effects of Buildings

Urban features such as buildings can change the direction of the wind significantly. For example, if the wind intensity is high then the velocity increases around the edges of tall buildings. Stagnation points may also be produced in the wake where there is no wind flow depending on the geometry of the structure. Eddies induced by the wind shear may cause the wind to flow in the reverse direction. This *down-draught* phenomenon can then trap pollutants in these low-pressure cavities, so producing high localized concentrations. During *down-wash*, the pollutant plume is sucked into the lee of the chimney stack of a building if the pressure is low enough in this region. As the steady-state Gaussian plume model (Equation (14.15)) is only applicable to dispersion over flat terrains, it, therefore, has to be modified for use in urban regions or areas that have complex terrains.

When a plume approaches a hillside, the plume centre line will be closer to the ground surface and will result in high ground-level concentrations. High ground deposition results, particularly when atmospheric conditions are stable. Furthermore, complex three-dimensional airflow patterns within the atmosphere can result is a highly complex mathematical problem (Dupuis and Lipfert, 1986). Procedures for accounting for buildings and complex terrains have been examined by Dupuis and Lipfert (1986) and Venkatram *et al.* (2001). An in-depth treatment of the status of complex terrain models used by the US EPA is provided by Schiermeier (1984), while a more recent study has been reported by ADMLC (2000).

14.4.4 Removal by Dry and Wet Deposition

Radioactive particles in the plume can be removed either by gravimetrical settling or when the plume impacts with the surface of the earth. This is known as *dry deposition or fall-out*. The dry deposition rate (D) can be estimated from the following relationship:

$$D = v_g X \tag{14.20}$$

where v_g is the deposition velocity of the particles and X is the air concentration. The deposition velocity is a function of the physical characteristics of the particle and the meteorological conditions, as well as the surface type (Simmonds *et al.*, 1995; ADMLC, 2000).

Precipitation (such as rain and snow) will also remove particles from the air through 'scavenging'. This process can be modelled by a simple exponential law as follows:

$$X_t = X_o \exp(-\Lambda t) \tag{14.21}$$

where X_o is the original concentration, X_t is the concentration after time t and Λ is the wash-out coefficient which is a function of rainfall and particle size. For further details, the interested reader is referred to Jones (1986) and ADMLC (2000) which provide a thorough treatment of this subject.

14.4.5 Resuspension of Particles

Radioactive particles that settle on the ground can be resuspended, especially if the conditions are dry and windy. A simple approach to take this into account is reported by IAEA (1994d). The concentration of the radionuclide in air ($X_a(t)$, Bq m^{-3}) is related to the inventory of the radionuclide on the soil surface ($C_s(t)$, Bq m^{-2}) through a time-dependent resuspension factor ($K(t)$, m^{-1}), as follows:

$$X_a(t) = K(t)C_s \tag{14.22}$$

where t is the time. The resuspension factor can be calculated from the following relationship:

$$K(t) = A \exp(-Bt) \tag{14.23}$$

where A (m^{-1}) and B (d^{-1}) are empirical constants. Further details on such approaches can be found in Nair *et al.* (1997).

14.4.6 Transport over Coastal regions

Most nuclear reactors are situated near coastlines and as a consequence the emissions of radioactive gases will inevitably, for some of the time at least, travel over water. Coastal regions exhibit very different surface characteristics and complex temporal and spatial variations in their meteorological conditions and diffusion processes. In such cases, the general Gaussian plume model cannot be used and modifications are required to account for these variations and their effects on the plume dispersion. For example, complexities can arise as a result of the land–sea breezes which change depending on many parameters, including air temperature. In these circumstances, one encounters internal boundary-layer turbulence effects which affect the vertical diffusions rates of the radioactive pollutants. As a result of sea breezes, it is feasible that the plume can be 'trapped' by the flow and be transported inland for several kilometres before travelling back towards the coast at higher altitudes as its temperatures rises. This is likely to happen during daytimes, while the reverse can happen during night-times.

Many reactors are situated in river valleys which introduce their own special effects (Van den Hoven, 1982). In such cases, the air flow can channel along the valley carrying the plume, with the diurnal heating cycle causing a reversal in the flow patterns. The 'micro-climate', as a result, can be very different to that outside the geographical domain of the valley. These effects restrict the dispersion of the plume and have to be taken into account in a predictive model.

14.5 LAGRANGIAN TRAJECTORY PARTICLE MODELS

The use of trajectories of particles and random-walk approaches is also common for describing the transport of radioactivity in the atmosphere. An example of a Lagrangian approach is the model of Izraehl *et al.* (1990) which has been used to investigate the intermediate and regional scale fall-out of the Chernobyl nuclear accident. This model has been used to calculate ground level concentrations of ^{137}Cs and ^{131}I at a range of European sites. Figure 14.8 shows the concentration contours for ^{137}Cs estimated with the model. Another example of a Lagrangian model is that reported by Bonelli *et al.* (1992). These authors have used the STRALE model to simulate dispersion of ^{137}Cs in three dimensions and compared the results with actual measurements. The scatter plot comparing the measured and modelled data is shown in Figure 14.9. A correlation coefficient of 0.63 is cited, which implies a good agreement for a model of this type. Desiato (1992) has reported the use of the Lagrangian particle model APOLLO for estimating ^{137}Cs resulting from the Chernobyl accident. When compared to the actual measurements, the model performs well and exhibits agreement with field data similar to that of the STRALE model discussed above. The UK NAME model (Nuclear Accident Dispersion Model) is a further example of a Lagrangian model (Maryon and Best, 1995). This model simulates the movement of a large number of passive air parcels ('particles') which contain a proportion of the released radioactivity. At each time step, the parcel or particle can move ('walk') randomly. Such models allow the trajectories of the plume to be mapped as shown in Figure 14.10.

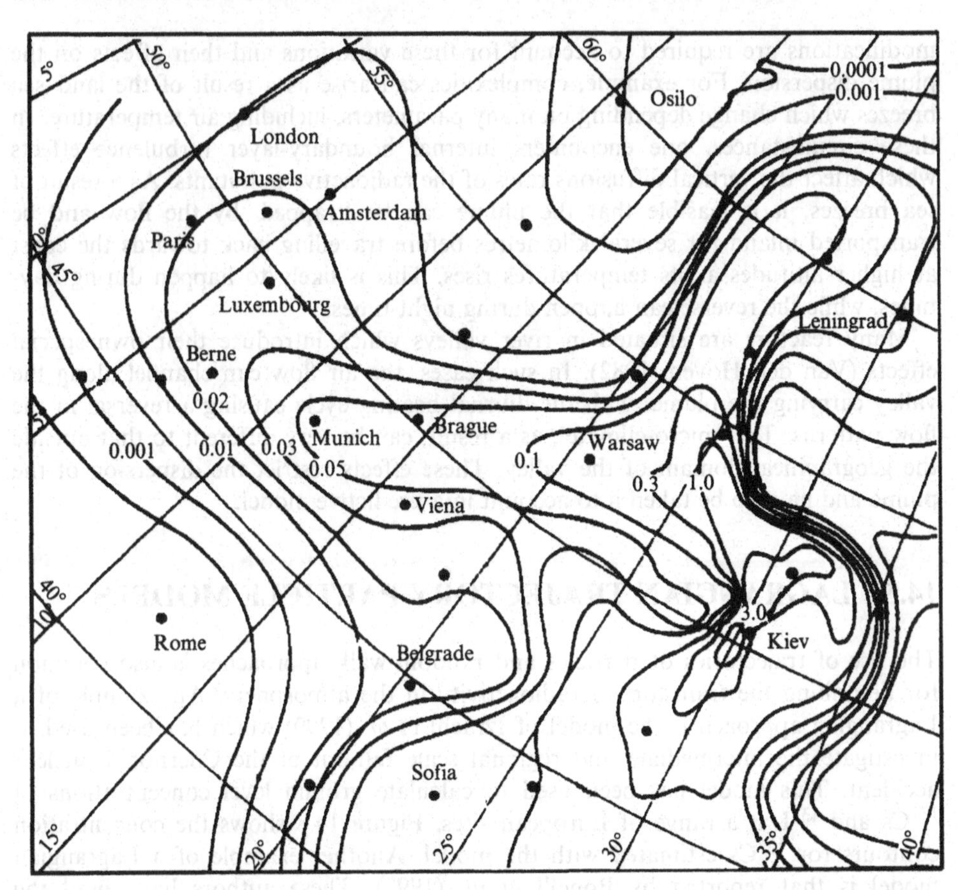

Figure 14.8 Contour map of cumulative ^{137}Cs fall-out density (Ci km^{-2}) calculated from a Lagrangian model (1 Ci = 3.7×10^{10} Bq). Reprinted from Izraehl, Y. A., Petrov, V. N. and Severov, D. A., 'Modelling of the transport and fallout of radionuclides from the accident at the Chernobyl nuclear power plant', in *Environmental Contamination following a Major Nuclear Accident*, Vol. 1, Symposium Proceedings, IAEA, 1990, 85–98. Copyright in these materials is vested in the International Atomic Energy Agency, Vienna, Austria from which permission for the republication must be obtained

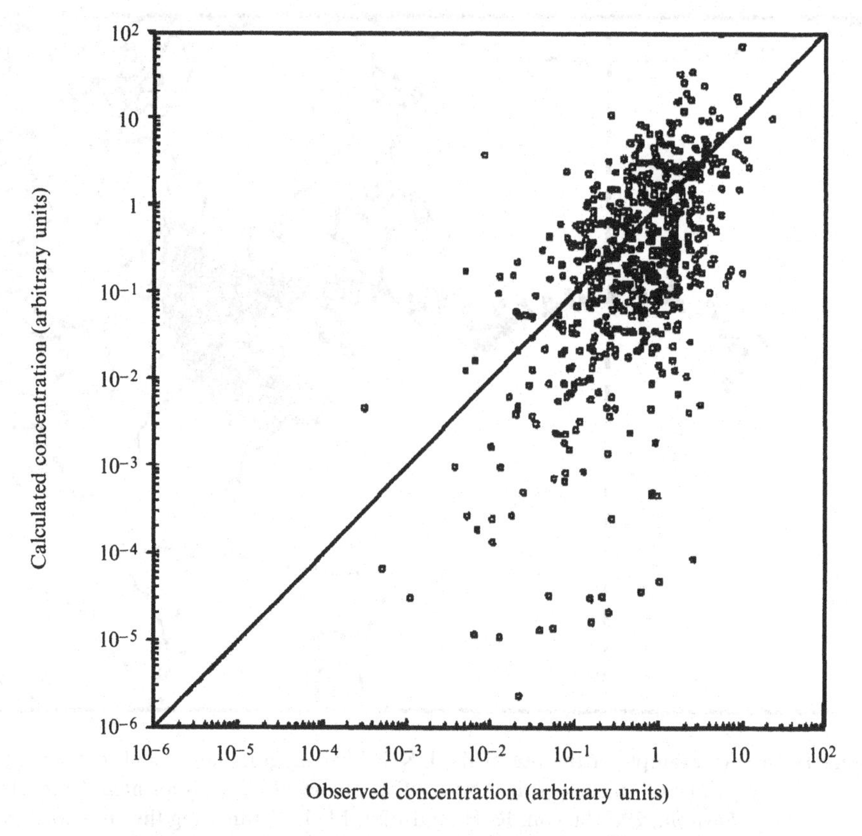

Figure 14.9 A scatter plot of modelled and observed data for ^{137}Cs 12 h air concentrations. Reprinted from *Atmos. Environ.*, **26**, Bonelli, P., Calori, G. and Finzi, G., 'A fast long-range transport model for operational use in episode simulation. Application to the Chernobyl accident', 2523–2535, Copyright (1992), with permission from Elsevier Science

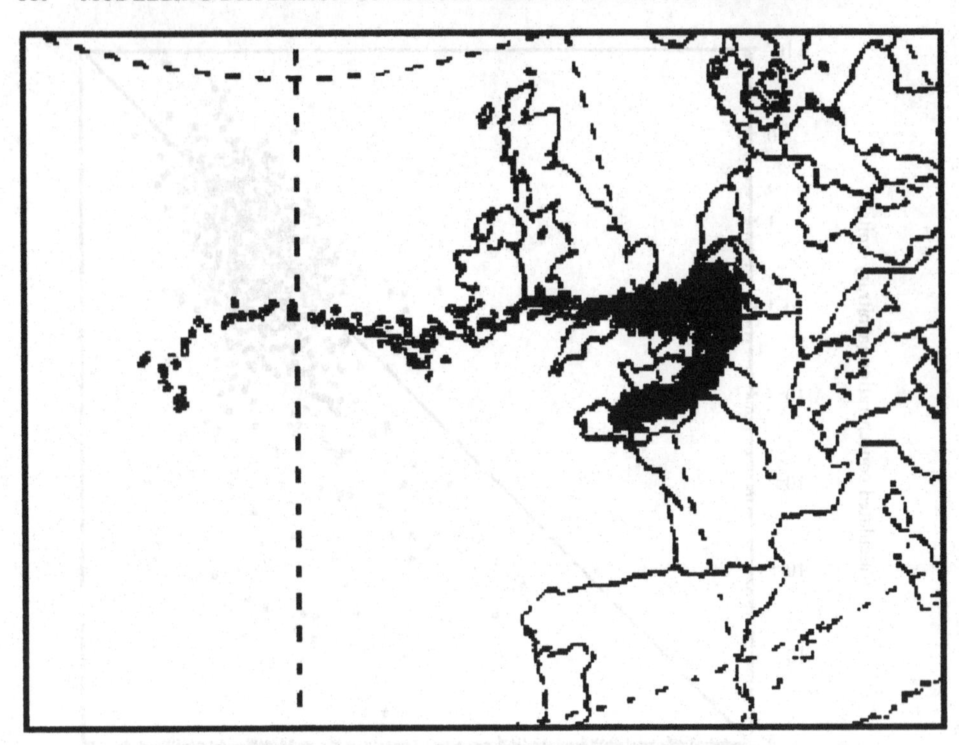

Figure 14.10 An example of the use of the UK NAME Lagrangian model to calculate end points of particle trajectories of a plume at 72 h. Reprinted from *Atmos. Environ.*, **29**, Maryon, R. H. and Best, M. J., 'Estimating the emissions from a nuclear accident using observations of radioactivity with dispersion model products', 1853–1869, Copyright (1995), with permission from Elsevier Science

15

Modelling Dispersion of Radionuclides in Aquatic and Terrestrial Environments

In many respects, the processes that affect radionuclide transport in the aquatic environment are more difficult to characterize and describe mathematically than the processes that affect atmospheric dispersion. Numerous factors come into play in addition to fluid dynamics and these include physical characteristics of the aquatic system, meteorology, biological processes and chemical processes. Mixing of the pollutants in the aquatic environment is generally slower than that in the atmosphere and hence residence times can be long. In addition, pollutants discharged into a river can be transported into the estuarine and coastal environments and eventually into the open seas with each system having its own special characteristics.

In the case of the terrestrial system, the characteristics of soil play a crucial role as this will determine whether the radionuclides are re-suspended or transferred into plants and animals. As in the case of the atmosphere (see Chapter 14), models that can be used to predict the behaviour of radionuclides in the aquatic and terrestrial environments can be complex or simple, depending on the processes included and the approach followed. This chapter provides an overview of the modelling approaches and their applications to predict concentrations of radionuclides in aquatic and terrestrial environments. The transfer of radioactivity from the physical environment to the biological environment will be addressed in Chapter 16, in relation to exposure pathways and dose estimation. This present chapter focuses on the dispersion of the radionuclides.

15.1 TRANSPORT OF RADIOACTIVITY IN AQUATIC SYSTEMS

In this chapter, the transport of radioactivity in rivers, lakes, estuaries and coastal areas is considered. Radioactive effluent, if discharged into rivers, will eventually flow into lakes and estuaries and finally into the coastal regions. The boundary between rivers and estuaries can be defined in terms of salinity or tidal influences. Salinity will increase as the river merges into the estuary. In terms of tidal influences, the river can be thought to start at the tidal limit. The boundary between estuaries and the coast is normally defined by geographical characteristics, with tidal influences being strong in coastal regions. Within lakes, wind influences are also important in dispersing and mixing the radioactivity. In addition, stratification can affect the movement of the pollutants. It is also important to recognize that as radioactive material enters the aquatic environment it will disperse in the aqueous phase and may also be transferred onto the physical and biological material such as sediment, vegetation and fish.

The three main processes that need to be considered when modelling radioactivity in water systems are as follows:

- physical mixing and dispersion
- interaction with sediments
- transfer to and from the biota

Although radioactivity can be transferred into the biota, and hence is important in terms of exposure in relation to the total inventory, most of the radioactivity resides in the water and sediment phases (IAEA, 1982a). The major processes that affect radionuclides in the water column are shown schematically in Figure 15.1.

15.1.1 Movement of Radioactivity in Riverine Systems

A simple empirical model that uses mixing ratios, transit times and decay of the radionuclide for estimating concentrations in the water phase is given by the following:

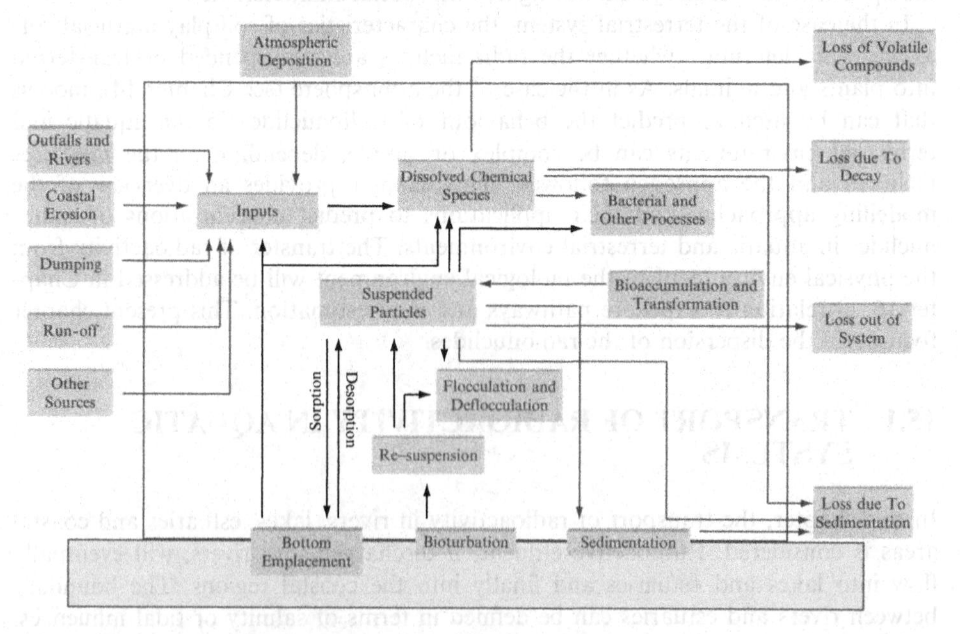

Figure 15.1 A schematic representation of the major processes that affect radionuclides in the water column (adapted from Eisenbud, 1987)

$$C_{w,i} = P\frac{Q_i}{F}M_p \exp\left(-\lambda_i t_p\right) \tag{15.1}$$

where $C_{w,i}$ is the concentration of radionuclide i in water (Bq l^{-1}), P is the conversion factor to relate units used for release rate river flow to units of concentration in water, Q_i is the input rate of radionuclide i into the river (Bq per year), F is the effluent flow (m^3 per year), M_p is the mixing ratio, a dimensionless reciprocal of the dilution factor, λ_i is the radioactive decay constant of radionuclide i (s^{-1}), and t_p is the average transit time for the radionuclides to reach the location of interest (s).

This steady-state model is suitable for continuous releases and cannot be applied to estimate concentrations resulting from isolated, short-term emissions. The input is taken as the annual total from a continuous release. The dimensionless mixing ratio, M_p, has a value of F/R, where R is the annual average river flow. For undiluted effluent, the value of M_p is unity. In practice, the value for M_p depends on the hydrology and the location of the points of interest. As the flows and releases are averaged over a year, this modelling approach will inevitably underestimate peak concentrations, for example, during periods of low flows. Detailed discussion of models of this nature can be found in USNRC (1977) and IAEA (1982b)

A simple model for estimating radioactivity in riverine situations has also been reported by Vorobiova and Degteva (1999). These authors have considered long-term releases and long-term averaged activity in river water. The main processes that have been considered in their stationary model are advection, sorption and decay. An analytical solution to the model equation is given by the following:

$$C(R) = C_0\frac{q_0}{q(R)}\exp\left(-kR - \frac{\lambda}{V}R\right) \tag{15.2}$$

where $C(R)$ and C_0 are the radionuclide concentrations at distances R and $R = 0$, respectively, $q\ (R)$ and q_0 are the volume flow rates of the river at distances R and $R = 0$, respectively, $k = \lambda_s V^{-1}$ (in which λ_s is the fractional rate of loss of activity by sorption onto sediments), V is the river velocity, and λ is the physical decay constant. An analytical solution to calculate the content of radioactivity, $S(R,t)$, in the bottom sediments per unit of river bed length at a distance R and time t is given by the following:

$$S(R,t) = \frac{A_0 k}{\lambda}\exp\left(-kR - \frac{\lambda}{V}R\right)\left[1 - \exp\left(-\lambda t\right)\right] \tag{15.3}$$

where A_0 is the activity at $R = 0$.

Vorobiova and Degteva (1999) have employed this model to study the dispersion of radioactivity in the Techa River (Southern Urals, Russia) resulting from contamination from a plutonium production plant. Concentrations of radionuclides such as Sr-90 and Ru-106 were calculated and were used to describe the extent of contamination in the region.

More complex models based on descriptions of physical processes, such as SHETRAN (Ewen, 1995), are available and provide detailed contaminant spatial information over river catchments. Due to their complex nature, numerical techniques are required to arrive at the concentrations and hence tend to be computationally intensive. A much simpler approach of using 'lumped' catchment scale models have been used over longer timescales, for example, by Cosby *et al.* (1985) and Whitehead *et al.* (1998). These models are simple but only give coarse scale information on the contamination. Sloan and Ewen (1999) have proposed an intermediate approach, which employs the concept of the catchment being described as a number of water storage compartments. This modelling approach provides finer spatial resolution information without the price of high computation times and has been developed to simulate long-term movement of radionuclides in the near-surface of a river catchment.

Figure 15.2 shows the model predictions compared to experimental data for a study of the Jucar River in Spain (Rodriquez-Alvarez and Sanchez, 2000). The study involved a multi-compartmental model which considered the water column as well as suspended and bed sediments. Results of the modelling runs of ^{238}U for two separate days are shown in Figure 15.2, illustrating overall good agreement with the measured data.

15.1.2 Movement of Radioactivity in Lakes

When describing movement of radionuclides in lakes, special factors have to be taken into account as there is normally limited or no uni-directional flow (IAEA, 1982a). These include the following:

(i) residence times of the radionuclide;

(ii) sedimentation rate of the insoluble material;

(iii) dilution by the input fresh water;

(iv) removal of the radionuclides by water currents.

As discharges are made into a lake system, equilibrium will eventually be attained as the inputs are balanced by the losses. The time taken to reach steady-state conditions will vary depending on the physical and environmental factors. In order to simplify the calculations of radionuclide concentrations, complete mixing can be assumed in such steady-state conditions. For small lakes, we can use the following expression to estimate the concentrations in the water column:

$$C_{w,i} = \frac{Q_i}{K_e V} \tag{15.4}$$

where $C_{w,i}$ is the steady-state concentration of the radionuclide i in the water fraction (Bq m^{-3}), Q_i is the annual input of the radionuclide i (Bq per year), and K_e is the effective removal rate constant. The latter is given by the following:

$$K_e = \lambda_i + \frac{r}{V} \tag{15.5}$$

where r is the rate of output of fresh water (m^3 per year), V is the lake volume (m^3), and λ_i is the decay constant of the radionuclide (per year).

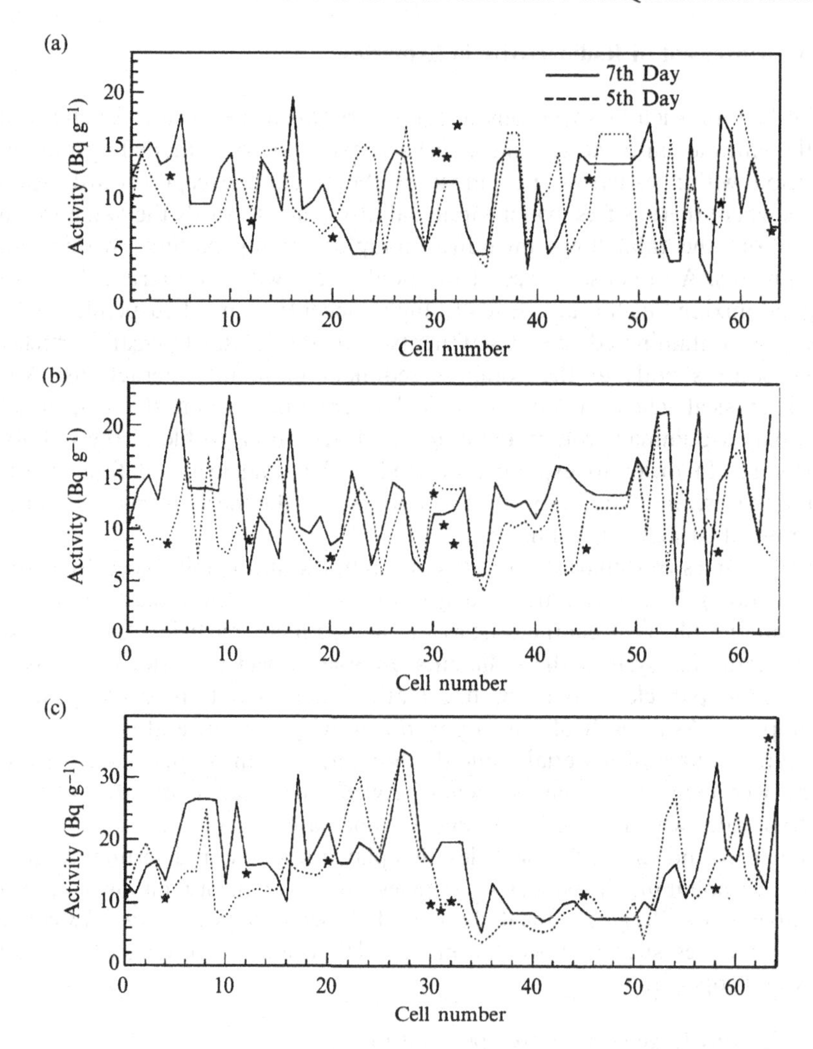

Figure 15.2 Model predictions for two days for the spatial distribution of ^{238}U activity in (a) suspended matter, (b) bottom sediments, and (c) water: (......), 5th day; (—), 7th day; ⋆, experimental values. Reprinted from *Appl. Math. Model.*, **25**, Rodriguez-Alvarez, M.J. and Sanchez, F., 'Modelling of U, Th, Ra and ^{137}Cs radionuclides behaviour in rivers. Comparison with field observations', 57–77, Copyright (2000), with permission from Elsevier Science

Whereas Equation (15.4) is applicable for small and well-mixed systems, lakes often exhibit stratification and can be also be connected to rivers. In situations such as these, it is important to take into account factors such as the rate of fresh water input, exchange rates between the stratified layers and seasonal mixing variations. In the case of large lakes, local gradients can occur in the vicinity of the point of release. The nature and extent of these gradients will inevitably depend on a range of environmental factors and physical and hydrological characteristics of the lake itself. Further considerations of such models can be found in USNRC (1977).

15.1.3 Movement of Radioactivity in Estuaries

Special factors, such as variations of salinity between the freshwater end of a river and the entrance into the sea, have to be considered when modelling radioactivity dispersion within estuaries. For most of the estuary, except towards the upper reaches where there is freshwater input, salinity tends towards the value for the sea. In addition, the tidal flows introduce complex mixing patterns within estuarine environments. As a consequence of the tidal forces within estuaries, there is generally good mixing but this also leads to high concentrations of suspended sediments. Mixing of contaminated fluvial sediments with the relatively clean particles from marine sources adds to the complex sediment–pollutant interactions. Although several physical, chemical and biological interactions occur, the suspended sediments play a dominant role in defining the distribution profile of the pollutants in the estuary. In order to explain and predict the transport and fate of estuarine pollutants it is, therefore, important to understand the mechanisms that control the behaviour of suspended sediments.

The dynamics of estuarine sediments is complex and is influenced by many variables including hydrodynamics and geomorphology of the estuary, mixing conditions and the physical characteristics of the sediment particles. Once transported into the estuarine system, the sediments become subject to various processes, such as recycling, particle formation, flocculation, deflocculation, decomposition and consumption. As a result of anthropogenic activity, run-off and waste disposal act as an input source of material while dredging acts as an output from the estuarine system. Recycling of sediments occurs between the estuary and the other sources over timescales which range from almost immediate to geological.

The nature and size of the particles also play a key role in determining the fate of radioactive pollutants in estuaries. Particles that have an organic origin are usually less stable and undergo chemical changes. The size of very *fine particles* can change due to processes such as flocculation and deflocculation. Fine particles have the following properties:

(i) relatively large surface area per unit mass;
(ii) easily transported in suspension;
(iii) contain clay minerals;
(iv) relatively high organic-matter content.

The large surface area of fine particles play an extremely important role in estuarine biogeochemical processes. They are a good source of planktonic and benthic organisms because they are rich in organic matter. Pollutants, such as radionuclides become sorbed onto the surfaces of these fine cohesive sediments. In addition, as a result of bi-directional tidal flows in estuarine systems, the exchange of, or removal of, radionuclides from the estuary is a complex phenomenon. Much of the sediments settle on mudflats during low tides, so exposing people and pets to radiation.

There are various sediment transport parameters that characterize the dynamics in terms of erosion, accretion and transport (Dyer, 1997). Some of the most important ones are as follows:

(i) settling velocity and concentration – for calculating sediment fluxes;

(ii) depositional stress – for predicting sediment location and deposition rates;

(iii) erosion stress – basic source parameter.

Turbulence fluctuations play an important role when describing the transport of suspended sediments in estuaries. These fluctuations can be generated by the channel bed, free shear, internal waves or wind-generated waves (West, 1990). Before a mathematical model of suspended sediment transport can be developed, it is, therefore, necessary to understand the mechanisms that generate the turbulence.

Simple models for estimating the dissolved fraction concentrations in estuaries assume a region of complete mixing around the vicinity of discharge and are of the form of the steady-state Equation (15.4) (IAEA, 1982a). The loss term incorporates radioactive decay as well as removal by exchange with the sea. In practice, considerable removal takes place through adsorption. Such simple models can provide a reasonable estimate of the maximum values of concentrations. An example of a more complex model is that reported by Perianez *et al.* (1996) who have applied a numerical model to study the dispersion of Ra-226 in suspended matter and dissolved phases in the Odiel River (in the south-west of Spain).

15.1.4 Movement of Radionuclides in Coastal Seas

There are several specific factors which have to be considered when predicting concentrations in and around coastal seas, as follows:

(i) geographical characteristics, ranging from nearly enclosed seas and experiencing limited flushing to those that are more open and form part of the oceans;

(ii) large variations in salinity especially in and near land-locked seas;

(iii) a major geographical boundary, namely the coastline, limits the mixing and dispersion of the radionuclides;

(iv) major tidal forces which influence the mixing processes;

(v) large variations in suspended sediments (although the sediment load is less than that in estuaries).

In addition to these factors, there can also be significant transfer of radioactivity to bed sediments which results in low concentrations in the water fraction. The simple model described by Equation (15.4) for lakes can also be applied to seas provided that a steady state is assumed. In this case, V is the mixed volume of the receiving water mass and K_e is the removal constant incorporating radioactive decay and loss of water from the mixed volume (r/V).

An example of a more complex modelling approach is that of Perianez (2000a) who reports on a numerical model to simulate the dispersion of ^{137}Cs and ^{240}Pu in the English Channel. Figure 15.3 illustrates contours of the radioactivity levels from ^{137}Cs in the English Channel generated from the computational values. The contours

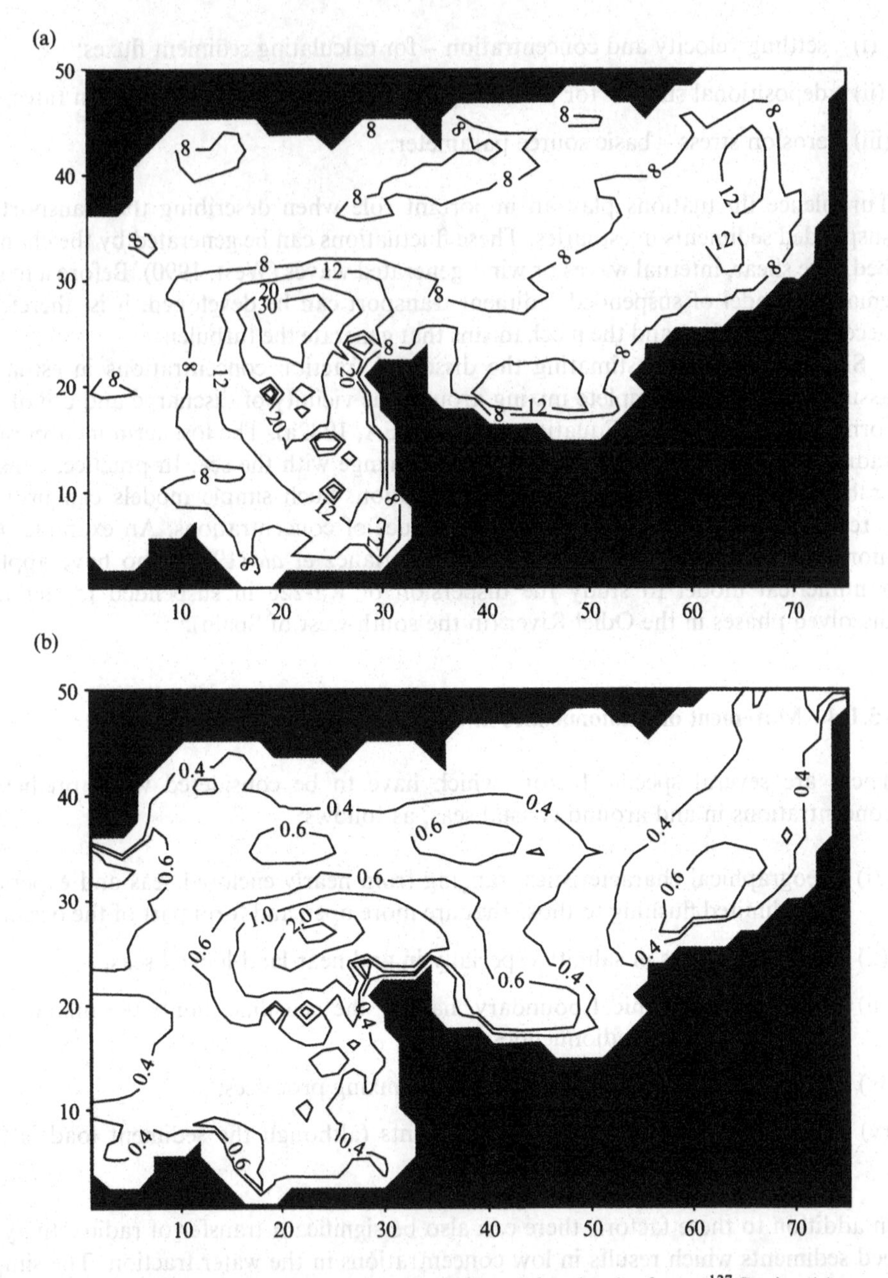

Figure 15.3 Computed contour maps of radioactivity levels from ^{137}Cs in (a) water (mBq l^{-1}), and (b) suspended matter (mBq g^{-1}) in the English Channel. Reprinted from *J. Environ. Radioact.*, **49**, Perianez, R., 'Modelling the tidal dispersion of Cs-137, Pu-239 and Pu-240 in the English Channel', 259–277, Copyright (2000), with permission from Elsevier Science

show that the radioactivity levels are generally higher near the French coast than the English side. In addition to hydrodynamics, the model takes account of suspended matter, deposition and re-suspension. The same author has also reported, in another study, the application of a numerical model but this time applied to the Irish Sea (Perianez, 2000b). Figure 15.4 shows the modelled values for $^{239,240}Pu$ compared to measured data for surface waters, suspended sediments and bottom sediments sampled at various distances from the Sellafield nuclear plant. The results demonstrate that the model is reasonably successful in reproducing the behaviour of the radionuclides in the three phases. The interested reader is also directed to other examples, such as du Bois and Guegueniat (1999), Osvath *et al.* (1999) and Perianez (1999).

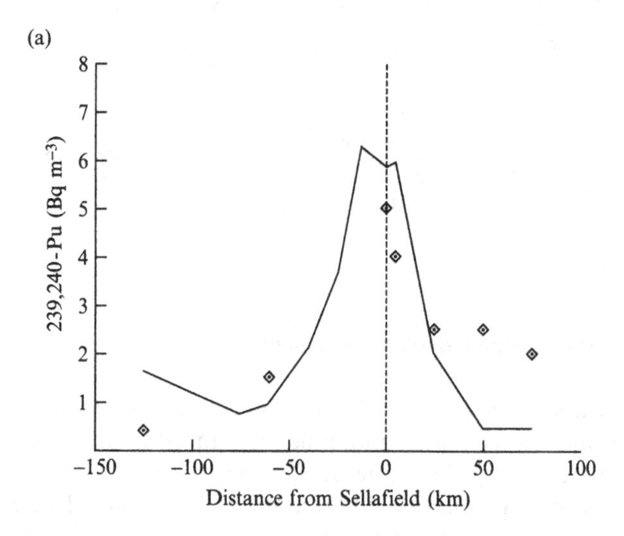

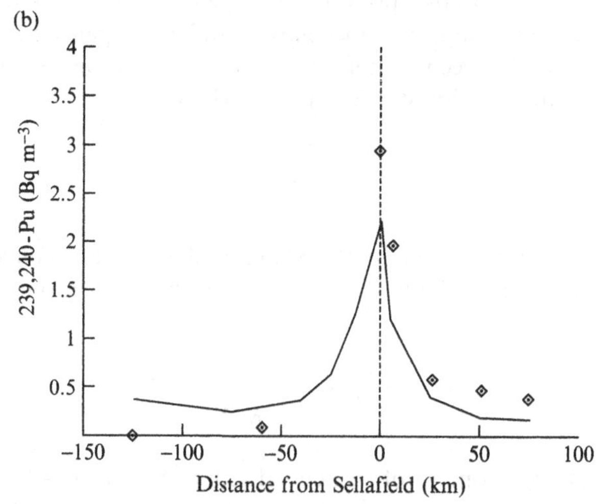

Figure 15.4 Modelled (—) and measured (◇) values for $^{239,240}Pu$ specific activities in (a) water, (b) suspended matter, and (c) bottom sediments in the Irish Sea. The negative and positive distances indicate south-west and north-west directions from the Sellafield nuclear plant. Reprinted from *J. Environ. Radioact.*, **49**, Perianez, R., 'Modelling the physico-chemical speciation of plutonium in the eastern Irish Sea', 11–33, Copyright (2000), with permission from Elsevier Science

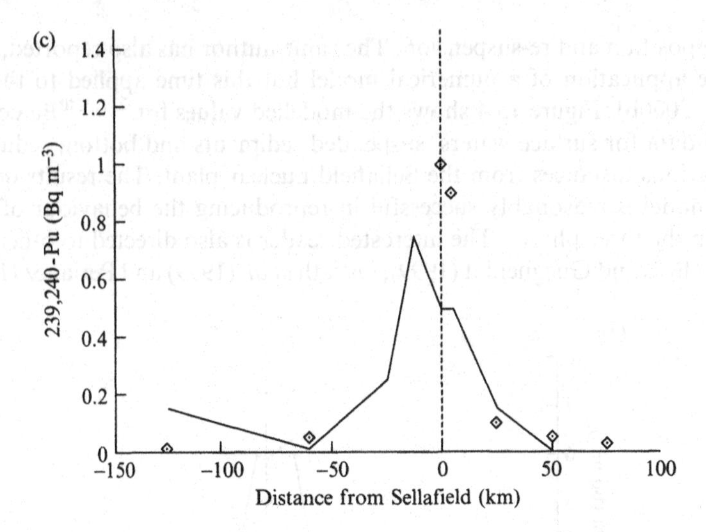

Figure 15.4 *Continued*

15.1.5 Transfer of Radioactivity to Sediment

As already discussed above, the role of sediments can be vital in determining the concentration and dispersion of radionuclides. In addition, as stated earlier, fine sediments in particular, offer large surface areas for adsorption and hence are especially important in estuarine environments. The removal of radionuclides by sediments can be through adsorption onto suspended and bed sediments involving processes such as sedimentation, flocculation and re-suspension. The partitioning of the radioactivity between water and sediment can be estimated simply by making use of the distribution coefficient, K_d, where:

$$K_d = \frac{C_s}{C_w} \tag{15.6}$$

where C_w is the concentration in the water and C_s is the concentration in the sediments. The reduction of the concentration in water is given by the following factor:

$$F = \frac{1}{1 + K_d X} \tag{15.7}$$

where X represents the content of suspended sediments and F is the factor by which the water concentration is reduced due to depletion by sediment. As K_d is in the denominator, a reduction in the concentration of radionuclides in the water is only significant when K_d and/or suspended sediment concentration is high. Values of K_d can be found in Eisenbud (1987) and IAEA (1982a). Table 15.1 shows some typical values for K_d for selected radionuclides. It should be noted that the values of K_d depend on the nature of the radionuclide and of the sediment and hence can be

Table 15.1 Typical value ranges for the distribution coefficient, K_d (ml g^{-1}) (adapted from IAEA, 1982a).

Radionuclide	K_d
Cs	10^2–10^4
I	10^2
Pb	10^4
Pu	10^4–10^5
Ra	10^2–10^3
U	10^2–10^3

site-specific. The dependence of K_d on particle size, for example, is discussed by Laissaoui and Abril (1999). A partitioning model to study sediments in an estuarine environment has been reported by Perianez and Martines-Aguirre (1997).

As an upper limit the concentration in the sediment or the water can be estimated from the following equation, assuming that an equilibrium exists between the water and sediment fractions:

$$C_s = C_w K_d \tag{15.8}$$

where C_s and C_w represent the concentrations of the radionuclide in the sediment and water, respectively. This simple model does not take into account the gradual buildup of radioactivity in the sediment fraction with time and hence it should be used to provide the maximum concentrations. The temporal dependence of the radioactivity buildup in the sediments can be described by the following expression:

$$C_{s,i} = \frac{[1 - \exp(-\lambda_i t_b)]}{\lambda_i} C_{w,i} K_{c,i} \tag{15.9}$$

where $C_{s,i}$ is the concentration of radionuclide i in the sediment (Bq kg^{-1}), $C_{w,i}$ is the concentration of radionuclide i in the water (Bq l^{-1}), $K_{c,i}$ is the transfer coefficient from water to sediment (l kg^{-1} h^{-1}), t_b is the length of time the sediment is exposed to contaminated water (taken nominally as 15 years by USNRC (1977), and λ_i is the decay constant of radionuclide i (per year) or, if available, the 'true' environmental decay constant, which takes into account removal processes. The time-dependent transfer coefficient, K_c, is site-specific. Values can be found in USNRC (1977). Time-dependent models can also be found in IAEA (1982b).

15.1.6 Movement of Radioactivity in Groundwater

There are several mechanisms by which radionuclides released into the environment can contaminate groundwater. These include seepage through lakes, streams and wells and transfer through soil (USEPA, 1996b). The movement of groundwater can be described by Darcy's equation (Kathren, 1991), as follows:

$$Q = uiA \tag{15.10}$$

where Q is the rate of seepage through the soil medium, u is the permeability of the soil, i is the hydraulic gradient and A is the area of the surface through which the contaminated water is seeping. This equation can be modified with a multiplicative factor to take into account radioactive decay and absorption/removal from the soil, to give the following:

$$Q = uiAe^{-kd\lambda t} \tag{15.11}$$

where k is the coefficient of absorption or removal from the soil, d is the depth of soil, λ is the the radioactivity decay constant, and t is the time. The term k varies with soil type and vegetation and hence takes into account processes such as adsorption and ion-exchange, as well as uptake by plants. Although Equation (15.11) describes the seepage of radionuclides to groundwater, it does not model the transport within the sub-surface water.

15.2 MOVEMENT OF RADIOACTIVITY IN THE TERRESTRIAL ENVIRONMENT

Radionuclides enter the terrestrial environment either directly by discharge or indirectly, for example, by deposition from the atmosphere. The obvious reason for being concerned about contamination of land is because of the possibility of exposure to humans. Soil and vegetation need to be considered when predicting radioactivity in the terrestrial environment. Models such as those reported in IAEA (1982a) consider both of these sub-systems, as well as the human food chain for radiological assessment. Processes such as deposition (dry and wet), as well as re-suspension, are important, along with interactions between soil and vegetation. In addition, meteorological conditions, topography, nature of soil and the type and extent of vegetation and animals are all important. Once the radioactivity reaches plants, it can be passed on to grazing animals and hence to humans.

15.2.1 Radioactivity in Soil

Soil consists of a complex mixture of minerals and organic matter, in addition to water and air. The size of the soil particles range from clay particles of sub-micron diameter to sand particles of dimensions of around 2 mm. The surface soil consists of a topsoil layer (about 0.5 m thick) which is rich in organic matter. Below this lies the sub-soil layer (about 0.3 m thick), consisting mainly of inorganic matter. The next level down is the pervious rock layer, before meeting the impervious bedrock.

Sand particles are generally chemically inert and because of their large sizes water travels through them with ease. Silt particles, which are smaller in size, offer larger surface areas and can be transported with contaminated water. Clay particles offer the largest surface area per unit mass and thus hinder water flow. Soluble radionuclides can be adsorbed onto the reactive surfaces of fine soil particles. They can also react with organic matter, precipitate, for example as oxides, or undergo ion-exchange. Depending on the time that a radionuclide remains in the soil, it can partition into various fractions through these processes. The extent to which any of these processes become important depends on the radionuclide itself and if there are any removal mechanisms, such as by root uptake. The movement of the radionuclide through the soil is hence determined to a large extent by these partitioning processes. If the radionuclide is moving with water, then, as it becomes adsorbed on to soil particles, its transport is retarded relative to the movement of the water. The partitioning of the radionuclide between the soil and the interstitial water is determined by the distribution coefficient (K_d), defined as follows:

$$K_d = \frac{\text{Concentration of radionuclide in the solid phase (Bq kg}^{-1}\text{)}}{\text{Concentration of the radionuclide in the soluble phase (Bq l}^{-1}\text{)}}. \quad (15.12)$$

The units of K_d are hence l kg^{-1} or ml g^{-1}. The reader should note the similarity here with Equation (15.6). From Equation (15.12), one can deduce that large K_d values imply strong adsorption onto the particle surfaces. Typical K_d value ranges for some common radionuclides are given above in Table 15.1. For fine clay particles, values are usually in the range of 10^3–10^5 for radionuclides such as ^{137}Cs and ^{60}Co.

Sorption of ions in the water flowing through the soil depends on the physical and chemical properties of the radionuclide and the soil material. In general, cations sorb more strongly because of the net negative charge on soil particles. As numerous ions are already present in soil, ion-exchange can be significant when radionuclides in liquid discharges travel in soil. In addition to the radionuclide itself and the physical and chemical properties of the soil, other factors that influence ion-exchange include pH and precipitation. Organic matter aids ion-exchange between the soluble radionuclides and the soil particles. This type of exchange is strongest in clay and weakest in sandy soils. Once radionuclides are in the soil, therefore, their spatial dispersion is determined to a large extent by the range of factors mentioned above. In sand, the movement is rapid, whereas in clay the movement is slow. Although vertical transport is aided by rainfall, the time taken for the radionuclides to reach the water table can be of the order of months, and even years.

15.2.2 Role of Vegetation – Uptake and Deposition

The uptake processes of the plant roots also retard the movement of radionuclides in soil. These processes are dependent on the physio-chemical characteristics of the soil, the chemical properties of the radionuclide, value of K_d and plant metabolic factors. Values for the uptake factors for dose calculations can be found in NCRP (1984), Eisenbud (1987) and Renaud *et al.* (1999). It is obvious from these sources that the range of these uptake values is considerable, implying that one has to be cautious when employing such data for dose calculations and also be familiar with the associated uncertainties, which can be considerable. During the process of uptake and ion-exchange, the radioactive nature of the pollutant is not important – it is the chemical properties of the radionuclide that are important. In practice, therefore, the plant roots do not distinguish between a radionuclide and its stable counterpart. A comprehensive compilation of soil-to-plant transfer factors is given in IAEA (1994e). Table 15.2 lists the ranges of values for some selected radionuclides.

Direct deposition of radionuclides onto foliage is an important mechanism for transferring radioactivity to vegetation from the atmosphere. This is normally expressed as activity per unit area per unit time and is given by the product of the deposition velocity v_d(m s^{-1}), and the concentration of the radionuclide in the air (Bq m^{-3}). The resulting units of deposition of radionuclides onto foliage are, therefore, Bq m^{-2} s^{-1}. Empirical expressions for calculating v_d in terms of particle size and wind speed can be found in various literature sources, e.g. Kathren (1991). Once the deposition has occurred, various processes, in addition to decay, can remove or deplete the radioactivity from the plant surfaces. These include evaporation, wash-off and removal by wind. When the plant eventually dies or is uprooted or disturbed, the radioactivity will also transfer to the soil.

In an event of an accident, foliar deposition and transfer is more important than root transfer. Renaud *et al.* (1999) have proposed a simple empirical model to estimate the contamination due to these two processes, as follows:

$$Cv_f(t) = DFT_f(0)e^{-(\lambda_b+\lambda_r)\Delta t} \tag{15.13}$$

$$Cv_r(t) = DFT_r e^{-\lambda_r \Delta t} \tag{15.14}$$

where Cv_f is the concentration due to foliar transfer (Bq kg^{-1}), Cv_r is the concentration due to root transfer (Bq kg^{-1}), D is the deposit (Bq m^{-2}), $FT_f(0)$ is the direct foliar transfer factor (m^2 kg^{-1}) for $t = 0$, FT_r is the root transfer factor (m^2 kg^{-1}), λ_b is the biological decay constant (d^{-1}), λ_r is the radioactive decay constant (d^{-1}), and Δt is the period of time since the deposit (d).

This simple model enables initial contamination of vegetation to be assessed which can be useful as an early evaluation or screening method following an accident. There is scarcity of data, however, in relation to foliar transfer to vegetables, and hence, this will limit the precision and accuracy of any model that relies on such data.

15.2.3 Re-suspension of Radionuclides

As radionuclides can deposit onto surfaces from the atmosphere, they can also be re-suspended back into the air through various processes. These include wind

Table 15.2 Soil-to-plant transfer factor ranges for some selected radionuclides (adapted from IAEA, 1994e).

Radionuclide	Transfer factor
Sr	0.01–1
Cs	0.001–0.1
Pb	0.001–0.01
U	0.001–01
Pu	10^{-5}–10^{-3}
Am	10^{-5}–10^{-3}

effects and other forms of physical disturbance. Once radioactivity has been re-suspended, this can lead to re-deposition and hence further contamination of the land and vegetation (Simmonds *et al.*, 1995). The transport caused by re-suspension will depend on factors such as particle size, wind speed, nature of the deposit and surface, and precipitation. Larger particles ($\sim$ 1–2 mm) may roll off the plant surfaces as a result of wind action, whereas finer particles may be transported into the air before settling gravimetrically. The very small particles (a few microns and less in size) can remain in the atmosphere for days and weeks. Re-suspension will also be affected by land-use practices and thus areas of different land-use types, such as agricultural and urban, will respond differently. Consequently, one can also classify the main processes that lead to re-suspension as either anthropogenic or natural. Anthropogenic activities, such as road traffic, farming, landscaping and construction, can all lead to localized re-suspension.

There are various methods for estimating re-suspension. A simple method makes use of the re-suspension factor, k, defined as follows:

$$k = C_a / D_s \tag{15.15}$$

where C_a is the concentration of the radionuclide in the air due to re-suspension (Bq m^{-3}) and D_s is the total surface contamination (Bq m^{-2}); hence, the units of k are m^{-1}. There are various limitations with this simple approach. For example, the expression assumes that the air concentration is due to a local source. Furthermore, as the value of k is normally time-averaged, it should be used with caution when the release is instantaneous, for example, in the case of an accident. Reviews of particle re-suspension can be found in Sehmel (1980), Nicholson (1988) and Simmonds *et al.* (1995). The data given in the literature indicate that values for k range from 10^{-4} to 10^{-9} m^{-1}. More recently, measurements of re-suspension as a function of particle size and the influence of parameters such as soil moisture have been reported by Wagenpfeil *et al.* (1999).

An alternative expression for re-suspension is given in terms of a re-suspension rate (Λ) (Horst, 1978; Simmonds *et al.*, 1995), as follows:

$$\Lambda(s^{-1}) = F_r / D_s \tag{15.16}$$

where F_s is the re-suspension flux (Bq m^{-2} s^{-1}) and D_s is the surface deposit (Bq m^{-2}). Another useful method reported in the literature (Simmonds *et al.*, 1995) is employed in cases where the deposit is attached to the soil and hence has aged somewhat, namely:

$$C_a = C_s S_E \tag{15.17}$$

where C_a and C_s are the air and soil concentrations (Bq m^{-3}), respectively, and S_E (kg m^{-3}) is the equivalent soil concentration or dust loading. Typical values for S_E are around 50–100 kg m^{-3} but can be several hundred kg m^{-3}.

A time-dependent model for re-suspension, reported by Garland and Pomery (1994), has also been discussed by Simmonds *et al.* (1995) and takes the following form:

$$k = (1.2 \times 10^{-6})t^{-1} \qquad (15.18)$$

where k is the re-suspension factor (m^{-1}) and t is the time (d). This equation is valid for $t \geq 1$, and can be modified to take into account radioactive decay, as follows:

$$k = (1.2 \times 10^{-6}) \exp\ (-\lambda t)/t \qquad (15.19)$$

where λ is the decay constant.

16

Assessment of Radiation Doses

People can receive radiation doses from various different sources and in many different ways. Earlier chapters have shown that most radiation doses are unavoidable in that they are caused in the main by natural phenomena (see Chapter 1). Nevertheless, it is necessary to estimate radiation doses in situations where they may be high enough to cause concern, and so protective actions might be necessary, or where they result from deliberate human actions, such as the introduction of a practice using radioactivity (see Section 4.3). It is impossible to measure radiation doses directly. Rather, they have to be inferred from measurements of, say, radionuclide concentrations in environmental materials or estimated by using models of radionuclide behaviour in the environment or derived using a combination of the two.

The doses individuals receive depend very largely on two factors – the levels of radioactivity in the environment (the 'Exposure') and how the individuals interact with that environment (the 'Habits'). Other factors, such as age, may also play a part, particularly in the case of intakes of radionuclides. An example illustrates how these factors interact. In the case of radionuclides in agricultural soil, an individual could be exposed to radionuclides in crops grown on the soil but the radiation dose will also depend upon the amount of the crops that are eaten, i.e. the habits. It will be appreciated that it is often easier to quantify exposure than habits. The issue is further complicated by the fact that the presence of radionuclides in environmental materials is rarely homogeneous; so, it is not just an individual's general habits that are important but more often the habits that are related to a particular area of contamination. Thus, in the example of crop consumption, it is the quantity of crops consumed from a particular location, i.e. the contaminated area, that is needed in order to assess doses.

It is obviously preferable, and usually more accurate, to base estimates of radiation doses on measurements of either dose rates from external irradiation or of radionuclide concentrations in environmental materials. The first step in assessing radiation doses to people is usually to identify the routes by which people are exposed to radiation or radionuclides. These include inhalation of radionuclides in the atmosphere, ingestion of radionuclides in foodstuffs and external exposure from radionuclides in environmental materials. Examples of these exposure pathways are illustrated in Figures 16.1 and 16.2.

This chapter outlines the principles for estimating doses. Detailed guidance on estimating doses for typical European situations is given in Simmonds *et al.* (1995). Guidance on estimating doses for the purpose or regulating discharges to the environment is given in IAEA (2001b).

16.1 GENERAL ISSUES IN CALCULATING DOSES

In evaluating doses, two factors at least have to be estimated, namely the level of exposure and the appropriate habits. The way that both of these factors are assessed will depend largely on the reason for calculating the doses. This is called the *assessment context* and should be established at the outset.

Very broadly, the context of an assessment of doses can be broken down into the following categories:

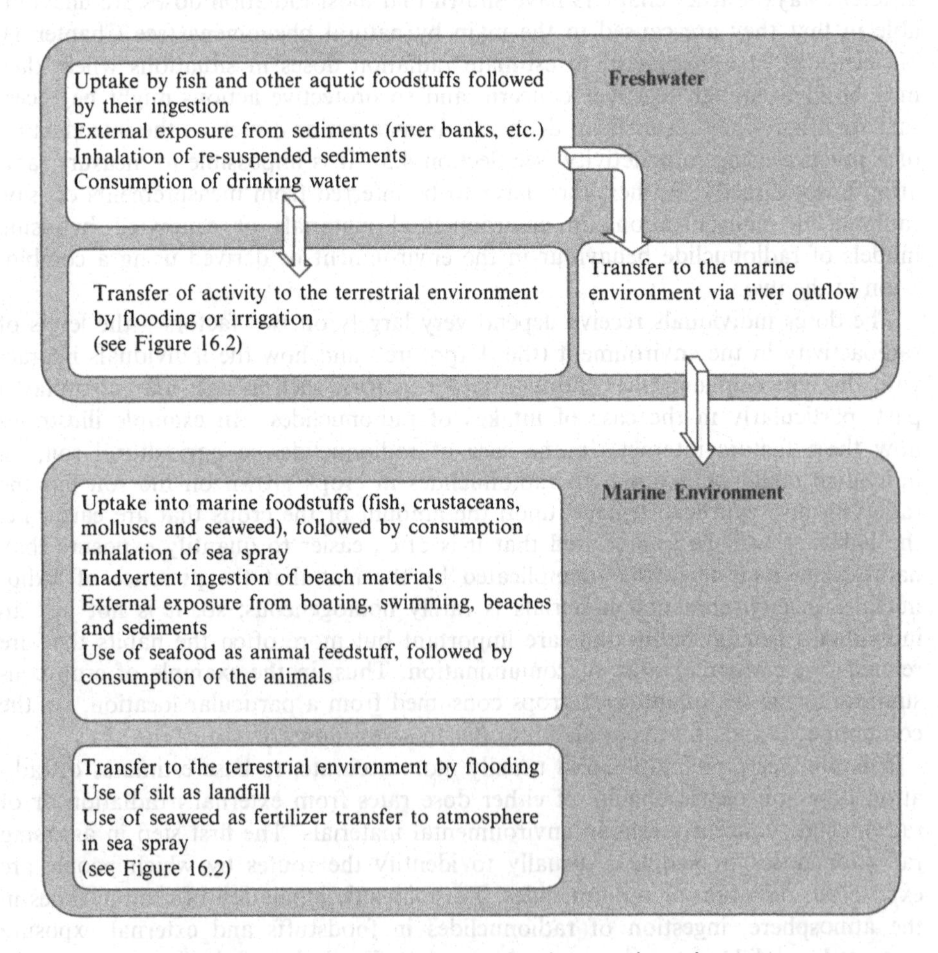

Freshwater
Uptake by fish and other aqutic foodstuffs followed by their ingestion
External exposure from sediments (river banks, etc.)
Inhalation of re-suspended sediments
Consumption of drinking water
Transfer of activity to the terrestrial environment by flooding or irrigation
(see Figure 16.2)
Transfer to the marine environment via river outflow
Marine Environment
Uptake into marine foodstuffs (fish, crustaceans, molluscs and seaweed), followed by consumption
Inhalation of sea spray
Inadvertent ingestion of beach materials
External exposure from boating, swimming, beaches and sediments
Use of seafood as animal feedstuff, followed by consumption of the animals
Transfer to the terrestrial environment by flooding
Use of silt as landfill
Use of seaweed as fertilizer transfer to atmosphere in sea spray
(see Figure 16.2)

Figure 16.1 Exposure pathways for the aquatic environment

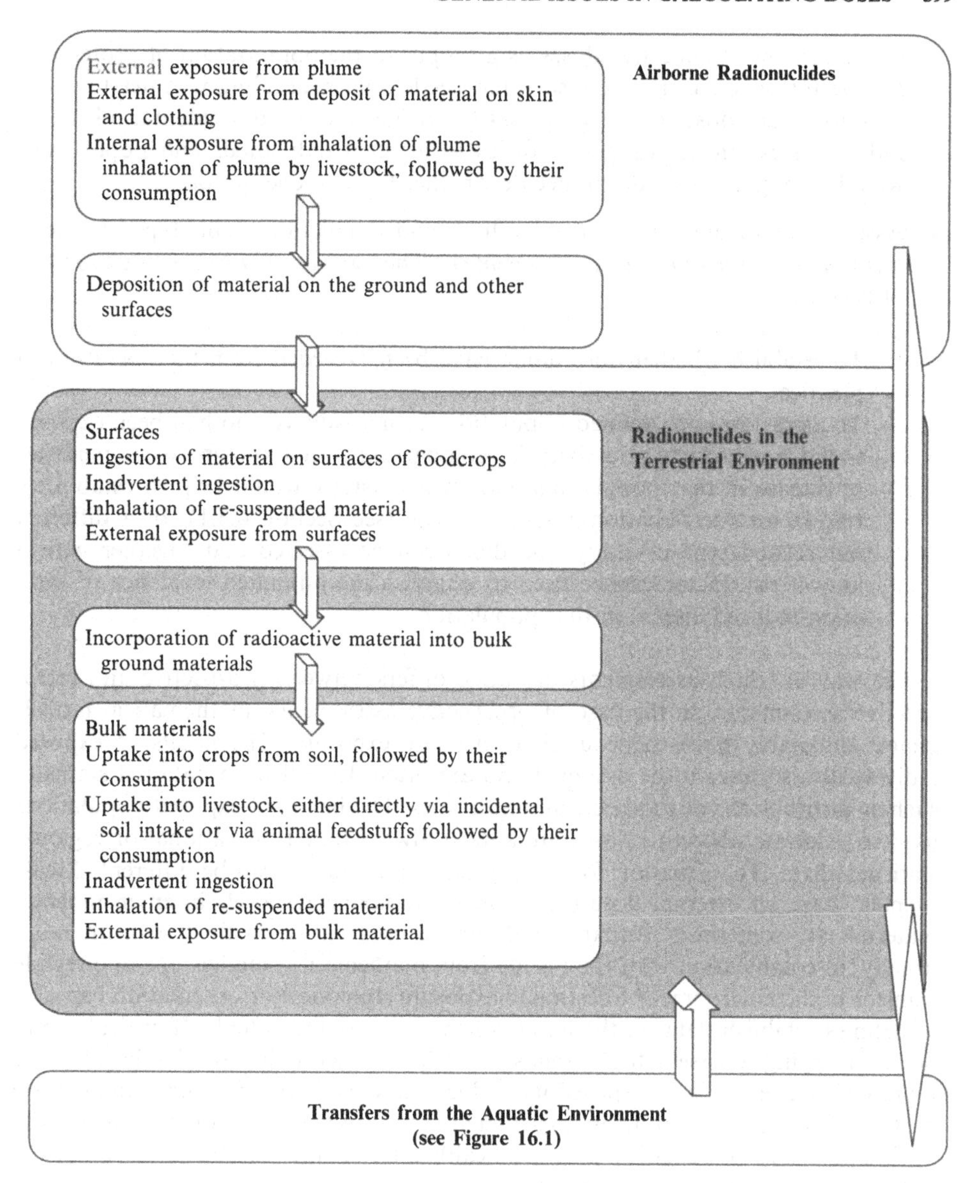

Figure 16.2 Exposure pathways for the atmospheric and terrestrial environments

- Assessments of doses to future individuals and populations. This type of assessment is termed a *prospective assessment*. Such assessments are done when new practices (see Chapter 4) are being considered for introduction as an input to judging their acceptability. An example would be a new hospital that is going to discharge radionuclides into the environment following the use of radiopharmaceuticals in the treatment or diagnosis of disease. A decision on whether these discharges are acceptable would involve an assessment of critical group doses for comparison with the appropriate constraint (see Chapter 5) and, possibly, an assessment of collective doses as an input to the optimization of protection. Prospective assessments are also often needed when deciding whether to intervene to reduce doses (see Section 4.4.1). In the case of an accidental release of radionuclides, an assessment of doses to affected individuals and populations would be required in order to decide whether to intervene or not.

- Assessments of doses to past individuals and populations. This type of assessment is termed a *retrospective assessment*. They are usually carried out for one of two reasons:

  - To establish whether dose limits have been exceeded from the operation of practices.
  - To assess doses to defined populations of individuals who had been exposed to a defined source of activity for comparison with a known excess incidence of disease in that population. This type of study would help to establish a risk factor for radiation-induced disease (see Section 4.3.1). In a different but related type of study, the doses can be assessed and together with a known risk factor can be used to establish the estimated incidence of radiation-induced disease in that population.

The way in which assessments are done differs between prospective and retrospective assessments. In the case of prospective assessments, mathematical models almost invariably have to be used to evaluate exposure. The habits of exposed individuals also have to be assumed and are often derived from national statistics such as distributions of food consumption rates. The habits of exposed populations used to calculate collective doses are almost always based on national or regional statistical data. The situation for retrospective assessments can be different. Measurement data on external dose rates and on radionuclide levels in environmental materials are sometimes available and can be used either on their own or, more usually, in combination with the results from mathematical models of radionuclide transfer in the environment to estimate exposure. Interview or observation can also sometimes establish data on the habits of the exposed individuals. If critical group doses are being assessed, the emphasis will be on identifying the habits of those representative of the most exposed individuals, whereas if the purpose is to estimate past doses to a population from a particular source of activity in order to assess the attributable risk of radiation-induced health effects, then the habits of a typical individual in the population would be taken.

Having established the context of the assessment, the next stage is usually to identify the exposure pathways. It is important to take account of all likely exposure pathways or the calculation of doses may be incomplete.

16.2 EXPOSURE PATHWAYS

Each environmental material, e.g. soil, air, water, etc., has a range of associated exposure pathways. The importance of any one of these will depend upon the particular radionuclide or radionuclides present.

Typically, radionuclides in the atmosphere, from, say, routine discharges to the atmosphere, can cause radiation doses to individuals by external exposure from radionuclides outside the body and internal exposure from radionuclides taken into the body. People can receive doses from gamma- or beta-emitting radionuclides in the plume while the radionuclides themselves remain outside the body – this is an example of external irradiation. Individuals may also inhale or ingest radionuclides; this causes internal irradiation.

Radionuclides present in the atmosphere as particulate material (dusts, etc.) may eventually deposit on the ground by, among other mechanisms, gravitational settling, leading to radiation exposure after the airborne plume of activity has passed. The exposure pathways for deposited radioactive material include external exposure from gamma- and beta-emitters and internal exposure from consumption of radionuclides incorporated into foodstuffs. Transfer of radionuclides through the foodchain to people can involve several stages depending upon the radionuclide concerned and the complexity of the foodchain. These stages can include uptake of radionuclides by vegetation, either by foliar absorption or by root absorption, and then translocation to the edible parts of the plant. The situation becomes even more complex when animal foodstuffs are considered; in such cases, the uptake of radionuclides by the animal after ingestion of contaminated feedstuff has to be allowed for, as does the behaviour of the radionuclide in the animal's body and the eventual translocation to edible parts. Foodchain models are described in more detail in Section 16.3.4. Material deposited on the ground may be re-suspended due to the action of winds, etc., and subsequently inhaled. Radionuclides that are isotopes of biologically available elements, i.e. carbon, sulfur and tritium, may also become directly incorporated into foodstuffs even when discharged as gases or vapours.

Radionuclides in the freshwater environment can cause internal irradiation from the consumption of abstracted drinking water and from consumption of freshwater fish that have incorporated radionuclides. Some radionuclides readily bind to sediments due to their chemical properties and may thus become bound to river beaches and banks, thus causing doses by external exposure or inhalation of re-suspended material. Freshwater may be used for irrigation of crops and this can lead to doses to people from foodchain pathways. Doses from discharges to the marine environment arise from similar exposure pathways to those for freshwater except that drinking water and irrigation pathways will not occur unless, of course, some form of desalination is undertaken.

Depending upon the chemical properties of the parent element, different exposure pathways will be important for different radionuclides. For example, external exposure from a plume of gas in the atmosphere will be important for beta- or gamma-emitting isotopes of chemically inert noble gases, e.g. ^{85}Kr and ^{41}Ar, whereas inhalation of the plume and of re-suspended material will be significant for alpha-emitters such as ^{239}Pu. Consumption of fish is an important pathway for marine discharges of ^{137}Cs and external exposure from radionuclides bound to beach sediments is important for discharges of ^{95}Zr. A summary of the important exposure pathways is given in Table 16.1, while examples of important exposure pathways for particular radionuclides are given in Table 16.2. In general, it can be seen that exposure pathways leading to intakes into the body are important for alpha-emitting radionuclides, whereas for gamma-emitters it is the external exposure pathways that are important. Pure beta-emitters generally need to be taken into the body to deliver a radiation dose but can also deliver significant doses to the skin if outside the body and in close contact with the body.

In estimating doses to people, it is important to add the doses from all the relevant exposure pathways. For example, a release of ^{239}Pu to atmosphere could cause radiation doses from inhalation, from ingestion of activity deposited on foodstuffs and from inadvertent ingestion of dusts, etc.

16.3 ESTIMATION OF EXPOSURE

This section describes the basis of how to estimate the exposure of individuals and of populations from radionuclides in the three main types of environmental materials, i.e. soil, water and air. The starting point is taken to be concentrations of radionuclides in environmental materials. In all cases, measured radionuclide concentrations or estimates of concentrations from models could equally well be used as the starting point for the calculation. In many assessments of past exposures, a combination of measurements and model predictions has been used. These three environmental media cannot, however, be thought about in isolation. Radionuclides in air can transfer to soils or water. Similarly, radionuclides in water can transfer to soils. Terrestrial foodchains are particularly noteworthy in this respect

Table 16.1 Summary of the important exposure pathways.

Pathway	Source
Atmospheric discharges	
Inhalation	Air; re-suspended deposits
Ingestion	Beef; liver and other offal; milk and milk products; mutton and lamb; grain; green vegetables; root vegetables;
External	Gamma-exposure from plume and deposited radionuclides; beta exposure from the plume
Liquid discharges	
Inhalation	Re-suspended soil following deposition of radionuclides; in sea spray; re-suspended beach material; sea spray
Ingestion	Drinking water; freshwater fish Meat, offal and meat products vegetables[a] Fish; crustaceans; molluscs; seaweed
Inadvertent ingestion	Soil following deposition of radionuclides in sea spray; beach material
External	Gamma-exposure from radionuclides deposited on land via sea spray; beaches; fishing gear – beta- and gamma-exposure

[a]Following deposition in sea spray or in irrigation water.

Table 16.2 Important exposure pathways for some radionuclides.

Radionuclide	Important exposure pathway
Discharges to atmosphere	
^{3}H	Ingestion of food (milk) and inhalation of plume
^{14}C	Ingestion of foodstuffs
^{32}P	Ingestion of foodstuffs
^{41}Ar	External irradiation from plume
$^{57-60}CO$	External irradiation from deposited activity
^{89}Kr	External irradiation from plume
^{131}I	Ingestion of foodstuffs (milk)
^{137}Cs	External irradiation from deposited activity; ingestion of foodstuffs
^{238}U	Inhalation of plume
$^{238-241}Pu$	Inhalation of plume
Discharges to marine environment	
^{3}H	Ingestion of seafood (fish, crustaceans and molluscs)
^{14}C	Ingestion of seafood (fish, crustaceans and molluscs)
^{32}P	Ingestion of seafood (fish)
^{60}Co	Ingestion of seafood and external irradiation (beaches, etc.)
^{90}Sr	Ingestion of seafood (fish and crustaceans)
^{106}Ru	Ingestion of seafood (crustaceans, molluscs and seaweed); external irradiation (beaches)
^{131}I	Ingestion of seafood (fish and crustaceans)
^{137}Cs	Ingestion of seafood (fish); external irradiation from deposited activity (beaches)
^{239}Pu	Ingestion of seafood (molluscs)

as radionuclides in the atmosphere can deposit directly on crops or be inhaled by animals used subsequently for foods. Irrigation can also lead to direct contamination of foodstuffs by radionuclides in freshwater as can ingestion of water by farm animals. There is, therefore, a separate discussion on estimating doses from complex terrestrial foodchains (Section 16.3.4).

16.3.1 Soil

Soil is an environmental material that is involved in many exposure pathways and measurements of radionuclide concentrations in this material or estimates from models can be used as a starting point for calculating doses. Examples of how to calculate doses via three exposure pathways are given below.

Ingestion of foodstuffs

Crops grown on contaminated soils can take up activity via the roots and from absorption from deposits on leaves and stems. Numerous experimental studies and observations on contaminated areas have shown that there is a relationship between the concentration of radionuclides in soil and the corresponding concentration in a foodcrop grown in that soil. This relationship is variously described as a *soil-to-plant concentration factor* or a *root-uptake factor*. There is sometimes confusion between the use of these two terms. The distinction usually made is that a soil-to-plant concentration factor refers to an observed relationship between the concentration of a radionuclide in soil and the corresponding concentration in the plant, whereas a root-uptake factor refers specifically to the concentration of radionuclide in a plant which is a result of root uptake in a plant grown on that soil. The proportion of activity in a plant that results from root uptake depends upon the chemical and physical properties of the radionuclide in question. The numerical value of the root-uptake factor depends upon the radionuclide, the plant species, and, to some extent, the soil type. Compilations of root-uptake factors have been published (IUR, 1989; Ng, 1982). In estimating the concentration of radionuclide in a plant using a root-uptake factor, allowance should be made for the amount of radionuclide present via other routes. For example, activity could be present in the plant from direct deposition onto the plant from the atmosphere, followed by translocation to edible parts. Soil contamination may also lead to the presence of activity in edible produce, although food preparation techniques may remove most of this. The suggested values for soil contamination range from 0.01 % per dry weight of grain seed to 0.1 % per dry weight of leafy green vegetables such as lettuce (Simmonds *et al.*, 1995). Thus, the concentration of radionuclides in a plant can be represented by the following relationship:

$$SC_i(RF_i + Y) = PC_i \tag{16.1}$$

where SC_i is the concentration of radionuclide i in the soil, RF_i is the root-uptake factor for radionuclide i, Y is a factor to allow for contamination by the soil, and PC_i is the concentration of radionuclide i in the plant.
This should be summed over all of the radionuclides of interest.

Published root-uptake factors assume equilibrium between the soil and the plant, and, as such, should not be used in circumstances where the concentrations in soil vary substantially with time during the growing season.

In many cases, the situation is more complex, particularly if animals are involved, and foodchain models are required. These are described later in Section 16.3.4.

Re-suspension of soil

Soil and dust can be re-suspended by the action of winds or by man-made disturbances such as ploughing. The material in the air could be inhaled and so this represents a possible exposure pathway to radionuclides in soils. The intake of radionuclides by this route (I, Bq) can be estimated from a knowledge of the concentration of radionuclides in soil (SC_i, Bq kg^{-1}), the concentration of soil particles in air (CA, kg m^{-3}), the breathing rate (Z, m^{-3} h^{-1}) and the time that an individual is exposed to the contaminated air (T, h), as follows:

$$SC_i \times CA \times Z \times T = I \tag{16.2}$$

The concentration of particles in air, CA, is often referred to as a *dust-loading factor*. It is a simple matter to calculate the resultant dose by use of the appropriate dose coefficient. The concentration of dusts, etc. in air depends upon the type of environment, e.g. urban or rural, and the presence of agents that could re-suspend particles in air. Example values for dust loadings are around 5 to 50 $\mu g\ m^{-3}$ in rural areas and between 100 and 800 $\mu g\ m^{-3}$ for large industrial areas (Simmonds *et al.*, 1995). In arid regions, human activities such as digging can cause dust loadings of up to 2000 $\mu g\ m^{-3}$.

In estimating doses by using this method, it is implicitly assumed that the concentration of radionuclides in re-suspended particles is the same as that measured in the bulk soil. This may not be the case, particularly if there has been a recent deposit of radionuclides onto the soil.

An appropriate approach in such situations is to use a *re-suspension factor*. Here, a relationship is assumed between the concentration of a radionuclide in air and its density on the underlying surface. This relationship has the following form:

$$k\ (m^{-1}) = \frac{\text{Concentration in air due to re-suspension (Bq m}^{-3}\text{)}}{\text{Surface deposit (Bq m}^{-2}\text{)}} \tag{16.3}$$

where k is the re-suspension factor.

In using the re-suspension-factor approach there is the implicit assumption that the airborne activity arises solely from the local surface. When k is estimated from measurements, its value depends upon the depth of surface material sampled; reported re-suspension factors range from 10^{-2} to $10^{-10} m^{-1}$ (Linsley, 1983).

There is some evidence that re-suspension factors decrease with time following an initial deposit of activity. Evidence comes from studies on fall-out from the Chernobyl accident (Garland *et al.*, 1992), from experimental studies under temperate conditions (Garland, 1983) and from studies in the Palomares region of Spain where there was a nuclear weapons accident in 1966. This accident lead to the dispersion of plutonium over an area of several square kilometres. Measurement data on concentrations of plutonium in air show that the re-suspension factor decreased from an initial value of around 10^{-7} per month to about 10^{-9} per month after a few months and then to 10^{-10} per month after a few years (Iranzo *et al.*, 1987). This decline is due to chemical and physical changes within the soil which result in radionuclides becoming associated with progressively larger soil particles and also mixing to greater depths. There is no evidence in particular that this phenomenon depends upon the specific radionuclide under consideration.

The value for the re-suspension factor for wind-driven re-suspension in northern European conditions at various times following deposition can be estimated from the following equation (Garland *et al.*, 1992):

$$k_{(T)} = k_{(T=0)} T^{-1} (T \geq 1\text{day}) \tag{16.4}$$

where $k_{(T)}$ is the re-suspension factor (m^{-1}), $k_{(T=0)} = 1.2 \times 10^{-6}$, and T is the time following deposition (days).

This formula is applicable to situations where there is wind-driven re-suspension in rural environments in northern Europe and probably also to semi-arid regions, such as the Palomares, and urban environments (see Simmonds *et al.*, 1995). However, where there is man-made re-suspension, from tractor driven ploughing, etc., a dust-loading approach is more appropriate.

External exposure

Radionuclides in soils can give rise to external irradiation. The distribution of radionuclides with depth in the soil is important as radiation can be absorbed by soil. It is generally assumed that irradiation at the surface from radionuclides in soil at depths greater than 30 cm can be ignored; this will be zero for β-emitting radionuclides and small for γ-emitting radionuclides when compared with the external irradiation due to radionuclides in the upper 30 cm of soil.

It is conventional to take the dose at 1 m above the surface as being representative of the dose to a person standing on that surface. Computer codes have been used to estimate the absorbed dose rate in air from the radionuclides in the slab of soil at a particular depth and then summing the contributions from radionuclides at other depths. The estimates take account of buildup in the soil and air, together with attenuation in soil and air. A description of models for calculating dose rates from external irradiation from radionuclides in soils is given in Simmonds *et al.* (1995). Alternatively, if an approximation is adequate the simple model described by Hunt (1984) for estimating dose rates over beaches could be used (see Section 16.3.3).

When estimating doses from external exposure using data on radionuclide concentrations in soil, it is important to take account of the time people spend on the soil itself and also of the time spent in buildings constructed on the soil. Buildings will shield the people inside from radiation emitted in the soil. The degree of shielding will depend, among other things, on the type of building, its construction, and the location of the individuals inside the building. Modelling studies suggest that typical dose rates indoors from gamma-emitting radionuclides in soils are about 10% of those experienced outdoors in a rural environment, all other things being the same (Crick and Brown, 1990).

16.3.2 Air

Very low levels of radionuclides are discharged to atmosphere from the nuclear industry (see Chapter 6) and from some other industries (see Chapter 7). The resulting levels of radionuclides in environmental materials are often so low that doses are estimated by using mathematical models of radionuclide behaviour in the environment. Accidents can cause radionuclides to be released to atmosphere; the major nuclear accidents have all involved such releases (see Chapter 6). If an accident happens, it is important to have an estimate of the doses when deciding on whether emergency countermeasures should be taken. (See Chapter 4 for a description of the radiological protection principles that would be applied in this type of situation.) In the initial stages of an accident, models will have to be used in order to estimate doses but later, measurement data will become available that will enable improved estimates to be made.

Doses from inhalation can usually be estimated in a straightforward manner from a knowledge of the average radionuclide concentration in air over the period of interest, CA (Bq m^{-3}), the breathing rate (m^3 per unit time) and the dose coefficients for inhalation for the radionuclides present.

Measured concentrations of radionuclides in particulate form in air can be obtained by passing large volumes of air through an air filter using a high-volume air sampler. Typically, these will trap particulate material of size greater than around 0.5 μm and will filter around 10 m^3 of air per minute. They will not trap radionuclides that are in the form of gases, such as ^{85}Kr, or, in many situations, isotopes of iodine, although the latter can be trapped in activated carbon filters. Returning to particulate material, the results from an air sampler will enable an estimate to be made of the average concentration in air over the period that the air sampler was operated. If the air sampler is operating during an accidental release of radionuclides, multiplication of this average value by the estimated release duration will give the *time-integrated air concentration* (Bq s m^{-3}). This is an important intermediate quantity in calculating doses from accidental releases to atmosphere. Doses from intakes of radionuclides by inhalation can be estimated by multiplying the time-integrated air concentration by the breathing rate (m^3 s^{-1}) to give the intake of radionuclide and then by the appropriate dose coefficient to yield the dose as follows:

$$\text{Dose} = \text{Time-integrated air concentration (Bq s m}^{-3}) \times \text{breathing rate (m}^3\text{ s}^{-1}) \times \text{dose coefficient (Sv Bq}^{-1}) \qquad (16.5)$$

This simple calculation assumes that only one radionuclide is present. If there is a mixture of different radionuclides, radiochemical analysis or gamma-spectrometry will have to be performed on the filter in order to establish the amounts of the different radionuclides present. The calculation would have to be performed for each radionuclide separately and the results then summed.

Radionuclides present as particulate material in the atmosphere will deposit on the ground and give rise to doses from, for example, soil-associated pathways (see Section 16.3.1). Two mechanisms cause particulate material to deposit, i.e. impaction on the underlying surface, termed *dry deposition*, and the action of rain, termed *wet deposition*.

Deposits of radionuclides on the ground can be estimated from the time-integrated air concentration by multiplication by a deposition velocity appropriate for that form of material. The *deposition velocity* is defined as the ratio of the amount of material deposited on the surface per unit area per unit time, to the concentration in air per unit volume at the surface. Typical deposition velocities for dry deposition of micron-sized particulates are around 10^{-3} m s^{-1}. An appropriate deposition velocity for reactive gases, such as iodine, would be 10^{-2} m s^{-1}. Iodine in an organic form has a lower deposition velocity of around 10^{-5} m s^{-1}. Noble gases do not deposit on the ground and therefore have a deposition velocity of zero.

Thus, the calculation becomes:

$$\text{Time-integrated air concentration (Bq s m}^{-3}\text{)} \times \text{deposition velocity (m s}^{-1}\text{)} = \text{ground deposit (Bq m}^{-2}\text{)} \tag{16.6}$$

If the time-integrated air concentration in the above equation is replaced by an air concentration, the result would be a deposition rate.

Under most circumstances in northern European temperate environments, at least, wet deposition is an important mechanism for transferring particulate material and reactive gases from the plume to the ground, especially in the case of continuous discharges over time periods greater than a few weeks or months. In such circumstances, wet deposition can be taken into account in an approximate manner by using an overall deposition velocity of about 10^{-2} m s^{-1} for particulate material and reactive gases (IAEA, 2001b). A more detailed consideration of wet deposition is given in Simmonds *et al.* (1995).

Carbon-14 and tritium do not generally deposit in the simple manner described above. The transfer of these radionuclides from atmosphere to the terrestrial environment is a complex process but can often be represented by a relatively simple specific activity approach. In this approach it is assumed that the specific activity of the ^{14}C and ^{3}H in foodstuffs grown in the terrestrial environment is the same as that in the atmosphere. The specific activity in the atmosphere is estimated from the results from the dispersion models, using a value for the carbon concentration in the atmosphere of 0.15 g m^{-3} with the corresponding value for water being 8 g m^{-3} (Simmonds *et al.*, 1995).

In theory, the deposition to other surfaces such as buildings or even skin can be estimated if the appropriate deposition velocity is available. The estimate of deposition on the ground can be used as described in the preceding section to calculate doses via soil-associated pathways.

Doses from external exposure to airborne radionuclides are estimated by taking account of the concentration of radionuclides in the plume, together with the size, shape and height of the plume at the location of interest. There are two approaches to this.

The first approach is the *semi-infinite cloud model*. In this approach, it is assumed that the concentration of radionuclides is uniform over the volume of the plume from which photons can reach the point at which the dose is being delivered and that the plume is in radiative equilibrium, i.e. the amount of energy absorbed by a given element of the plume is equal to that released by the same element. The absorbed dose rate in air is given by the following relationship:

$$D_\gamma = k_l X \sum_{j=1}^{n} I_j E_j \tag{16.7}$$

where D_γ is the absorbed dose rate in air (Gy per year), X is the annual average concentration of the particular radionuclide in air (Bq m^{-3}), E_j is the initial energy of the photon (MeV), I_j is the fraction of photons of initial energy E_j emitted per disintegration, n is the number of photons of particular energies emitted per disintegration, and k_l is a constant equal to 2.0×10^{-6} (Gy per year per MeV m^{-3} s^{-1}).

The mean free path length of photons of energy less than about 20 keV is sufficiently short that the assumptions inherent in this model are likely to be adequate in all situations. However, for photons of higher energy the assumptions made in this model are likely to be invalid near to the release point and the contributions to the dose rate at a particular location from points in the plume should be modelled explicitly by using a *finite-cloud model*. This involves simulating the plume by a number of small volume sources and integrating over these sources. A detailed description is beyond the scope of this present book and the interested reader is referred to Simmonds *el al.* (1995).

In estimating these doses, account should be taken of the time likely to be spent indoors. Generally, models estimate dose rates in air in the open. Individuals in buildings will be shielded to certain extent, depending upon the type of structure, from photons emitted from the part of the plume that is outside the building. The shielding provided by buildings can reduce the dose rate indoors to between 0.01 and 0.9 of that outdoors (Brown, 1988). Typical values for European-type buildings are 0.2 for family houses and 0.07 for multi-storey buildings (Brown and Jones, 1993). The doses are also likely to be lower to people outdoors in an urban environment from those calculated for outdoors in rural areas by the models because of the shielding afforded by neighbouring buildings; a factor of 0.7 has been proposed to take account of this phenomenon (Brown and Jones, 1993).

Alternatively, an estimate of the annual effective dose from external irradiation from airborne gamma-emitters can be estimated from the following equation (the appropriate dose-conversion factors, calculated using a semi-infinite cloud model, are given in an IAEA publication (IAEA, 2001b)):

$$D_G = C_A D_{FI} O \tag{16.8}$$

where D_G is the annual effective dose from immersion in the plume of gamma-emitters, and C_A is the annual average concentration in air of gamma-emitting radionuclides (calculated using models described in Chapter 14, in Bq m^{-3}). The parameter D_{FI} is the dose coefficient relating unit annual average concentration of a particular radionuclide in air to annual dose (Sv per year per Bq m^{-3}). The factors for common radionuclides are given in Table 16.3. (Note that the factors given in this table are valid for semi-infinite cloud conditions.) In addition, O is an occupancy factor to allow for the fraction of the year that an individual is in a plume of a particular average concentration.

External doses from beta-particles emitted by radionuclides in the plume can be calculated in a way similar to that described above for photons. However, as the distance travelled by beta-particles, several metres at most, is generally much smaller than the dimensions of the plume, an infinite-cloud model can be used in all circumstances. Shielding by buildings can also be treated in a simplified way. In circumstances where radionuclides are being released to atmosphere for extended periods of time, e.g. routine releases, and thus the air inside buildings has reached equilibrium with that outside, the effects of building shielding can be ignored when estimating doses from beta-particles. This is because both indoors and outdoors most of the dose will come from the volume of air within a few metres of the individual. In situations where the air inside the building is not in equilibrium with that outside,

Table 16.3 Various coefficients for calculating doses from external exposure from radionuclides in a plume in the atmosphere[a] (Sv per year per Bq m^{-3}).

Nuclide	Effective dose
^{60}Co	4.0×10^{-6}
^{137}Cs[b]	8.7×10^{-7}
^{129}I	1.2×10^{-8}
^{131}I	5.8×10^{-7}
^{35}S	1.0×10^{-10}
^{131m}Te[c]	2.5×10^{-6}

[a]Data taken from IAEA (2001b). Note that this table only presents some selected examples – the reference should be consulted for a comprehensive list.
[b]Includes a contribution from progeny ^{137m}Ba.
[c]Includes contributions from progeny ^{131}Te and ^{131}I.

such as may be the case for a very short-term release, the building could shield an individual from almost all of the potential dose from the beta-particles.

Typical ranges of beta-particles in tissue are of the order of a few millimetres at most and so the only body organ that needs to be considered in estimating doses from external irradiation is the skin. The surface layer of the skin is dead tissue; radiosensitive cells are located in the basal layer at a typical depth of 70 μm. Thus, the dose rate in skin is estimated from the absorbed dose rate in air by allowing for the absorption of energy in the skin's surface 70 μm layer as follows:

$$H_\beta = 0.5e^{-\mu d} D_\beta w_r \tag{16.9}$$

where H_β is the equivalent dose rate in skin (Sv per year), w_r is the radiation weighting factor for beta-particles (1, see Section 4.3), μ is the absorption coefficient in tissue and is inversely proportional to the range in tissue of the mean energy of the beta-particles being considered (m^{-1}), d is the thickness of the epidermal layer (70 μm), and D_β is the absorbed dose rate in air (Gy per year).

Dose factors enabling annual skin doses to be estimated by using the following equation are given in Table 16.4 (IAEA, 2001b):

Table 16.4 Various coefficients for calculating doses to the skin from radionuclides in a plume in the atmosphere[a] (Sv per year per Bq m^{-3}).

Nuclide	Skin dose
^{137}Cs[b]	1.4×10^{-6}
^{32}P	1.4×10^{-6}
^{35}S	9.2×10^{-9}
^{90}Sr[c]	2.9×10^{-7}
^{60}Co	4.6×10^{-6}

[a]Data taken from IAEA (2001b). Note that this table only presents some selected examples – the reference should be consulted for a comprehensive list.
[b]Includes a contribution from progeny ^{137m}Ba.
[c]Includes a contribution from progeny ^{90}Y.

$$D_S = C_A \times D_{SI} \times O \qquad (16.10)$$

where D_S is the annual equivalent dose to the skin (Sv per year), and D_{SI} is the dose factor relating unit annual average concentration in air of a particular radionuclide to skin dose (Sv per year per Bq m^{-3}). The other symbols are as given above for gamma-emitting radionuclides.

The average annual dose equivalent to the skin is multiplied by the weighting factor for this tissue, i.e. 0.01, in order to estimate the contribution to effective dose (see Chapter 4).

16.3.3 Water

The aquatic environment represents surface waters, including streams, rivers and lakes, together with estuaries and the marine environment. Various types of installations, including electricity-generating power reactors, hospitals and research laboratories, make discharges of very low levels of radioactive materials into this environment, mainly in the first instance to rivers.

The first issue in estimating doses from measurements of radionuclides in water is whether the measurement has been taken on filtered water or not. This can make a considerable difference to the result as some radionuclides, including isotopes of plutonium, can bind strongly, but usually reversibly, to suspended sediments.

The fraction of the total concentration of a radionuclide that is present in solution, i.e. in filtered seawater, F_w, is given by the following equation:

$$F_w = \frac{1}{1 + K_d \alpha} \qquad (16.11)$$

where α is the suspended sediment load (t m^{-3}), and K_d is the distribution coefficient (Bq t^{-1} per Bq m^{-3}).

The distribution coefficient is defined as the ratio of the amount of radionuclide per unit weight of dry sediment to the amount per unit volume of water. Values of K_d for various elements in coastal and marine waters have been published by the IAEA (IAEA, 1985). The corresponding values for the freshwater environment are often different; one reason for this is the competition from ions in the marine environment. Values for some parent elements of radionuclides of interest are given in Table 16.5 for both the freshwater and marine environments. Typical suspended sediment loads in ocean waters far from land are around 1×10^{-7}t m^{-3}; the corresponding values for coastal seas are usually between one and two orders of magnitude higher. Thus for ocean waters, those radionuclides with values of K_d higher than 10^7 (Bq t^{-1} per Bq m^{-3}) will be predominantly associated with sediment. For coastal waters, the corresponding K_d is around 10^5 to 10^6 (Bq t^{-1} per Bq m^{-3}).

The concentration of radionuclides in aquatic foods (fish, crustaceans, molluscs and seaweed) can be estimated from the concentration of radionuclide in filtered seawater by using the appropriate concentration factor. Once again, values for individual elements can be different in the freshwater and marine environments. Concentration factors for various radionuclides in seafoods have been published by the IAEA (IAEA, 1985) and a selection is reproduced in Table 16.6, together with values for freshwater foodstuffs.

Table 16.5 Distribution coefficients (K_d) for elements in the aquatic environment ($Bq\ t^{-1}$ per $Bq\ m^{-3}$).

Element	K_d		
	Coast[a]	Ocean[a]	Freshwater[b]
H	1	1	3×10^{-2}
C	2×10^3	2×10^3	2×10^3
Co	2×10^5	1×10^7	2×10^4
Ru	3×10^2	1×10^3	7×10^3
Cs	3×10^3	2×10^3	2×10^3
Pu	1×10^5	1×10^5	1×10^5
Am	2×10^6	2×10^6	4×10^5

[a]Data taken from IAEA (1985). Note that this table only presents some selected examples – the reference should be consulted for a comprehensive list.
[b]Data taken from Simmonds *et al.* (1995). Note that this table only presents some selected examples – the reference should be consulted for a comprehensive list.

Table 16.6 Concentration factors for elements in the aquatic environment ($Bq\ t^{-1}$ per $Bq\ m^{-3}$).

Element	Concentration factor				
	Marine[a]			Freshwater[b]	
	Fish	Molluscs	Seaweed	Fish	Plants
H	1	1	1	9×10^{-1}	9×10^{-1}
C	2×10^4	2×10^4	1×10^4	4.6×10^3	4.6×10^3
Co	1×10^3	5×10^3	1×10^4	300	470
Ru	2	2×10^3	2×10^3	10	200
Cs	100	30	50	2×10^3	500
Pu	40	3×10^3	2×10^3	3.5	350
Am	50	2×10^4	8×10^3	25	5×10^3

[a]Data taken from IAEA (1985).
[b]Data taken from Simmonds *et al.* (1995).

Radionuclides in the aquatic environment can result in radiation doses from a number of exposure pathways in addition to the consumption of foods. The most important group of additional pathways is that arising from the presence of people on beaches – referred to as *beach occupancy* – or on riverbanks. The presence of radionuclides in these materials can result in external exposure and doses from re-suspension of activity and inadvertent ingestion; the manner in which these doses are calculated is very similar to the way they are calculated in the case of soils (see Section 16.3.1).

The starting point is an estimate of the concentration of radionuclides in beach or riverbank materials. If an existing situation is being investigated, then direct measurements can be made but it is also possible to arrive at a rough estimate of radionuclide concentrations beach materials from measurements on water taken from the neighbouring sea area or river or from measurements on the suspended sediments. As a rule of thumb, the concentration of radionuclides in coastal beaches will be around 10 to 100 times lower than the concentration in the suspended sediment. This reduction in concentration arises because the radionuclides bind to the surface of the sediment particles and, as beach particles are larger than suspended particles, the surface area of beach particles is lower per unit mass than that of sediment particles.

Doses from external exposure can be calculated in a similar way to that described for soils (see Section 16.3.1). However, beaches tend to be better mixed over depth than do soils and Hunt (1984) has developed a simple model to estimate external exposure from radionuclides in beaches. This model assumes that the radionuclides are homogeneously mixed down to a depth of 30 cm in wet sand of density 1.6 g cm^{-3}, as follows:

$$D = 0.1584 \times C \times E \qquad (16.12)$$

where D is the absorbed dose rate in air at a height of 1 m (μGy h^{-1}), C is the concentration of a particular radionuclide in wet sand (Bq cm^{-3}), and E is the mean gamma-energy per decay of the radionuclide (MeV).

The parameter C can be replaced by the radionuclide concentration by weight in dry sediment (Bq g^{-1}). The absorbed gamma-dose rates in air can be converted to effective dose rates by multiplying by 0.87.

The estimates of dose rates have to be combined with estimates of the time individuals spend on beaches in order to calculate their doses.

People on beaches or riverbanks may breathe in or inadvertently ingest material. Doses from the former pathway can be estimated by using a dust-loading approach (see Section 16.3.1); in the absence of specific information, it can be assumed that the airborne concentration of beach materials is 100 μg m^{-3} (Wilkins *et al.*, 1994). Inadvertent ingestion of beach material can occur following contamination of hands by foodstuffs. This will obviously vary considerably from person to person and will also be different for different age groups; values of 42 mg h^{-1} for adults and 10 year olds, and 84 mg h^{-1} for one year old children have been proposed (Wilkins *et al.*, 1994).

Freshwater is used as drinking water. Doses from this pathway are simply estimated from the radionuclide content of the water, the amount of water drunk and the appropriate dose coefficient. Where water has been treated prior to drinking, removal of radionuclides by the treatment may need to be taken into account.

Radionuclides in the aquatic environment can be transferred to the terrestrial environment. Flooding of rivers and the use of river water for irrigation and for animals' drinking water are obvious examples. Estimating the doses delivered from these pathways will probably involve using the procedures described in the previous two sections. In the marine environment, radionuclides can be transferred on an ongoing basis via sea spray. This is probably only significant for areas close to points where radionuclides are being discharged in the marine environment, for example, from nuclear installations situated on the coast. Details of how to calculate doses from this exposure pathway are described in Simmonds *et al.* (1995).

Other exposure pathways that deserve consideration include external exposure from handling of fishing gear and external exposure from swimming in water containing radionuclides. Simple models for estimating gamma-dose rates in these situations have been described by Hunt (1984). The model for estimating dose rates from radionuclides in sands entrained in fishing gear is a simple modification of the model described above for exposure from beaches, with a factor of 0.1 being introduced to allow for the lower solid angle subtended at the person by the fishing gear than by the beach, and a further factor of 0.1 introduced to convert from radionuclide concentration in beach material to radionuclide concentration per unit volume of fishing gear. The model for estimating the gamma-dose rate to a swimmer or diver is described by the following relationship:

$$D = 0.576 \times C \times E \tag{16.13}$$

where D is the dose rate in water (μGy h^{-1}), C is the concentration of a particular radionuclide in seawater, including that on suspended sediment (Bq g^{-1}), and E is the mean gamma-energy per decay of the radionuclide (MeV).

The dose rate in tissue can be taken to be the same as that in water. This equation only gives the doses from gamma-exposure. Eckerman and Ryan (1993) include doses from beta-emitting radionuclides.

Small users of radioactive materials, such as hospitals and research laboratories, often discharge into drains, which pass their effluent to sewage systems before release into rivers. Thus, pathways such as external exposure and inhalation of resuspended material could expose workers at sewage works. Furthermore, in some cases the treated solids are used for land conditioning and this could lead to additional exposures from, say, ingestion of radionuclides from foodstuffs grown on the land. Doses from these pathways can be evaluated by using methods described in this present section.

16.3.4 Terrestrial Foodchains

Some examples of how to calculate doses from radionuclides in foodstuffs have been given in earlier sections. Those examples covered simple situations. However, the processes leading to the presence of radionuclides in human foodstuffs can be complex and in order to understand the radiological implications of releases of radionuclides to the environment such situations may need to be considered more fully.

Releases of radionuclides to air or water can lead to the presence of radionuclides in terrestrial foodstuffs. The basic processes are represented in Figure 16.3. Radionuclides in air can deposit onto agricultural land. Part of the activity will deposit directly onto the ground and the foliage of any vegetation growing on the land may intercept a part. Radionuclides can be deposited in a similar manner from water used for irrigation, although the ratio of the activity intercepted by foliage versus that deposited directly onto the ground will not necessarily be the same as for a deposition from atmosphere.

Radionuclides are removed from the surfaces of plants by weathering and other natural processes, with half-lives ranging from a few days to several tens of days. A fraction of the surface deposit may be absorbed and translocated to other parts of the plant – a process known as *translocation*. The significance of this process depends, among other things, upon the chemical characteristics of the radionuclide and, as

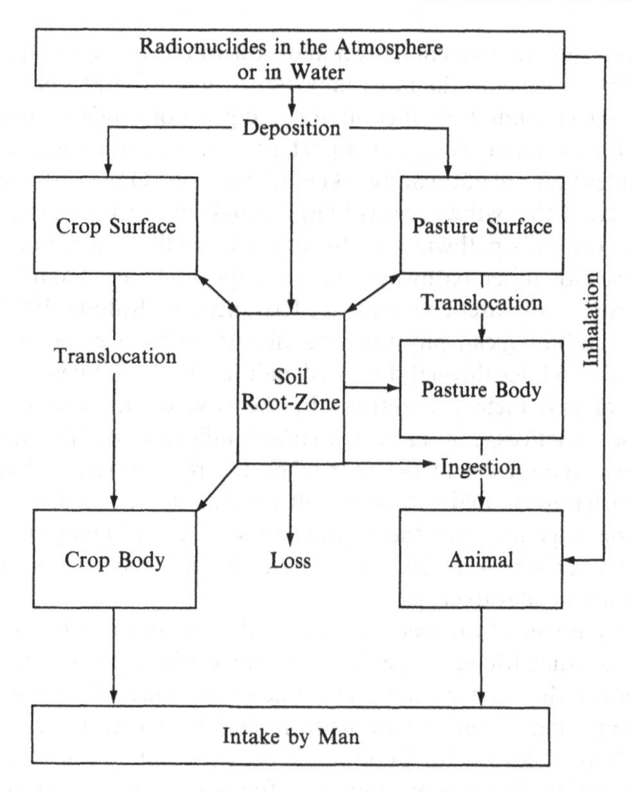

Figure 16.3 Illustration of the basic foodchain processes (adapted from Simmonds *et al.*, 1995)

such, is more important for some radionuclides, notably isotopes of caesium, than for others, such as isotopes of plutonium. If the plant is ingested immediately following the deposition without any preparation, such as washing, this can in itself be an important pathway of human exposure, either directly or via animal foodstuffs.

Radionuclides in soil may, as described in Section 16.3.1, be absorbed into plants via the roots and subsequently transfer to the edible parts of the plant. Following a accidental, or short-term release, root uptake becomes important when direct surface contamination has declined due to weathering, etc. In such situations, the process is significant for the longer-lived, biologically available radionuclides, such as ^{90}Sr and ^{137}Cs. Re-suspension processes and splashing during rainfall can lead to the contamination of plant surfaces by radionuclides in soils. Such processes tend to be important only in the long term and for radionuclides that are relatively poorly taken up by roots, e.g. isotopes of plutonium.

Radionuclides are lost from the rooting zone by migration down the soil column. In some cases, biogeochemical changes may also occur that alter the availability of radionuclides for root uptake. A well-known example is the fixation of caesium by clay particles in some soils – a process that can markedly reduce root uptake in some soils (Wilkins *et al.*, 1985).

The transfer of radionuclides to animal food products is another important pathway for human exposure. Animals can take up radionuclides by inhalation, from drinking water, from inadvertent ingestion of soil and from ingestion of pasture grass or other fodder. In general, the most important route of intake of radionuclides by animals is via the consumption of contaminated grass or fodder. However, inadvertent consumption of soil may become important after deposition has stopped but only for radionuclides that are not readily taken up via roots. One of the most well known and well studied of the pathways involving animals in northern temperate countries is the pasture–cow–milk pathway (see, for example, Fulker and Grice, 1989). In other areas of the world, other pathways can be important; an example is the lichen–reindeer pathway in sub-arctic regions (see Rissanen and Rahola, 1989).

As well as the biological, physical and chemical processes described above, the transfer of radionuclides through the foodchain to man is influenced by the prevailing agricultural and dietary practices. In the case of accidental and short-term releases, these practices can have a marked influence on the transfer to man depending upon the season of the year when the release occurs. For example, the extent of transfer to animal products will depend upon whether the animals are grazing outside, together with the amount of stored feed being provided. Clearly, the time of year at which different crops are planted and harvested will influence transfer of radionuclides to crops.

There are a number of different models of the transfer of radionuclides through terrestrial foodchains. Models differ in their degree of complexity and, more importantly, in whether they can estimate the time-dependence of radionuclide transfer. This is very important when attempting to model the consequences of an accidental release of radionuclides as, for example, it is important to know whether and for how long restrictions on the consumption of foodstuffs are likely to be in place.

Models of transfer through the foodchain can be divided into two broad categories: *multiplicative* and *dynamic compartmental.* The first type comprises a series of factors that link different parts of the foodchain (see, for example Till and Meyer, 1983). Dynamic compartmental models are more complex. Transfer of radionuclides through the foodchain is modelled by a series of interconnected compartments. Thus, the process shown in Figure 16.3 could be represented in this manner. Examples of compartment models are given in Brown and Simmonds (1995) and Pröhl (1991).

16.4 ESTIMATION OF INDIVIDUAL AND COLLECTIVE DOSES

Information on the exposure of individuals and populations can be combined with data on habits to estimate doses. Before describing these procedures, some general points should be made about the use of measurements and models in estimating doses.

In using measurement data on radionuclides in environmental materials or measured gamma- or beta-dose rates to calculate doses, the following points should be remembered:

1. How representative are the measurements, i.e. do they adequately represent typical radionuclide concentrations in the material from which people could receive radiation doses?

2. Do the measurements cover the important exposure pathways or can doses via such pathways be adequately estimated from the measurements using models?

When using models, the important questions are as follows:

1. Are the models validated for the intended purpose?
2. Is the spatial and temporal resolution of the model appropriate?
3. Does the model adequately represent the important pathways?

16.4.1 Doses to Individuals

The dose to an individual is a function of his or her exposure to radionuclides and his or her habits, together with a radionuclide-dependent constant. The following example illustrates the basic principles.

Doses from the ingestion of radionuclides in foodstuffs can be calculated from measured or estimated radionuclide concentrations in the food. Supposing measurements show the presence of three radionuclides in potatoes, i.e. ^{239}Pu, ^{90}Sr and ^{137}Cs. The measured concentrations are, respectively, 2, 1 and 10 Bq kg^{-1}. This is the exposure. The dose to an individual consuming 10 kg of these potatoes – the habits – would be estimated in a fairly straightforward way by calculating the amount of each radionuclide in 10 kg of potatoes and multiplying this by a radionuclide-dependent factor, which in this case is the appropriate dose coefficient.[1] The dose coefficients for ingestion by an adult member of the public of ^{239}Pu, ^{90}Sr and ^{137}Cs are, respectively, 2.5×10^{-7}, 2.8×10^{-8} and 1.3×10^{-8} Sv Bq^{-1}. It should also be remembered that ^{90}Sr will be present in foods in secular equilibrium with its daughter ^{90}Y, a short-lived radionuclide with a half-life of 2.67 d. The dose coefficient calculated for ^{90}Sr takes account of ingrowth of ^{90}Y after intake into the body but not of the fact that the daughter will be present in the intake itself. Similarly, ^{137}Cs will be in equilibrium with its daughter ^{137m}Ba but in this case the daughter is so short-lived that its presence in the potato can be ignored for this purpose.

Thus, the calculation becomes as shown in the following table:

Radionuclide	Concentration (Bq kg^{-1})	Activity ingested (Bq)	Dose coefficient (Sv Bq^{-1})	Effective dose (Sv)
^{239}Pu	2	20	2.5×10^{-7}	5.0×10^{-6}
^{90}Sr	1	10	2.8×10^{-8}	2.8×10^{-7}
^{90}Y	1	10	2.7×10^{-9}	2.7×10^{-8}
^{137}Cs	10	100	1.3×10^{-8}	1.3×10^{-6}

which gives a total dose of 6.5×10^{-6} Sv.

[1] A dose coefficient is the radiation dose from an intake of 1 Bq of a radionuclide; internationally agreed values have been published for each radionuclide. The values are usually different for intakes by individuals in different age categories and for intakes by inhalation or ingestion (see Chapter 4).

This example illustrates a particular problem in carrying out this type of calculation, namely, the distribution of a particular radionuclide in a foodstuff may not be uniform. Studies on potatoes grown in plots of contaminated soil in the UK show that whereas ^{137}Cs is distributed fairly uniformly throughout the peel and flesh of the potato, about half of the ^{90}Sr is found in the peel together with almost all of the ^{239}Pu and ^{241}Am (Green *et al.*, 1997). This has implications for dose calculations in that, obviously, if measurements are made on whole potatoes but peeled potatoes are eaten, the dose resulting from any ^{239}Pu or ^{241}Am will be substantially overestimated. Thus, in estimating doses from measurements on foodstuffs it is important that the measurement data refer to the edible portions.

The calculation of doses to individuals is illustrated in a general way in Figures 16.4 and 16.5 for intakes of radionuclides and external exposure, respectively. The overall equations can be expressed as follows:

(a) For intakes of activity, we use the following equation:

$$\text{Dose per unit time} = \text{dose coefficient} \times \text{activity concentration} \times \text{intake rate} \times \text{fraction of time exposed} \quad (16.14)$$

(b) For external exposure, we use the following equation:

$$\text{Dose per unit time} = \text{dose rate per unit concentration} \times \text{activity concentration} \times \text{fraction of time exposed} \quad (16.15)$$

Individuals may receive doses from both types of exposure and so it may be necessary to sum the doses. As in all cases where doses are summed, care should be taken that the summation is realistic. In other words, make sure that ridiculous combinations of habits are not being assumed: the same individuals will not be high-rate consumers of all types of foodstuffs. Some information on human habits for characterizing critical groups is given in Table 16.7. The data are based on UK experience (see, for example, NRPB, 1998). Generalized data for several areas of the world for use in calculating critical group doses are given in IAEA (2001b).

One last point to note is that an important implication of the fact that doses are a function of exposure and habits is that the location where the exposures are highest need not be where the doses to individuals are highest. This is more likely to occur in the case of external exposure than for most other pathways.

16.4.2 Calculating Collective Doses

The collective dose (see Section 4.1.5) to a particular population from a specified source of radiation is the sum of all of the doses to that population from that particular source. The calculation normally involves dividing the geographical area around the discharge location into a number of areas, within each one of which radionuclides are assumed to be uniformly distributed. Exposure can be estimated by using the equations described above in Section 16.3 and combining with the population characteristics to provide a value for the collective dose for that area.

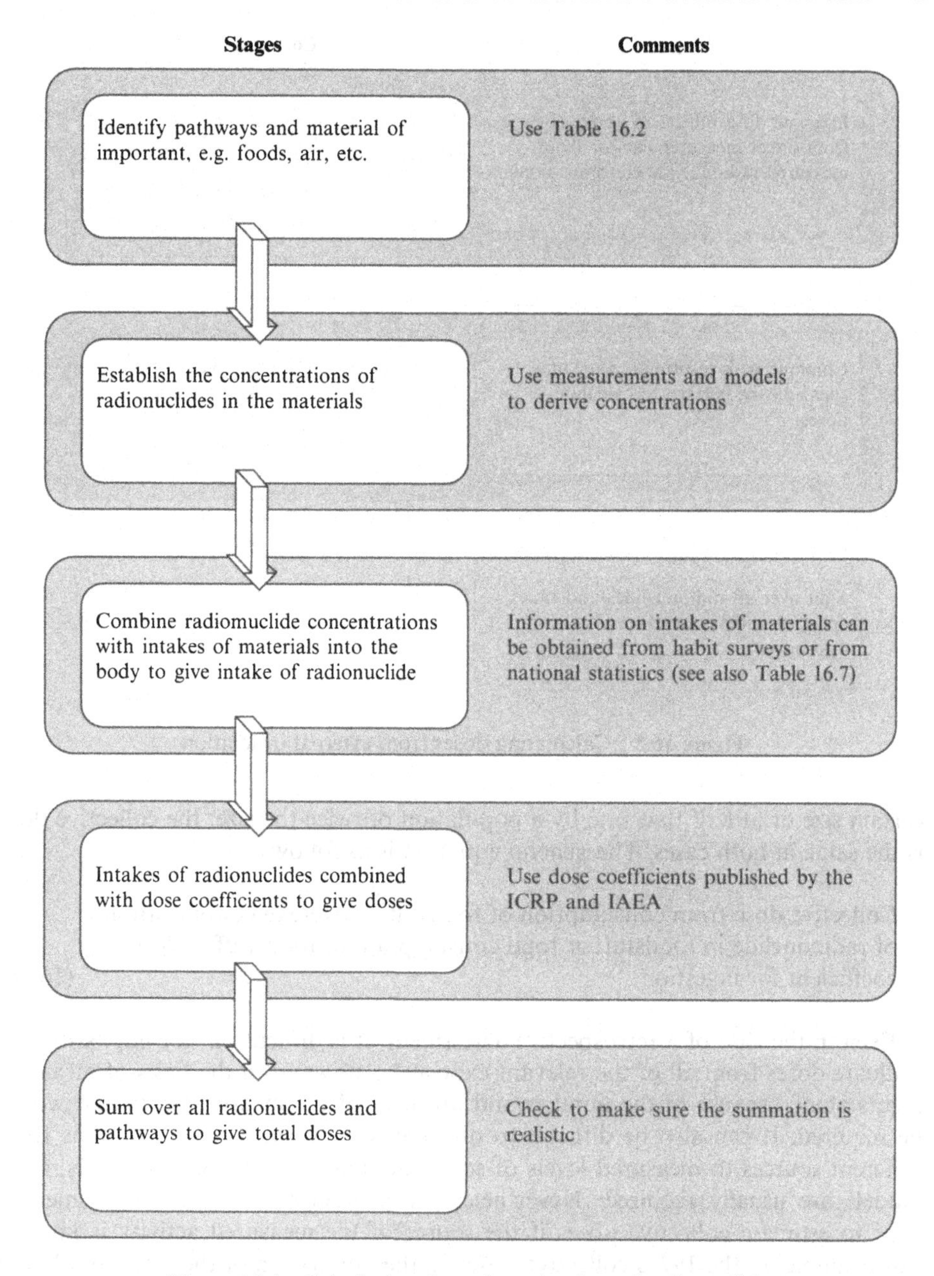

Figure 16.4 Calculating doses from intakes of radionuclides

The total collective dose would be obtained by summing across areas and exposure pathways. A knowledge of individual habits is often not required in this calculation. In particular, in estimating collective doses from the consumption of foodstuffs, the required parameters are the average radionuclide concentration in the foodstuff and the amount that is eaten by the exposed population. It doesn't matter whether the foodstuff is eaten at a particular rate by a population of a

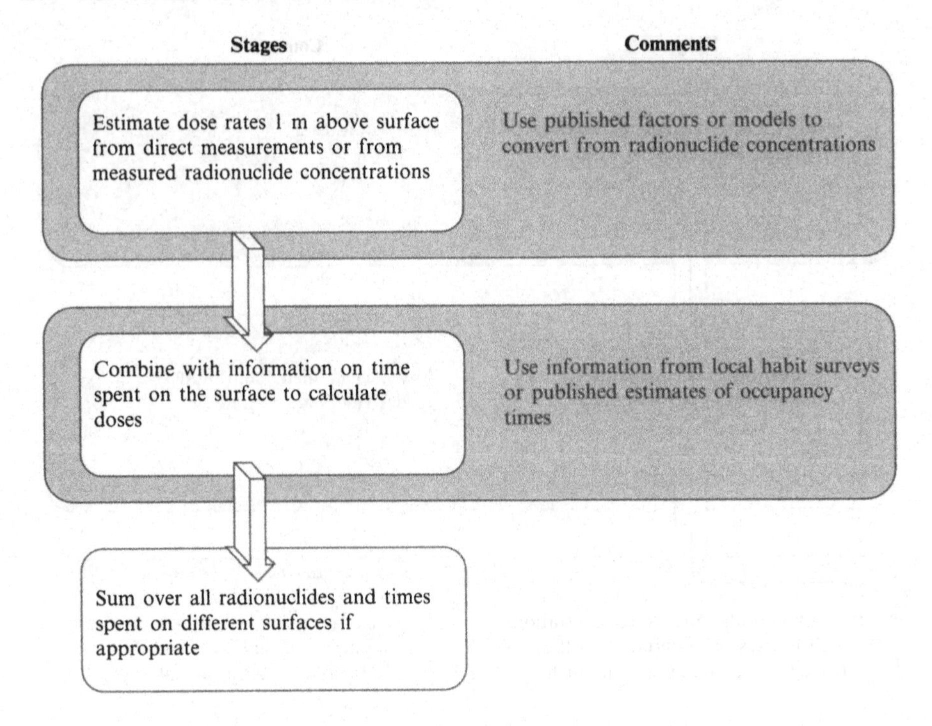

Figure 16.5 Calculating doses from external irradiation

certain size or at half that rate by a population of twice the size: the collective dose is the same in both cases. The general equation is as follows:

Collective dose from consumption of foodstuff = average concentration of radionuclide in foodstuff × total consumption of foodstuff × dose coefficient for ingestion (16.16)

Even in the case of a retrospective assessment, it is difficult, if not impossible, to evaluate doses from all of the relevant exposure pathways on the basis of measurements alone because of the number and diversity of the measurements that would be required. It can also be difficult to distinguish between the contributions from different sources to measured levels of some radionuclides in some materials; thus, models are usually required. Nevertheless, measurement data can sometimes be used to estimate collective doses if the source of the measured activity is known. For example, in the 1970s collective doses to the population of the UK from liquid discharges from the Sellafield reprocessing plant were estimated from measurements of ^{137}Cs concentrations in fish.

Measured levels of ^{137}Cs in fish from northern European marine waters can, potentially, contain contributions from discharges from the reprocessing plants at Sellafield and La Hague, weapons test fall-out and, latterly, the Chernobyl accident. In the 1970s, the dominant source of this radionuclide in these waters was discharges from Sellafield; the contribution from weapons test fall-out was corrected by using measurements on fish from locations distant from Sellafield. Knowing

Table 16.7 Example critical group habits.
Data on intakes[a]

Source	Annual consumption rates (kg per year)	
	Infants	Adults
Milk and milk products	365	300
Meat and meat products	10	75
Green vegetables	20	80
Root vegetables	50	130
Cereals	30	110
Fruit and fruit juices	50	60
Freshwater fish	1	10
Sea fish	5	100
Crustaceans	0	20
Molluscs	0	20
Drinking water	250	600
Inhalation rate (m^3 per year)	1900	7300

Data on occupancy

Parameter	Infants	Adults
Distance from discharge point (m)	1000	1000
Proportion of time outside (%)	10	30
River bank occupancy (h per year)	30	500
Beach occupancy (h per year)	30	2000
Shielding afforded by habitation[b]		
cloud gamma	0.2	0.2
deposited gamma	0.1	0.1

[a]Food, drinking water and inhalation rate.
[b]Unit-less.

the average concentrations of ^{137}Cs in the edible fraction of fish consumed in the UK, all that is required is information on the total amount of the fish consumed by the population in the relevant time period in order to calculate the corresponding collective dose.

16.5 APPLYING MATHEMATICAL MODELS OF ENVIRONMENTAL TRANSFER OF RADIONUCLIDES IN RADIOLOGICAL ASSESSMENTS OF RELEASES OF RADIONUCLIDES

Estimation of the radiological impact of *releases* of radionuclides to the environment will usually involve mathematical modelling of radionuclide *transfer* in the environment. A detailed description of such models is given in Chapters 13 to 15. There are, however, some important factors that have to be taken into account when applying these models to releases to the environment. These factors are listed here. The degree that they need to be addressed depends to some extent upon the 'assessment context'.

- The characteristics of the release.

– *Radionuclide content.* The radionuclide composition of the release should be characterized in terms of the discharge rate for each radionuclide or group of radionuclides. If radionuclides are grouped, it should be on the basis of radiological impact; for assessment purposes, one radionuclide will have to be taken as representative of the group.
– *Release location.* For atmospheric releases, this will include the release height and also factors such as buoyancy of the released gases and the influence of the immediate surroundings, including 'building-wake' effects. The discharge point for liquid releases will often be a purpose-built pipeline, but in the case of small users of radionuclides could be an entrance point into a sewerage system.

- Characterization of the receiving environment.

– This will include factors such as local topographic features, local population distribution, meteorological characteristics, information on local food production, and information on local habits. If the purpose of the assessment is the calculation of critical group doses, the location and habits of the group should be established.

Following characterization of the release and of the receiving environment, the next stage is to ensure that appropriate parameter values or ranges of values are used within the chosen model. It is during this stage that the modelling system could be validated to ensure that it represents the situation of interest. It should be checked that the modelling system can represent all of the exposure pathways of interest and has the appropriate degree of spatial and temporal resolution.

Appendix 1 Acronyms Used in the Text

ADC	Analogue-to-Digital Convertor
ADMLC	Atmospheric Dispersion Modelling Liaison Committee
ADMS	Atmospheric Dispersion Modelling System
ADR	Adsorbed Dose Rate
AERMIC	American Meteorological Society/Environmental Protection Agency Regulatory Model Improvement Committee
AERMOD	AERMIC Dispersion Model
ATMES	Atmospheric Model Evaluation Study
BEIR	Biological Effects of Ionizing Radiation Committee – A Committee of the US National Academy of Sciences
BGO	Bismuth Germanate
BIOMOVS	Biospheric Model Validation Study
BWR	Boiling-Water Reactor
CAMAC	Computer Automated Measurement and Control
CMB	Chemical Mass Balance
DDREF	Dose and Dose Rate Effectiveness Factor
DNA	Deoxyribonucleic acid
EC	European Commission
EPA	Environmental Protection Agency (USA)
ETEX	European Tracer Experiment
FA	Factor of Agreement
FAO	Food and Agriculture Organization
FB	Fraction Bias
FBR	Fast Breeder Reactor
FOEX	Factor of Exceedance
FWHM	Full-Width at Half-Maximum
GCR	Gas-Cooled Reactor
GPS	Global Positioning System
HLW	High-Level Waste
IAEA	International Atomic Energy Agency
ICRP	International Commission on Radiological Protection
ICRU	International Commission on Radiation Units and Measurements

IUPAC	International Union of Pure and Applied Chemistry
IUR	International Union of Radioecologists
Kerma	Kinetic Energy Released per Unit Mass
LET	Linear Energy Transfer
LILW	Low- and Intermediate-Level Waste
LWR	Light-Water Reactor
MCA	Multi-Channel Analyser
MCB	Multi-Channel Buffer
MDA	Minimum Detectable Activity
MLRA	Multiple Linear Regression Analysis
MOX	Mixed-Oxide Fuel
NAS	National Academy of Sciences (USA)
NCRP	National Council on Radiation Protection and Measurements
NEA	Nuclear Energy Agency of the Organization for Economic Co-operation and Development (OECD)
NIM	Nuclear Instrument Standard
NIST	National Institute of Standards and Technology
NMSE	Normalized Mean Square Error
NORM	Naturally Occurring Radioactive Material
NPL	National Physical Laboratory (UK)
NRC	Nuclear Regulatory Commission (USA)
NRPB	National Radiological Protection Board (UK)
PCA	Principal Component Analysis
PIPS	Passivated Implanted Planar Silicon
PVC	Poly(vinyl chloride)
PWR	Pressurised-Water Reactor
QA	Quality Assurance
QC	Quality Control
RMS	Royal Meteorological Society (UK)
SCA	Single-Channel Analyser
SHV	Super High Voltage
SSNTD	Solid-State Nuclear-Track Detectors
SURRC	Scottish Universities Research and Reactor Centre
UNSCEAR	United Nations Scientific Committee on the Effects of Atomic Radiation
VAMP	Validation of Environmental Model Predictions
WHO	World Health Organization
WL	Working Level
WMO	World Meteorological Organization

Appendix 2 Units, Terms and Conversion Factors for Radiation, Radioactivity and Related Areas

It is now accepted that, as far as possible, all measurements are expressed in Système International (SI) units – the International System of Units. In some areas of science (including radiation and radioactivity), units other than SI units can sometimes be used. In the Table 2 below, these will be designated as SI*. In addition, certain non-SI units are still (incorrectly) widely used and found in many, older textbooks. Some of these are also included in the table below but should now only be regarded as being of historical interest.

The SI system is based on a number of fundamental units of length, mass, time, etc. Thus, the basic unit of length is the *metre* (m), the basic unit of mass is the *kilogram* (kg), the basic unit of time is the *second* (s), the basic unit of (thermodynamic) temperature is the *kelvin* (K), and the basic unit of current is the *ampere* (A). There are other base units but these are not of interest here. The interested reader should consult books such as Drazil (1983) or reports such as CODATA (Cohen and Barry, 1987).

All other units are derived from these base units. A few of these derived units are given special names and those relevant to radiation or radioactivity are listed below.

Table 1 Derived SI units with special names.

Unit	Definition
becquerel (Bq)	The unit of activity of a radionuclide. The becquerel is equal to one reciprocal second. Since activity is in disintegrations per second, 1 Bq is equal to one disintegration per second
coulomb (C)	The quantity of electricity carried in one second by a current of one ampere
degree Celsius (°C)	The unit 'degree Celsius' is equal to the unit 'kelvin'. The zero of the Celsius scale is the temperature of the ice point (273.15 K)
gray (Gy)	The unit of absorbed dose of ionizing radiation, equal to one joule per kilogram. The absorbed dose is the mean energy imparted by ionizing radiation to matter, per unit of mass of irradiated material, at the place of interest. The gray may also be used as the unit of Kerma (see text and Appendix 1 for further details)
joule (J)	The work done when a material subject to a force of one newton is displaced through a distance of one metre in the direction of the force $= kg\ m^2\ s^{-2} = N\ m = W\ s = 6.2415 \times 10^{18}\ eV = 1.0 \times 10^7$ erg
newton (N)	The force, which when applied to a mass of one kilogram, gives it an acceleration of one metre per second squared $= kg\ m\ s^{-2}$
volt (V)	The difference of electrical potential between two points of a conducting wire carrying a constant current of one ampere when the power dissipated between these points equals one watt
watt (W)	A watt is the power which in one second gives rise to an energy of one joule

Table 2 Selected units and definitions.

Unit	Definition
angstrom (Å)	Unit of wavelength $= 10^{-10}\ m = 10^{-1}$ nm
atomic mass unit (u) [SI*]	1/12th of the rest mass of a neutral atom of ^{12}C in the ground state $= 1.660\,57 \times 10^{-27}$ kg
barn (b)	Unit of nuclear cross-section (used as a unit in nuclear reactions and indicates the probability of a reaction occurring) $= 10^{-28}\ m^2 = 10^{-24}\ cm^2$
becquerel per kilogram ($Bq\ kg^{-1}$)	SI unit of specific activity of a radionuclide $= 2.702\,70 \times 10^{-11}\ Ci\ kg^{-1} (= kg^{-1}\ s^{-1})$;1 $Ci\ kg^{-1} = 3.7 \times 10^{10}\ Bq\ kg^{-1} = 37\ GBq\ kg^{-1}$
coulomb per kilogram ($C\ kg^{-1}$)	SI unit of exposure (e.g. ionizing radiation) $= 1\ kg^{-1}\ s\ A = 1\ A\ m^2\ J^{-1}\ s^{-1} = 3.875\,97 \times 10^3$ R (roentgen)
coulomb per kilogram second ($C\ kg^{-1}\ s^{-1}$)	SI unit of exposure rate $= kg^{-1}\ A = 3.875\,97 \times 10^3\ R\ s^{-1}$
curie (Ci)	Unit of activity of a radionuclide $= 3.7 \times 10^{10}$ Bq = 37 GBq
curie per kilogram ($Ci\ kg^{-1}$)	Unit of specific activity of a radionuclide $= 3.7 \times 10^{10}\ Bq\ kg^{-1} = 37\ GBq\ kg^{-1}$
electronvolt per metre ($eV\ m^{-1}$) [SI*]	unit of linear stopping power and linear energy transfer $= 1.602\,19 \times 10^{-19}\ J\ m^{-1}$

electronvolt square metre per kilogram ($eV\ m\ kg^{-1}$) [SI*]	Unit of mass stopping power $= 1.602\,19 \times 10^{-19}\ J\ m^2\ kg^{-1}$
electronvolt (eV) [SI*]	Unit of energy – the energy an electron acquires when falling through a potential of of one volt $= 1.602\,19 \times 10^{-19}$ J
erg (erg)	CGS unit of work $= 10^{-7}$ J
fermi	Unit of length used for nuclear distances $= 10^{-15}$ m
gram-rad	Unit of integral absorbed dose $= 10^{-5}\ J = 10\ \mu J$
gray per second ($Gy\ s^{-1}$)	SI unit of absorbed dose rate and Kerma rate $= 1\ m^2\ s^{-3} = 1\ W\ kg^{-1} = 10^2\ rad\ s^{-1}$
litre (L) [SI*]	One decimetre cubed $= 1\ dm^3$
manSv	SI unit of collective effective dose (the sum of all doses to all the individuals in a given population) $= m^2\ s^{-2} = 1\ J\ kg^{-1} = 1\ Nm\ kg^{-1}$
rad (rad)	Unit of absorbed dose $= 10^{-2}\ Gy = 10^2\ erg^{-1}$
rad per second ($rad\ s^{-1}$)	Unit of absorbed dose rate $= 10^{-2}\ Gy\ s^{-1}$
rem	Originally defined as 'roentgen equivalent man' – unit of dose equivalent $= 10^{-2}$ Sv (now obsolete)
roentgen (R)	Unit of exposure $= 2.58 \times 10^{-4}\ C\ kg^{-1}$
roentgen per second ($R\ s^{-1}$)	Unit of exposure rate $= 2.58 \times 10^{-4}\ C\ kg^{-1}\ s^{-1} (A\ kg^{-1})$
roentgen metre squared per curie hour ($R\ m^2\ Ci^{-1}\ h^{-1}$)	Unit of a specific gamma ray constant $= 1.936\,94 \times 10^{-18} C\ m^2\ kg^{-1}$
sievert (Sv)	SI unit of dose equivalent (the absorbed dose with a weighting factor depending on the biological effect of the radiation) $= 1\ m^2\ s^{-2} = 1\ J\ kg^{-1} = 1\ Nm\ kg^{-1} = 100$ rem

Appendix 3 Data for the Most Important Environmental Radionuclides

Cosmogenic Nuclides

These are nuclides produced by nuclear reactions between cosmic-rays and the nuclei of atoms in the atmosphere, rocks, oceans, etc. Details are shown in Table 1 below.

Terrestrial Radionuclides

This component of background radioactivity comes from nuclides that occur naturally in the earth. There are in fact some several dozen naturally occurring radionuclides with half-lives comparable to the age of the earth ($\sim 4.5 \times 10^9$ years) and which are believed to be primordial, i. e. present in the material from which the earth was formed. Of these, however, only four are important in environmental monitoring, namely ^{40}K, ^{232}Th, ^{238}U and ^{235}U. The first of these, ^{40}K, simply decays to a stable nuclide. The latter three decay to a stable isotope of lead through a sequence of intermediate radionuclides of varying half-lives and decay modes that form a decay chain. If there is no chemical or physical separation of these intermediates, then a given series attains secular radioactive equilibrium and all of the members of the series may be detected. In Table 2 below, only the most important decay modes are listed. In addition only gamma-rays with energies greater than 20 keV and intensities greater than 1% are usually listed. For some nuclides emitting a large number of gamma-rays, only the most prominent are given. Intensities are *absolute*, i.e. gammas per 100 parent decays, unless otherwise indicated. The decay chains for ^{232}Th, ^{238}U and ^{235}U are shown in Figures 1–3, respectively.

Anthropogenic Radionuclides

The final component of environmental radioactivity originates from human activity. These radionuclides may be very locally confined but, on the other hand, may be distributed very widely, particularly those resulting from nuclear weapon tests

Table 1

Nuclide	Half-life[a]	Mode of decay (energy of emitted particle)[b,c]	Associated γ-ray energy (keV) (intensity (%))[d]	Production rate		
				Air[e]	Igneous rocks[f]	Limestone[f]
^{3}H	12.3 y	β^- (18.59 keV)	No γ-rays	0.25	2.1	2.0
^{7}Be	53.29 d	Electron capture	477.6 (10.4)	0.08	0.43	0.48
^{10}Be	1.6×10^6 y	β^- (555.9 keV)	No γ-rays	0.05	0.06	0.06
^{14}C	5730 y	β^- (156.5 keV)	No γ-rays	2.5	0.36	0.37
^{22}Na	950.8 d[g]	β^+ (500, 1800 keV)	1275 (99.9)[g]	8.6×10^{-5}	0.72	0.03
^{26}Al	7.4×10^5 y	β^+ (1067, 2196 keV) – 82.1%; electron capture – 17.9%:	1130 (2.5); 1809 (99.8); 2938 (0.25)	1.4×10^{-5}	1.18	0.04
^{36}Cl	3.01×10^5 y	β^- (709.6 keV) – 98.1%: β^+ /electron capture – 1.9%	No γ-rays	1.1×10^{-3}	0.20	0.84
^{37}Ar	35.04 d	Electron capture	No γ-rays	—	0.6	4.1
^{39}Ar	269 y	β^- (565 keV)	No γ-rays	5.6×10^{-3}	0.32	0.8
^{81}Kr	2.29×10^5 y	Electron capture	276.0 (0.3)	10^{-6}	—	—

[a]Data from Firestone (1996): d, days; y, years.
[b]The beta-energy given is the end-point energy of the beta-spectrum.
[c]Data from OECD/NEA Data Bank (1993).
[d]Data from Browne and Firestone (1986).
[e]Data from Kathren (1984): units of atoms s^{-1} cm^{-1}.
[f]Data from Yokoyama *et al.* (1977): units of atoms min^{-1} kg^{-1}.
[g]Data from IAEA (1991).

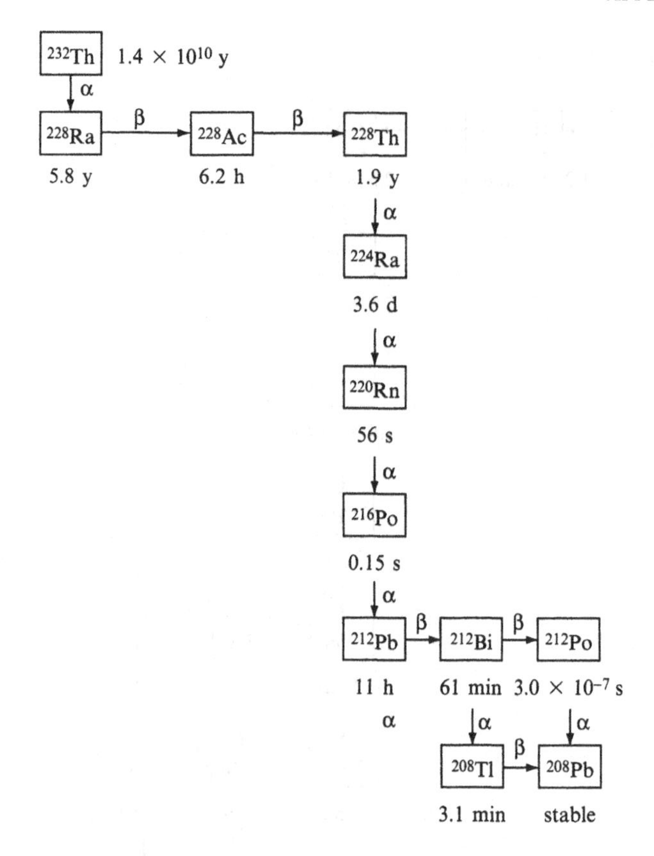

Figure 1 Decay chain for ^{232}Th: s, seconds: min, minutes: h, hours; d, days; y, years

and nuclear accidents. The main sources of anthropogenic radioactivity are weapon tests and industrial and medical applications of radionuclides. Routine discharges from nuclear power plants and reprocessing plants or from nuclear accidents make up only a small fraction of the total in this category, except on a local scale. The only exception to this is the radioactivity resulting from the Chernobyl accident. This was the only accident which resulted in significant amounts of activity appearing outside the boundaries of the country in which it occurred.

Two anthropogenic radionuclides, i.e. ^{3}H and ^{14}C, are also produced cosmogenically (see Table 1). This anthropogenic component may affect some types of measurement, e.g. ^{14}C dating. It then becomes necessary to separate the anthropogenic from the cosmogenic contribution.

In Table 3 below, only those nuclides are listed with half-lives greater than 10 days and which are readily measurable or contribute a significant fraction to the overall environmental dose to man.

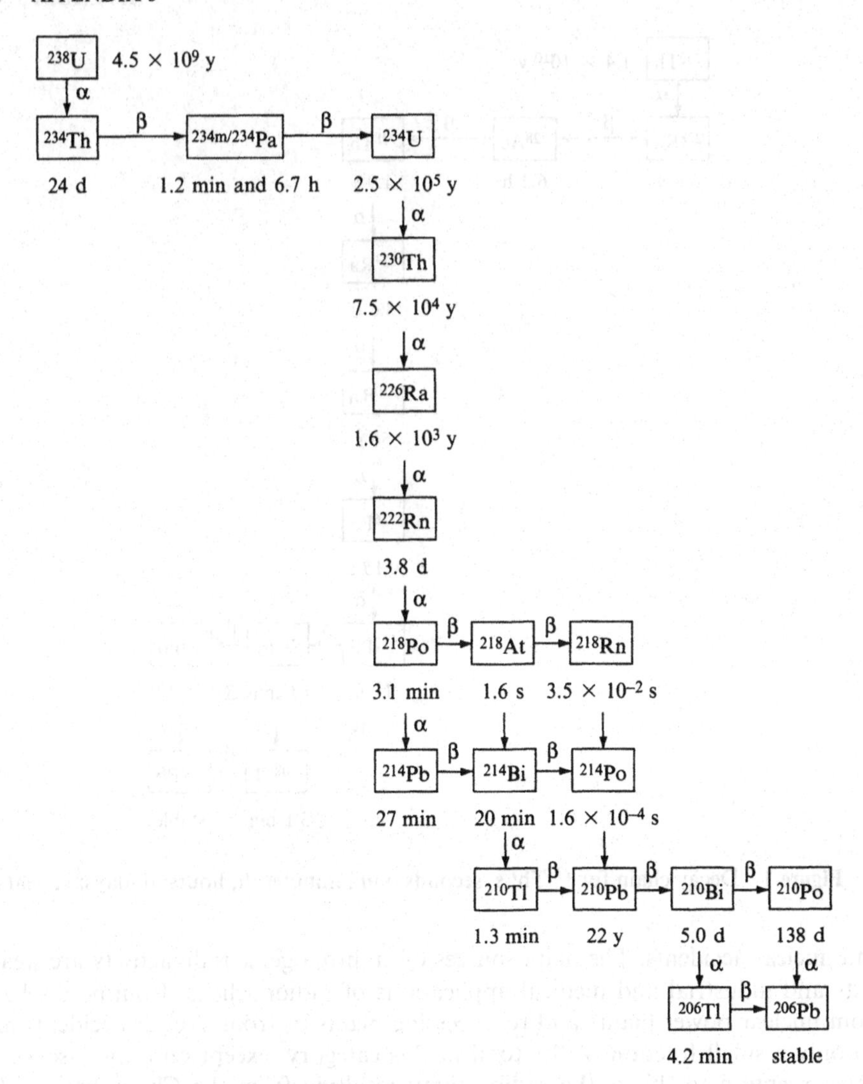

Figure 2 Decay chain for ^{238}U: s, seconds; min, minutes; h, hours; d, days; y, years

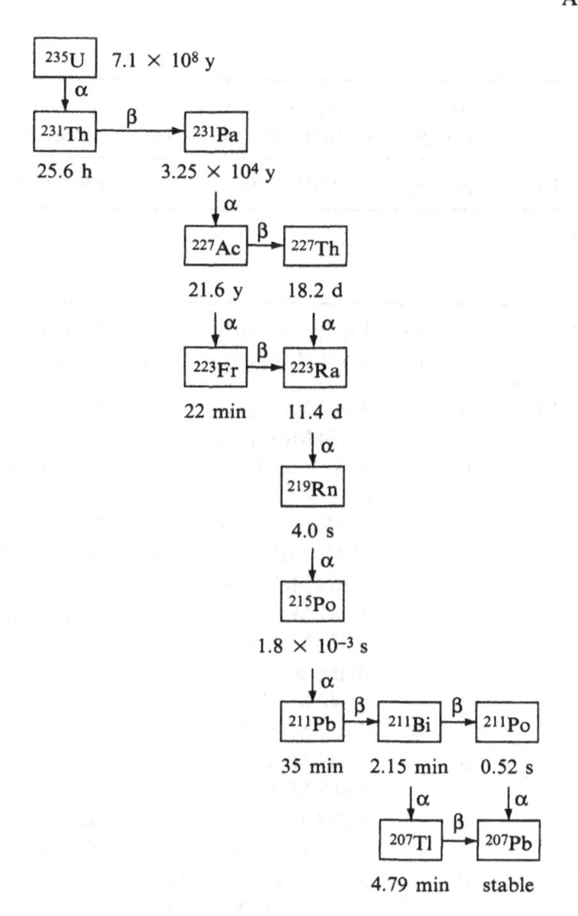

Figure 3 Decay chain for ^{235}U: s, seconds; min, minutes; h, hours; d, days; y, years

Table 2

Nuclide	Half-life[a]	Mode of decay	Energy of particle emitted (intensity (%))[b]	Associated gamma-ray energy (keV) (intensity (%))[c]
^{40}K	1.277×10^9 y	β^- EC[d]	1.33 MeV (89.3) (10.7)	1460.8 (10.7)

^{232}Th Decay Series

Nuclide	Half-life[a]	Mode of decay	Energy of particle emitted (intensity (%))[b]	Associated gamma-ray energy (keV) (intensity (%))[c]
^{232}Th	1.405×10^{10} y	α	4.013 MeV (77); 3.954 MeV (23)	No γ-rays
^{228}Ra	5.75 y	β^-	39 keV (56); 15.5 keV (35)	No γ-rays
^{228}Ac	6.15 h	β^-	491 keV (5); 606 keV (8.6); 1.014 MeV (6.7); 1.168 MeV (40); 1.741 MeV (6.5); 2.079 MeV (9.3) [plus many others]	93.4 (5.6); 338.4 (12.4); 463.1 (4.6); 794.8 (4.6); 911.2 (29); 964.6 (5.8); 969.0 (17.4) [plus several others]
^{228}Th	1.913 y	α	5.340 MeV (27.6); 5.423 MeV (71.7)	84.4 (1.3)
^{224}Ra	3.66 d	α	5.449 MeV (5.5); 5.685 MeV (94)	240.8 (3.9)
^{220}Rn	55.6 s	α	6.288 MeV (99.9)	No γ-rays
^{216}Po	0.145 s	α	6.779 (> 99.9)	No γ-rays
^{212}Pb	10.64 h	β^-	334 keV (82.6); 458 keV (5.1); 573 keV (12.3)	238.6 (43.6)
^{212}Bi	60.55 min	β^- (64.0 %)	1.519 MeV (4.5); 2.246 MeV (55.2) [plus 5 others]	39.8 (1.1); 727.2 (6.7)
		α (35.9 %)	6.051 MeV (25.2); 6.090 MeV (9.7) [plus 9 others]	785.5 (1.1); 1620.7 (1.5)
^{212}Po	2.99×10^{-7} s	α	8.785 MeV (100)	No γ-rays
^{208}Tl	3.05 min	β^-	1.286 MeV (24.4); 1.519 MeV (22.1); 1.796 MeV (49.4) [plus 12 others]	277.3 (6.8); 510.6 (21.6); 583.0 (86); 860.3 (12.0); 2614.4 (99.8) [plus several others]

^{238}U Decay Series

Nuclide	Half-life[a]	Mode of decay	Energy of particle emitted (intensity (%))[b]	Associated gamma-ray energy (keV) (intensity (%))[c]
^{238}U	4.468×10^9 y	α	4.150 MeV (23); 4.199 MeV (77)	— (< 1)

^{234}Th	24.10 d	β^-	105.8 keV (6.2); 106.2 keV (18); 198.5 keV (72.5) [plus 4 others]	63.3 (3.8); 92.4 (2.7); 92.8 (2.7)
^{234m}Pa	1.17 min	β^- (99.8 %) IT (0.16 %)[e]	2.197 MeV (98.2) [plus 21 others]	— (< 1)
^{234}Pa	6.7 h	β^-	483.7 keV (24.5); 484.4 keV (11.2); 654 keV (15.4) [plus 38 others]	131.3 (20); 569.5 (10.7); 883.2 (12.0); 925.7 (11) [plus many others]
^{234}U	2.455×10^5 y	α	4.723 MeV (28.4); 4.775 MeV (71.4) [plus 3 others]	— (< 1)
^{230}Th	7.538×10^4 y	α	4.621 MeV (23.4); 4.688 MeV (76.3) [plus 5 others]	— (< 1)
^{226}Ra	1600 y	α	4.785 MeV (94.5) [plus 4 others]	186.1 (3.3)
^{222}Rn	3.82 d	α	5.490 MeV (> 99.9)	No γ-rays
^{218}Po	3.10 min	α (> 99.9 %)	6.003 (100)	No γ-rays
^{214}Pb	26.8 min	β^-	678 keV (48.2); 735 keV (43); 1030 keV (5.6) [plus 8 others]	241.9 (7.5); 295.1 (19.2); 351.9 (37.1) [plus several others]
^{214}Bi	19.9 min	β^- (> 99.9 %)	1.505 MeV (18); 1.540 MeV (18); 3.270 MeV (16.1) [plus 62 others]	609.3 (46.1); 1120.3 (15.0); 1238.1 (5.9); 1764.5 (15.9) [plus many others]
^{214}Po	164.3 µs	α	7.687 MeV (> 99.9)	— (< 1)
^{210}Pb	22.3 y	β^-	16.5 keV (82); 63 keV (18)	46.5 (4.1)
^{210}Bi	5.013 d	β^-	1.162 MeV (> 99.9)	No γ-rays (< 0.1)
^{210}Po	138.38 d	α	5.305 MeV (100)	No γ-rays (< 0.1)

^{235}U Decay Series

Nuclide	Half-life[a]	Mode of decay	Energy of particle emitted (intensity (%))[b]	Associated gamma-ray energy (keV) (intensity (%))[c]
^{235}U	7.038×10^8 y	α	4.396 MeV (58); 4.368 MeV (15.8) [plus 15 others]	143.8 (10.5); 163.4 (4.7); 185.7 (53); 205.3 (4.7) [plus several others]
^{231}Th	25.52 h	β^-	206 keV (12.4); 287 keV (11.4); 288 keV (36.5); 305 keV (33.9) [plus 9 others]	84.2 (6.6)

Nuclide	Half-life[a]	Mode of decay	Energy of particle emitted (intensity (%))[b]	Associated gamma-ray energy (keV) (intensity (%))[c]
^{231}Pa	3.276×10^4 y	α	4.951 MeV (22.9); 5.014 MeV (25.4); 5.030 MeV (20)	283.7 (1.6); 300.1 (2.4); 330.1 (1.3)
^{227}Ac	21.773 y	α (1.38%)	4.938 MeV (0.52); 4.951 MeV (0.65); [plus 17 others]	— (< 1)
		$β^-$ (98.62 %)	20.5 keV (9.9); 35.7 keV (34.5); 45 keV (54.2)	— (< 1)
^{227}Th	18.72 d	α	5.757 MeV (20.5); 5.978 MeV (23.4); 6.038 MeV (24.5) [plus 55 others]	50.1 (8.5); 236.0 (11.2); 256.2 (6.7) [plus several others]
^{223}Fr	21.8 min	$β^-$	913 keV (12.0); 1068 keV (14.0); 1098V (61.0) [plus 18 others]	50.1 (33); 79.7 (8.9); 85.4 (2.4); 88.5 (4.0); 99.9 (1.4); 205.0 (11.1); 234.8 (3.7)
^{223}Ra	11.435 d	α	5.607 MeV (24.2); 5.716 MeV (52.5) [plus 23 others]	122.3 (1.2); 144.2 (3.3); 154.2 (5.6); 269.4 (13.6); 323.9 (3.9); 338.3 (2.8); 444.9 (1.3)
^{219}Rn	3.96 s	α	6.553 MeV (11.5); 6.819 MeV (80.8); [plus 9 others]	271.1 (9.9); 401.7 (6.6)
^{215}Po	1.781 ms	α	7.386 (> 99.9)	No γ-rays (< 0.1)
^{211}Pb	36.1 min	$β^-$	541 keV (6.6); 1373 keV (90.8) [plus 9 others]	404.9 (3.8); 427.0 (1.7); 831.9 (3.8)
^{211}Bi	2.14 min	α	6.279 MeV (16.4); 6.623 MeV (83.4)	350.1 (12.8)
^{207}Tl	4.77 min	$β^-$	1.422 MeV (99.8) [plus 2 others]	— (< 1)

[a]Data from Firestone (1996): s, seconds; min, minutes; h, hours; d, days; y, years.
[b]Data from OECD/NEA Data Bank (1993).
[c]Data from Browne and Firestone (1986).
[d]EC, electron capture.
[e]IT, isomeric transition.

Table 3

Nuclide	Half-life[a]	Mode of decay	Energy of particle emitted (intensity (%))[b]	Associated gamma-ray energy (keV) (intensity (%))[c]
^{54}Mn	312.5 d	EC (100%)[d]	—	834.8 (> 99.9)
^{55}Fe	2.73 y	EC (100%)[d]	—	No γ-rays
^{85}Kr	10.76 y	β^-	687 keV (99.6) [plus one other]	— (< 1)
^{89}Sr	50.50 d	β^-	1.492 MeV (> 99.9)	No γ-rays
^{90}Sr	28.78 y	β^-	546 keV (100)	No γ-rays
^{91}Y	58.51 d	β^-	338 keV (0.3); 1.543 MeV (99.7)	— (< 1)
^{95}Zr	63.98 d	β^-	365.2 keV (55); 397.7 keV (44.6); 887.2 keV (0.7)	724.2 (44.1); 756.7 (54.5)
^{103}Ru	39.35 d	β^-	112 keV (6.5); 225 keV (90); 723.1 keV (3.5) [plus 2 others]	497.1 (88.7); 610.3 (5.6)
^{106}Ru	373.6 d	β^-	39.4 keV (100)	No γ-rays
^{125}Sb	2.758 y	β^-	95.4 keV (13.6); 130.8 keV (18.1); 303.4 keV (40.2); 622.0 keV (13.5) [plus 4 others]	427.9 (29.4); 463.4 (10.5); 600.5 (17.8); 606.6 (5.0); 635.9 (11.3) [plus several others]
^{125}I	60.14 d	EC (100%)[d]	—	35.5 (6.7)
^{129}I	1.57×10^7 y	β^-	150.4 keV (100)	39.6 (7.5)
^{134}Cs	2.062 y	β^-	89 keV (27.1); 658 keV (70.3); [plus 3 others])	563.2 (8.4); 569.3 (15.4); 604.7 (97.6); 795.9 (85.4); 802.0 (8.7); 1365.2 (3.0) [plus several others]
^{137}Cs	30.07 y	β^-	512 keV (94.6); 1173 keV (5.4)	661.7 (85.2)
^{141}Ce	32.50 d	β^-	436 keV (70.5); 580 keV (29.5)	145.4 (48.4)
^{144}Ce	284.89 d	β^-	184.7 keV (19.6); 318.2 keV (76.5)	80.1 (1.1); 133.5 (11.1)
^{238}Pu	87.7 y	α	5.456 MeV (28.8); 5.499 MeV (71.0) [plus 13 others]	— (< 1)
^{239}Pu	2.411×10^4 y	α	5.106 MeV (11.5); 5.144 MeV (15.1); 5.156 MeV (73.3) [plus 66 others]	— (< 1)

Table 3 (*continued*)

Nuclide	Half-life[a]	Mode of decay	Energy of particle emitted (intensity (%))[b]	Associated gamma-ray energy (keV) (intensity (%))[c]
^{240}Pu	6563 y	α	5.124 MeV (27); 5.168 MeV (72.9) [plus 4 others]	— (< 1)
^{241}Pu	14.35 y	β^-	20.8 keV (> 99.99)	— (< 1)
^{242}Cm	162.8 d	α	6.070 MeV (25.9); 6.113 MeV (74) [plus 6 others]	— (< 1)

[a]Data from Firestone (1996): d, days; y, years.
[b]Data from OECD/NEA Data Bank (1993).
[c]Data from Browne and Firestone (1986).
[d]EC, electron capture.

Bibliography

Adams, J.A.S. and Lowder, W.M. (Eds), 1964. *The Natural Radiation Environment*, The University of Chicago Press, Chicago, IL.

Allen, S.E. (Ed.), 1989. *Chemical Analysis of Ecological Materials*, Blackwell Scientific, London.

Anderson, D.R., Sweeney, D.J. and Williams, T.A., 1981. *Introduction to Statistics – An Applications Approach*, West Publishing Company, St. Paul, MN.

Bevington, P.R., 1969. *Data Reduction and Error Analysis for the Physical Sciences*, McGraw-Hill, New York.

Birks, J.B. and Poullis, G.C., 1972. Liquid scintillators. In: Crook, M.A., Johnson, P. and Scales, B. (Eds), *Liquid Scintillation Counting*, Heyden and Son, London.

Cooper, P.N., 1986. *Introduction to Nuclear Radiation Detectors*, Cambridge University Press, Cambridge, UK.

Crouthamel, C.E., 1970. *Applied Gamma-Ray Spectrometry*, 2nd Edn (revised and enlarged by F. Adams and R.Dams), International Series of Monographs in Analytical Chemistry, Vol. 41, Pergamon Press, Oxford, UK.

Debertin, K. and Helmer, R.G., 1988. *Gamma- and X-Ray Spectrometry with Semiconductor Detectors*, North-Holland, Oxford, UK.

Department of the Environment, 1989. Sampling and Measurement of Radionuclides in the Environment, Report by the Methodology Sub-Group to the Radioactivity Research and Environmental Monitoring Committee (RADREM), HMSO, London.

Durrani, S.A. and Bull, R.K., 1987. *Solid State Nuclear Track Detection*, Pergamon Press, Oxford, UK.

Dyer, A., 1974. *An Introduction to Liquid Scintillation Counting*, Heyden, London.

Environmental Measurements Laboratory, 1992. Chieco, N.A., Bogen, D.C. and Knutson, E.O. (Eds), *Procedures Manual (HASL-300)*, 27th Edn, Vol. 1, US Department of Energy, New York.

EU Website (The European Union Online) [http://europa.eu.int/].

Geary, W., 1986. *Radiochemical Methods*, ACOL Series, Wiley, Chichester, UK.

Gilbert, R.O., 1987. *Statistical Methods for Environmental Pollution Monitoring*, Van Nostrand Reinhold, New York.

Harshaw Radiation Detectors, 1984. Harshaw Booklet, Harshaw/Filtrol, Solon, OH.

Hendee, W.R. and Edwards, F.M. (Eds), 1996. *Health Effects of Exposure to Ionising Radiation*, Institute of Physics, Bristol, UK.

Horrocks, D.L., 1974. *Applications of Liquid Scintillation Counting*, Academic Press, London.

International Atomic Energy Agency, 1970. Reference Methods for Marine Radioactivity Studies, International Atomic Energy Agency Technical Report Series No. 168, International Atomic Energy Agency, Vienna.

International Atomic Energy Agency, 1973. Reference Methods for Marine Radioactivity Studies, International Atomic Energy Agency Technical Report Series No. 169, International Atomic Energy Agency, Vienna.

International Atomic Energy Agency, 1974. Recommended Instrumentation for Uranium and Thorium Exploration, International Atomic Energy Agency Technical Report Series No. 158, International Atomic Energy Agency, Vienna.

International Atomic Energy Agency, 1975. Reference Methods for Marine Radioactivity Studies II, International Atomic Energy Agency Technical Report Series No. 169, International Atomic Energy Agency, Vienna.

International Atomic Energy Agency, 1989. Construction and Use of Calibration Facilities for Radiometric Field Equipment, International Atomic Energy Agency Technical Report Series No. 309, International Atomic Energy Agency, Vienna.

Knoll G.F., 1979 (1st Edn), 1989 (2nd Edn). *Radiation Detection and Measurement*, Wiley, New York.

Kratochvil, B., 1985. Sampling for chemical analysis of the environment: statistical considerations. In: *Trace Residue Analysis*, ACS Symposium Series, Vol. 284, American Chemical Society, Washington, DC, pp. 5–23.

Liquid Scintillation Counting, Proceedings of Symposia on Liquid Scintillation Counting: Vol. 1 (1971) Dyer, A. (Ed.); Vol. 2 (1972), Crook, M.A., Johnson, P. and Scales, B. (Eds); Vol. 3 (1974), Vol. 4 (1976) and Vol. 5 (1978), Crook M.A. and Johnson, P. (Eds): Heyden & Son, London.

Lodge, J.P., Jr (Ed.) 1989. *Methods of Air Sampling and Analysis*, 3rd Edn, Lewis Publishers, Chelsea, MI.

Low Level Measurements of Radioactivity in the Environment, Proceedings of International Summer Schools: First (1988) and Second (1991), Garcia-Leon, M. and Madurga, G. (Eds); Third (1994), Garcia-Leon, M. and Garcia-Tenorio, R. (Eds): World Scientific, Singapore.

McDowell, W.J., 1986. Alpha Counting and Spectrometry Using Liquid Scintillation Methods, Report NAS-NS-3116, Technical Information Center, US Department of Energy, Oak Ridge, TN.

McKeever, S.W.S., 1985. *Thermoluminescence of Solids*, Cambridge University Press, Cambridge, UK.

Mullard Ltd, 1977. Geiger–Müller Tubes, Technical Information Document, Mullard Ltd, London.

Pearson, E.S. and Hartley, H.O. (Eds), 1954, *Biometrika Tables for Statisticians*, Vol. 1, Cambridge University Press, Cambridge, UK.

Perry, R. and Young, R.J. (Eds), 1977. *Handbook of Air Pollution Analysis*, Chapman & Hall, London.

Reid, J.M., 1972, *The Atomic Nucleus*, Penguin Books, Harmondsworth, Middlesex, UK.

Sampling and Measurement of Radionuclides in the Environment, 1989. A Report by the Methodology Sub-Group to the Radioactivity Research and Environmental Monitoring Committee, (RADREM), HMSO, London.

Tölgyessy, J. and Bujdosó, E., 1991. *Handbook of Radioanalytical Chemistry*, Vols I and II, CRC Press, Boca Raton, FL.

Wilkes, R.J., 1987. *Principals of Radiological Physics*, Churchill Livingstone, Edinburgh, UK.

References

Aarkrog, A., 1996. Inventory of nuclear releases in the world. In: Luykx, F.F. and Frissel, M. J. (Eds), *Radioecology and the Restoration of Radioactive-Contaminated Sites*, NATO ASI Series, Kluwer Academic Publishers, Dordrecht, The Netherlands.

Aarkrog, A., Dahlgaard, H., Frissel, M., Foulquier, L., Kulikov. N.V., Molchanova, I.V., Myttenaere, C., Nielsen, S.P., Polikarpov, G.C. and Yushkov, P.I., 1992. Sources of anthropogenic radionuclides in the Southern Urals. *J. Environ. Radioact.*, **15**, 69–80.

ADMLC, 2000. Atmospheric Dispersion Modelling Liaison Committee Annual Report 1998/1999, National Radiological Protection Board, Chilton, UK.

Al-Khayat, T.A.H., Van Eygen, B., Hewitt, C.N. and Kelly, M., 1992. Modelling and measurement of the dispersion of radioactive emissions from a nuclear fuel fabrication plant in the UK. *Atmos. Environ.*, **26A**, 3079–3087.

Allisy, A., 1996. Henri Becquerel: the discovery of radioactivity. *Radia. Protect. Dosim.*, **68**, 3–10.

AMC Technical Brief, 2000. Is My Calibration Linear?, AMC Technical Brief No. 3, December 2000. The Royal Society of Chemistry, Cambridge, UK.

AMC Technical Brief, 2001. Robust Statistics: A Method of Coping With Outliers, AMC Technical Brief No. 6, April 2001, The Royal Society of Chemistry, Cambridge, UK.

Ames, B.N., 1989. Endogenous DNA damage as related to cancer and ageing. *Mutat. Res.*, **214**, 41–46.

ApSimon, H.A., 1986. The use of computers in emergency situations and the choice of dispersion model. In: *Emergency Planning and Preparedness for Nuclear Facilities*, Symposium Proceedings, Rome, 4–8 November 1985, International Atomic Energy Agency, Vienna, pp. 213–223.

Arlett, C.F., 1992. Human cellular radiosensitivity – the search for the holy grail or poisoned chalice. *Adv. Radiat. Biol.*, **16**, 273–292.

Attix, F.H., 1986. *Introduction to Radiological Physics and Radiation Dosimetry*, Wiley, New York.

Baetsle, L.H., 1991. Study of the Radionuclides Contained in Wastes Produced by the Phosphate Industry and their Impact on the Environment, European Commission EUR 13262EN, European Commission, Luxembourg.

Barnett, V., 1974. *Elements of Sampling Theory*, Hodder and Stoughton, London.

Barnett, V. and Lewis, T., 1985. *Outliers in Statistical Data*, 2nd Edn, Wiley Chichester, UK.

Basabikov, O.T., Gabbasov, M.N., Zelenov, V.I., Loborev, V.M., Markovtsev, A.S. and Sudakov, V.V., 1994. Altai Territory radioactive contamination levels and population effective doses exposure evaluation as a result of nuclear blasts at the Semipalatinsk test site. In: *Remediation and Restoration of Radioactive-Contaminated Sites in Europe*, Symposium Proceedings, Antwerp, 11–15 October 1993, European Commission, Luxembourg, RP **74**, pp. 403–420.

Baxter, M.S., MacKenzie, A.B., East, B.W. and Scott, E.M., 1996. Natural decay series radionuclides in and around a large metal refinery. *J. Environ. Radioact.*, **32**, 115–133.

Beck, M.B., Jakeman, A.J. and McAleer, M.J., 1993. Construction and evaluation of models of environmental systems. In: Jakeman, A.J., Beck, M.B. and McAleer, M.J. (Eds), *Modelling Change in Environmental Systems*, Ch. 1, Wiley, Chichester, UK, pp. 3–35.

Becker, P., 1989. Phosphates and Phosphoric Acid: Raw Materials, Technology and Economics of the Wet Process, Fertilizer Science and Technology Series Vol. 6, 2nd Edn, Marcel Dekker, New York.

BEIR IV, 1988. Health Risks from Radon and other Internally Deposited Alpha-Emitters, National Academy Press, Washington, DC.

Berger, M.J., Coursey, J.S. and Zucker, M.A., 1999. ESTAR, PSTAR and ASTAR: Computer Programs for Calculating Stopping-Power and Range Tables for Electrons, Protons and Helium Ions (Version 1.21) [Online]. Available: http://physics.nist.gov/Star [Accessed, 7 May 2002]. National Institute of Standards and Technology, Gaithersburg, MD.

Beychok, M.R., 1979. *Fundamentals of Stack Gas Dispersion*, 3rd Edn (updated to 1994), Milton R. Beychok, Irvine, CA.

BIOMOVS, 1993. Biospheric Model Validation Study BIOMOVS Technical Report 15, Final Report, Swedish Radiation Protection Institute, Stockholm.

Bodmer, W.F., 1991. Inherited susceptibility to cancer. In: Franks, L.M. and Teich, N.M. (Eds), *Introduction to the Cellular and Molecular Biology of Cancer*, Oxford University Press, Oxford, UK, pp. 98–124.

Bonelli, P., Calori, G. and Finzi, G., 1992. A fast long-range transport model for operational use in episode simulation. Application to the Chernobyl accident. *Atmos. Environ.*, **26**, 2523–2535.

Botezatu, E., Grecea, C., Botezatu, G., Capitanu, O., Sandor, G. and Peic, T., 1996. Radiation exposure potential from coal-fired power plants in Romania. In: *Proceedings of the 1996 International Conference on Radiological Protection*, Vol. 2, April 14–19 1996, Vienna, Austria, pp. 196–198.

Bradley, E.J., 1993. Natural Radionuclides in Environmental Materials, UK HM Inspectorate of Pollution, Report RR 93/063.

Bridges, O. and Bridges, J.W., 1995. Radioactive waste problems in Russia. *J. Radiol. Protect.*, **15**, 223–234.

Briggs, G.A., 1973. Diffusion Estimates for Small Emissions. Environmental Research Laboratory, Air Resources, Atmospheric Turbulence and Diffusion Laboratory Report ATDL 106, USAEC, Oak Ridge, TN.

Bristow, Q., 1978. The Application of Airborne Gamma-Ray Spectrometry in the Search for Radioactive Debris from the Russian Satellite Cosmos 954 ('Operation Morning Light'), Current Research, Part B, Paper 1978–1B, Geological Survey of Canada, Ottawa, pp. 151–162.

Brown, J., 1988. The Effectiveness of Sheltering as a Countermeasure in the Event of an Accident, Radiological Protection Bulletin No. 97, National Radiological Protection Board, Chilton, UK.

Brown, J. and Jones, J.A., 1993. Location Factors for Modification of External Radiation Doses, Radiological Protection Bulletin No. 144, National Radiological Protection Board, Chilton, UK.

Brown, J. and Simmonds, J.R., 1995. A Dynamic Model for the Transfer of Radionuclides Through Terrestrial Foodchains, National Radiological Protection Board, Chilton, Report No. NRPB-R273, HMSO, London.

Browne, E. and Firestone, R.B., 1986. In: Shirley, V.S. (Ed.), *Tables of Radioactive Isotopes*, Wiley, New York.

Bualert, S., 2001. Development and Application of an Advanced Gaussian Urban Air Quality Model, PhD Thesis, University of Hertfordshire, Hatfield, UK.

Bunzl, K., 1997. Probability for detecting hot particles in environmental samples by sample splitting. *Analyst*, **122**, 653–656.

Burch, P.R.J., Duggleby, J.C., Oldroyd, B. and Spiers, F.W., 1964. Studies of environmental radiation at a particular site with a static γ-ray monitor. In: Adams, J.A.S. and Lowder, W.M. (Eds), *The Natural Radiation Environment*, The University of Chicago Press, Chicago, IL, pp. 767–779.

Burcham, W.E. and Jobes, M., 1995. *Nuclear and Particle Physics*, Longman Scientific and Technical, London.

Burnett, W.C., Schultz, M.K. and Hull, C.D., 1996. Radiological flow during the conversion of phosphogypsum to ammonium sulphate. *J. Environ. Radioact.*, **32**, 33–51.

Carmichael, J.B., 1988. Worldwide production and utilization of phosphogypsum. In: *Proceedings of the Second International Symposium on Phosphogypsum*, Miami, FL, December, 1986, Publication No. 01–037–055, Florida Institute of Phosphate Research, Bartow, FL, pp. 105–116.

Cawse, A., 1988. Environmental radioactivity in Caithness and Sutherland. Part 1: radionuclides in soil, peat and crops in 1979. *Nucl. Energy*, **27**, 193–213.

Cawse, P.A. and Baker, S.J., 1990. A Survey of Radioactive Caesium in British Soils: Comparison of Accumulations Pre- and Post-Chernobyl, AEA-EE-0047, Atomic Energy Authority, Harwell, UK.

Chang, W.P., Chan, C.-C. and Wang J.-D., 1997. ^{60}Co contamination in recycled steel resulting in elevated civilian radiation doses: causes and challenges. *Health Phys.*, **73**, 465–472.

Chatfield, C., 1988. *Problem Solving – A Statisticians Guide*, Chapman & Hall, London.

Cimorelli, A.J., Perry, S.G., Venkatram, A., Weil, J.C., Paine, R.J., Wilson, R.B., Lee, R.F. and Peters, W.D., 1998 AERMOD Description of Model Formulation, AERMIC, USEPA, Research Triangle Park, NC, 113 pp.

Clarke, R.H., 1979. The First Report of a Working Group on Atmospheric Dispersion: A Model for Short and Medium Range Dispersion of Radionuclides Released into the Atmosphere, National Radiological Protection Board, Harwell, Report No. NRPB-R91, HMSO, London.

Clarke, R.H., 1989. Current radiation risk estimates and implications for the health consequences of Windscale, TMI and Chernobyl accidents. In: *Proceedings of the UKAEA Conference on Medical Response to Effects of Ionising Radiation*, London, 28–30 June 1989, pp. 102–118.

CODEX, 1989. Contaminants: Guideline Levels for Radionuclides in Food following Accidental Nuclear Contamination for Use in International Trade, Supplement to CODEX Alimentarious, Volume XVII, CODEX Alimentarious Commission, World Health Organization, Geneva.

Cohen, E.R. and Barry, N.T., 1987. The 1986 CODATA recommended values of the fundamental physical constants. *J. Res. Nat. Bur. Stand.*, **92**, 85–95.

Colls, J., 1997. *Air Pollution – An Introduction*, E & FN Spon, London.

Cooper, J.R., 1992. The radiological impact of routine radioactive discharges from nuclear sites. *J. Radiol. Protect.*, **12**, 9–16.

Cooper, J.R., 2000. Principles for removing material from regulatory control. In: *Safety of Radioactive Waste Management*, Proceedings of International Conference, Cordoba, Spain, 13–17 March 2000, International Atomic Energy, Vienna, pp. 197–208.

Cooper, J.R., Mobbs, S.F. and Barraclough, I., 1992. Radiological criteria for solid waste disposal: historical developments and new ideas. *Mater. Res. Soc. Symp.*, **257**, 25–34.

Cosby, B.J., Hornberger, G.M., Galloway, J.N. and Wright, R.F., 1985. Modelling the effects of acid deposition. Assessment of a lumped parameter model of soil water and stream chemistry. *Water Resources Res.*, **21**, 51–63.

Cox, R., 1996. Fundamental biological processes in radiation tumorigenesis. *Radiat. Protect. Dosim.*, **68**, 105–110.

Crick, M.J. and Brown, J., 1990. EXPURT: A Model for Evaluating Exposure from Radioactive Material Deposited in the Urban Environment, National Radiological Protection Board, Chilton, Report No. NRPB-R235, HMSO, London.

Crick, M.J. and Linsley, G.S., 1984. An assessment of the radiological impact of the Windscale reactor fire, October 1957. *Int. J. Radiat. Biol.*, **46**, 479–506.

Croarkin, C., 1985. National Assurance Programs Part II: Development and Implementation, National Bureau of Standards Special Publication 676-II, National Bureau of Standards, Washington, DC.

Cross, W.G., Arneja, A. and Ing, H., 1986. The response of chemically etched CR-39 to protons of 10 keV to 3 MeV. *Nucl. Tracks*, **9**, 649–652.

Croudace, I.W., 1991. A reliable and accurate procedure for preparing low-activity efficiency calibration standards for germanium gamma-ray spectrometers. *J. Radioanal. Nucl. Chem. Lett.*, **153**, 151–162.

Currie, L.A., 1968. Limits for qualitative detection and quantitative determination. *Anal. Chem.*, **40**, 586–593.

D'Amours, R., 1998. Modelling the ETEX plume dispersion with the Canadian Emergency Response Model. *Atmos. Environ.*, **32**, 4335–4341.

Dalheimer, A. and Henrichs, K., 1994. Monitoring of workers occupationally exposed to thorium in Germany. *Radiat. Protect. Dosim.*, **53**, 207–209.

Davis, J.P., Barraclough, I.M. and Mobbs, S.F., 1991. Methodology for Assessing Suitable Systems for Management of Reactor Decommissioning Wastes, EUR 12701 EN, European Commission, Luxembourg.

de Soete, D., Gijbels, R. and Hoste, J., 1972. *Neutron Activation Analysis*, Series of Monographs on Analytical Chemistry and its Applications, Vol. 34, Wiley-Interscience, London.

Deal, L.J., Doyle, J.F., Burson, Z.G. and Boyns, P.K., 1972. Locating the lost Athena missile in Mexico by the Aerial Radiological Measuring System (ARMS). *Health Phys.*, **23**, 95–98.

Desiato, F., 1992. A long-range dispersion model evaluation study with Chernobyl data. *Atmos. Environ.*, **26A**, 2805–2820.

Desiato, F., Anfossi, D., Trini Castelli, S., Ferrero, E. and Tinarelli, G., 1998. The role of wind field, mixing height and horizontal diffusivity investigated through two Lagrangian particle models. *Atmos. Environ.*, **32**, 4157–4165.

Dickson, B.H., Bailey, R.C. and Grasty, R.L., 1981. Utilizing multi-channel airborne gamma-ray spectra. *Can. J. Earth. Sci.*, **18**, 1793–1801.

Draxler, R.R., 1976. Determination of atmospheric diffussion parameters. *Atmos. Environ.*, **10**, 99–105.

Drazil, J.P., 1983. *Quantities and Units of Measurements: A Dictionary and Handbook*, Mansell (an Alexandrine Press Book), London.

du Bois, P.B. and Gueguenit, P., 1999. Quantitative assessment of dissolved radiotracers in the English Channel: Sources, average impact of la Hague reprocessing plant and conservative behaviour. *Continental Shelf Res.*, **19**, 1977–2002.

Dubasov, Yu., Krivohatskii, A., Kharitonov, K. and Gorin, V., 1994. Radioactive contamination of the Semipalatinsk testing ground and adjacent territories in consequence of atmospheric nuclear tests in 1949–1962. In: *Remediation and Restoration of Radioactive-Contaminated Sites in Europe*, Symposium Proceedings, Antwerp, 11–15 October 1993, European Commission, Luxembourg, RP **74**, pp. 369–382.

Dupuis, L.R. and Lipfert, F.W., 1986. Estimating the Cost of Uncertainty in Air Quality Modelling, Electric Power Research Institute Report, Doc. No. DE86 013935 (EPRI-EA-4707), Research Project 2301-1, Electric Power Research Institute, Palo Alto, CA.

Dyer, K.R., 1997. *Estuaries: A Physical Introduction*, Wiley, Chichester, UK.

Dyson, N.A., 1990. *X-Rays in Atomic and Nuclear Physics*, Cambridge University Press, Cambridge, UK.

EC, 1990a. The Radiological Exposure of the Population of the European Community from Radioactivity in North European Marine Waters: Project MARINA, European Commission RP-47 EUR 12483 EN, European Commission, Luxembourg.

EC, 1990b. PAGIS: Proceedings of the PAGIS Information Day, European Commission EUR 12676 EN, European Commission, Luxembourg.

EC, 1994. The Radiological Exposure of the Population of the European Community to Radioactivity in the Mediterranean Sea: MARINA-MED Project, European Commission RP-70 EUR 15564 EN, European Commission, Luxembourg.

EC, 1995. Radioactive Effluents from Nuclear Power Stations and Nuclear Fuel Reprocessing Plants in the European Community, 1977–1986, European Commission, EUR 15928 EN, European Commission, Luxembourg.

EC, 1998. *Recommended Radiological Protection Criteria for the Recycling of Metals from the Dismantling of Nuclear Installations*, RP **89**, European Commission, Luxembourg.

EC, 1999. *Radiation Protection Principles Concerning the Natural Radioactivity of Building Materials*, RP **112**, European Commission, Luxembourg.

EC, 2000a. *Practical Use of the Concepts of Clearance and Exemption – Part 1*, RP **122**, European Commission, Luxembourg.

EC, 2000b. The Radiological Exposure of the Population of the European Community to Radioactivity in the Baltic Sea: MARINA-BALT Project, European Commission RP-110 EUR 19200 EN, European Commission, Luxembourg.

Eckerman, K.F. and Ryan, J.C., 1993. External Exposure to Radionuclides in Air, Water and Soil, Federal Guidance Report No. 12, US Environmental Protection Agency, Washington, DC.

Eisenbud, M., 1987. *Environmental Radioactivity from Natural, Industrial and Military* Sources, Academic Press, New York.

Environmental Measurements Laboratory, 1992. Chieco, N.A., Bogen, D.C. and Knutson, E.O. (Eds), *Procedures Manual (HASL-300)*, 27th Edn, Vol. 1, US Department of Energy, New York.

Ettenhauber, E. and Lehmenn, R., 1986. The collective dose equivalent due to the naturally occurring radionuclides in building materials in the German Democratic Republic – Part 1. External exposure. *Health Phys.*, **50**, 49–56.

Ewen, J. 1995. Contaminant transport component of the catchment modelling system SHETRAN. In: Trudgill, S.T. (Ed.), *Solute Modelling in Catchment Systems*, Wiley, Chichester, UK, pp. 417–441.

Eyre, B.L., 1996. Industrial applications of radioactivity. *Radiat. Protect. Dosim.*, **68**, 63–72.

Farris, W.T., Napier, B.A., Ikenberry, T.A. and Shipler, D.B., 1996. Radiation doses from Hanford site releases to the atmosphere and the Columbia River. *Health Phys.*, **71**, 588–601.

Ferguson, T.S., 1961. On the rejection of outliers. In: Neyman, J. (Ed.), *Proceedings of the 4th Berkeley Symposium on Mathematics, Statistics and Probability*, Vol. 1 Berkeley, CA, University of California Press, Berkeley, CA, pp. 253–287.

Fetisov, V.I., Romanov, G.N. and Drozhko, E.G., 1994. Practice and problems of environmental restoration at the location of the Industrial Association MAYAK. In: *Remediation and Restoration of Radioactive-Contaminated Sites in Europe*, Symposium Proceedings, Antwerp, 11–15 October 1993, European Commission, Luxembourg, RP **74**, pp. 507–522.

Firestone, R.B., 1996. *Tables of Isotopes*, 8th Edn, Vol. 1, Wiley-Interscience, New York.

Fleischer, R.L., Price, P.B. and Walker, R.M., 1975. *Nuclear Tracks in Solids*, University of California Press, London.

Foti, S.C., 1977. Ashing of vegetation for the determination of ^{131}I. *Health Phys.* **33**, 387–391.

Fulker, M.J. and Grice, J.M., 1989. Transfer of radiocaesium from grass and silage to cow's milk. *Sci Total Environ.*, **85**, 129–138.

Garland, J.A., 1983. Some recent studies of the resuspension of deposited material from soil and grass. In: Pruppacher, H.R., Semonin, R.G. and Slinn, W.G.N. (Eds), *Precipitation Scavenging, Dry Deposition and Resuspension*, Vol. 2, Elsevier, Amsterdam, pp. 1087–1097.

Garland, J.A. and Pomery, I.R., 1994. Resuspension of fall-out material following the Chernobyl accident. *J. Aerosol. Sci.*, **25**, 793–806.

Gibbs, K.J., 1994. Radium contamination: an overview of UK Ministry of Defence Experience. In *Remediation and Restoration of Radioactive-Contaminated Sites in Europe*, Symposium Proceedings, Antwerp, 11–15 October 1993, European Commission, Luxembourg, RP **74**, pp. 281–293.

Gifford, F.A., 1976. Turbulent diffusion-typing schemes: A review. *Nucl. Safety*, **17**, 68–86.

Gilbert, R.O., 1987. *Statistical Methods for Environmental Pollution Monitoring*, Van Nostrand Reinhold, New York.

Glasstone, S., 1979. *Sourcebook on Atomic Energy*, Krieger, New York.

Golder, A., 1972. Relations among stability parameters in the surface layer. *Boundary Layer Met.*, **3**, 47–58.

Grasty, R.L., 1978. Estimating the fallout on Great Slave Lake from Cosmos 954. *Trans. Am. Nucl. Soc.*, **30**, 116–118.

Grasty, R.L., Kosanke, K.L. and Ford, R.S., 1978. Fields of view of airborne gamma-ray detectors. *Geophysics*, **44**, 1447–1457.

Green, N., Wilkins, B.T. and Poultney, S., 1997. Distribution of radionuclides in potato tubers: implications for dose assessments. *J. Radioanal. Nucl. Chem.*, **226**, 75–78.

Hart, I.R. and Saini, L., 1992. Biology of tumour metastasis. *Lancet*, **339**, 1453–1457.

Harvey, D.S., 1998. Natural radioactivity in iron and steel production. In: *Proceedings of NORM II Symposium*, Krejfeld, Germany, pp. 62–66.

Harvey, M.P., Hipkin, J., Simmonds, J.R., Mayall, A., Cabianca, T., Fayers, C. and Haslam, I., 1994. Radiological Consequences of Waste arising with Enhanced Natural Radioactivity Content from Special Metal and Ceramic Processes, EUR 15613 EN, European Commission, Luxembourg.

Haywood, S.M. and Smith, J.G., 1990. Assessment of the Radiological Impact of the Residual Contamination in the Maralinga and Emu Areas, National Radiological Protection Board, Chilton, Report No. NRPB-R237, HMSO, London.

Haywood, S.M. and Smith, J.G., 1992. Assessment of the potential doses at the Maralinga and Emu test sites. *Health Phys.*, **63**, 624–630.

Hedvall, R. and Erlandsson, B., 1996. Radioactivity concentrations in non-nuclear industries. *J. Environ. Radioact.*, **32**, 19–31.

Henderson-Sellers, B., 1984. *Pollution of our Atmosphere*, Adam-Hilger, Bristol, UK.

Hipkin, J. and Paynter, R.A., 1991. Radiation exposures to the workforce from naturally occurring radioactivity in industrial processes. *Radiat. Protect. Dosim.*, **36**, 97–100.

Horst, T.W., 1978. Estimation of air concentrations due to the suspension of surface contamination. *Atmos. Environ.*, **12**, 797–802.

Horwitz, W., 1990. Nomenclature for sampling in analytical chemistry (Recommendations 1990). *Pure Appl. Chem.*, **62**, 1193–1208.

Huber, A.H., Snyder, W.H., Thompson, R.S. and Lawson, R.E. Jr, 1980. The Effects of Squat Building on Short Stack Effluents – A Wind Tunnel Study, EPA Report 600/4-80-055, USEPA, Research Triangle Park, NC.

Hughes, J.S., 1999. Ionising Radiation Exposure of the UK Population: 1999. Review, National Radiological Protection Board, Chilton, NRPB-R311, HMSO, London.

Hughes, J.S. and O'Riordan, M.C., 1994. Radiation Exposure of the UK Population: 1993. Review, National Radiological Protection Board, Chilton, NRPB-R263, HMSO, London.

Hunt, G.J., 1984. Simple models for prediction of external radiation exposure from aquatic pathways. *Radiat. Protec. Dosim.*, **8**, 215–220.

IAEA, 1982a. Generic Models and Parameters for Assessing the Environmental Transfer of Radionuclides from Routine Releases, International Atomic Energy Agency Safety Series No. 57, STI/PUB/611, International Atomic Energy Agency, Vienna.

IAEA, 1982b. Hydrological Dispersion of Radioactive Material in Relation to Nuclear Power Plant Siting: A Safety Guide, International Atomic Energy Agency Safety Series No. 50-SG-S6, International Atomic Energy Agency, Vienna.

IAEA, 1985. Sediment K_ds and Concentration Factors for Radionuclides in the Marine Environment, International Atomic Energy Agency Technical Report Series No. 247, International Atomic Energy Agency, Vienna.

IAEA, 1986. *Emergency Planning and Preparedness for Nuclear Facilities*, Symposium Proceedings, Rome, 4–8 November 1985, International Atomic Energy Agency, Vienna.

IAEA, 1988. The Radiological Accident in Goiania, International Atomic Energy Agency, Vienna.

IAEA, 1989. Construction and Use of Calibration Facilities for Radiometric Field Equipment, International Atomic Energy Agency Technical Report Series No. 309, International Atomic Energy Agency, Vienna.

IAEA, 1990. The Use of Gamma-Ray Data to Define the Natural Radiation Environment, IAEA-TECHDOC-586, International Atomic Energy Agency, Vienna.

IAEA, 1991. X-Ray and Gamma-Ray Standards for Detector Calibration, IAEA-TECHDOC-619, International Atomic Energy Agency, Vienna.

IAEA, 1993. Validation of Environmental Model Predictions (VAMP). A Program for Testing and Improving Biospheric Models using Data from the Chernobyl Fallout, International Atomic Energy Agency STI/PUB/932, International Atomic Energy Agency, Vienna.

IAEA, 1994a. Safety Indicators in Different Time Frames for the Safety Assessment of Underground Radioactive Waste Repository, IAEA-TECHDOC-767, International Atomic Energy Agency, Vienna.

IAEA, 1994b. Classification for Radioactive Waste: A Safety Guide, International Atomic Energy Agency Safety Series No. 111-G-1.1, International Atomic Energy Agency, Vienna.

IAEA, 1994c. *Nuclear Safety Review 1994, Part D, IAEA Yearbook 1994*, International Atomic Energy Agency, Vienna.

IAEA, 1994d. Modelling of Resuspension, Seasonality, and Losses during Food Processing, First Report of the VAMP Terrestrial Working Group, International Atomic Energy Agency IAEA-TECHDOC-647, International Atomic Energy Agency, Vienna.

IAEA, 1994e. Handbook of Parameter Values for the Prediction of Radionuclide Transfer in Temperate Environments, International Atomic Energy Agency Technical Report No. 364, International Atomic Energy Agency, Vienna.

IAEA, 1995a. The Principles of Radioactive Waste Management, International Atomic Energy Agency Safety Series No. 111–F, International Atomic Energy Agency, Vienna.

IAEA, 1995b. *IAEA Yearbook 1995*, International Atomic Energy Agency, Vienna.

IAEA, 1996a. International Basic Safety Standards for Protection against Ionizing Radiation and for the Safety of Radiation Sources, International Atomic Energy Agency Safety Series No. 115, International Atomic Energy Agency, Vienna.

IAEA, 1996b. Clearance Levels for Radionuclides in Solid Materials: Application of the Exemption Principles – Interim Report for Comment, IAEA-TECHDOC-855, International Atomic Energy Agency, Vienna.

IAEA, 1996c. *IAEA Yearbook 1996*, International Atomic Energy Agency, Vienna.

IAEA, 1996d. Lessons Learned from Accidents in Industrial Irradiation Facilities, International Atomic Energy Agency, Vienna.

IAEA, 1997. *IAEA Yearbook 1997*, International Atomic Energy Agency, Vienna.

IAEA, 1998. Clearance of Materials Resulting from the Use of Radionuclides in Medicine, Industry and Research, IAEA-TECHDOC-1000, International Atomic Energy Agency, Vienna.

IAEA, 1999a. Protection of the Environment from the Effects of Ionising Radiation: A Report for Discussion, IAEA-TECHDOC-1091, International Atomic Energy Agency, Vienna.

IAEA, 1999b. Inventory of Radioactive Waste Disposals at Sea, IAEA-TECHDOC-1105, International Atomic Energy Agency, Vienna.

IAEA, 2000. Regulatory Control of Radioactive Discharges to the Environment: Safety Guide, International Atomic Energy Agency Safety Standard Series No. WS-G-2.3, International Atomic Energy Agency Vienna.

IAEA, 2001a. International Atomic Energy Agency Bulletin, Vol. 43, No. 3, International Atomic Energy Agency, Vienna.

IAEA, 2001b. Generic Models for Use in Assessing the Impact of Discharges of Radioactive Substances to the Environment, International Atomic Energy Agency Safety Report Series No. 19, International Atomic Energy Agency, Vienna.

IAEA Managing Radioactive Waste Fact Sheet, IAEA Website [www.iaea.org/worldatom/Periodicals/Factsheets/English/manradwa.html].

IARC, 1994. Direct estimates of cancer mortality due to low doses of ionising radiation: an international study. *Lancet*, **344**, 1039–1043.

ICRP, 1977. Recommendations of the International Commission on Radiological Protection, ICRP Publication 26, Ann ICRP 1, No. 3, Pergamon Press, Oxford, UK.

ICRP, 1991a. 1990 Recommendations of the International Commission on Radiological Protection, ICRP Publication 60, Ann ICRP 21, Nos 1–3, Pergamon Press, Oxford, UK.

ICRP, 1991b. Principles for Intervention for Protection of the Public in a Radiological Emergency, ICRP Publication 63, Ann 1CRP 22, No. 4, Pergamon Press, Oxford, UK.

ICRP, 1994a. Dose Coefficients for Intakes of Radionuclides by Workers, ICRP Publication 68, Ann ICRP 24, No. 4, Pergamon Press, Oxford, UK.

ICRP, 1994b. Protection against Radon-222 at Home and at Work, ICRP Publication 65, Ann ICRP 23, No. 2, Pergamon Press, Oxford, UK.

ICRP, 1996. Age Dependent Doses to Members of the Public from Intake of Radionuclides: Part 5. Compilation of Ingestion and Inhalation Dose Coefficients, ICRP Publication 72, Ann ICRP 26, No. 1, Pergamon Press, Oxford, UK.

ICRP, 1997. Radiological Protection Policy for the Disposal of Radioactive Waste, ICRP Publication 77, Ann ICRP 27 (Supplement 1997), Pergamon Press, Oxford, UK.

ICRP, 1998a. Radiation Protection Recommendations as Applied to the Disposal of Long-Lived Solid Radioactive Waste, ICRP Publication 81, Ann ICRP 28, No. 4, Pergamon Press, Oxford, UK.

ICRP, 1998b. Genetic Susceptibility to Cancer, ICRP Publication 79, Ann ICRP 28, Nos 1–2, Pergamon Press, Oxford, UK.

ICRP, 1999a. The ICRP Database of Dose Coefficients: Workers and Members of the Public, CD-ROM (Version 1.0), Elsevier, Amsterdam.

ICRP, 1999b. Protection of the Public in Situations of Prolonged Radiation Exposure, ICRP Publication 82, Ann ICRP 29, Nos 1–2, Pergamon Press, Oxford, UK.

ICRU, 1984. Stopping Powers for Electrons and Positrons, ICRU Report 37, International Commission on Radiation Units and Measurements, Bethesda, MD.

ICRU, 1993. Stopping Powers and Ranges for Protons and Alpha Particles, ICRU Report 49, International Commission on Radiation Units and Measurements, Bethesda, MD.

ICRU, 1998. Fundamental Quantities and Units for Ionizing Radiation, ICRU Report 60, International Commission on Radiation Units and Measurements, Bethesda, MD.

Ingamells, C.O. and Switzer, P., 1973. A proposed sampling constant for use in geochemical analysis. *Talanta*, **20**, 547–568.

Institution of Environmental Health Officers, 1988. *Environ. Health Monit.*, February [complete issue].

Iranzo, E., Salvador, S. and Iranzo, C.E., 1987. Air concentrations of Pu-239 and Pu-240 and potential radiation doses to persons living near Pu contaminated areas in Palomares (Spain). *Health Phys.*, **52**, 453–461.

IUR, 1989. Sixth Report of the Working Group on Soil-to-Plant Transfer Factors, International Union of Radioecologists, RIVM, Bilthoven, The Netherlands.

Izraehl, Y.A., Petrov, V.N. and Severov, D.A., 1990. Modelling of the transport and fallout of radionuclides from the accident at the Chernobyl nuclear power plant. In: *Environmental Contamination Following a Major Nuclear Accident*, Symposium Proceedings, Vol. 1, Vienna, 16–20 October 1989, International Atomic Energy Agency, Vienna, pp. 85–98.

Jacob, P., Kenigsberg, Y., Zvonova, I., Goulko, G., Buglova, E., Heidenreich, W.F., Golovneva, A., Bratilova, A.A., Drozdovitch, V., Kruk, J., Pochtennaja, G.T., Balonov, M., Demidchik, E.P. and Paretzke, H.G., 1999. Childhood exposure due to the Chernobyl accident and thyroid cancer risk in contaminated areas of Belarus and Russia. *Br. J. Cancer*, **80**, 1461–1469.

Jacobson, M.Z., 1999. *Fundamental of Atmospheric Modelling*, Cambridge University Press, Cambridge, UK.

Jones, J.A., 1981. The Third Report of a Working Group on Atmospheric Dispersion – The Estimation of Long Range Dispersion and Deposition of Continuous Releases of Radionuclides to the Atmosphere, National Radiological Protection Board, Harwell, Report No. NRPB-R123, HMSO, London.

Jones, J.A., 1986. The Seventh Report of a Working Group on Atmospheric Dispersion – The Uncertainty in Dispersion Estimates Obtained from the Working Group Models, National Radiological Protection Board, Harwell, Report No. NRPB-R199, HMSO, London.

Kaplan, I., 1963. *Nuclear Physics*, 2nd Edn, Addison-Wesley, Reading, MA.

Kathren, R.N., 1984. *Radioactivity in the Environment: Sources, Distribution and Surveillance*, Harwood Academic, London.

Kathren, R.N., 1991. *Radioactivity in the Environment*, Harwood Academic, London.

Kelly, G.N., Jones, A.J., Bryant, P.M. and Morley, F., 1975. The Predicted Radiation Exposure of the Population of the European Community Resulting from Discharges of Krypton-85, Tritium, Carbon-14 and Iodine-129, CEC Doc. No. V/2676/75, Commission of the European Communities, Luxembourg.

Kinner, N.E., Malley, J.P. Clement, J.A., Quern, P.A., Schell, G.S. and Lessard, C.E., 1991. Effects of sampling technique, storage, cocktails, sources of variation and extraction on the liquid scintillation technique for radon in water. *Environ. Sci. Technol.*, **25**, 1165–1171.

Kip, A., Bousquet, A., Evans, R. and Tuttle, W., 1946. Design and operation of an improved counting rate meter. *Rev. Sci. Instr.*, **17**, 323–333.

Kirchner, T.B., Whicker, F.W., Anspaugh, L.R. and Ng, Y.C., 1996. Estimated internal dose due to ingestion of radionuclides from the Nevada test site fallout. *Health Phys.*, **71**, 487–496.

Kleeman, A.W., 1967. Sampling error in the chemical analysis of rocks. *J. Geol. Soc. Aust.*, **14**, 43–47.

Klug, W., Graziani, G., Grippa, G., Pierce, D. and Tassone, C., 1992. *Evaluation of Long-Range Atmospheric Models Using Environmental Radioactivity Data from the Chernobyl Accident*, Elsevier, Barking, UK.

Knoll, G.F., 1989. *Radiation Detection and Measurement*, 2nd Edn, Wiley, New York.

Kocher, D.C., 1979. A Dynamic Model of the Global Iodine Cycle for the Estimation of Doses to the World Population from Releases of Iodine-129 into the Environment, Report ORNL/NUREG-59, Oak Ridge National Laboratories, Oak Ridge, TN.

Laissaoui, A. and Abril, J.M., 1999. A theoretical technique to predict the distribution of radionuclides bound to particles in surface sediments. *J. Environ. Radioact.*, **44**, 71–84.

Lane, D.P., 1992. P-53, Guardian of the genome. *Nature (London)*, **358**, 15–16.

Langner, J., Robertson, L., Persson, C. and Ullerstig, A., 1998. Validation of the operational emergency model at the Swedish Meteorological and Hydrological Institute using data from ETEX and the Chernobyl accident. *Atmos. Environ.*, **32**, 4325–4333.

Lauritzen, B. and Mikkelsen, T., 1999. A probabilistic dispersion model applied to the long range transport of radionuclides from the Chernobyl accident. *Atmos. Environ.*, **33**, 3271–3279.

Linsley, G.S., 1983. Resuspension in vegetated environments and its radiological significance. In: *Proceedings of Seminar on the Transfer of Radioactive Materials in the Terrestrial Environment Subsequent to an Accidental Release to Atmosphere*, Vol. 1, Dublin, April 11–13 1983, CEC Doc. No. V/3004/83, Commission of the European Communities, Luxembourg, pp. 79–91.

Livingstone, D.A., 1963. Chemical composition of rivers and lakes. In: *Data of Geochemistry*, 6th Edn, US Geological Survey Professional Paper 440–G, US Geological Survey, Washington, DC, Ch. 6, pp. G40–G51.

Longworth, G., (Ed.), 1998. *The Radiochemical Manual*, AEA Technology, Harwell, UK.

Lubenau, J.O. and Yusko, J.G., 1998. Radioactive materials in recycled metals – an update. *Health Phys.*, **74**, 293–299.

Luckey, T.D., 1982. Physiological benefits from low levels of ionising radiation. *Health Phys.*, **43**, 781–789.

Lyons, T.J. and Scott, W.D., 1990. *Principles of Air Pollution Meteorology*, Belhaven, London.

Mahesh, K., Weng, P.S. and Furetta, C., 1989. *Thermoluminescence in Solids and its Application*, Nuclear Technology Publications, Ashford, UK.

Martin, A., Mead, S. and Wade, B.O., 1997. Materials Containing Natural Radionuclides in Enhanced Concentrations, European Commission EUR 17625 EN, European Commission, Luxembourg.

Maryon, R.H. and Best, M.J., 1995. Estimating the emissions from a nuclear accident using observations of radioactivity with dispersion model products. *Atmos. Environ.*, **29**, 1853–1869.

Matthews, I.P., Kouris, K., Jones, M.C. and Spyrou, N.M., 1980. Theoretical and experimental investigations on the applicability of the Poisson and Ruark–Devol statistical density functions in the theory of radioactive decay and counting. *Nucl. Instrum. Methods*, **171**, 369–375.

McDonald, P., Baxter, M.S. and Scott, E.M., 1996. Technological enhancement of natural radionuclides in the marine environment. *J Environ Radioact.*, **32**, 67–90.

McHugh, C.A., Carruthers, D.J. and Edmunds, H.A., 1997. ADMS and ADMS-urban. *Int. J. Environ. Pollution*, **8**, 438–440.

Mellander, H., 1989. Airborne Gamma Spectrometric Measurements of the Fall-Out over Sweden after the Nuclear Reactor Accident at Chernobyl, USSR, International Atomic Energy Agency, Internal Report IAEA/NENF/NM-89-1, International Atomic Energy Agency, Vienna.

Miller, C.W. and Denham, L.S., 1994. An approach to dose reconstruction. In: *Assessing the Radiological Impact of Past Nuclear Activities and Events*, IAEA-TECHDOC-755, International Atomic Energy Agency, Vienna, pp. 79–85.

Miller, C.W. and Smith, J.M., 1996. Why should we do environmental dose reconstructions?. *Health Phys.*, **71**, 420–424.

Mobbs, S.F., Harvey, M.P., Martin, J.S., Mayall, A. and Jones, M.E., 1991. Comparison of the Waste Management Aspects of Spent Fuel Disposal and Reprocessing: Post-Disposal Radiological Impact, European Commission EUR 13561 EN, European Commission, Luxembourg.

Monin, A.S. and Yaglom, M., 1971. *Statistical Fluid Mechanics*, MIT Press, Cambridge, MA.

Mosca, S., Graziani, G., Klug, W., Bellasio, R. and Biaconi, R., 1998. A statistical methodology for the evaluation of long-range dispersion models: an application to the ETEX exercise. *Atmos. Environ.*, **32**, 4307–4324.

Muirhead, C.R., Goodill, A.A., Haylock, R.G.E., Vokes, J., Little, M.P., Jackson, D.A., O'Hagan, J., Thomas, J.M., Kendall, G.M., Silk, T.J., Bingham, D. and Berridge, G.L.C., 1999. Second Analysis of the National Registry for Radiation Workers: Occupational Exposure to Ionising Radiation and Mortality, National Radiological Protection Board, Chilton, Report No. NRPB-R307, HMSO, London.

Müller, J.W., 1979. Some second thoughts on error statements. *Nucl. Instrum. Methods*, **163**, 241–251.

Nair, S.K., Miller, C.W., Thiessen, K.M., Garger, E.K. and Hoffman, F.O., 1997. Modelling the resuspension of radionuclides in the Ukrainian Regions impacted by Chernobyl fallout. *Health Phys.*, **72**, 77–85.

National Group for Studying the Radiological Implications of the Use of Zircon Sand, 1985. Radiation protection aspects of the use of zircon sand. *Sci. Total Environ.*, **45**, 135–142.

NCRP, 1979. Tritium in the Environment, NCRP Report 62. National Council on Radiation Protection and Measurements, Washington, DC.

NCRP, 1984. Predicting the Transport, Bioaccumulation and Uptake by Man of Radionuclides Released into the Environment, NCRP Report 76, National Council on Radiation Protection and Measurements, Bethesda, MA.

NCRP, 1987. Radiation Exposure of the US Population from Consumer Products and Miscellaneous Sources, NCRP Publication 95. National Council on Radiation Protection and Measurements, Washington, DC.

NEA, 1985. Review of the Continued Suitability of the Dumping Site for Radioactive Waste in the North-East Atlantic, Nuclear Energy Agency, Paris.

NEA, 1987. The Radiological Impact of the Chernobyl Accident in OECD Countries, Nuclear Energy Agency, Paris.

NEA, 1996. Chernobyl – Ten Years On: Radiological and Health Impact, Nuclear Energy Agency, Paris.

NEA, 2000. Radiological Impacts of Spent Nuclear Fuel Management Options: A Comparative Study, Nuclear Energy Agency, Paris.

Ng, Y.C., Colsher, C.S. and Thompson, S.E., 1982. Soil-to-Plant Concentration Factors for Radiological Assessments, Report NUREG/CR-2975, Lawrence Livermore National Laboratory, Livermore, CA.

Nicholson, K.W., 1988. A review of particle resuspension. *Atmos. Environ.*, **22**, 2639–2651.

NRC, 1995a. *Technical Basis for Yucca Mountain Standards*, National Research Council, National Academy Press, Washington, DC.

NRC, 1995b. *Radiation Dose Reconstruction for Epidemiological Uses*, National Research Council, National Academy Press, Washington, DC.

NRPB, 1992. Board Statement on Radiological Protection Objectives for the Land-Based Disposal of Solid Radioactive Wastes, Documents of the NRPB, Vol. 3, No. 2, National Radiological Protection Board, HMSO, London.

NRPB, 1993. Occupational, Public and Medical Exposure, Documents of the NRPB, Vol. 4, No. 2, National Radiological Protection Board, HMSO, London.

NRPB, 1995. Risk of Radiation-Induced Cancer at Low Doses and Dose Rates for Radiation Protection Purposes, Documents of the NRPB, Vol. 6, No. 1, National Radiological Protection Board, HMSO, London.

NRPB, 1996a. Risk from Deterministic Effects of Ionising Radiation, Documents of the NRPB, Vol. 7, No. 3, National Radiological Protection Board, HMSO, London.

NRPB, 1996b. Generalised Derived Limits for Radioisotopes of Strontium, Iodine, Caesium, Plutonium, Americium and Curium, Documents of the NRPB, Vol. 7, No. 1, National Radiological Protection Board, HMSO, London.

NRPB, 1998. Revised Generalised Derived Limits for Radioisotopes of Strontium, Ruthenium, Iodine, Caesium, Plutonium, Americium and Curium, Documents of the NRPB, Vol. 9, No. 1, National Radiological Protection Board, HMSO, London.

Nuclear Engineering International, 2001. *World Nuclear Industry Handbook*, Nuclear Engineering International, Dartford, UK.

OECD/NEA Data Bank, 1993. JEF-2.2 Decay Data, Issy-Les-Moulineaux, France.

Oliveira, A.R., Hunt, J.G., Valverde, N.J.L., Brandao-Mello, C.E. and Farina, R., 1991. Medical and related aspects of the Goiania accident: an overview. *Health Phys.*, **60**, 17–24.

Oliveira, A.P., Soares, J., Tirabassi, T. and Rizza, U., 1998. A surface energy-budget model coupled with a Skewed Puff Model for investigating the dispersion of radionuclides in a sub-tropical area of Brazil. *Il Nuovo Cimento*, **21**, 631–646.

Open University Course Team, 1989. *Seawater: Its Composition, Properties and Behaviour*, Pergamon Press, Oxford, UK.

Osvath, I., Povinec, P.P. and Baxter, M.S., 1999. Kara Sea radioactivity assessment. *Sci. Total Environ.*, **237–238**, 167–179.

Pan, Z., 1993. Radiological Impact of Coal-Fired Energy in China, China National Nuclear Corporation, Beijing, China.

Pan, Z., Wang, Z., Chen, Z., Zhang, Y. and Xie, J., 1996. Radiological environmental impact of the nuclear industry in China. *Health Phys.*, **71**, 847–862.

Panofsky, H.A. and Dutton, J., 1984. *Atmospheric Turbulence*, Wiley, New York.

Pasquill, F., 1961. The estimation of the dispersion of windborne material. *Meteorol. Magazine*, **90**, 33–49.

Pasquill, F. and Smith, F.B., 1983. *Atmospheric Dispersion*, 3rd Edn, Ellis Horwood, Chichester, UK.

Pasternack, B.S. and Harley, N.H., 1971. Detection limits for radionuclides in the analysis of multi-component gamma ray spectrometer data. *Nucl. Instrum. Methods*, **91**, 533–540.

Pentreath, R.J., 2002. Radiation protection of people and the environment: developing a common approach. *J. Radiol. Protect.*, **22**, 45–56.

Perianez, R., 1999. Three-dimensional modelling of the tidal dispersion of non-conservative radionuclides in the marine environment. Application to Pu-239, Pu-240 dispersion in the eastern Irish Sea. *J. Marine Systems*, **22**, 37–51.

Perianez, R., 2000a. Modelling the tidal dispersion of Cs-137 and Pu-239, Pu-240 in the English Channel. *J. Environ. Radioact.*, **49**, 259–277.

Perianez, R., 2000b. Modelling the physico-chemical speciation of plutonium in the eastern Irish Sea. *J. Environ. Radioact.*, **49**, 11–33.

Perianez, R. and Martinez-Aguirre, A., 1997. Uranium and thorium concentrations in an estuary affected by phosphate fertilizer processing: experimental results and a modelling study. *J. Environ. Radioact.*, **35**, 281–304.

Perianez, R., Abril, J.M. and Garcia Leon, M., 1996. Modelling the dispersion of non-conservative radionuclides in tidal waters. 2. Application to Ra-226 dispersion in an estuarine system. *J. Environ. Radioact.*, **31**, 253–272.

Peterson, S.-R. and Kirchner, T.B., 1998. Data quality and validation of radiological assessment models. *Health Phys.*, **74**, 147–157.

Pierce, D.A., Shimizu, Y., Preston, D.L., Vaeth, M. and Mabuchi, M., 1996. Studies of the mortality of atomic bomb survivors. Report 12, part 1. Cancer: 1950 to 1990. *Radiat. Res.*, **146**, 1–27.

Pröhl, G., 1991. Modellierung der Radionuklidausbreitung in Nahrungsketten nach Deposition von Sr-90, Cs-137 und I-131 auf landwirtschaftlich genutzte Flächen, GSF-Bericht 29/90, GSF, Neuherberg, Germany.

Randle, K., 1967. Radiochemical Studies of Ion Exchange, PhD Thesis, University of Durham, Durham, UK.

Renaud, P., Real, J., Maubert, H. and Roussel-Debet, S., 1999. Dynamic modelling of the cesium, strontium and ruthenium transfer to grass and vegetables. *Health Phys.*, **76**, 495–501.

Rissanen, K. and Rahola, T., 1989. Cs-137 concentration in reindeer and its fodder plants. *Sci. Total Environ.*, **85**, 199–206.

RMS, 1995. Atmospheric Dispersion Modelling: Guidelines on the Justification of Choice and Use of Models and the Communication and Reporting of Results, Royal Meteorological Society, Reading, UK.

Robinson, C.A., Mayall, A., Attwood, C.A., Cabianca, T., Dodd, D.H., Fayers, C.A., Jones, K.A. and Simmonds, J.R., 1994. Critical Group Doses around Nuclear Sites in England and Wales, National Radiological Protection Board, Chilton, Report No. NRPB-R271, HMSO, London.

Rochedo, E.R.R., 2000. The radiological accident in Goiania: environmental aspects. In: *Restoration of Environments with Radioactive Residues*, International Atomic Energy Agency, Vienna, pp. 365–384.

Rodriguez-Alvarez, M.J. and Sanchez, F., 2000. Modelling of U, Th, Ra and ^{137}Cs radionuclides behaviour in rivers. Comparison with field observations. *Appl. Math. Model.*, **25**, 57–77.

Roek, D.R., Reavey, T.C. and Hardin, J.M., 1987. Partitioning of natural radionuclides in the waste streams of coal-fired utilities. *Health Phys.*, **52**, 311–323.

Romanov, G.N. and Drozhko, Ye.G., 1996. Ecological consequences of the activities of the MAYAK plant. In Luykx, F.F. and Frissel, M.J. (Eds), *Radioecology and the Restoration of Radioactive-Contaminated Sites*, NATO ASI Series, Kluwer Academic Publishers, Dordrecht, The Netherlands.

Rowland, R.E., 1994. Radium in Humans. A Review of US Studies, Report ANL/ER-3 UC-408, Argonne National Laboratory, Argonne, IL.

Ryall, D.B. and Maryon, R.H., 1998. Validation of the UK Meterological. Office's NAME model against the ETEX dataset. *Atmos. Environ.*, **32**, 4265–4276.

Sanderson, D.C.W., Scott, E.M., Baxter, M.S., Martin, E. and Ni Riain, S., 1993. The Use of Aerial Radiometrics for Epidemiological Studies of Leukaemia, SURRC Report, Scottish Universities Research and Reactor Centre, East Kilbride, Scotland, UK.

Sanderson, D.C.W., Allyson, J.D., Tyler, A.N. and Scott, E.M., 1995. Environmental applications of airborne gamma spectrometry. In: *Applications of Uranium Exploration Data and Techniques in Environmental Studies*, IAEA-TECHDOC-827, International Atomic Energy Agency, Vienna, pp. 71–93.

Sandor, G.N., Peic, T., Peic, R., 1996. Radioactive polluting potential of coal-fired power plants from Romania. In: *Proceedings of the 1996 International Conference on Radiological Protection*, Vol. 2, April 14–19 1996, Vienna, pp. 229–231.

Schaller, K., Dalrymple, G.J., Dodd, R., Malherbe, J., Mehling, O. and Mobbs, S.F., 1991. Assessment of waste management scenarios for light water reactor spent fuel. In: Cecille, L. (Ed.), *Radioactive Waste Management and Disposal*, Elsevier Applied Science, London, pp. 53–69.

Schiermeier, F.A., 1984. Scientific Assessment Document on Status of Complex Terrain Models for EPA Regulatory Applications, EPA Report No. EPA-600/3-84-103, USEPA, Research Triangle Park, NC.

Scholten, L.C., 1996. Approaches for Regulating Management of Large Volumes of Waste Containing Natural Radionuclides in Enhanced Concentrations, European Commission EUR 16956EN, European Commission, Luxembourg.

Scholten, L.C., Roelofs, L.M.M. and van der Steen, J., 1993. A Survey of Potential Problems for Non-Nuclear Industries Posed by Implementation of New EC Standards for Natural Radioactivity, KEMA, Report 40059-NUC, The Netherlands.

Schull, W.J., 1996. *Effects of Atomic Radiation: A Half-Century of Studies from Hiroshima and Nagasaki*, Wiley-Liss, New York.

Scott, B.R., 1993. Early-occurring and continuing effects. In: *Modifications of Models Resulting from Addition of Effects of Exposure to Alpha-Emitting Nuclides*, US Nuclear Regulatory Commission, Report NUREG/CR-4214 Rev1, Part 11, Addendum 2 (LMF-136), Nuclear Regulatory Commission, Washington, DC, pp. 7–27.

Sehmel, G.A., 1980. Particle resuspension: a review. *Environ. Int.*, **4**, 107–127.

Seibert, P., Beyrich, F., Gryning, S.-E., Joffre, S., Rasmussen, A. and Tercier, P., 1998. Mixing height determination for dispersion modelling, (Report of Working Group 2). In: Fisher, B.E.A., Erbrink, J.J., Finardi, S., Jeannet, S., Joffre, S., Morselli, M.G., Pechinger, U., Seibert, P. and Thomson, D.J. (Eds), COST Action 710 – Final Report, *Harmonisation of the Pre-Processing of Meteorological Data for Atmospheric Dispersion Models*, European Commission EUR 18195 EN, European Commission, Luxembourg, Part 3, pp. 1–120.

Seinfeld, J.H., 1986. *Atmospheric Chemistry and Physics of Air Pollution*, Wiley, New York.

Seltzer, S.M., Inokuti, M., Paul, H. and Bichsel, H., 2001. Response to the commentary by J F Ziegler regarding ICRU Report 49: stopping powers and ranges for protons and alpha particles. *Radiat. Res.*, **155**, 378–381.

Semenov, B., Oi, N., Grigoriev, A. and Takats, F., 1995. Overview of spent fuel management. In: *Safety and Engineering Aspects of Spent Fuel Storage*, Symposium Proceedings, Vienna, 10–14 October 1994, International Atomic Energy Agency, Vienna, pp. 3–14.

Sheih, C.M., Wesley, M.L. and Hicks, B.B., 1979. Estimated dry deposition velocities of sulfur over the eastern US and surrounding regions. *Atmos. Environ.*, **13**, 361–368.

Shipler, D.B., Napier, B.A., Farris, W.T. and Freshley, M.D., 1996. Hanford dose reconstruction – an overview. *Health Phys.*, **71**, 532–544.

Simmonds, J.R., Lawson, G. and Mayall, A., 1995. Methodology for Assessing the Radiological Consequences of Routine Releases of Radionuclides to the Environment, Radiation Protection 72, European Commission EUR 15760 EN, European Commission, Vienna.

Sjoblom, K.-L., Salo, A., Bewers, J.M., Cooper, J., Dyer, R.S., Lynn, N.M., Mount, M.E., Povinec, P.P., Sazykina, T.G., Schwarz, J., Scott, E.M., Sivintsev, Y.T., Tanner, J.E., Warden, J.M. and Woodhead, D., 1999. International Arctic Seas Assessment Project. *Sci. Total Environ.*, **237/238**, 153–166.

Sloan, W.T. and Ewen, J., 1999. Modelling long term contaminant migration in a catchment at fine spatial and temporal scale using the UP system. *Hydrol. Process.*, **13**, 823–846.

Smith, G.M. and White, I.F., 1983. A Revised Global-Circulation Model for I-129, National Radiological Protection Board, Chilton, Report No. NRPB-M81, HMSO, London.

Smith, K.R., Crockett, G.M., Oatway, W.B., Harvey, M.P., Penfold, J.S.S. and Mobbs, S.F., 2001. Radiological Impact on the UK Population of Industries which Use or Produce Materials Containing Enhanced Levels of Naturally Occurring Radionuclides. Part 1: Coal-Fired Electricity Generation, National Radiological Protection Board, Chilton, Report No. NRPB-R327, HMSO, London.

Snedcor, G.W. and Cochran, W.G., 1980. *Statistical Methods*, 7th Edn, Iowa State University Press, Ames, IA.

Sokhi, R.S., San Jose, R., Moussiopoulos, N. and Berkowicz, R., 2000. *Urban Air Quality: Measurement, Modelling and Management*, Kluwer Academic Publishers, Dordrecht, The Netherlands.

Stevens, W., Thomas, D.C., Lyon, J.L., Till, J.E., Kerber, R.A., Simon, S.L., Lloyd, R.D., Elghany, N.A. and Preston-Martin, S., 1990. Leukemia in Utah and radioactive fallout from the Nevada test site: a case control study. *J. Am. Med. Assoc.*, **264**, 585–590.

Stradling, G.N., Stather, J.W., Gray, S.A., Moody, J.C., Ellender, M., Pearce, M.J. and Collier, C.G., 1992. Radiological implications of inhaled plutonium-239 and americium-241 in dusts at the former nuclear test site in Maralinga. *Health Phys.*, **63**, 641–650.

Stumm, W., 1992. *Chemistry of the Solid–Water Interface*, Wiley-Interscience, New York.

TAG, 1990. Rehabilitation of Former Nuclear Test Sites in Australia, Report by the Technical Assessment Group, Department of Primary Industries and Energy, Australian Government Publishing Service, Canberra, Australia.

Tertian, R. and Claisse, F., 1982. *Principles of Quantitative X-Ray Fluorescence Analysis*, Heyden, London.

Testa, C., Desideri, D., Meli, M.A., Roselli, C., Bassignani, A., Finazzi, P.B., 1993. Radium, uranium and thorium concentrations in low specific activity scales and waters of some oil and gas production plants. *J Radioanal. Nuc. Chem.*, **170**, 117–124.

Testa, C., Desideri, D., Meli, M.A., Roselli, C., Bassignani, A., Colombo, G. and Fresca Fantoni, R. 1994. Radiological protection and radioactive scales in oil and gas production. *Health Phys.*, **67**, 34–38.

Theodorsson, P., 1993. Systems for low-level beta- and gamma-counting. In: *Low Level Measurements of Radioactivity in the Environment*, Proceedings of the Third International Summer School, World Scientific, Singapore, pp. 33–52.

Till, J.E. and Meyer, H.R. (Eds), 1983. *Radiological Assessment. A Textbook on Environmental Dose Analysis*, US Nuclear Regulatory Commission, NUREC/CR-3332, (ORNL-5968), Nuclear Regulatory Commission, Washington, DC.

Till, J.E., Simon, S.L., Kerber, R., Lloyd, R.D., Stevens, W., Thomas, D.C., Lyon, J.L. and Preston-Martin, S., 1995. The Utah thyroid cohort study: analysis of the dosimetry results. *Health Phys.*, **68**, 473–483.

Timmermans, C.W.M. and van der Steen, J., 1996. Environmental and occupational impacts of natural radioactivity from some non-nuclear industries in the Netherlands. *J. Environ. Radioact.* **32**, 97–104.

Titley, J.G., Cabianca, T., Lawson, G., Mobbs, S.F. and Simmonds, J.R., 1995. Improved Global Dispersion Models for Iodine-129 and Carbon-14, European Commission Report EUR 15880 EN, European Commission, Luxembourg.

Tracy, B.L. and Prantl, F.A., 1982. Radiological Implications of Thermal Power Production, IAEA-SM-254/6, International Atomic Energy Agency, Vienna.

Trapeznikov, A.V., Pozolotina, V.N., Chebotina, M.Ya., Chukanov, V.N., Trapeznikova, V.N., Kulikov, N.V., Nielsen, S.P. and Aarkrog, A., 1993. Radioactive contamination of the Techa River, The Urals. *Health Phys.*, **65**, 481–488.

Trapeznikov, A.V., Aarkrog, A., Yekidin, A., Karavaeva, Ye., Kulikov, N., Lisovskikh, V., Mikhailovskaya, L., Molochanova, I., Chebotina, M., Chukanov, V. and Yushkov, P., 1994. Radiological investigation of the Techa River (Urals) and of the soil and vegetation cover in its flood plain. In: *Remediation and Restoration of Radioactive-Contaminated Sites in Europe*, Symposium Proceedings, Antwerp, 11–15 October 1993, European Commission, Luxembourg, RP **74**, pp. 485–503.

Tsaturov, Yu. S. and Anisimova, L.I., 1994. Radionuclide contaminated territories of Russia: identification, restoring and rehabilitation aspects. In: *Remediation and Restoration of Radioactive-Contaminated Sites in Europe*, Symposium Proceedings, Antwerp, 11–15 October 1993, European Commission, Luxembourg, RP **74**, pp. 309–324.

Turner, D.B., 1994. *Workbook of Atmospheric Dispersion Estimates: An Introduction to Dispersion Modelling*, CRC/Lewis Publishers, Boca Raton, FL.

Turner, J.E., 1995. *Atoms, Radiation and Radiation Protection*, Wiley, New York.

UNSCEAR, 1982. United Nations Scientific Committee on the Effects of Atomic Radiation, 1982 Report to the General Assembly, with Scientific Annexes, United Nations, New York.

UNSCEAR, 1988. United Nations Scientific Committee on the Effects of Atomic Radiation, 1988 Report to the General Assembly, with Scientific Annexes, United Nations, New York.

UNSCEAR, 1993. United Nations Scientific Committee on the Effects of Atomic Radiation, 1993 Report to the General Assembly, with Scientific Annexes, United Nations, New York.

UNSCEAR, 2000a. United Nations Scientific Committee on the Effects of Atomic Radiation, 2000 Report to the General Assembly, with Scientific Annexes, Volume I: Sources, United Nations, New York.

UNSCEAR, 2000b. United Nations Scientific Committee on the Effects of Atomic Radiation, 2000 Report to the General Assembly, with Scientific Annexes, Volume II: Effects, United Nations, New York.

USEPA, 1992. Protocols for Defining the Best Performing Model, USEPA Publication EPA-450/4–92–008b, Office of Air Quality Planning and Standards Emissions, Technical Support Division, US Environmental Protection Agency, Research Triangle Park, NC.

USEPA, 1995. User's Guide for the Industrial Source Complex (ISC2) Dispersion Models: Volume II – Description of Model Algorithm, USEPA Publication No. EPA-454/4–95–003b, Office of Air Quality Planning and Standards Emissions, Monitoring and Analysis Division, US Environmental Protection Agency, Research Triangle Park, NC.

USEPA, 1996a. Meteorological Processor for Regulatory Models (MPRM) User's Guide, USEPA Publication No. EPA-454/B-96-002, Office of Air Quality Planning and Standards Emissions, Monitoring and Analysis Division, US Environmental Protection Agency, Research Triangle Park, NC.

USEPA, 1996b. Documenting Groundwater Modelling at Sites Contaminated with Radioactive Substances, USEPA Publication No. EPA 540-R-96-003, Office of Air Quality Planning and Standards Emissions, Monitoring and Analysis Division, US Environmental Protection Agency, Research Triangle Park, NC.

USEPA, 1999. PCRAMMET User's Guide, Office of Air Quality Planning and Standards Emissions, Monitoring and Analysis Division, US Environmental Protection Agency, Research Triangle Park, NC.

USNRC, 1977. Calculation of Annual Doses to Man from Routine Releases of Reactor Effluents for the Purpose of Evaluating Compliance with 10 CFR Part 50, Regulatory Guide 1.109, US Nuclear Regulatory Commission, Washington, DC.

Van den Hoven, I., 1982. Meteorological Considerations in the Development of a Real time Atmospheric Dispersion Model for Reactor Effluent Exposure Pathway, NRC FIN B7120 (Contract No: NRC 01–81–017), Division of Health, Siting and Waste Management, Office of Nuclear Regulatory Research, US Nuclear Regulatory Commission, Wasington, DC.

van Dop, H. and Nodop, K. (Eds), 1998. ETEX, A European Tracer Experiment. *Atmos. Environ.*, **32**(24) 4098–4375 [complete issue].

van Dop, H., Addis, R., Fraser, G., Girardi, F., Graziani, G., Inoue, Y., Kelly, N., Klug, W., Kumala, A., Nodop, K. and Pretel, J., 1998. ETEX: A European tracer experiment: observations, dispersion modelling and emergency response. *Atmos. Environ.*, **32**, 4089–4094.

Vandenhove, H. 2000. European sites contaminated by residues from the ore extracting and processing industries. In: *Restoration of Environments with Radioactive Residues*, International Atomic Energy Agency, Vienna, pp. 61–89.

Vanmarcke, H. 1996. Exhalation of radon and thoron from phosphogypsum used as building material. In: *Proceedings of the 1996 International Congress on Radiological Protection*, Vol. 2, April 14–19 1996, Vienna, pp. 38–40.

Vanmarcke, H. and Zeevaert, T., 2000. Restoration of the areas environmentally contaminated by the Olen radium facility. In: *Restoration of Environments with Radioactive Residues*, International Atomic Energy Agency, Vienna, pp. 517–539.

Vargo, G.J. (Ed.), 2000. *The Chernobyl Accident: A Comprehensive Risk Assessment*, Battelle Press, Richland, WA.

Venkatram, A., Brode, R., Cimorelli, A., Lee, R., Paine, R., Perry, S., Peters, W., Weil, J., Wilson, R., 2001. A complex terrain dispersion model for regulatory applications. *Atmos. Environ.*, **35**, 4211–4221

Verplancke, J. 1992. Low level gamma spectroscopy: low, lower, lowest. *Nucl. Instrum. Methods Phys. Res.*, **A312**, 174–182.

Vo, D.T., 1999. Extended Evaluations of the Commercial Spectrometer Systems for Safeguards Applications, Los Alamos National Laboratory Report LA-13604-MS, Los Alamos, NM.

Vo, D.T., Russo, P.A. and Sampson, T.E., 1998. Comparison between Digital Gamma-Ray Spectrometer (DSPEC) and Standard Nuclear Instrumentation Methods (NIM) Systems, Los Alamos National Laboratory Report LA-13393-MS, Los Alamos, NM.

Vorobiova, M.I. and Degteva, M.O., 1999. Simple model for the reconstruction of radionuclide concentrations and radiation exposure along the Techa River. *Health Phys.*, **77**, 142–149.

Wade Patterson, H., 1997. Setting standards for radiation protection: the process appraised. *Health Phys.*, **72**, 450–457.

Wagenpfeil, F., Paretzke, H.G., Peres, J.M. and Tschiersch, J.T., 1999. Resuspension of coarse particles in the region of Chernobyl. *Atmos. Environ.*, **33**, 3313–3323.

Wampach, R., Bisa, R., Pflugrad, K. and Simon, R. (Eds), 1995. *Decommissioning Nuclear Installations*, European Commission, Luxembourg.

Webster, R., 1977. *Quantitative and Numerical Methods in Soil Classification and Survey*, Clarendon Press, Oxford, UK.

Weil, J.C., 1985. Updating applied diffusion-models. *J. Climate Appl. Meteorol.*, **24**, 1111–1130.

Weinberg, R.A., 1991. Tumour suppressor genes. *Science*, **254**, 1138–1146.

Wendum, D., 1998. Three long range transport models compared to the ETEX experiment: a performance study. *Atmos. Environ.*, **32**, 4297–4305.

West, M.S., 1990. Aspects of Intertidal Sediment Dynamics in the Severn Estuary, PhD Thesis, University of Birmingham, Birmingham, UK.

Whicker, F.W., Kirchner, T.B., Anspaugh, L.R. and Ng, Y.C., 1996. Ingestion of Nevada test site fallout: internal dose estimates. *Health Phys.*, **71**, 477–486.

Whitehead, P.G., Wilson, E.J. and Butterfield, D., 1998. A semi-distributed integrated nitrogen model for multiple source assessment in catchments (INCA): Part I – model structure and process equations. *Sci. Total Environ.*, **210**, 547–558.

WHO, 1988. Derived Intervention Levels for Radionuclides in Food: Guidelines for Application after Widespread Contamination Resulting from a Major Nuclear Accident, World Health Organization, Geneva.

WHO, 1993. *Guidelines on Drinking Water Quality*, 2nd Edn, World Health Organization, Geneva.

Wilkins, B.T., Green, N., Stewart, S.P. and Major, R.O., 1985. Factors that affect the association of radionuclides with soil phases. In: Bulman, R.A. and Cooper, J.R. (Eds), *Speciation of Fission and Activation Products in the Environment*, Elsevier, Oxford, UK.

Wilkins, B.T., Simmonds, J.R. and Cooper, J.R., 1994. An Assessment of the Present and Future Implications of Radioactive Contamination of the Irish Sea Coastal Region of Cumbria, National Radiological Protection Board, Chilton, NRPB-R267, HMSO, London.

Wilson, D.J. and Netterville, D.D.J., 1978. Interaction of a roof-level plume with a downwind building. *Atmos. Environ.*, **12**, 1051–1059.

Woodruff, M., 1988. Tumour clonality and its biological consequences. *Adv. Cancer Res.*, **50**, 197–229.

Yamamoto, M., Yamamori, S., Komura, K. and Sakanoue, M., 1980. Behaviour of plutonium and americium in soils. *J. Radioanal. Res.*, **21**, 204–212.

Yokoyama, Y., Reyss, J.L. and Guichard, F., 1977. Production of radionuclides by cosmic rays at mountain altitudes. *Earth Planet Sci Lett.*, **36**, 44–50.

Zannetti, P., 1990. *Air Pollution Modelling: Theories, Computational Methods and Available Software*, Van Nostrand Reinhold, New York.

Ziegler, J.F., 1999. Commentary: comments on ICRU Report 49: stopping powers and ranges for protons and alpha particles. *Radiat. Res.*, **512**, 219–222.

Index